Methods and Protocols in Food Science

Series Editor
Anderson S. Sant'Ana
University of Campinas
Campinas, Brazil

For further volumes:
http://www.springer.com/series/16556

Methods and Protocols in Food Science series is devoted to the publication of research protocols and methodologies in all fields of food science.

Volumes and chapters will be organized by field and presented in such way that the readers will be able to reproduce the experiments in a step-by-step style. Each protocol will be characterized by a brief introductory section, followed by a short aims section, in which the precise purpose of the protocol will be clarified.

Novel Food Analysis

Techniques and Applications

Edited by

Tanmay Sarkar

Department of Food Processing Technology, West Bengal State Council of Technical Education, Malda, West Bengal, India

Humana Press

Editor
Tanmay Sarkar
Department of Food Processing
Technology
West Bengal State Council of Technical Education
Malda, West Bengal, India

ISSN 2662-950X ISSN 2662-9518 (electronic)
Methods and Protocols in Food Science
ISBN 978-1-0716-4786-8 ISBN 978-1-0716-4787-5 (eBook)
https://doi.org/10.1007/978-1-0716-4787-5

This Humana imprint is published by the registered company Springer Science+Business Media, LLC, part of Springer Nature.
The registered company address is: 1 New York Plaza, New York, NY 10004, U.S.A.

Dedication

To my beloved mother, Arati Sarkar, and father, Bholanath Sarkar, and my sister, Tanusree Sarkar, whose unwavering love, support, and encouragement have shaped my journey.

And to God, for the strength and guidance throughout every step of the way.

Preface to the Series

The Methods and Protocols in Food Science series is devoted to the publication of research protocols and methodologies in all fields of food science. The series is unique as it includes protocols developed, validated, and used by food and related scientists, and theoretical basis is also provided for each protocol. Aspects related to improvements in the protocols and adaptations and further developments in the protocols may also be approached.

The Methods and Protocols in Food Science series aims to bring the most recent developments in research protocols in the field as well as very well-established methods. As such the series targets undergraduates, graduates, and researchers in the field of food science and correlated areas. The protocols documented in the series will be highly useful for scientific inquiries in the field of food sciences and presented in such a way that the readers will be able to reproduce the experiments in a step-by-step style.

Each protocol will be characterized by a brief introductory section, followed by a short aims section, in which the precise purpose of the protocol is clarified. Then, an in-depth list of materials and reagents required for employing the protocol is presented, followed by comprehensive and step-by-step procedures on how to perform that experiment. The next section brings the dos and don'ts when carrying out the protocol, followed by the main pitfalls faced and how to troubleshoot them. Finally, template results will be presented and their meaning/conclusions addressed.

The Methods and Protocols in Food Science series will fill an important gap, addressing a common complain of food scientists, regarding the difficulties in repeating experiments detailed in scientific papers. With this, the series has a potential to become a reference material in food science laboratories of research centers and universities throughout the world.

Campinas, Brazil *Anderson S. Sant'Ana*

Preface

Food analysis has witnessed a wonderful transformation over the past decade, pushed with the aid of the convergence of cutting-edge technologies, growing customer demand for transparency, and the growing complexity of the global meals supply chain. *Protocols in Novel Food Analysis: Techniques and Applications* is a well-timed contribution to this evolving landscape, presenting a complete and realistic manual to advanced methodologies for detecting, profiling, and authenticating food additives and contaminants.

This book brings collectively a numerous range of progressive protocols that span molecular biology, spectroscopy, nanotechnology, metabolomics, and computational intelligence—each contributing uniquely to the safety, nice, and traceability of food products. Techniques that include CRISPR-Cas systems and DNA barcoding provide exceptional precision in identifying foodborne pathogens and verifying species authenticity. At the same time, nano-sensors, lab-on-a-chip structures, and surface plasmon resonance (SPR) are revolutionizing speedy, real-time food testing abilities.

The book is structured to serve as a reference and as a practical toolkit for researchers, analytical chemists, food technologists, and regulatory scientists. Each chapter presents a protocol, combining theoretical underpinnings with step-via-step programs, demanding situations, and predicted results. Topics range from spectroscopic and electrochemical techniques to lipidomics, proteomics, and the mixing of artificial intelligence and machine learning in food statistics interpretation.

Particular interest has been given to emerging and interdisciplinary regions which include tunable resistive pulse sensing (TRPS) for nanoparticle analysis, hyperspectral imaging for non-negative assessment, and digital nose and tongue technologies for sensory profiling. Moreover, the combination of isotopic and metabolomic gear allows a holistic know-how of meals origin and purposeful attributes.

As food systems become more intricate and client expectations extra stringent. By bridging innovation and practical application, it supports the development of meals protection, authenticity, and nutritional perception.

This collection of protocols will not only serve as a valuable reference for researchers but also act as a catalyst for advancing research and fostering innovation in the field of food analysis.

Malda, West Bengal, India ***Tanmay Sarkar***

Contents

Contributors

AMINU ABDULLAHI • *Department of Biotechnology, Vel Tech Rangarajan Dr. Sagunthala R&D Institute of Science and Technology, Chennai, India; Centre for Biomaterials and Environmental Biotechnology Vel Tech Rangarajan Dr. Sagunthala Institute of Science and Technology, Chennai, India; Department of Biotechnology, Modibbo Adama University, Yola, Nigeria*

MOHAMED MUJAMIL ABUTHAHIR • *Department of Biotechnology, Faculty of Engineering and Technology, SRM Institute of Science and Technology, Tiruchirappalli, Tamil Nadu, India*

EMMANUEL IBUKUN ADULOJU • *Department of Toxicology, Advanced Medical and Dental Institute (AMDI), Universiti Sains Malaysia, Kepala Batas, Penang, Malaysia; Department of Biochemical and Chemical Sciences, The Federal Polytechnic Offa, Offa, Kwara State, Nigeria*

KHALED AHMADAULI • *Campus of Agriculture and Natural Resources, Faculty of Natural resources, University of Tehran, Karaj, Iran*

EMMANUEL AYORINDE AJIBOYE • *Centre for Critical Minerals, Institute of Energy and Mineral Science, The Pennsylvania State University, State College, University Park, PA, USA*

V. C. ANITHA KRISHANAN • *Department of Food Technology, Nehru Institute of Technology, Coimbatore, Tamil Nadu, India*

AKRAMUL ANSARY • *Department of Pharmaceutical Chemistry, NETES Institute of Pharmaceutical Science, NEMCARE Group of Institutions, Mirza, Assam, India*

USAMA ANWAR • *Laboratory of Materials Science and Nanotechnology (LMNT), CR-INSTM, Department of Biomedical Sciences, Viale S. Pietro 43/B, University of Sassari, Sassari, Italy*

JOBIL J. ARACKAL • *Department of Food Technology, Saintgits College of Engineering, Kottayam, Kerala, India*

MUHAMMAD TAYYAB ARSHAD • *Institute of Food Science and Technology, The University of Lahore, Lahore, Pakistan*

MUHAMMAD ASIF • *National Institute of Food Science and Technology, University of Agriculture Faisalabad, Faisalabad, Pakistan; Department of Food Science and Technology, University of Agriculture Faisalabad-Subcampus Burewala, Burewala, Pakistan*

ELHAM AZARPAZHOOH • *Agricultural Engineering Research Department, Khorasan Razavi Agricultural and Natural Resources Research and Education Center, AREEO, Mashhad, Iran*

AYESHA BAIG • *Department de Quimica Analitica, Facultad de Quimica, Universidad de Sevilla, Sevilla, Spain*

HOMA BEHMADI • *Agricultural Engineering Research Institute, Agricultural Research, Education and Extension Organization (AREEO), Karaj, Iran*

ABHILASH BHARADWAJ • *Department of Pharmacognosy, NETES Institution of Pharmaceutical Science, NEMCARE Group of Institution, Mirza, Assam, India*

FAKHRA BIBI • *Department of Chemistry, University of Gujrat, Gujrat, Pakistan*

AMIR BOUALLEGUE • *Laboratory of Food Oral Processing, School of Food Science and Biotechnology, Zhejiang Gongshang University, Hangzhou, China*

Amrita Chakraborty • *Assistant Professor, Mount Carmel College, Bengaluru, Karnataka, India*

Ayan Chatterjee • *School of Health Science and Technology, Medhavi Skills University, Topakhani, Sikkim, India*

Amit Kumar Das • *Department of Pharmaceutics, NETES Institute of Pharmaceutical Science, NEMCARE Group of Institutions, Mirza, Assam, India*

Siddhartha Sankar Das • *Department of Pharmaceutical Chemistry, NETES Institute of Pharmaceutical Science, NEMCARE Group of Institutions, Mirza, Assam, India*

Suchandra Datta • *Guest Faculty, Department of Home Science, University of Calcutta, Kolkata, West Bengal, India*

José Antônio de Queiroz Lafetá Junior • *Cândido Tostes Dairy Institute, Agricultural Company of Minas Gerais (EPAMIG), Juiz de Fora, MG, Brazil*

Ragunathan Devendran • *Charles Tanford Protein Center, Martin Luther University, Halle (Saale), Germany*

Harmeet Singh Dhillon • *Dairy Microbiology Division, ICAR- National Dairy Research Institute, Karnal, Haryana, India*

Estela F. Diaz • *Universidad de Manila, Manila, Philippines*

Koushik Nandan Dutta • *Department of Pharmacognosy, NETES Institute of Pharmaceutical Science, NEMCARE Group of Institutions, Mirza, Assam, India*

Sumitha Elayaperumal • *Department of Biotechnology and Bioinformatics JSS Academy of Higher Education and Research, Mysore, Karnataka, India*

Walid Elfalleh • *Department of Biology, College of Science, Imam Mohammad Ibn Saud Islamic University (IMSIU), Riyadh, Saudi Arabia*

Mujahid Farid • *Department of Environmental Science, University of Gujrat, Gujrat, Pakistan*

Shumaila Fayyaz • *Department of Chemistry, The University of Lahore, Lahore, Pakistan*

Lorena Evangelista Fernandes • *Cândido Tostes Dairy Institute, Agricultural Company of Minas Gerais (EPAMIG), Tenente Luiz de Freitas, Juiz de Fora, MG, Brazil*

Gnanaselvan Gnanasekaran • *Department of Bioproducts and Biosystems Engineering, University of Minnesota, Minneapolis, MN, USA*

Rashmita Gogoi • *Department of Pharmacognosy, NETES Institution of Pharmaceutical Science, NEMCARE Group of Institution, Mirza, Assam, India*

Rajpreet Kaur Goraya • *Advanced Post-Harvest Technology Centre, Centre for Applied Research, Innovation and Entrepreneurship, Lethbridge Polytechnic, Lethbridge, AB, Canada*

T. Harshitha • *Department of Biotechnology and Bioinformatics, JSS Academy of Higher Education and Research, Mysore, Karnataka, India*

Darshana Hazarika • *Department of Community Health Nursing, Nemcare Institute of Nursing Sciences, Kamrup, Assam, India*

Haider Ijaz • *Department of Chemistry, University of Gujrat, Gujrat, Pakistan*

Zimal Ijaz • *Department of Chemistry, University of Gujrat, Gujrat, Pakistan*

Rukshana Irani • *Department of Food & Nutrition, Raidighi College, Raidighi, West Bengal, India*

Syed Tohshin Ishtiyak • *Department of Pharmacognosy, NETES Institution of Pharmaceutical Science, NEMCARE Group of Institution, Mirza, Assam, India*

Beenish Israr • *Institute of Home Sciences, University of Agriculture Faisalabad, Faisalabad, Punjab, Pakistan*

NIZWA ITRAT • *Institute of Home Sciences, University of Agriculture Faisalabad, Faisalabad, Punjab, Pakistan*

KHWALA BEN JEDDOU • *Laboratory of Plant Improvement and Valorization of Agricultural Resources, ENIS, University of Sfax, Sfax, Tunisia; Université de Sfax, ENIS, Unité de Service Bioréacteur couplé à un ultrafiltre, Sfax, Tunisia; High Institute of Applied Biology of Medenine, University of Gabes, Medenine, Tunisia*

SUFIYANU JIGA ABUBAKAR • *Marine Biodiversity centre, University of Aberdeen, Scotland, United kingdom*

B. K. JNANESH • *Department of Biotechnology and Bioinformatics JSS Academy of Higher Education and Research, Mysore, Karnataka, India*

ANN JUSTIN KAPPEN • *Department of Food Technology, Saintgits College of Engineering, Kottayam, Kerala, India*

ROBIN KAURA • *Dairy Engineering Division, ICAR- National Dairy Research Institute, Karnal, Haryana, India*

TARANGINI KORUMILLI • *Department of Biotechnology, Vel Tech Rangarajan Dr. Sagunthala R& D Institute of Science and Technology, Chennai, India; Centre for Biomaterials and Environmental Biotechnology Vel Tech Rangarajan Dr. Sagunthala Institute of Science and Technology, Chennai, India*

SUMAN LAMICHHANE • *Department of Pharmacy, Shree College of Technology, Chitwan, Nepal; Department of Pharmacy, Chitwan Medical College, Chitwan, Nepal*

MUHAMMAD RIZWAN MAJEED • *Department of Nutrition, University of Tennessee, Knoxville, TN, USA*

SAMMRA MAQSOOD • *National Institute of Food Science and Technology, University of Agriculture Faisalabad, Faisalabad, Pakistan*

KHADIJA MUNAWAR • *Department of Chemistry, University of Gujrat, Gujrat, Pakistan*

PRAMILA MURUGESAN • *Department of Biotechnology, Faculty of Engineering and Technology, SRM Institute of Science and Technology, Tiruchirappalli, Tamil Nadu, India*

MONDHER MZOUGHI • *Department of Food Engineering, Faculty of Chemical and Metallurgical Engineering, Istanbul Technical University, Maslak, Istanbul, Türkiye*

BALAKRISHNAN NAVINA • *Department of Post-Harvest Technology, ETSEA-University of Lleida, Lleida, Spain*

A NIKHILESH • *Department of Food Technology, Saintgits College of Engineering, Kottayam, Kerala, India*

ZAINAB NOOR • *Department of Biochemistry and Biotechnology, University of Gujrat, Gujrat, Pakistan*

BOSEDE ALICE OMACHI • *Department of Dietetics and Human Nutrition, University of KwaZulu-Natal, Pietermaritzburg, KwaZulu-Natal, South Africa*

ANA FLÁVIA COELHO PACHECO • *Cândido Tostes Dairy Institute, Agricultural Company of Minas Gerais (EPAMIG), Tenente Luiz de Freitas, Juiz de Fora, MG, Brazil*

FLAVIANA COELHO PACHECO • *Department of Food Technology (DTA), Federal University of Viçosa (UFV), Viçosa, MG, Brazil*

PAULO HENRIQUE COSTA PAIVA • *Cândido Tostes Dairy Institute, Agricultural Company of Minas Gerais (EPAMIG), Tenente Luiz de Freitas, Juiz de Fora, MG, Brazil*

JITENDRA PANDEY • *Department of Chemistry, University of Hawaii at Manoa, Honolulu, HI, USA*

POLLOBI PORASAR • *Department of Pharmacognosy, NETES Institution of Pharmaceutical Science, NEMCARE Group of Institution, Mirza, Assam, India*

G. Rajeshwari • *Department of Biotechnology, Faculty of Engineering and Technology, SRM Institute of Science and Technology, Tiruchirappalli, Tamil Nadu, India*
Hosahalli S. Ramaswamy • *Department of Food Science and Agricultural Chemistry, Macdonald Campus of McGill University. Anne de Bellevue, Sainte-Anne-de-Bellevue, QC, Canada*
H. P. Vasantha Rupasinghe • *Department of Plant, Food, and Environmental Sciences, Faculty of Agriculture, Dalhousie University, Halifax, Canada*
Kübra Sağlam • *Department of Food Processing, Food Technology Program, Istanbul Gelisim Vocational School, Istanbul Gelisim University, Avcilar, Istanbul, Türkiye*
Supriya Sahani • *Department of Pharmacognosy, NETES Institution of Pharmaceutical Science, NEMCARE Group of Institution, Mirza, Assam, India*
R. Sanjana • *Department of Biotechnology and Bioinformatics, JSS Academy of Higher Education and Research, Mysore, Karnataka, India*
T. D. Sanjana • *Department of Biotechnology and Bioinformatics, JSS Academy of Higher Education and Research, Mysore, Karnataka, India*
S. Saranyadevi • *Department of Biotechnology, Faculty of Engineering (FoE), Karpagam Academy of Higher Education, Coimbatore, Tamil Nadu, India*
T Selva Kumar • *Vel Tech Rangarajan Dr. Sagunthala R&D Institute of Science and Technology, Chennai, Tamil Nadu, India*
Javeria Shabbir • *Department of Nutrition and Dietetics, The University of Faisalabad, Punjab, Pakistan*
Sadaf Shakoor • *Department of Food Science and Technology, University of Agriculture Faisalabad-Subcampus Burewala, Burewala, Pakistan*
Dipjyoti Sharma • *Department of Pharmacognosy, NETES Institution of Pharmaceutical Science, NEMCARE Group of Institution, Mirza, Assam, India*
B. S. Shrinidhi • *Department of Biotechnology and Bioinformatics JSS Academy of Higher Education and Research, Mysore, Karnataka, India*
Ayesha Siddiqa • *Aquatic Ecology and Water Management, Wageningen University and Research, Wageningen, Netherlands*
Mohit Singla • *Advanced Post-Harvest Technology Centre, Centre for Applied Research, Innovation and Entrepreneurship, Lethbridge Polytechnic, Lethbridge, AB, Canada*
Chitranayak Sinha • *Dairy Engineering Division, ICAR- National Dairy Research Institute, Karnal, Haryana, India*
Yuvaraj Sivamani • *Crescent School of Pharmacy, B.S. Abdur Rahman Crescent Institute of Science and Technology, Chennai, Tamilnadu, India*
Saranyadevi Subburaj • *Department of Biotechnology Karpagam Academy of Higher Education, Coimbatore, India*
Akshay Sunil • *Department of Biotechnology and Bioinformatics JSS Academy of Higher Education and Research, Mysore, Karnataka, India*
K. Surendra • *Department of Physics, Centurion University of Management and Technology, Vizianagaram, Andhrapradesh, India*
Nishchal Thakur • *Dairy Microbiology Division, ICAR- National Dairy Research Institute, Karnal, Haryana, India*
Aman Ullah • *College of Natural and Applied Sciences (CNAS), Department of Agricultural, Food & Nutritional Science, University of Alberta, Edmonton, AB, Canada*
Mudassar Zafar • *Department of Chemistry, The University of Lahore, Lahore, Pakistan*

MUHAMMAD NADEEM ZAFAR • *Department of Chemistry, University of Gujrat, Gujrat, Pakistan*
ASIF ZARNAB • *University of South Australia, UniSA Clinical and Health Sciences, Adelaide, Australia*
MUHAMMAD ZUBAIR • *Department of Chemistry, University of Gujrat, Gujrat, Pakistan*

Chapter 1

CRISPR-Cas Systems for Foodborne Pathogen Detection

T. D. Sanjana, R. Sanjana, T. Harshitha, Yuvaraj Sivamani, K. Surendra, Ragunathan Devendran, and Sumitha Elayaperumal

Abstract

Foodborne pathogens are microorganisms such as bacteria, fungi, and viruses that cause significant disease outbreaks worldwide. Even though there are many recent advancement in techniques in food preservation and food safety, these pathogens still cause substantial impacts on public health due to contaminated food. This chapter shows different methods like culture-based, Polymerase Chain Reaction (PCR), immunoassays, and next-generation sequencing (NGS)-based methods but mainly focuses on the recently developed gene-editing technique. CRISPR Cas9 system is an advanced and currently emerging scientific technique that helps alter an organism's DNA as it is more accurate and speed than other techniques. This review shows how CRISPR-Cas9 technology helps to detect foodborne pathogens.

Key words Foodborne, CRISPR Cas9, Foodborne Illnesses, Biological materials

1 Introduction

Foods are the basic essential nutrients for all organisms, which are essentially obtained from plants and animals and consist of proteins, minerals, vitamins, and other organic compounds. It is necessary to consume the proper quality of food to maintain proper health and wellness. Food preservation is a method to maintain the shelf life and quality of food for a longer period. There will be spoilage of food, which will be caused by microorganisms due to the availability of nutrients for their growth [1]. According to the World Health Organization, approximately 600 million cases of foodborne illnesses and 420,000 deaths could occur in a year, among them 40% of foodborne diseases are carried by children under the age of 5 years and cause 125,000 deaths every year. This impacts the economy, causes health issues, and increases mortality. Foodborne pathogens are characterized by bacteria, viruses, and parasites that contaminate the food and cause foodborne illness. It is necessary to maintain proper food safety management to protect public health

Tanmay Sarkar (ed.), *Novel Food Analysis: Techniques and Applications*, Methods and Protocols in Food Science,
https://doi.org/10.1007/978-1-0716-4787-5_1,

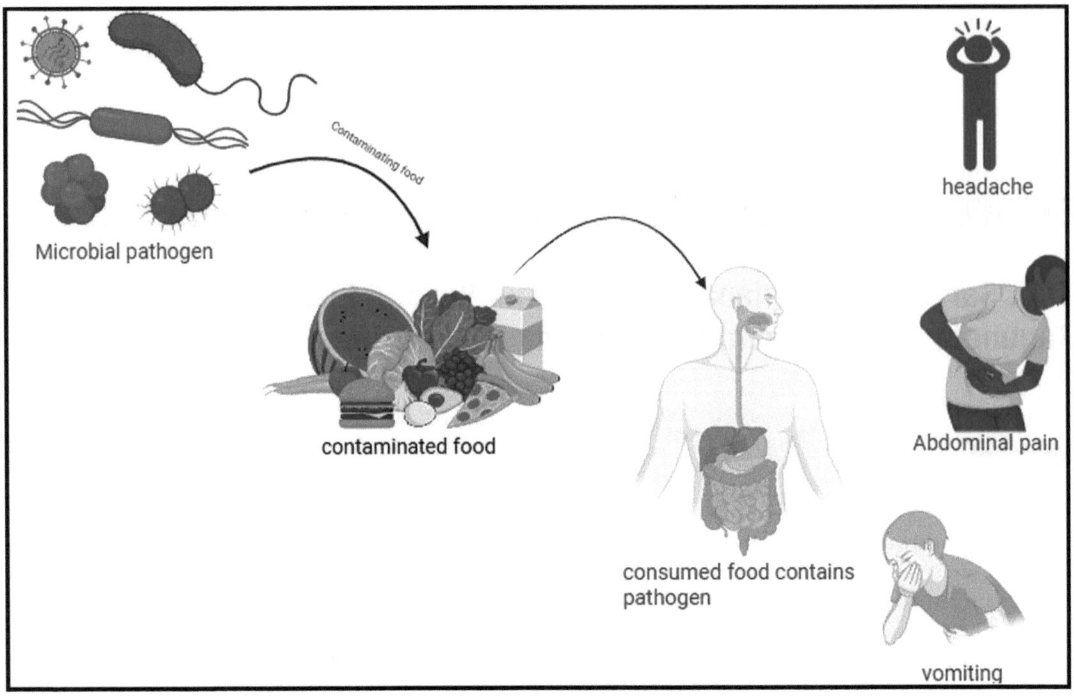

Fig. 1 Effect of microbial pathogens on human health

from foodborne pathogens by implementing strategies to control the risk of contaminating the food. These microorganisms contaminate food and can cause mild to severe diseases in humans [2]. Figure 1 describes the impact of microbial pathogens on human health. They include bacteria such as Salmonella and *E. coli*, viruses like Norovirus and Hepatitis A, as well as parasites such as *Toxoplasma gondii* and *Giardia lamblia*. Foodborne illnesses are classified into three categories based on how they affect the body: food intoxication, food infection, and food toxi-infection. Each category contains different toxins and pathogens and causes different symptoms based on the mechanism of action, which requires specialized treatment and prevention [3].

Food Intoxication It is a common and dangerous illness caused by the consumption of food that is contaminated with toxins produced by pathogens such as bacteria, fungi, and parasites. This food intoxication causes mild symptoms (stomach upset) to severe life-threatening conditions (hospitalization and death), based on the amount of toxins taken by the person. So, it is essential to prevent food intoxication by proper storage, handling, and preparation of food, which helps in reducing the risk of contamination and maintaining proper health. Some of the pathogens that cause food intoxication are *Clostridium botulinum*, *Staphylococcus aureus*, etc. [3].

Table 1
Overview of foodborne pathogens: onset time, symptoms, and contamination sources

Organisms	Incubation time to cause symptoms	Predominant symptoms	Sources
Salmonella	12–72 h	High fever, diarrhea, abdominal cramps, headache, blood in urine or stool	Raw or undercooked meat, poultry, eggs, seafood, and unpasteurized dairy products
E. coli	10 days	Diarrhea, loss of appetite, abdominal or pelvic pain, cloudy, foul-smelling urine	Raw fruits and vegetables, undercooked meat, and unpasteurized beverages, drinking contaminated water
Norovirus (stomach flu)	12–48 h	Nausea, vomiting, diarrhea, stomach pain, headache, fever, and body aches. Complications: Feeling weak or dizzy, fussiness, sleeping throughout the day	Food and water that are contaminated with norovirus
Hepatitis A	14–28 days	Malaise, fever, abdominal discomfort, nausea, dark-colored urine, and jaundice	Lack of safe water, poor sanitation
Toxoplasma gondii	1–5 days	Swollen lymph nodes, inflammation of the lungs, heart muscles, and eyes, numbness, seizures, toxoplasmic encephalitis, and coma	Eating undercooked meat, contaminated food or water, blood transfusion, or organ transplantation, mother to fetus

Food Infection It is an illness caused by consuming food that contains viable microorganisms. During this infection, the microorganisms multiply within the host organism. It requires several hours to several days for the incubation period, which takes place within the intestinal tract before symptoms develop. Food infection is classified into two types: enterotoxigenic type of food infection and invasive type of food infection [4]. It may cause fever and vomiting. Some food infection pathogens include *Vibrio cholerae*, Shigella, etc. Table 1.

Food Toxi-Infection Food toxi-infections are those caused by food being contaminated by biological agents or toxins. Their dependency is based on temperature and moisture conditions that occur significantly during the summer season. They have a specialized character that can reach large populations within one to two hours. This type of infection occurs on the food which are prepared ahead of time and consumed hours later without storing it in a proper condition. It may cause mild to severe symptoms, but they will disappear after a few days. Some of the pathogens that cause food poisoning are *Bacillus cereus* and *E. coli* [5].

1.1 Microbial Pathogens Causing Foodborne Illness

Bacteria

Salmonella are bacteria which are gram-negative bacterial pathogens that belong to the Enterobacteriaceae family. Salmonella causes diseases like gastroenteritis, bacteremia, and typhoid fever; the most common is gastroenteritis. About 95% of humans are affected by Salmonellosis. The consumption of contaminated products like meat, poultry, eggs, milk, and seafood, as well as uncooked or improperly cooked meat, causes foodborne pathogen diseases in humans [6]. *E. coli* is one of the main foodborne pathogens, types like enteropathogenic, and enter-invasive are type of foodborne pathogens that cause diarrhea. Shiga toxin-producing Escherichia coli (STEC) is a major cause of foodborne illness. This Shiga toxin *E. coli* causes bloody diarrhea and can even cause death. Among *E. coli* bacteria the E. coli 0157:H7 is the most dangerous strain that causes foodborne pathogen diseases. *E. coli* also acts as an indicator bacterium, indicating fecal contamination, which indicates the presence of coliform and other enteric bacteria that contaminate the food products [6].

Viruses

Norovirus is a major foodborne pathogen that is highly transmissible. A single norovirus can cause approximately 0.5% infection (Table 1). It is highly contagious and causes gastroenteritis with symptoms like vomiting and diarrhea. Norovirus infection can cause severe infection in children and older persons; the persons who are affected by Norovirus may require hospitalization. Sanitation is one of the most important causes of norovirus to cause infection in humans, they spread from person to person and even from surface to person modes [7]. It spreads through the fecal-oral route, contaminating food and water and causing foodborne pathogens. Food can be contaminated at the source or even during preparation. Hepatitis A virus is a foodborne pathogen that can be transmitted through contaminated food, water, environmental surfaces, and even through direct or indirect contact between person and persons. It is commonly transmitted through the fecal-oral route. Hepatitis A is stable under environmental conditions like freezing, heat, chemicals, and desiccation [6]. The consumption of shellfish also causes the foodborne pathogen caused by the hepatitis A virus. Shellfish growing waters are monitored to observe the presence of the virus in the water, as shellfish can cause this virus to form a foodborne pathogen. *Toxoplasma gondii* is a parasitic protozoan that causes toxoplasmosis, and foodborne pathogen illness. *Toxoplasma gondii* uses cats as its primary reservoir and warm-blooded organisms. In humans, *Toxoplasma gondii* is acquired through the ingestion of contaminated food and water containing the oocyst, contaminated blood transfusion, and organ transplantation [8]. *Toxoplasma gondii* is caused by the uptake of

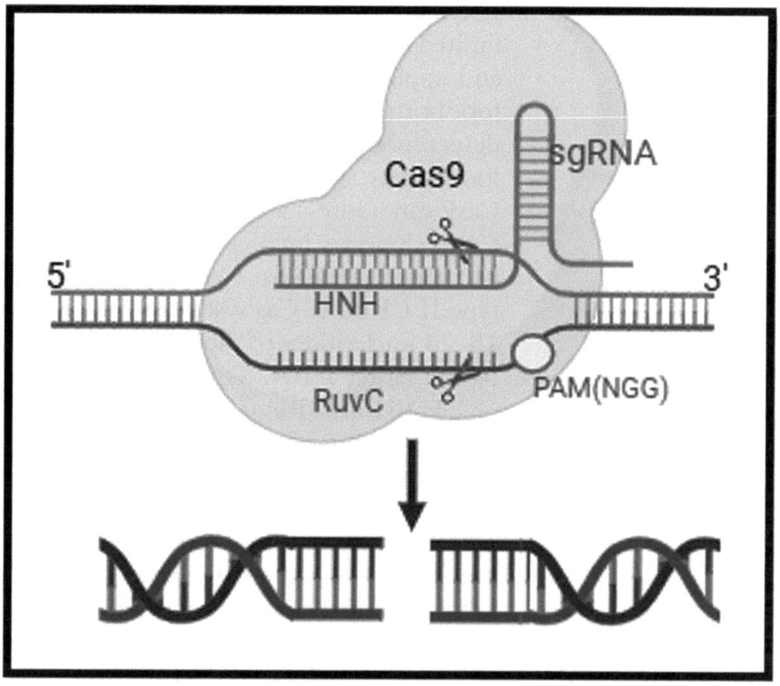

Fig. 2 Mechanism of CRISPR-Cas9-mediated genome editing

cysts in raw or undercooked meat, with fresh pork and beef being the primary source for *Toxoplasma gondii*. *Giardia lamblia* is a parasite that causes an intestinal infection called Giardiasis. It is one of the significant foodborne pathogens. It is primarily transmitted by the fecal-oral route. Giardia cysts are ingested by a person when present in contaminated food and water. It can be found in food sources like meat or uncooked meat, and it can even be found in the water used to irrigate crops [9].

This chapter particularly concentrates on the CRISPR-CAS9-based detection of foodborne pathogens. The recently emerged technology Clustered Regularly Interspaced Short Palindromic Repeats (CRISPR)-Cas9 has opened new opportunities for the detection and identification of foodborne pathogens (Fig. 2). This technology brings revolutionary opportunities in molecular biology, particularly in gene editing. It is a natural, programmable technique that is highly specific, and it is a powerful tool that can target nucleic acid sequences. It is not only an important tool in genetic engineering, but it is also important in diagnostic applications. In the context of pathogen detection, CRISPR-Cas9 is highly specific and sensitive, and it is a rapidly performing tool that identifies the genetic markers of various foodborne microorganisms [10]. This CRISPR technology typically uses isothermal application methods that operate at a constant temperature and avoid the cost and complexity of thermal cyclers required for

Polymerase Chain Reaction (PCR), enabling detection in resource-limited settings. This chapter explains the principles, advancements, and applications of CRISPR-Cas9 technology in the detection of foodborne pathogens. There are a wide variety of techniques to detect foodborne pathogens, each playing a pivotal role in ensuring food safety. While these techniques exhibit drawbacks, CRISPR Cas9 can address and overcome them. There are several CRISPR-Cas techniques involved in the detection of foodborne pathogens like Cas12a, Cas13a, Cas14a, and Cas9 [11]. Cas9 is a part of the type II CRISPR-Cas system, where the guide RNA has a length of 18–24 nucleotides of the spacer sequence, which is complementary to the target DNA. This Cas9 system has two nuclease domains, i.e., HNH and RuvC, where HNH cuts the DNA strand that is complementary to the guide RNA (gRNA) and RuvC cuts the noncomplementary strand. This guide RNA/single-guide RNA (sgRNA) is a synthetic RNA that combines target recognition and a scaffold for the Cas9 binding. This requires a protospacer adjacent motif (PAM), which is essential for Cas9 binding and cleaving DNA; this makes this system ideal for targeting pathogen-specific genes [12]. This Cas9 targets and cuts the double-stranded DNA (dsDNA), which results in the blunt ends. As this does not perform trans cleavage, unlike the other Cas proteins like Cas12 and Cas13, this Cas9-based protein cuts only the guided, specific DNA sequence and does not cut the surrounding sequences.

The CRISPR-Cas9 detection provides superior specificity and rapid results compared to culture, ELISA, and PCR-based methods. These traditional approaches have suffered from long turnaround times or cross-reactivity, whereas this system operates under isothermal conditions, and also requires minimal instrumentation and high sensitivity. The table explains how various traditional detection methods differ from one another, along with CRISPR-Cas9, and compares their effectiveness, specificity, and sensitivity in detecting foodborne pathogens [12].

Method	Specificity and sensitivity	Time and conditions	Limitations/advantages
Culture	High specificity Moderate sensitivity	Slow (24–72 h) Requires incubation	Time-consuming Requires the presence of viable organisms
ELISA	Moderate specificity Moderate sensitivity	Moderate time (2–6 h) Non-isothermal	Requires labelled antibodies Highly prone to cross reactivity
PCR	High specificity High sensitivity	Moderate time (3–5 h) Requires thermal cycling	Requires expensive equipment Sensitive to inhibitors

(continued)

Method	Specificity and sensitivity	Time and conditions	Limitations/advantages
CRISPR-Cas9	Superior specificity Very high sensitivity	Rapid (<1 h) Isothermal conditions	Minimal instrumentation Field-deployable Low cross-reactivity

2 Materials

The detection of foodborne pathogens has been revolutionized with the introduction of CRISPR-Cas9 technology, enabling the rapid, sensitive, and accurate identification of microbial contaminants in food products. This technology employs the Cas9 enzyme guided by sgRNA to process specific DNA sequences of pathogens such as Salmonella, *Escherichia coli*, and *Listeria monocytogenes* [13]. Detection systems based on CRISPR-Cas9 offer many materials and methods tailored to increasing sensitivity, shortening detection time, and allowing testing to be done at the site of sample collection. A vital part of these detection systems is the design of specific sgRNAs, which determine the area of the pathogen's DNA that Cas9 is to cut. Cas9, upon binding to the target site, will make double-strand breaks, which can be used to generate signals in many types of assays. As an illustration, in lateral flow tests, gold nanoparticles (AuNPs) undergo a colorimetric change that indicates the presence of the target DNA-Cas9 complex, providing simple and fast detection [14]. For better sensitivity in detection, CRISPR-Cas9 assays routinely use isothermal amplification approaches such as loop-mediated isothermal amplification (LAMP) and recombinase polymerase amplification (RPA) for amplifying the target DNA. Further developments in signal transduction have improved the efficiency of CRISPR-Cas9-based biosensors. As an example, for the detection of infections at extremely low concentrations, the detection signal can be amplified through strand displacement actions and hybridization chain reactions (HCR). Moreover, other research has been conducted to use metal-organic frameworks (MOFs) as scaffolds for CRISPR-Cas9 assays in order to increase the stability and sensitivity of the detection systems [15]. In the efforts to make CRISPR-Cas9 diagnostic tools more cost-effective and portable, there have been developments such as color-changing paper-based assays. These tests, which were designed for the detection of Salmonella, produce results in a few hours and are considerably cheaper than traditional methods. These innovations can greatly aid in monitoring food safety, especially in less economically developed regions.

2.1 Core Biological Components

The innovation of CRISPR-Cas9 technology brought rapid and precise detection of foodborne pathogens. Target biological components of detection systems based on CRISPR-Cas9 include: the Cas9 protein, sgRNA, target DNA sequences, and amplification enzymes. This review delineates every compound, explaining its role and contribution to pathogen detection. The protein Cas9 serves as an RNA-guided DNA endonuclease and is a central feature of CRISPR-Cas9 systems [16]. Cleavage of dsDNA is enabled as Cas9, an enzyme from *Streptococcus pyogenes*, cuts dsDNA at regions indicated by sgRNA. With two nucleases, RuvC domain and HNH domain, Cas9 is the second component of CRISPR-Cas9. The HNH domain cuts the strand holding a copy of sgRNA, while RuvC domain cuts the strand that does not hold a complementary strand. Their joint action at the target region results in a double-strand break. Detection of foodborne pathogens such as *Salmonella enterica*, *Escherichia coli*, and *Listeria monocytogenes* is made possible through Cas9's specificity and the use of advanced food diagnostics technologies [17]. In one study, hemolysin (hly) gene detection from *L. monocytogenes* was achieved following the use of Cas9 and isothermal exponential amplification reaction (EXPAR) with a sensitivity of 0.82 amol. sgRNA is produced from the fusion of two RNA forms, the trans-activating crRNA (tracrRNA) and the CRISPR RNA (crRNA). Alongside the cleaved DNA strand, the chimeric RNA guides Cas9 to the designated double-strand break (DSB) site of interest. A 20-nucleotide stretch within the sgRNA delineated the target DNA, ensuring binding occurs. The PAM sequence (usually 5'-NGG-3′) is equally necessary for site recognition by Cas9 since it provides ample opportunity for binding alongside the target sequence. Unlike other methods, sgRNAs are meant to identify genes exclusive to pathogens. As an illustration, antifood poisoning detection systems based on CRISPR-Cas9 have deployed sgRNAs directed against the invA gene from Salmonella for identification of the pest in contaminated food samples [18]. Detection systems that utilize CRISPR-Cas9 have a particular focus on the specificity and sensitivity related to the selection of target DNA sequences, which sit at the very core of genetic detection. The target sequences are usually some forms of distinct genomic markers of the disease in question. The sensitivity for detection systems with CRISPR-Cas9 at their core is mainly boosted through the use of nucleic acid amplifying techniques. The ease of detection is achieved by increasing the amount of target DNA. RPA does not require thermal cycling as it is an isothermal amplification method that takes place at a constant temperature. It is effective for point-of-care (POC) use, fast, and sensitive. A variety of foodborne pathogens, such as Salmonella and Listeria, have been identified using RPA coupled with CRISPR-Cas9 systems [19].

2.2 Isothermal Amplification Assisted by Loops (LAMP)

LAMP is another isothermal technique that is capable of amplifying DNA to a very high degree. When paired with CRISPR-Cas9, LAMP significantly augments detection abilities and enables detection of infections at lower levels. The gold standard method for amplifying DNA remains the PCR. In the case of detection using CRISPR-Cas9, the target sequence is first amplified using PCR, which increases the sensitivity of the assay. In CRISPR-Cas9 detection, multiple functions are attributed to nanoparticles (NPs) within CRISPR-Cas9 systems that function as biosensors [20]. They include the following: signal amplification, target enrichment, and provision for multisensory detection of the desired analyte. The unique optical properties of AuNPs, particularly localized surface plasmon resonance (LSPR), which enables colorimetric detection, make them popular. In CRISPR-AWARD experiments, ssDNA probes can be used to functionalize AuNPs. The presence of target DNA causes the probes to aggregate, which leads to a distinct color change. Cleavage of ssDNA linkers in the detection of Salmonella using the CRISPR/Cas12 system results in preventing AuNP aggregation, maintaining the red colored solution, which indicates passing of the test. MNPs facilitated the extraction and purification of target nucleic acids from complex biological mixtures [21]. The attachment of specific probes or antibodies to MNPs allows for the magnetic separation of pathogens or target DNA. This enrichment step improves the detection sensitivity of CRISPR-based detection. For example, Salmonella DNA has been isolated by MNPs and detected using the CRISPR/Cas12a system. Semiconductor nanoparticles that display high intensity, scotopic stability, and fluorescence that is size-dependent are called quantum dots (QDs). When QDs are used in CRISPR assays, their fluorescence changes because of nucleic acid probe and Cas enzyme interaction, resulting in enzymatic target cleavage. Such features enable sensitive and multiplexed detection of pathogens [22]. Nanozymes are materials on a nanoscale that exhibit properties of catalysts akin to those found in enzymes. They can be integrated into CRISPR-based biosensors to boost the detection signals. For example, detection of *Burkholderia pseudomallei* DNA can be achieved with high sensitivity using PtPd@PCN-224 nanozymes with CRISPR/Cas14a systems. The electrochemical signal that is caused by the nanozyme breakdown of H2O2 correlates with the target DNA. GO is a two-dimensional oxide of carbon that has a large surface area and the ability to quench fluorescence. GO can adsorb ssDNA probes with fluorescein tags in CRISPR-based sensors. The probes are released and fluorescence is restored when the target is recognized and cut by the Cas enzymes [23]. This technique has been employed to capture microorganisms leading to sepsis. These reporters transform the events of molecular recognition within the CRISPR-Cas9 systems into an output measurable in value. Depending on the type of detection sensor used, different reporters for CRISPR-Cas9 signal reporters need to be used. These reporters are usually adopted because sensitivity and real-time monitoring are their forte. An ssDNA probe is linked to a fluorophore and a quencher. Cleavage of the target DNA causes loss of fluorescence. Once the Cas enzyme gets to the target, the probe gets cleaved and the quencher is separated from the fluorophore, leading to the emission of fluorescence [24]. This method has been used for the detection of several foodborne pathogens. Its simplicity, as well as the lack of sophisticated equipment for colorimetric detection, makes it quite appealing. An increase in AuNPs' aggregates caused the red color to change to purple/blue, which makes them colorimetric reporters. In CRISPR-based assays, target DNA holds its red color, while in the absence of target DNA, aggregation occurs along with a color shift to purple and or blue. This method makes Salmonella and other bacteria visible to the naked eye. Detection with SERS reporters is more sensitive than regular Raman scattering, as SERS enhances the signals [25]. Changes in current, voltage, or impedance are measured, as in electrochemical detection, the target is detected when there is a change. The electrochemical signal of CRISPR-based electrochemical biosensors is altered as the Taq enzymes cut DNA probes. For instance, there is a gFET biosensor where the output signal is changed for detecting electrical signals and a dCas9-sgRNA complex to target DNA hybridize. Using Raman-sensitive reagents, gold or silver nanoparticles are functionalized with DNA probes in CB SERS tests. The subsequent changes in SERS signal after the target identification and cleavage done by Cas enzymes signify the negative presence of the pathogen that the signal targets [25].

Molecular diagnostics have been revolutionized by the development of CRISPR-Cas9 technology since it can pinpoint nucleic acids very precisely. Combining CRISPR-Cas9 with various consumables and assay platforms has facilitated swift and accurate detection of foodborne pathogens in the food safety industry. This integration serves a crucial role in public health and epidemic surveillance. Microfluidic devices, or "lab-on-a-chip" devices, have emerged as powerful tools in CRISPR-based diagnostics [26]. These systems enable automation, downsizing, and high-throughput analysis by combining several laboratory operations onto a single chip. They are made from widely used materials such as glass, quartz, hydrogels, polymers, and paper. Microfluidic devices that are paper-based (mUPAD) are particularly advantageous because of their low-cost and simple fabrication, as well as the ability to employ capillary-driven flow, which eliminates the need for external pumps.

2.3 Lateral Flow Assays (LFA)

Due to their simplicity and rapid turnaround times, lateral flow assays (LFAs) are routinely used for POC testing. In CRISPR-based LFA systems, the Cas9 enzyme is guided by a specific sgRNA to bind to the target DNA and undergoes cleavage. The presence of the pathogen is confirmed by the creation of a visible line on a test strip, which is the result of the complex being detected by the labeled probes, most commonly gold nanoparticles (AuNPs). Wang et al. developed a CRISPR-Cas9-based lateral flow nucleic acid assay (CASLFA) specifically designed for Listeria monocytogenes. This assay showed high specificity and practicality for field testing by achieving a limit of detection of 150–200 copies per reaction within one hour. Control of the fluidic system is taken a step further with DMFs, which allow precise manipulation of individual droplets on a surface using electric signals [27]. The integration of CRISPR/Cas systems into DMFs enables fully automated, high-throughput detection with minimal reagents. A DMF chip-CRISPR/Cas12a integration system for *Staphylococcus aureus* detection was developed by Lu et al. Working with only 6.6 μL of reagents, the system achieved a lower limit of detection of 32 CFU/mL within 55 min. The translation of biological interactions into electric signals provides a high level of sensitivity and even the possibility of further miniaturization of electrochemical biosensors. When integrated with CRISPR/Cas systems, these biosensors can detect specific DNA sequences through changes in voltage or current [28]. The important enzymes that perform target recognition and cleavage of nucleic acids are the different forms of the Cas9 protein and the Cas9 protein itself. Other components that are also of necessity include nucleotides, the buffers designed for enzyme activity, and sgRNAs corresponding to specific pathogen genes. Other techniques of DNA amplification, like LAMP and RPA, are routinely employed in CRISPR-based tests to enhance the tests' sensitivity. These kits add value to CRISPR detection by providing enzymes and primers for rapid isothermal DNA amplification [24, 25]. The results can be viewed using colorimetric or fluorescein probes. In fluorophore assays, the activation of Cas9 results in the destruction of a covalently linked quencher, releasing a signal. A colorimetric assay could employ AuNPs that aggregate and change color in the presence of the target DNA. To prevent cross-contamination, consumable microfluidic chips and test strips must be designed as disposables. These parts are designed from PDMS, paper, or plastic to make them user-friendly because they come preloaded with the necessary reagents. As stated before, having a precise sample to work on is crucial for accuracy. The elimination and purification of complex food matrices containing DNA require the use of filters, lysis buffers, and nucleic acid extraction kits, which are compatible with the following steps in the CRISPR-based testing process.

3 Methods

3.1 Cas9-Mediated Fluorescence Detection

Eating healthy is good for one's overall well-being; however, if the food is contaminated with pathogens, it can lead to severe health risks. Public health is of utmost priority in order to avoid adverse epidemiological events. Relying on culture-based techniques often leads to an elaborate setup, which can cost time and money to get it working. Thanks to CRISPR-Cas9 technology, the detection and identification of foodborne pathogens is now easy, efficient, and quick. Precise recognition of different types of DNA allows differentiation of nonpathogenic and pathogenic operators. Researchers have made extraordinary advancements with fluorescent tags, enabling expedient identification of pathogenic DNA sequences using remarkable detection techniques [29].

3.2 Working Principle

As regards the Cas9 integration fluorescence-based bio-sensors, the concept is based principally on the sequence-specific recognition and cutting of the DNA. Here, a sgRNA pair leads the Cas9 enzyme to the target region within the genome of the organism of interest. After binding, Cas9 cleaves both strands of the DNA at the specific site. A chemical reaction that results from this interaction is made evident when a probe with a fluorophore and a quencher is added. If the target DNA is absent, the probe is intact, and quenching is maintained [28, 29]. When the probe is displaced or cut by the action of the binding and cleavage of the DNA by Cas9, the quenching and unquenching result in the separation of the amino acid and fluorophore. This displacement results in a detectable signal. The decrease in intensity of fluorescence in the probe results from the presence of nucleotides, specifically in the relative amount of the pathogen DNA hybridized to the probe. The process begins with the design of an sgRNA that is complementary to a unique and conserved region of the target pathogen's DNA [30]. This ensures specificity in detection and minimizes off-target effects. The synthesized sgRNA is incubated with the Cas9 enzyme to form a ribonucleoprotein complex. This complex is essential for the subsequent DNA recognition and cleavage steps. DNA is extracted from food samples that have the potential to be contaminated. Processes such as homogenization, lysis, and purification may be required to obtain DNA of adequate quality for analysis. The reaction mixture is increased with a DNA probe that has been specifically designed and tagged with a fluorophore and a quencher [30]. This probe is designed to bind near the target region, which is recognized by the sgRNA. Optimal conditions include combining the Cas9-sgRNA complex with target DNA and a fluorescent probe. If the target DNA is present, the fluorescent probe will be displaced or cut because Cas9 binds and cuts at the exact region. The increase in fluorescence intensity resulting from the dissociation of the fluorophore from the quencher can be quantified using a fluorometer or comparable instrument [31]. The analysis of data involves checking the acquired fluorescence data for the presence of the targeted pathogen. It is possible to evaluate both qualitative and quantitative measures due to the relationship between the intensity of the fluorescence signal and the concentration of pathogenic DNA in the sample. The technique of detecting Cas9-mediated fluorescence is a sensible extension to surveillance in public health and monitoring food safety since it allows the accurate detection and identification of foodborne pathogens rapidly and sensitively [32].

3.3 CRISPR-Cas9-Based Colorimetric Detection

Prompt and accurate detection of foodborne microorganisms is crucial for food safety and public health. Detection methods like PCR and culturing techniques usually require specialized equipment, expert personnel, and time-consuming processes. Conversely, colorimetric detection methods provide a noninstrumental interpretation by displaying a change that is easily detectable at a glance [32]. Incorporating CRISPR-Cas9 technology into colorimetric assays provides a new avenue for pathogen detection. The system's ability to recognize specific sequences within the target DNA enables the development of assays designed to undergo clear chromatic alterations in the presence of specific infections. This strategy particularly caters to the needs of POC testing for low-resource settings because it combines the features of precision molecular diagnostics with effortless visual inspection [33]. The colorimetric detection approach based on CRISPR-Cas9 technology relies on specific DNA recognition and cleavage, coupled with the production of a colorimetric signal. In this approach, a sgRNA brings the Cas9 protein to the target pathogen genome's complementary DNA sequence. The Cas9 protein, once binding, performs double-strand termination of the targeted region [5]. To achieve this goal, the AuNPs that are customized with shape-specific DNA probes are also employed to yield the expected color change based on the reporter device responsible for responding to the cleavage event. One of the approaches is using AuNPs functionalized with DNA probes. When the said DNA is absent, the AuNPs remain in solution and preserve their red color. However, upon cleavage of the target DNA by the Cas9-sgRNA complex, the remaining DNA fragments increase in AuNPs and subsequently decrease in color to blue or purple. This shift in color, which can be observed without the assistance of a microscope, indicates the pathogen's presence [34].

The very first task within the experiment is synthesizing an sgRNA that pairs with vernacular, highly conserved regions in the sequence of the pathogen's DNA. This enhances detection specificity and eliminates any off-target interactions. The detected food samples, which are contaminated are further processed to extract

DNA. This process involves homogenization, lysis, and purification to obtain high-quality DNA suitable for analysis. The sgRNA is incubated with the Cas9 enzyme to form a ribonucleoprotein complex. The complex is responsible for the cleavage and subsequent recognition of the pliable DNA [35]. Potentially contaminated food samples are the source of DNA used. Purification, lysis, homogenization, or other procedures to generate DNA suitable for analysis are what these steps require. AuNPs are functionalized using target DNA sequence complementary DNA probes. These probes are designed to assist in the aggregation of fragments of DNA when they hybridize with the broken pieces of DNA. Participants respond to and identify the AuNP probes to which the DNA fragments hybridize in the event that the Cas9-sgRNA complex is formed [36]. If present, Cas9 will bind and cleave the DNA edge with a foresight to the most posterior (5′) of the fragments, which will then complement with AuNP probes. During this process, the AuNPs will undergo capping, resulting in a color change from red to blue or purple, which can be recognized with the naked eye. The target pathogen is qualitatively assessed based on the colorimetric shift, as a visually noted color change signifies pathogenic presence. Quantitative analysis can be computed using a spectrophotometer or other optical devices to quantify the intensity of color. This manipulation allows easy and rapid identification of foodborne pathogens. Its prominent feature lies in capturing the pathogen on site, in resource-limited settings, making it a preferred choice [37].

3.4 CRISPR-Cas9 Coupled with PCR for Detection of Foodborne Pathogens

The detection of foodborne pathogens is crucial for ensuring safety and public health. Conventional techniques like culture methods and standard PCR have been accepted for use; however, they are often limited in sensitivity, specificity, and the time taken. Applying CRISPR-Cas9 technology with PCR seems to resolve these issues. CRISPR-Cas9, which was first identified as an adaptive immune system present in bacteria, is a technology that can precisely mark and cleave DNA strands. When it is mobilized alongside PCR, this system can identify pathogens quickly and precisely measure foodborne pathogen detection, even at trace amounts. This technique takes advantage of both PCR's amplificatory strength and CRISPR-Cas9's recognition specificity thus giving food diagnostics a new hope in pathogen detection [25].

3.5 CRISPR-Cas9-Assisted Lateral Flow Assay (CASLFA)

The safety of food and public health greatly relies on the detection of any foodborne pathogens. Established diagnostic techniques like cultures and even PCR usually need an embedded facility, skilled personnel, and considerable time for results, which increases the period for food to be available for consumption. These limitations called for the creation of POC diagnostic tools, which are sensitive and very simple to operate [38]. Merging CRISPR-Cas9 technology with lateral flow assays (LFAs) has shown great potential for POC diagnostics. CRISPR-Cas9 systems guarantee precise detection and discrimination of biological target sequences while LFAs ensure fast and easy-to-interpret results interpreted visually. This permits the creation of tests that identify specific DNA profiles linked to foodborne illnesses without high laboratory-grade facilities. The CASLFA relies on the combination of lateral flow detection and the more advanced sequencing set recognition provided by the CRISPR-Cas9 system. The procedure is performed in the following stages [5]. Initially, the food's DNA must undergo amplification with biotin-labeled primers through PCR or one of the isothermal amplification techniques. An sgRNA is created such that it will hybridize with one of the subset regions of the DNA sequence that has already undergone amplification. The corresponding sgRNA with Cas9 protein creates a complex with the corresponding target DNA, without making a double-strand break in the DNA. A lateral flow strip is prepared with Cas9-sgRNA-DNA combinational treatment [39]. Target DNA can hybridize with AuNPs bonded to specific DNA probes that are part of the strip. As the sample flows along the strip, streptavidin binds with biotin on the amplified DNA, capturing the complexes at the test line; this occurs as the sample moves along the strip. Significant accumulation of AuNPs at the test line indicates the presence of the target pathogen due to the conspicuous color change that occurs [40]. The results of this procedure can be observed roughly within an hour, which enables the rapid and accurate detection of foodborne pathogens. An assay design guarantees high specificity due to the CRISPR-Cas9 system's ability to precisely target regions. Potentially contaminated food samples are collected and put through DNA extraction procedures to obtain whole genomic DNA. A set of biotinylated primers for the specific target pathogen's genome region is synthesized. PCR or isothermal amplification can be used for target sequence amplification. SgRNA specific to the particular DNA sequence is synthesized and cloned [41]. The assembled sgRNA is combined with Cas9, and the complex is incubated to form the Cas9-sgRNA complex. Upon addition of biotinylated amplified DNA, the biotin-labeled complex can be captured and the Cas9-sgRNA complex bound. The sample is placed onto the sample pad containing the AuNP-DNA probe. During sample flow through the strip, the complexes are captured by streptavidin at the test line and produce a shift in color that can be seen.

3.6 Cas9-Triggered Enzyme-Free DNA Signal Amplification

Traditional detection methods, such as PCR and culture-based methods, often take too long to process. These methods require particular instruments, trained personnel, and take time for sample processing. In the current scenario of public health and safety, foodborne pathogens need to be detected in a timely and precise manner. To try and alleviate these problems in detection approaches, strategies that combine molecular biology technologies have been developed. With the combination of CRISPR-Cas9 techniques with signal amplification methods, detection becomes more efficient and sensitive without needing sophisticated enzymatic reactions. Due to its inherent precision, these enzymatic reactions are rendered useless for target recognition. The use of these technologies can greatly benefit areas with low resources, as well as POC areas [42].

3.6.1 Principle of Operation

A nonenzymatic signal amplification procedure is carried out after a sequence-specific recognition and cleavage mechanism in the Cas9-triggered enzyme-free DNA signal amplification method: An engineered sgRNA saws by hand traces to the nucleotide sequence of organisms within a scope, and this sgRNA attaches to the sequence of interest, which is exclusive to a certain foodborne pathogen. The double-strand break at the cleavage site is enabled by the attachment of Cas9. The cutting action laser targets the boundaries of the linkage structure (DNA fragment), which yields fragments suitable for a hybridization reaction. They participate in a series of hybridization reactions that form DNA constructs, which can be detected through colorimetric alternations or fluorescence [43]. Because this method takes advantage of nucleic acids and their inherent ability to selectively self-organize and self-interact, without the need for any additional enzymes, it increases the simplicity and reduces the cost of the test. Collect and procure food samples that might be contaminated with pathogens. Through the use of standard extraction methods suitable for the food matrix, retrieve all the DNA. Guide sgRNA was designed and synthesized to curate oligonucleotides with homologous regions in the genome of interest [43]. The active CRISPR-Cas9 complex is formed when a reserved sgRNA is added to the Cas9 Enzyme. For optimal results, the targeted sequence of the extracted DNA should be appropriately bound and cleaved using the CRISPR-Cas9 complex. Present homologous probes are designed to hybridize specifically to the cleaved fragments of DNA. After undergoing a series of hybridization steps, those probes can either emit signal molecules or form certain visible constructs. Appropriate techniques, such as fluorescence spectroscopy or colorimetric assays, can be employed to estimate the rate of reaction of forming or cleaving signaling molecules. Analyze the outcomes in relation to whether or not the suspected pathogen's DNA is present. This technology allows rapid, precise, and cost-effective detection of foodborne pathogens in field settings or locations lacking laboratory facilities [44].

3.7 Cas9 and Dual Enzyme Electrochemical Biosensors

A global public health issue is presented by foodborne pathogens because it leads to increased morbidity and takes up healthcare system resources, and hence requires swift action using detection mechanisms that are both accurate and targeted. Traditionally, culture-based and molecular diagnostic methods can be very effective; however, they are often cumbersome and require sophisticated equipment, which does not allow for fieldwork [44]. As a result of the growing need for pathogen detection, electrochemical biosensors have emerged due to their high sensitivity, portability, and rapid response time. Molecular identification of biological agents along with CRISPR-Cas9 systems has enabled to improve further the specificity of electrochemical biosensors, aiding in the detection of specific pathogens. Pathogens of interest can now be accurately detected due to programmable RNA-guided endonucleases, which can cut defined portions of DNA. In conjunction with electrochemical transducers, the cutting action of Cas9 can be transformed into measurable electric signals, making it possible to detect harmful DNA present in food samples rapidly and accurately [45]. This combination can easily be translated into food safety because it provides effective detection along with molecular identification to develop a strong food safety tool. DNA targets, along with the transduction of its electrochemical signals, establish a flexible system enabling signal transduction pathways whereby the signals are capable of converting into measurable outputs. The combinatorial cleavage capability of the Cas9-sgRNA complex can be used to devise Cas9-mediated electrochemical biosensors [46]. The main steps are as follows:

A sgRNA is synthesized that can specifically recognize and bind to the unique sequence of DNA of the foodborne pathogen. The Cas9-sgRNA complex has the target sequence in the sample DNA and cuts it in the form of a double-strand break (DSB). The installation of a biosensor includes DNA probes (sDNA) that are complementary to the target or the products of its cleavage. As a result of Cas9 cleavage, the DNA layer on the electrode surface changes, which in turn alters the properties of charge carrier transport. These changes manifest as measurable changes in some of the electrochemical characteristics, like impedance, voltage, or current. For instance, in the case of cleavage, there is a possibility of liberation or rearrangement of redox-active labels or DNA strands, which result in quantifiable signals that are directly proportional to the DNA content of the pathogen [47].

3.8 Enrichment of Target Sequences for Sequencing Using CRISPR-Cas9 Methods

Correct genomic determination of foodborne pathogens is vital in studying their diversity, antibiotic resistance, and source outbreak tracking. One of the key obstacles that hinders the full utilization of next-generation sequencing (NGS) in complex food samples is the vast excess of host or background DNA. Sequencing can be preceded by DNA-based techniques that selectively capture pathogen-specific DNA, known as target enrichment, which increases sensitivity and reduces the cost of sequencing [48].

A novel CRISPR-Cas9 method enriches particular DNA fragments from a complex sample through cleavage by Cas9 and sequence-specific enzymatic activity. With the aid of guide RNAs (gRNAs), Cas9 can be programmed to target specific genomic regions of interest in pathogens, wherein it can either capture the required DNA or remove the excess DNA while enhancing sequencing accuracy and efficiency. This technique is useful in food safety for detecting low-abundance organisms more efficiently and accurately [48].

3.8.1 Working Principle

The CRISPR-Cas9 method of target enrichment has two components: cutting and retention/exclusion, each needed in sequence. At step one, they are marked for programmable cutting with exclusion on the lasso technique, removing area pieces of the fringe that encircle the selected region undergoing modification [49]. This level of programming allows steerable sgRNAs to home in target a range of pathogen genomic sequences of relevant interest, a process that is also referred to as "programmable targeting." sgRNAs act as guides for the Cas9 endonuclease, which performs double-strand breaks at select loci in the DNA sample. The enriched pathogen DNA fragments of interest are preserved when the DNA sample undergoes size selection, affinity purification, or exonuclease digestion that removes off-target DNA fragments [50]. Enriched DNA has already been purified and is ready for high-throughput sequencing. Its enhanced sensitivity also allows for deeper coverage of the pathogen genomes. This tailored enrichment allows complete genomic analysis of pathogens found in complex food matrices by enhancing detection.

Using different kits, harvest total DNA from food samples of potential microbial pathogens, containing high-quality, high-molecular-weight DNA suitable for advanced procedures. Maintain the design of essential genomic regions of the pathogens, including antibiotic resistance or virulence genes from components of interest [51]. Purified Cas9 protein is mixed with generated sgRNAs under controlled incubation conditions to form the Cas9-sgRNA complex. Incubate the extracted DNA with the Cas9-sgRNA complex to allow for specific targeted cuts in the DNA sequence. Depending upon the design of the experiment, cleavage results in fragments that can be selectively enriched or depleted with DNA. Either exonucleases or magnetic beads are used to remove the uncleaved fragments and selectively keep the cleaved fragments (for example, size selection or affinity capture). Enriched DNA now predominantly contains the target sequences of the pathogen. From the enriched DNA, create sequencing libraries using common protocols like fragmentation and adaptor ligation. The libraries can be sequenced using PacBio, Oxford Nanopore, or Illumina [52].

4 Notes

Every year, illness outbreaks from foodborne pathogens, which include *Toxoplasma gondii* and *Giardia lamblia* parasites, Norovirus and Hepatitis A viruses, and Salmonella and *E. coli* bacteria, result in millions of illnesses. Outbreaks continue even when food safety practices are enhanced due to food contamination. These bacteria and viruses have varying symptoms and modes of action that could lead to food intoxication, infection, or toxi-infection. The sensitivity, speed, and field applicability of traditional detection techniques like culture methods, PCR, and immunoassays are limited. Culture methods, although reliable, are time-consuming (24–72 h) [53]. PCR is sensitive and selective, but requires extensive equipment. Fast immunoassays like ELISA suffer from cross-reactivity, which can lead to incorrect positive identification. Viruses transmitted through food repeatedly remain a major danger to food safety as well as to the population's health due to the significant number of hospitalizations, deaths, and global diseases they inflict. There is always a need to focus on the prompt detection and control of foodborne illness outbreaks, even with better surveillance, hygiene, and food storage practices. While traditional detection methods such as PCR, culture-based techniques, and immunoassays have provided invaluable support, most of them suffer from a lack of sensitivity, rapidness, economic feasibility, and convenience for field applications [54]. The innovation of CRISPR-Cas9 as a gene-editing and diagnostic device has brought about new ways of detection and mitigation of microbial threats in foods. Starting from an immune system for bacteria, precision diagnostics have adapted CRISPR-Cas9 because it can. This means that pathogens such as *Salmonella enterica*, *E. coli* O157: H7, and *Listeria monocytogenes*, as well as many others, can be detected at unprecedented speed and accuracy in relation to food safety. Expensive machinery and specialized personnel are not used. The prospects of CRISPR-Cas9 technology for detecting foodborne pathogens are likely to be very driven and innovative in the future. With the food industry's focus on developing tools for accurate and cost-effective diagnosis at a faster pace, it is anticipated that CRISPR-Cas9 will become immensely useful owing to its specificity, programmability, and ease of modification. One of the main future expectations is that CRISPR-Cas9 will be integrated and miniaturized into portable, POC diagnostic devices. These portable platforms, which could be coupled with handheld or smartphone readers, will allow the detection of foodborne pathogens in farms, food processing units, markets, and even in kitchens [55]. This advancement will minimize reliance on central laboratories and support faster decision-making, recalls, and better outbreak containment. Moreover, there is active research aimed at increasing the stability of CRISPR-Cas9 reagents, specifically for the Cas9 protein and sgRNAs through lyophilization and encapsulation. This will strengthen the shelf life and thermal tolerance of the reagents, which will make CRISPR-based diagnostics possible in remote locations where cold chain logistics are not available.

There are projections that multiplexed detection systems will be constructed using CRISPR-Cas9 systems in the future. Scientists are attempting to develop sgRNA detection systems that can simultaneously identify multiple foodborne pathogens: bacteria, viruses, and even parasites, in one test. These multiplexed assays will significantly cut time and expenses without compromising the thoroughness of the risk assessment of food products. Efforts are also being made to merge the CRISPR-Cas9 framework with NGS to enable rapid target enrichment within NGS workflows. Sample preparation for NGS using CRISPR will increase the sensitivity and depth of sequencing as pathogen-specific DNA will be preferentially amplified, allowing the identification of even ultra-trace quantities of pathogens and their antimicrobial resistance genes. Genomic surveillance, monitoring outbreaks, and studying the evolution of virulent strains will greatly benefit from these applications [56].

References

1. Karmas E (1988) The major food groups, their nutrient content, and principles of food processing. Nutr Eval Food Process:7–19. https://doi.org/10.1007/978-94-011-7030-7_2
2. Stereotactic Ablative Radiotherapy for the Treatment of Oligometastatic Cancer: A Clinical Review as Part of a Health Technology Assessment, Version 2.0. Stereotact Ablative Radiother Treat Oligometastatic Cancer A Clin Rev as Part a Heal Technol Assessment, Version 20 Clin Rev 2022
3. Hernández-Cortez C, Palma-Martínez I, Uriel Gonzalez-Avila L, Guerrero-Mandujano A, Castro-Escarpulli RCS, Hernández-Cortez CG et al (2017) Food Poisoning Caused by Bacteria (Food Toxins). Poisoning—From Specif Toxic Agents to Nov Rapid Simpl Tech Anal. https://doi.org/10.5772/INTECHOPEN.69953
4. O'Shea H, Blacklaws BA, Collins PJ, McKillen J, Fitzgerald R (2019) Viruses associated with foodborne infections. Ref Modul Life Sci:B978-0-12-809633-8.90273-5. https://doi.org/10.1016/B978-0-12-809633-8.90273-5
5. Olson R, Mathews L, Liu M, Schellenberg D, Mou B, Berrang T et al (2020) Stereotactic ablative radiotherapy for the comprehensive treatment of 1-3 Oligometastatic tumors (SABR-COMET-3): study protocol for a randomized phase III trial. BMC Cancer 20:1–12. https://doi.org/10.1186/S12885-020-06876-4/FIGURES/2
6. Lamichhane B, Mawad AMM, Saleh M, Kelley WG, Harrington PJ, Lovestad CW et al (2024) Salmonellosis: an overview of epidemiology, pathogenesis, and innovative approaches to mitigate the antimicrobial resistant infections. Antibiot 13:76. https://doi.org/10.3390/ANTIBIOTICS13010076
7. Yuan L, Hensley C, Mahsoub HM, Ramesh AK, Zhou P (2020) Microbiota in viral infection and disease in humans and farm animals.

Prog Mol Biol Transl Sci 171:15–60. https://doi.org/10.1016/bs.pmbts.2020.04.005

8. S. Al-Malki E. (2020) Toxoplasmosis: stages of the protozoan life cycle and risk assessment in humans and animals for an enhanced awareness and an improved socio-economic status. Saudi J Biol Sci 28:962. https://doi.org/10.1016/J.SJBS.2020.11.007
9. Almeria S, Dubey JP (2021) Foodborne transmission of toxoplasma gondii infection in the last decade. An overview. Res Vet Sci 135:371–385. https://doi.org/10.1016/j.rvsc.2020.10.019
10. Ebrahimi S, Khosravi MA, Raz A, Karimipoor M, Parvizi P (2023) CRISPR-Cas technology as a revolutionary genome editing tool: mechanisms and biomedical applications. Iran Biomed J 27:219. https://doi.org/10.61186/IBJ.27.5.219
11. Li X, Zhong J, Li H, Qiao Y, Mao X, Fan H et al (2023) Advances in the application of CRISPR-Cas technology in rapid detection of pathogen nucleic acid. Front Mol Biosci 10:1260883. https://doi.org/10.3389/FMOLB.2023.1260883
12. Li T, Yang Y, Qi H, Cui W, Zhang L, Fu X et al (2023) CRISPR/Cas9 therapeutics: progress and prospects. Signal Transduct Target Ther 81:1–23. https://doi.org/10.1038/s41392-023-01309-7
13. Bhattacharjee R, Nandi A, Mitra P, Saha K, Patel P, Jha E et al (2022) Theragnostic application of nanoparticle and CRISPR against food-borne multi-drug resistant pathogens. Mater Today Bio 15:100291. https://doi.org/10.1016/J.MTBIO.2022.100291
14. Huang T, Zhang R, Li J (2023) CRISPR-Cas-based techniques for pathogen detection: retrospect, recent advances, and future perspectives. J Adv Res 50:69–82. https://doi.org/10.1016/J.JARE.2022.10.011
15. Matthews MC, van der Linden J, Robène I, Rozsasi S, Coetzee B, Campa M et al (2025) A combined recombinase polymerase amplification CRISPR/Cas12a assay for detection of Fusarium oxysporum f. sp. cubense tropical race 4. Sci Rep 15:2436. https://doi.org/10.1038/S41598-025-85633-8
16. Xu Y, Li Z (2020) CRISPR-Cas systems: overview, innovations and applications in human disease research and gene therapy. Comput Struct Biotechnol J 18:2401. https://doi.org/10.1016/J.CSBJ.2020.08.031
17. Panda G, Ray A (2022) Decrypting the mechanistic basis of CRISPR/Cas9 protein. Prog Biophys Mol Biol 172:60–76. https://doi.org/10.1016/J.PBIOMOLBIO.2022.05.001
18. Uniyal AP, Mansotra K, Yadav SK, Kumar V (2019) An overview of designing and selection of sgRNAs for precise genome editing by the CRISPR-Cas9 system in plants. 3 Biotech 9:223. https://doi.org/10.1007/S13205-019-1760-2
19. Son H (2024) Harnessing CRISPR/cas systems for DNA and RNA detection: principles, techniques, and challenges. Biosens 14:460. https://doi.org/10.3390/BIOS14100460
20. Kanitchinda S, Srisala J, Suebsing R, Prachumwat A, Chaijarasphong T (2020) CRISPR-Cas fluorescent cleavage assay coupled with recombinase polymerase amplification for sensitive and specific detection of Enterocytozoon hepatopenaei. Biotechnol Reports 27:e00485. https://doi.org/10.1016/J.BTRE.2020.E00485
21. Nanda BP, Rani P, Paul P, Aman SSG, Bhatia R (2024) Recent trends and impact of localized surface plasmon resonance (LSPR) and surface-enhanced Raman spectroscopy (SERS) in modern analysis. J Pharm Anal 14:100959. https://doi.org/10.1016/J.JPHA.2024.02.013
22. Tang C, He Z, Liu H, Xu Y, Huang H, Yang G et al (2020) Application of magnetic nanoparticles in nucleic acid detection. J Nanobiotechnology 181:1–19. https://doi.org/10.1186/S12951-020-00613-6
23. Chen K, Du Z, Zhang Y, Bai R, Zhu L, Xu W (2025) Exploring nucleic acid nanozymes: a new frontier in biosensor development. Biosens 15:142. https://doi.org/10.3390/BIOS15030142
24. Sobral AF, Dinis-Oliveira RJ, Barbosa DJ (2025) CRISPR-Cas technology in forensic investigations: principles, applications, and ethical considerations. Forensic Sci Int Genet 74:103163. https://doi.org/10.1016/J.FSIGEN.2024.103163
25. Kabiraz MP, Majumdar PR, Mahmud MMC, Bhowmik S, Ali A (2023) Conventional and advanced detection techniques of foodborne pathogens: a comprehensive review. Heliyon 9:e15482. https://doi.org/10.1016/J.HELIYON.2023.E15482
26. Ghouneimy A, Mahas A, Marsic T, Aman R, Mahfouz M (2023) CRISPR-based diagnostics: challenges and potential solutions toward point-of-care applications. ACS Synth Biol 12:1–16. https://doi.org/10.1021/ACSSYNBIO.2C00496/ASSET/IMAGES/LARGE/SB2C00496_0007.JPEG

27. Osborn MJ, Bhardwaj A, Bingea SP, Knipping F, Feser CJ, Lees CJ et al (2021) CRISPR/Cas9-based lateral flow and fluorescence diagnostics. Bioengineering 8:23. https://doi.org/10.3390/BIOENGINEERING8020023
28. Kim HS, Kweon J, Kim Y (2024) Recent advances in CRISPR-based functional genomics for the study of disease-associated genetic variants. Exp Mol Med 56:861–869. https://doi.org/10.1038/S12276-024-01212-3
29. de Ridder D, Kroese F, Evers C, Adriaanse M, Gillebaart M (2017) Healthy diet: health impact, prevalence, correlates, and interventions. Psychol Health 32:907–941. https://doi.org/10.1080/08870446.2017.1316849
30. Juskowiak B (2010) Nucleic acid-based fluorescent probes and their analytical potential. Anal Bioanal Chem 399:3157. https://doi.org/10.1007/S00216-010-4304-5
31. Marras SAE (2006) Selection of fluorophore and quencher pairs for fluorescent nucleic acid hybridization probes. Methods Mol Biol 335: 3–16. https://doi.org/10.1385/1-59745-069-3:3
32. Aladhadh M (2023) A review of modern methods for the detection of foodborne pathogens. Microorg 11:1111. https://doi.org/10.3390/MICROORGANISMS11051111
33. Frango A, Chen Q, Zhang H, Liu Z (2025) Rapid and specific colorimetric assay for detecting CRISPR-Cas9 in genome-edited plants using loop-mediated isothermal amplification. Food Anal Methods 18:798–806. https://doi.org/10.1007/S12161-024-02754-Y/FIGURES/7
34. Chen J, Chen Y, Huang L, Lin X, Chen H, Xiang W et al (2024) Trans-nuclease activity of Cas9 activated by DNA or RNA target binding. Nat Biotechnol 43:558–568. https://doi.org/10.1038/S41587-024-02255-7
35. Yan H, Qi M, Li H (2025) Characterization and full sequencing of 100 Nt sgRNA and large RNA using site-directed cleavage and liquid chromatography tandem mass spectrometry. Anal Chem 97:9228. https://doi.org/10.1021/ACS.ANALCHEM.4C06606/SUPPL_FILE/AC4C06606_SI_002.XLSX
36. Vidic J, Vizzini P, Manzano M, Kavanaugh D, Ramarao N, Zivkovic M et al (2019) Point-of-need DNA testing for detection of foodborne pathogenic bacteria. Sensors (Basel) 19:1100. https://doi.org/10.3390/S19051100
37. Ivanov IE, Wright AV, Cofsky JC, Palacio Aris KD, Doudna JA, Bryant Z (2020) Cas9 interrogates DNA in discrete steps modulated by mismatches and supercoiling. Proc Natl Acad Sci USA 117:5853–5860. https://doi.org/10.1073/PNAS.1913445117/SUPPL_FILE/PNAS.1913445117.SAPP.PDF
38. Moghaddam NS, Jahadakbar A, Amerinatanzi A, Elahinia M, Miller M, Dean D (2016) Metallic fixation of mandibular segmental defects: graft immobilization and orofacial functional maintenance. Plast Reconstr Surg—Glob Open 4:1–8. https://doi.org/10.1097/GOX.0000000000000859
39. Hadi J, Rapp D, Dhawan S, Gupta SK, Gupta TB, Brightwell G (2023) Molecular detection and characterization of foodborne bacteria: recent progresses and remaining challenges. Compr Rev Food Sci Food Saf 22:2433–2464. https://doi.org/10.1111/1541-4337.13153;SUBPAGE:STRING:FULL
40. Ding Y, Liu J (2024) Interfacing DNA and aptamers with gold nanoparticles: from fundamental colloid and Interface Sciences to biosensors. Chinese J Chem 42:2391–2400. https://doi.org/10.1002/CJOC.202400113;PAGE:STRING:ARTICLE/CHAPTER
41. Xu D, Zeng H, Wu W, Liu H, Wang J (2023) Isothermal amplification and CRISPR/Cas12a-system-based assay for rapid, sensitive and visual detection of Staphylococcus aureus. Foods 12:4432. https://doi.org/10.3390/FOODS12244432/S1
42. Lazcka O, Campo FJ Del, Muñoz FX (2007) Pathogen detection: a perspective of traditional methods and biosensors. Biosens Bioelectron 22:1205–1217. https://doi.org/10.1016/J.BIOS.2006.06.036
43. Wang H, Wang H, Willner I, Wang F (2020) High-performance biosensing based on autonomous enzyme-free DNA circuits. Top Curr Chem 378:1–32. https://doi.org/10.1007/S41061-020-0284-X/METRICS
44. Wilkinson RA, Martin C, Nemudryi AA, Wiedenheft B (2018) CRISPR RNA-guided autonomous delivery of Cas9. Nat Struct Mol Biol 261:14–24. https://doi.org/10.1038/s41594-018-0173-y
45. Wachholz Junior D, Kubota LT (2024) CRISPR-based electrochemical biosensors: an alternative for point-of-care diagnostics? Talanta 278:126467. https://doi.org/10.1016/J.TALANTA.2024.126467
46. Deroco PB, Wachholz Junior D, Kubota LT (2021) Recent advances in point-of-care biosensors for the diagnosis of neglected tropical diseases. Sensors Actuators B Chem 349: 130821. https://doi.org/10.1016/J.SNB.2021.130821

47. Chin S, Goyon A, Zhang K, Kurita KL (2023) Middle-out sequence confirmation of CRISPR/Cas9 single guide RNA (sgRNA) using DNA primers and ribonuclease T1 digestion. Anal Bioanal Chem 415:2809–2818. https://doi.org/10.1007/S00216-023-04693-9
48. Schadron T, van den Beld M, Mughini-Gras L, Franz E (2024) Use of whole genome sequencing for surveillance and control of foodborne diseases: status quo and quo vadis. Front Microbiol 15:1460335. https://doi.org/10.3389/FMICB.2024.1460335/XML/NLM
49. Wu X, Kriz AJ, Sharp PA (2014) Target specificity of the CRISPR-Cas9 system. Quant Biol 2:59. https://doi.org/10.1007/S40484-014-0030-X
50. Dong C, Gou Y, Lian J (2022) SgRNA engineering for improved genome editing and expanded functional assays. Curr Opin Biotechnol 75:102697. https://doi.org/10.1016/J.COPBIO.2022.102697
51. Gand M, Bloemen B, Vanneste K, Roosens NHC, De Keersmaecker SCJ (2023) Comparison of 6 DNA extraction methods for isolation of high yield of high molecular weight DNA suitable for shotgun metagenomics Nanopore sequencing to detect bacteria. BMC Genomics 24:1–14. https://doi.org/10.1186/S12864-023-09537-5/FIGURES/5
52. Mekler V, Minakhin L, Semenova E, Kuznedelov K, Severinov K (2016) Kinetics of the CRISPR-Cas9 effector complex assembly and the role of 3′-terminal segment of guide RNA. Nucleic Acids Res 44:2837. https://doi.org/10.1093/NAR/GKW138
53. Bintsis T (2017) Foodborne pathogens. AIMS Microbiol 3:529. https://doi.org/10.3934/MICROBIOL.2017.3.529
54. Tabatabaei MS, Islam R, Ahmed M (2021) Applications of gold nanoparticles in ELISA, PCR, and immuno-PCR assays: a review. Anal Chim Acta 1143:250–266. https://doi.org/10.1016/J.ACA.2020.08.030
55. Ding R, Long J, Yuan M, Jin Y, Yang H, Chen M et al (2021) CRISPR/Cas system: a potential technology for the prevention and control of COVID-19 and emerging infectious diseases. Front Cell Infect Microbiol 11:639108. https://doi.org/10.3389/FCIMB.2021.639108/XML/NLM
56. Wei C, Lei X, Yu S (2024) Multiplexed detection strategies for biosensors based on the CRISPR-Cas system. ACS Synth Biol 13:1633–1646. https://doi.org/10.1021/ACSSYNBIO.4C00161/ASSET/IMAGES/MEDIUM/SB4C00161_0008.GIF

Chapter 2

Nanosensors for Rapid Pesticide Residue Quantification

G. Rajeshwari, Mohamed Mujamil Abuthahir, Pramila Murugesan, and Gnanaselvan Gnanasekaran

Abstract

The overuse of pesticides in modern agriculture has prompted serious concerns about food safety, environmental sustainability, and public health. The rapid and reliable detection of pesticide residues is critical for avoiding dangerous exposure and maintaining regulatory compliance. Conventional detection techniques, such as gas chromatography (GC) and liquid chromatography-mass spectrometry (LC-MS), have excellent accuracy but are time-consuming, expensive, and necessitate complicated laboratory infrastructure. Nanosensors have emerged as a cutting-edge technology for quick, low-cost, and extremely sensitive pesticide residue quantification. These sensors use nanomaterials, including metal nanoparticles (NPs), carbon-based nanostructures, and quantum dots (QDs), to improve signal detection via electrochemical, optical, and biosensing approaches. Their unique qualities, including high surface area, tunability, and strong interaction with target molecules, allow for precise and real-time detection of ultra-low pesticide doses. Recent advances in nano-enabled electrochemical sensors, fluorescence-based platforms, and surface-enhanced Raman spectroscopy (SERS) have made it possible to detect pesticides in food, water, and environmental samples. This review delves into the most recent advances in nano-sensor technology for pesticide detection, with a focus on sensor design, operating principles, real-world applications, and future directions.

Key words Nanosensors, Nanomaterials, Rapid quantification, Pesticide residue detection, Environmental monitoring

1 Introduction

Pesticides are a numerous and diverse group of chemical compounds, which are used to eliminate pests in agriculture and households. They enable the quantities and the quality of crops and food to be controlled, and help to limit the many human diseases transmitted by insect or rodent vectors [1]. There is a huge body of evidence on the relation between exposure to pesticides and elevated rates of chronic diseases such as different types of cancers, diabetes, neurodegenerative disorders, including Parkinson's, Alzheimer's, and amyotrophic lateral sclerosis (ALS), birth defects, and reproductive disorders [2]. Therefore, strict regulations and

Tanmay Sarkar (ed.), *Novel Food Analysis: Techniques and Applications*, Methods and Protocols in Food Science,
https://doi.org/10.1007/978-1-0716-4787-5_2,

maximum residue limits (MRLs) have been established by regulatory bodies, such as the Food and Agriculture Organization (FAO) and European Food Safety Authority (EFSA), to protect consumers from the harmful effects of pesticide residues. Conventional methods for pesticide residue detection primarily rely on chromatographic techniques, such as gas chromatography (GC) and high-performance liquid chromatography (HPLC) coupled with mass spectrometry (MS) for identification and quantification [3]. Rapid pesticide residue detection technologies are quick and low-cost methods that can be used as a supplement to conventional technologies [4]. Moreover, the need for complex sample preparation further limits their application for rapid and on-site monitoring. This has intensified the search for alternative detection strategies that offer rapidity, affordability, sensitivity, and ease of operation. In this context, nanosensors have emerged as highly promising analytical tools due to their ability to detect trace levels of pesticides through enhanced physicochemical interactions provided by nanomaterials. These sensors integrate nanostructures such as gold nanoparticles (AuNPs), carbon nanotubes (CNTs), graphene, and quantum dots (QDs) with biological recognition elements such as enzymes, antibodies, and aptamers to achieve high sensitivity and selectivity [5]. Nanosensors can transduce biological recognition actions into measurable optical, electrochemical, or fluorescence signals, enabling rapid detection with minimal sample preparation. Multiple studies have demonstrated the successful application of nanosensors for detecting organophosphates, carbamates, pyrethroids, and neonicotinoids in fruits, vegetables, and cereal products. The versatility of these sensors lies in their tunable detection mechanisms, allowing the deployment of electrochemical, colorimetric, surface-enhanced Raman scattering (SERS), and fluorescence-based platforms according to the target analyte and sample matrix. This chapter aims to systematically review the principles, fabrication, and application of nanosensors for pesticide residue detection in food. The subsequent sections provide an in-depth discussion on nanomaterial synthesis, sensor construction, detection methodologies, performance evaluations, and the practical challenges associated with integrating these sensors into existing food safety frameworks.

2 Materials

2.1 Nanomaterials

Nanomaterials form the core of nano-sensor technology due to their remarkable physicochemical properties such as high surface-area-to-volume ratio, enhanced conductivity, unique opto-electronic features, and versatile surface modification potential. Among metallic nanomaterials, AuNPs have gained significant attention for colorimetric and electrochemical sensing applications.

Owing to their strong surface plasmon resonance (SPR) effect, AuNPs exhibit intense color changes upon aggregation, making them ideal candidates for visible pesticide detection [6]. Furthermore, their ease of synthesis via citrate reduction and excellent biocompatibility allow for functionalization with enzymes, antibodies, and aptamers. Carbon-based nanomaterials, including carbon nanotubes (CNTs) and graphene oxide (GO), are widely employed in electrochemical nanosensors. These materials provide exceptional electrical conductivity, large surface areas for immobilization of recognition elements, and chemical stability under varied conditions [7]. Single-walled and multi-walled carbon nanotubes (SWCNTs, MWCNTs) have been utilized to enhance the electron transfer kinetics of electrode surfaces, thereby improving the sensitivity and linear detection range of sensors. Graphene oxide, derived through the oxidation of graphite powder by Hummers' method, offers a highly functionalized surface rich in oxygenated groups, such as hydroxyl, carboxyl, and epoxy, facilitating covalent attachment of biomolecules. Its excellent dispersion in aqueous systems and robust electrochemical performance make it a valuable material for constructing pesticide detection platforms. QDs, typically cadmium selenide/zinc sulfide (CdSe/ZnS) core-shell nanocrystals, serve as fluorescent probes in sensors for ultra-trace pesticide analysis. These semiconductor nanoparticles (NPs) display size-dependent fluorescence emission and high quantum yields, which can be modulated by the interaction of target pesticides with surface-conjugated recognition molecules, leading to quenching or enhancement of emission signals. Magnetic NPs, such as Fe_3O_4, are often employed for sample pre-concentration and magnetic separation, reducing matrix interferences and simplifying sensor protocols [8].

2.2 Biological Recognition Elements

2.2.1 Enzymes

Enzymatic optical sensors provide high selectivity and sensitivity for identifying and quantifying the target analyte [9]. Insecticides have traditionally been utilized as enzyme inhibitors or substrates, which play a significant role in enzymatic reactions and indirectly induce signal response to the optical sensor. As expected, enzyme specificity can boost the efficiency of optical sensors in accurately detecting pesticides [10]. The most commonly used enzymes for the enzymatic detection of insecticides are acetylcholinesterase (AChE) and organophosphate hydrolase (OPH), with few records of other enzymes such as butyryl cholinesterase (BChE), alkaline phosphatase (ALP), choline oxidase (ChOx), horseradish peroxidase (HRP), tyrosinase (TYR), trypsin (TRY), and streptavidin.

2.2.2 Antibodies

Antibodies, as bio-recognition elements, are now commonly used as a potential alternative method for immunoassay for pesticide detection [11]. It is also utilized to produce immunosensors in the clinical and biochemical sectors [12]. Polyclonal, monoclonal, and recombinant antibodies are widely utilized for pesticide detection. Because of the extremely high equilibrium association constants, these antibodies can recognize antigens sensitively [13].

2.2.3 Aptamers

Aptamers are single-nucleic acid molecules with nucleobases synthesized in vitro without the necessity for animal or cell growth [14]. Compared to natural antibodies, aptamers are usually easier and less expensive to synthesize, as well as having a strong stability in sustained renaturation and denaturation [15]. Aptamers have a high affinity in the nanomolar to picomolar range for their targets, with dissociation constant (K_d) values [16]. Furthermore, aptamers typically recognize and bind to corresponding targets directly as "lock-and-key" through molecular shape complementarities, stacking of aromatic rings, electrostatic and van der Waals interactions, and hydrogen bonding, making their detection much more effective and convenient [17].

2.2.4 Molecularly Imprinted Polymers

Molecularly imprinted polymers (MIPs) are recognition elements known as plastic antibodies, having specialized recognition capacity [18]. MIPs are easily created by in situ copolymerization of functional monomers around a template molecule [19]. MIPs have been widely used in sensor development to boost separation efficacy and detection sensitivity due to their excellent stability and amazing mechanical qualities [20]. Furthermore, MIPs have a high potential for application as custom polymers in manufacturing biosensors, particularly when biological recognizers (enzymes and antibodies) are unavailable. Based on the benefits indicated, MIPs are receiving interest in the development of drug delivery, biosensors, and environmental cleanup [21].

2.3 Analytical Instrumentation

Several analytical techniques have been developed and applied for the detection of pesticides. However, they are all expensive and laborious, requiring complex pretreatments to assist detection. To eliminate these issues, several effortless, selective, cost-effective, rapid, and sensitive methodologies have been established for the detection of pesticides. Some of the techniques for the detection of pesticides are displayed in Table 1.

3 Methods

There are different methods, which have been adapted for the fabrication of nanomaterials such as chemical vapor deposition (CVD), Sol-Gel process, molecular self-assembly, lithography, fibre pulling, and chemical etching as illustrated in Fig. 1.

Table 1
Techniques for the detection of pesticide residues

S. no	Instruments	Advantages	Disadvantages
1	High-Performance Liquid Chromatography (HPLC)	High sensitivity, selectivity, and rapid detection	Time-consuming and expensive
2	Gas Chromatography-Mass Spectrometry (GC-MS)	Minimal sample Preparation and real-time detection	Requires complex sample preparation
3	Enzyme-Linked Immunosorbent Assay (ELISA)	Multiplex detection capability and portable systems	Limited to specific pesticide targets
4	Thin-Layer Chromatography (TLC)	Quantitative detection	Subjective interpretation of results
5	Nanosensor-based methods	Miniaturization, portability, and label-free detection	Complex fabrication processes, limited sample throughput

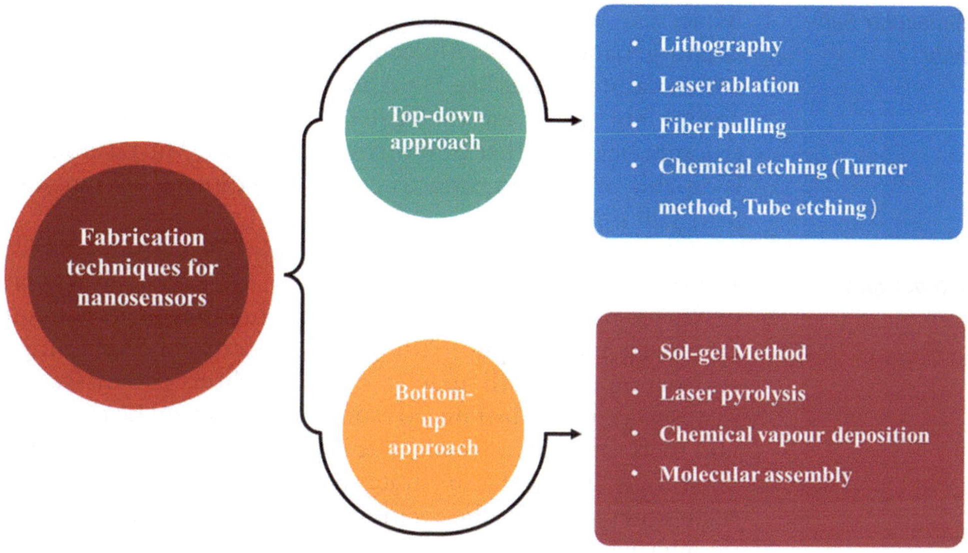

Fig. 1 Fabrication techniques for nanosensors

3.1 Chemical Vapor Deposition (CVD)

CVD utilizes substrates and volatile precursors. Precursors interact with the substrate at high temperatures, breaking it down and depositing crystals. This approach is widely utilized in the semiconductor industry [22]. CVD forms thin films of various materials onto substrates by chemically reacting gaseous precursors. In a heated reaction chamber, these precursors react and dissolve on a substrate to form a solid layer during CVD. CVD at high temperatures (750 °C) can create MWCNTs from carbon-based precursors. Uniform and controlled deposition is required for nanosensor property repeatability [23]. Furthermore, plasma enhancement (PECVD) decreases the reaction temperature and enables the production of more sensitive nanostructures with less energy. This version enables deposition at lower temperatures while maintaining high-quality material growth and thus boosting adaptability and industrial scalability [24]. Furthermore, the sensitivity and unique surface contacts of CVD-fabricated nanosensors make them ideal for food safety applications [25].

3.2 Sol-Gel Process

The sol-gel method involves the combination of sol and gel. A sol is made up of solid colloidal particles that are evenly distributed throughout a liquid, whereas a gel is a solid-like material generated by connected nanostructures in a liquid phase. This approach involves blending solid NPs in liquid media, resulting in the production of a gel with nanoscale pore diameters. Sol-gel properties are regulated by temperature, catalysts, pH, time, and solvent type [26]. The technique involves a succession of procedures including mixing, sol formation, aging, gelation, drying, dehydration, chemical stabilization, and densification [27].

3.3 Molecular Self-Assembly

Molecular self-assembly (MSA) is an extremely efficient way of creating nanosensors. Molecular self-assembly is the process by which natural things like DNA are created. This procedure includes assembling atoms and molecules to form a finely specified nanostructure held together by noncovalent bonds. The resulting gadget is capable of creating nanostructures with dimensions ranging from 1 to 100 nm [3].

3.4 Lithography

Lithography is one of the most popular top-down fabrication processes for nanosensors. Lithography techniques that are frequently employed include photolithography, electron beam lithography, and nanoimprint lithography. These methods use masks or templates to expose substrates to radiation or electron beams selectively and therefore pattern them. The exposed areas change physically or chemically, giving the substrate the required patterns. Lithography techniques allow for the creation of nanoscale objects with intricate geometries by offering high-resolution control over feature size and pattern shape. Their versatility enhances the functionality of nanosensors by enabling the integration of several sensing elements on a single chip [3].

3.5 Fibre Pulling

The pulling fibre method, which is frequently used in fibre optics technology, applies tensile strain to a fibre along its main axis in order to obtain nanoscale dimensions. The fibre is heated to promote elongation. Pulling fibre nanosensors may identify contaminants, pathogens, and chemical residues at low concentrations because of their high sensitivity and specific surface interactions. Specifically, volatile organic chemicals, which signal food spoiling, may be detected by fibre optic-based sensors with a nanofiber taper. These nanofibers' high transmission efficiency and sensitivity make them perfect for real-time food quality monitoring [3].

3.6 Chemical Etching

One of the most important top-down fabrication procedures for nanosensors is etching. Common etching methods include wet etching and dry etching. In wet etching, materials are selectively removed using chemical solutions, while in dry etching, the substrate surface is etched using plasma or reactive gases. Using etching processes, materials can be precisely removed to create structures and features at the nanoscale. In nanosensors, they are essential for defining the channels, patterns, and sensing elements. The etching parameters, including temperature, etching time, and etchant composition, can be carefully controlled to create nanosensors with the required surface characteristics and geometries [3].

3.7 Diverse Tuning Strategies for Enhancing the Sensing Performance of Nanomaterials

Many tuning strategies have been developed to improve the sensing performance of nanomaterials. These methods aim to modify the nanomaterial's surface characteristics and physicochemical properties. Nanostructure engineering, composite formation, surface functionalization, and doping or alloying are commonly adopted techniques for improving the sensing performance of NPs, as illustrated in Fig. 2. Nanostructure engineering is used to enhance the surface area and active sites to enhance their sensing activity with a short period of reaction time. A synergistic impact occurs by

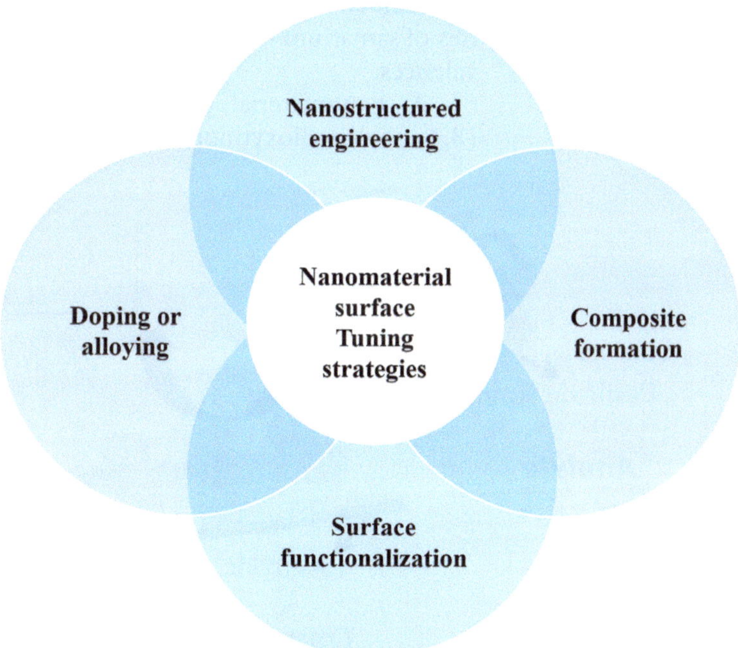

Fig. 2 Nanomaterial surface tuning strategies

hybridizing the benefits of various elements such as metal oxides, polymers, or carbon-based compounds. When compared to individual components, composites with other elements provide better overall performance by improving conductivity, stability, and sensitivity. Selectivity is increased by adding particular functional groups (organic molecules, polymers, and biomolecules) to the surface of nanomaterials. Functionalization improves selectivity and reduces interference from other species by increasing the affinity of NPs for target analytes. Doping nanomaterial-based sensors with transition metals can expand their range of applications by improving their optical property, conductivity, selectivity, and catalytic activity.

3.8 Applications of Nanosensors in Pesticide Residue Detection

3.8.1 Electrochemical-Based Detection

Pesticide electrochemical sensing techniques rely on the right working electrode (transducer), its functionalization and optimization, and electrical parameters like potential, charge, and current. A working electrode, a reference electrode, and a counter electrode are the three electrodes that make up a conventional electrochemical cell. The redox reaction between the target analyte (pesticide) and the working voltage determines how the electrode material is tuned (Fig. 3). The successful preparation of a samarium molybdate electrocatalyst obtained from sonochemistry allowed for the sensitive electrochemical detection of carbendazim in real environmental samples. Samarium molybdate nanomaterial is more sensitive, stable, and anti-interference for the electrochemical sensor's detection of carbendazim. The modified screen-printed electrode showed a current response (6.6 μA) that was twice as high as the unmodified screen-printed electrode due to the remarkable properties of samarium molybdate, such as its chemical stability and many valences.

A nanomaterial functionalized electrode based on poly (3,4-ethylenedioxythiophene) polymer and MWCNTs for the

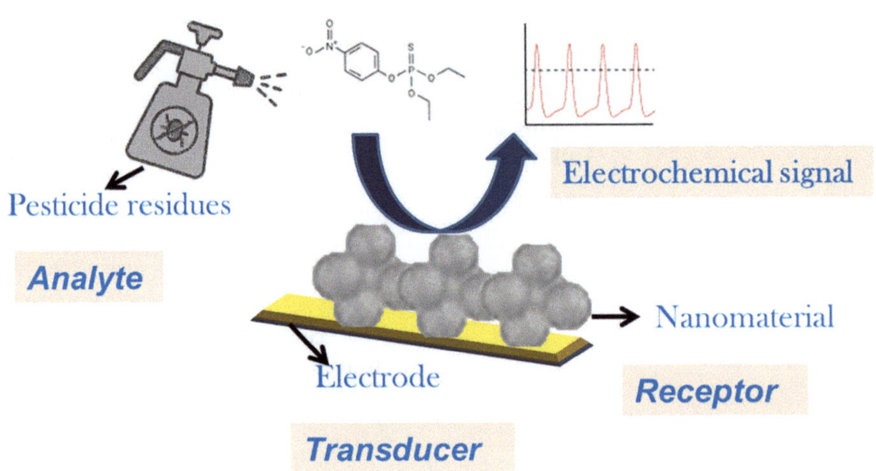

Fig. 3 Schematic representation of electrochemical-based detection of pesticide residues

measurement of carbamate levels (Mancozzeb-MCZ) in water. With a limit of detection (LOD) of 10 μM and a linear range of 25–150 μM, the electrochemical analysis of modified electrodes was investigated for a particular signal resulting from the interaction of MCZ and MWCNTs. Pesticides in a drinking water system might be easily and sensitively detected using the described technology [28]. Using a similar method reported that they detected carbofuran (CBF) and carbamates (isoprocarb, carbaryl, and fenobucarb) in wheat and maize samples using an electrochemical sensor-based screen-printed electrode modified with carbon nanospheres. With the lowest LOD for the pesticides in the nanomolar range, the proposed electrochemical sensor demonstrated a consistent and highly selective response [29].

3.8.2 Optical Detection

As an advanced analytical technology, spectral technology allows for the qualitative or quantitative analysis of target molecules based on their various spectral characteristics of target molecules. Spectral analysis is fast and simple, has wide applications, and is relatively low-cost. It has great potential in the field of rapid detection of pesticide residues. However, the sensitivity and selectivity of spectral analysis need improvement to meet the requirements of trace pesticide residue detection. In recent years, a variety of new spectral analysis methods have been developed for the detection of pesticide residues. Pesticide detection can differ significantly depending on whether it is done in the vapor phase, in solution, or in the solid form.

3.8.3 Fluorescence-Based Detection

Fluorescence technique is one of the most widely used methods to identify insecticides that are simple to use and have a high sensitivity and selectivity. In addition, it is also widely used for biomedical, environmental monitoring, food protection, and quality control. In this method, a spectrofluorophotometer can generate the signal shift and be observed by the naked eye on-site. The benefits of high specificity, excellent sensitivity, and rapid response have drawn a lot of interest in a variety of innovative fluorescence sensors. These nanosensors typically consist of three primary components: a fluorescent emitter (or fluorophore), a recognition element that interacts specifically with the target analyte, and a nanomaterial platform that supports and often enhances the signal response. Different types of recognition elements and materials have been used in the manufacturing of fluorescence sensors, including enzymes, semiconductor nanomaterials, metal nanomaterials, carbon materials, and noble metals, as they can enhance the sensitivity of the sensor [30]. The detection mechanism generally relies on fluorescence quenching or enhancement, depending on the nature of the interaction between the fluorophore and the pesticide. When pesticides are introduced, the fluorescence signal generally changes in "turn-off" manner due to the quenching of fluorescence caused by the

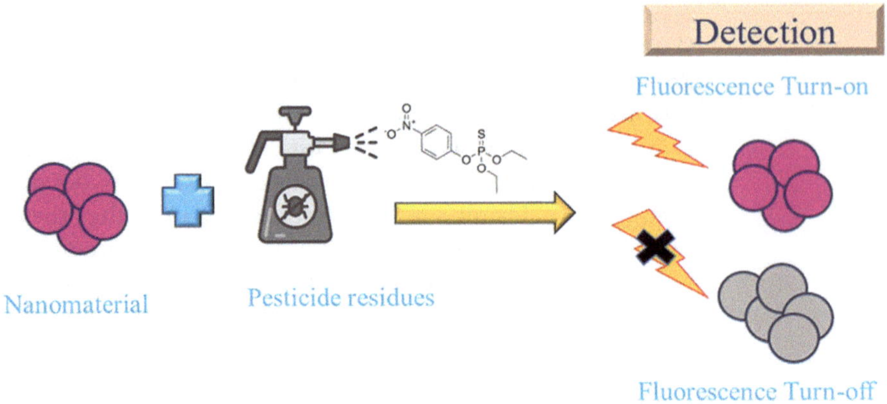

Fig. 4 Fluorescence-based detection of pesticide residues

pesticide (Fig. 4). This type of quenching happens when energy is transferred from the sensor to the pesticide, which can occur through an inner filter effect or photoinduced electron transfer electron exchange or Förster resonance energy transfer (FRET). For instance, carbon QDs coated with vitamin B12 showed dual fluorescence emission from excitation at 355 nm: radiation from the carbon dots at 417 nm and radiation from the energy transfer from the carbon dots to the coating of vitamin B12 at 550 nm. When the pesticide carbofuran was introduced, the energy transfer was disrupted, which caused the fluorescence emission at 550 nm to drop and the fluorescence emission at 355 nm to increase. The authors hypothesized that carbofuran forms a charge transfer complex on the surface of the QD, resulting in static quenching of the carbon dot emission and an inhibition of the energy transfer pathway [31].

In another study, a nanocrystal-based pesticide sensor was developed that used cadmium sulfide nanocrystals as energy donors and Eosin Y as an energy acceptor. When the nanocrystals were excited, energy was transferred to and emitted from the Eosin Y fluorophore. The fluorescence emission of the nanocrystals increased when chlorpyrifos was added to the solution because of pesticide-induced aggregation. However, there was no change in the energy transfer peak that was seen as Eosin Y dye emission. Lastly, the device detected ppb-level quantities of chlorpyrifos [32]. In addition, a significant number of fluorescent pesticide sensors based on nanomaterials employ NPs, including metals, semimetals, or non-metals, which are spherical particles with diameters in the nanometer range. One example of the use of metallic NPs for pesticide detection was the detection of organophosphorus and carbamate pesticides using AuNPs and rhodamine B. Without the use of pesticides, thiocholine is produced by active acetylcholinesterase (AChE), which attaches to the rhodamine on the surface of the NPs and promotes the development of aggregates with distinctive fluorescence emission. When pesticides known to be AChE inhibitors are introduced, the production of thiocholine and the resulting agglomeration of NPs are prevented. A dual-readout pesticide sensor is made possible by the formation of monomeric NPs with a distinct fluorescence signature. There have also been other reports of sensors based on AChE activity influencing metallic nanoparticle fluorescence emission [33]. Pesticides might potentially be detected by metallic NPs via fluorescence energy transfer. These systems worked especially well when the donor included a lanthanide that could produce visible light, absorb many photons, and be stimulated in the near-infrared spectrum [34].

There have also been reports of using semimetal NPs for the detection of fluorescent pesticides. One such example is the macrocycles that contain boronate displayed improved sensitivity toward the detection of 2,6-dichloro-4-nitroaniline [35]. Lastly, it has also been shown that metal-free NPs exhibit detectable fluorescence changes in response to the presence of pesticides. In one instance, fluorescence quenching was achieved by metal-free poly-fluorene-poly-benzothiadiazole NPs as energy donors to manganese oxide nanosheet acceptors. The intrinsic nanoparticle fluorescence was restored when pesticides were added to this design because they interfered with the energy transfer [36]. A high-performing sensor is frequently produced by combining numerous kinds of nanostructures. A magnetic molecularly imprinted Fe_3O_4@SiO_2@MIPIL displays a specific response to imidacloprid (IMD) and a detection limit (LOD) of 0.3 nM [37]. Metalloenzyme mimics of metal-bound peptides and paired them with QDs to develop a visible proportional fluorescence sensor (QDs-peptide) for the detection of organophosphate pesticides. The fluorescence of the sensor changed to blue from red as the concentration of pesticide increased, and the LOD of methyl paraoxon reached 35 nM. Samples of strawberries, blueberries, and jujubes showed good accuracy in detecting organophosphate pesticides [38].

3.8.4 Colorimetric Detection

Colorimetric sensors have been recognized as effective approaches due to their high sensitivity, ease of production, quick detection, and naked-eye or visual color change reaction phenomenon. Crucially, colorimetric techniques usually avoid the need for specific equipment or the expertise of a qualified technician. Generally, silver nanoparticles (AgNPs) and AuNPs have been commonly adopted nanomaterials for colorimetric-based sensing toward pesticides. Because AuNPs and AgNPs have large extinction coefficients, a suitable molecular contact between the NPs' surfaces and the target analytes leads to aggregation, which clearly changes color [31] (Fig. 5). Furthermore, the colorimetric method's selectivity

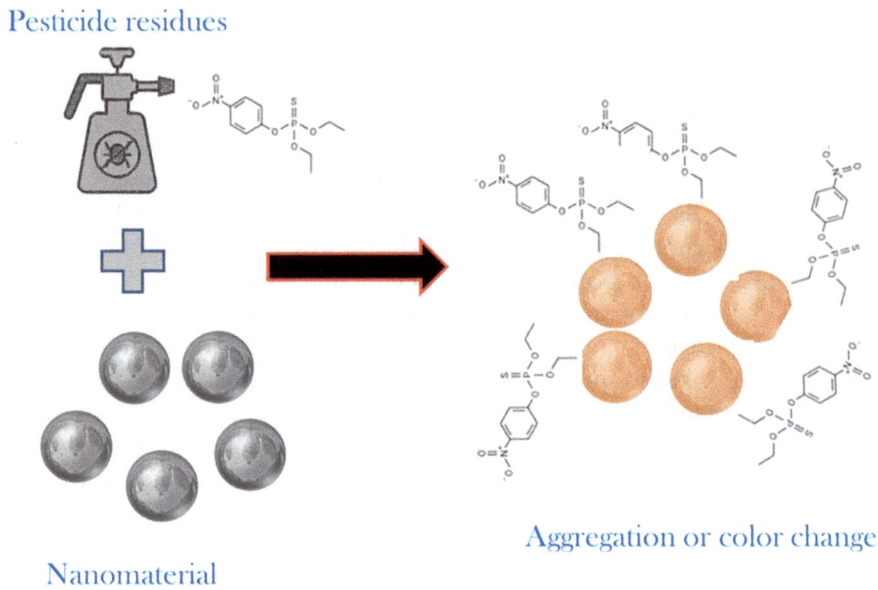

Fig. 5 Fluorescence-based detection of pesticide residues

and sensitivity are enhanced by the functionalization, integration, and modification of NPs with ligands. Bashir et al. (2024) described a colorimetric sensor for malathion based on gold nanorods' potential catalytic activity using 3,3′,5,5′tetramethylbenzidine/hydrogen peroxide (TMB/H_2O_2) as a substrate [39]. The assay revealed an inverse relationship between malathion concentration and gold nanorod activity. The sensor demonstrated the inherent catalytic activity of gold nanorods (nanozyme) comparable to the peroxidase enzyme, with a LOD of 1.78 μM for malathion. Chen et al. (2018) reported a new colorimetric sensor that uses AuNPs (citrate stabilized) to quantify dimethoate (DMT) and terbuthylazine (TBA). A colorimetric sensing mechanism was developed using the aggregation and dispersion of AuNPs with pesticide molecules.

The color shifts from blue to red for DMT insecticide, with a LOD of 0.02 μM. The sensor's only drawback is that it cannot be used for detection in samples containing both pesticides (TBA/DMT) because the detection methods are different [40]. A colorimetric sensor mechanism was recently constructed by directly determining the contact response of pesticide molecules with borohydride-stabilized AgNP, resulting in increased surface area for binding. Unlike chlorpyrifos and malathion, phorate responded selectively by changing color [41]. In order to quantitatively evaluate triazophos concentrations, Ma et al. developed a novel, easy, simple colorimetric detection technique that uses citrate-capped silver nanoparticle (Cit-AgNP) aggregation in On-Off mode. Without the need for further complex surface modifications, the

Cit AgNPs were successfully employed as a colorimetric probe. When these Cit AgNPs aggregated in the presence of the triazophos, substitution processes, p-p interactions, and hydrogen bonding caused the color to change from yellow to orange, pink, and eventually purple [42]. A colorimetric sensor was designed to detect dithiocarbamate pesticide using AgNPs stabilized by sodium dodecyl sulphate (AgNPs@SDS) with LOD of 0.0091 μM and AgNPs@CTAB with LOD of 0.0979 μM [43]. The sensor produced reliable results and can be used to quantify pesticides in environmental and fruit juice samples.

3.8.5 Surface-Enhanced Raman Scattering (SERS)

Surface-enhanced Raman spectroscopy (SERS) is an analytical technique based on the Raman scattering effect that can improve the Raman scattering signal of organic molecules by adsorbing them on a metal surface. This technology has excellent detection precision, simple pretreatment, good selectivity, rapid and nondestructive testing, and several advantages for determining trace contaminants. Nie et al. used Raman spectroscopy to detect malathion by targeting the aptamer pesticide structure [44]. Numerous chemical and biological sensing applications have led to the development of SERS-based sensors utilizing nanomaterials. NPs, Ag, and Au are usually functionalized with particular molecules or receptors that interact with target analytes in a selective manner. SERS, which uses the interaction between nanomaterials and pesticide molecules, has shown itself to be a useful technique for pesticide detection. Pesticides, including imidacloprid and acetamiprid, have been effectively identified by AuNPs functionalized SERS arrays in both aqueous solutions and real-world samples, demonstrating the improved detection made possible by specific functionalization [45].

The SERS signal is amplified by the properties of NPs, including their size, shape, and composition. For particular pesticide analytes, this amplification has been especially shown with Ag/rGO composites and Fe_3O_4/Ag nanocomposites under particular excitation lines, greatly enhancing detection limits. Despite its advantages, SERS has drawbacks, such as the inability to get reliable and consistent SERS readings. One of SERS's main drawbacks is that different substrates may have different enhancement factors and signal intensities, which makes cross-experimental comparisons difficult. Thus, consistency in the distribution of nanomaterials and repeatability in manufacturing techniques are crucial elements to take into account. Furthermore, not much is known about the amplification that results from the interaction of the analyte molecules with the nanostructures, such as semiconductors and materials.

4 Notes

The use of nanosensor technology in pesticide residue detection is quickly changing traditional analytical paradigms in food and environmental safety. With growing global awareness of the dangers of pesticide contamination, there is an urgent need for sensitive, fast, and dependable detection technologies. Despite their great accuracy, traditional techniques, including GC, LC-MS, and HPLC, are nonetheless limited by their operational complexity, labor-intensive protocols, and laboratory-bound nature. Nanosensors use nanomaterials' unique physicochemical features to provide a highly sensitive, selective, and portable solution for real-time pesticide monitoring in a variety of matrices. Nanomaterials such as AuNPs, carbon nanotubes, graphene oxide, and QDs have proven remarkable performance in boosting electrochemical, optical, and biosensing platforms due to their high surface area, variable surface chemistry, and efficient signal amplification abilities. Furthermore, biosensor setups that use enzymes, antibodies, and aptamers provide extremely specific biorecognition systems for pesticide detection at ultra-trace levels. This has resulted in considerable advances in detecting organophosphates, carbamates, and neonicotinoids in agricultural and environmental samples. Despite these advantages, issues remain regarding nanosensor stability, matrix interference, and the lack of regulatory validation for routine deployment. Future potential includes combining nanosensors with Internet of Things (IoT) systems and smartphone interfaces, as well as developing multiplexed sensors capable of simultaneous multi-residue detection. Furthermore, studies on eco-friendly, biodegradable nanomaterials and scalable sensor production techniques will improve their environmental sustainability and commercial viability. Finally, nanosensor-based pesticide detection has enormous potential to revolutionize food safety monitoring, protect consumer health, and promote sustainable farming practices.

References

1. Fenik J, Tankiewicz M, Biziuk M (2011) Properties and determination of pesticides in fruits and vegetables. TrAC Trends Anal Chem 30: 814–826
2. Mostafalou S, Abdollahi M (2013) Pesticides and human chronic diseases: evidence, mechanisms, and perspectives. Toxicol Appl Pharmacol 268(2):157–177
3. Joshi S, Upadhyay J, Mathur A, Bishi S, Rupanagunta G, Sarkar B (2024) Nanosensors in pesticide residue detection in grapes. Grape Insight:30–42
4. Xu L, Abd El-Aty AM, Eun JB, Shim JH, Zhao J, Lei X et al (2022) Recent advances in rapid detection techniques for pesticide residue: a review. J Agric Food Chem 70(41): 13093–13117
5. Mirres A, Silva B, Tessaro L, Galvan D, Andrade J, Aquino A, Joshi N, Junior C, Carlos. (2022) Recent advances in nanomaterial-based biosensors for pesticide detection in foods. Biosensors 12:572
6. Zhang C, Qiu M, Wang J, Liu Y (2023) Recent advances in nanoparticle-based optical sensors for detection of pesticide residues in soil. In: Biosensors, vol 13(4). MDPI. https://doi.org/10.3390/bios13040415

7. Cernat A, Tertiş M, Săndulescu R, Bedioui F, Cristea A, Cristea C (2015) Electrochemical sensors based on carbon nanomaterials for acetaminophen detection: a review. Anal Chim Acta 886:16–28. https://doi.org/10.1016/j.aca.2015.05.044
8. Tan, J., Wang, T., Li, Y., Xu, S., Chen, S., & Hao, H. (2021). Review on functionalized magnetic nanoparticles for the pretreatment of organophosphorus pesticides. In Green Processing and Synthesis (Vol. 10, 1, pp. 485–498). De Gruyter Open Ltd. https://doi.org/10.1515/gps-2021-0051
9. Amine A, Arduini F, Moscone D, Palleschi G (2016) Recent advances in biosensors based on enzyme inhibition. Biosens Bioelectron 76: 180–194. https://doi.org/10.1016/j.bios.2015.07.010
10. Chapman, J., Ismail, A. E., & Dinu, C. Z. (2018). Industrial applications of enzymes: recent advances, techniques, and outlooks. In: Catalysts, vol 8(6). MDPI. https://doi.org/10.3390/catal8060238
11. Farka, Z., Juřík, T., Kovář, D., Trnková, L., & Skládal, P. (2017). Nanoparticle-based immunochemical biosensors and assays: recent advances and challenges. In Chemical Reviews (Vol. 117, 15, pp. 9973–10042). American Chemical Society. https://doi.org/10.1021/acs.chemrev.7b00037
12. Li YF, Sun YM, Lei HT, Wang H, Shen YD, Yang JY, Xu ZL, Beier RC, Gee S, Hammock BD, Wang Z, Sun X (2017) Immunochemical techniques for multianalyte analysis of chemical residues in food and the environment: a review. In: TrAC - Trends in Analytical Chemistry, vol 88. Elsevier B.V., pp 25–40. https://doi.org/10.1016/j.trac.2016.12.010
13. Jia M, Zhongbo E, Zhai F, Bing X (2020) Rapid multi-residue detection methods for pesticides and veterinary drugs. In: Molecules, vol 25(16). MDPI AG. https://doi.org/10.3390/molecules25163590
14. Song KM, Lee S, Ban C (2012) Aptamers and their biological applications. In: Sensors, vol 12(1), pp 612–631. https://doi.org/10.3390/s120100612
15. Lan, L., Yao, Y., Ping, J., & Ying, Y. (2017). Recent Progress in nanomaterial-based optical Aptamer assay for the detection of food chemical contaminants. In ACS Applied Materials and interfaces (Vol. 9, 28, pp. 23287–23301). American Chemical Society. https://doi.org/10.1021/acsami.7b03937
16. Yao C, Zhu T, Qi Y, Zhao Y, Xia H, Fu W (2010) Development of a quartz crystal microbalance biosensor with aptamers as bio-recognition element. Sensors 10(6): 5859–5871. https://doi.org/10.3390/s100605859
17. Duan, N., Wu, S., Dai, S., Gu, H., Hao, L., Ye, H., & Wang, Z. (2016). Advances in aptasensors for the detection of food contaminants. In Analyst (Vol. 141(13), pp. 3942–3961). Royal Society of Chemistry. https://doi.org/10.1039/c6an00952b
18. Volkert AA, Haes AJ (2014) Advancements in nanosensors using plastic antibodies. Analyst 139(1):21–31. https://doi.org/10.1039/c3an01725g
19. Chen L, Xu S, Li J (2011) Recent advances in molecular imprinting technology: current status, challenges and highlighted applications. Chem Soc Rev 40(5):2922–2942. https://doi.org/10.1039/c0cs00084a
20. Algieri C, Drioli E, Guzzo L, Donato L (2014) Bio-mimetic sensors based on molecularly imprinted membranes. In: Sensors (Switzerland), vol 14(8), pp 13863–13912. MDPI AG. https://doi.org/10.3390/s140813863
21. Rico-Yuste A, Carrasco S (2019) Molecularly imprinted polymer-based hybrid materials for the development of optical sensors. In: Polymers, vol 11(7). MDPI AG. https://doi.org/10.3390/polym11071173
22. Gold K, Slay B, Knackstedt M, Gaharwar AK (2018) Antimicrobial activity of metal and metal-oxide based nanoparticles. Adv Therapeutics 1(3). https://doi.org/10.1002/adtp.201700033
23. Abd IK, Shano AM, Khodair ZT (2022) MWCNT thin films by CVD method and some applications. J Nano Electron Phys 14(6). https://doi.org/10.21272/jnep.14(6).06002
24. Yi K, Liu D, Chen X et al (2021) Plasma-enhanced chemical vapor deposition of two-dimensional materials for applications. Acc Chem Res 54(4):1011–1022
25. Awlqadr FH, Altemimi AB, Qadir SA, Hama Salih TA, Alkanan ZT, AlKaisy QH, Mohammed OA, Hesarinejad MA (2025) Emerging trends in nano-sensors: a new frontier in food safety and quality assurance. In: Heliyon, vol 11(1). Elsevier Ltd. https://doi.org/10.1016/j.heliyon.2024.e41181
26. Parashar, M., Shukla, V. K., & Singh, R. (2020). Metal oxides nanoparticles via sol–gel method: a review on synthesis, characterization and applications. In Journal of Materials Science: Materials in Electronics (Vol. 31, 5, pp. 3729–3749). Springer. https://doi.org/10.1007/s10854-020-02994-8
27. Baino F, Fiume E, Miola M, Verné E (2018) Bioactive sol-gel glasses: processing,

properties, and applications. Int J Appl Ceram Technol 15(4):841–860. https://doi.org/10.1111/ijac.12873

28. Zamora-Sequeira R, Starbird-Pérez R, Rojas-Carillo O, Vargas-Villalobos S (2019) What are the Main sensor methods for quantifying pesticides in agricultural activities? A review. Molecules 24(14):2659. https://doi.org/10.3390/molecules24142659. PMID: 31340442; PMCID: PMC6680408
29. Della Pelle F, Angelini C, Sergi M, Del Carlo M, Pepe A, Compagnone D (2018) Nano carbon black-based screen printed sensor for carbofuran, isoprocarb, carbaryl and fenobucarb detection: application to grain samples. Talanta 186:389–396. https://doi.org/10.1016/j.talanta.2018.04.082. Epub 2018 Apr 30
30. Fauzi NIM, Fen YW, Omar NAS, Hashim HS (2021) Recent advances on detection of insecticides using optical sensors. Sensors 21:3856. https://doi.org/10.3390/s21113856
31. Campos BB, Contreras-Cáceres R, Bandosz TJ, Jiménez-Jiménez J, Rodríguez-Castellón E, da Silva JCGE et al (2017) Carbon dots coated with vitamin B12 as selective ratiometric nanosensor for phenolic carbofuran. Sensor Actuator B Chem 239:553–561. https://doi.org/10.1016/j.snb.2016.08.055
32. Lasagna M, Hielpos MS, Ventura C, Mardirosian MN, Martin G, Miret N et al (2020) Chlorpyrifos subthreshold exposure induces epithelial-mesenchymal transition in breast cancer cells. Ecotoxicol Environ Safety 205:111312. https://doi.org/10.1016/j.ecoenv.2020.111312
33. Luo Q-J, Li Y-X, Zhang M-Q, Qiu P, Deng Y-H (2017) A highly sensitive, dual-signal assay based on rhodamine B covered silver nanoparticles for carbamate pesticides. Chin Chem Lett 28:345–349. https://doi.org/10.1016/j.cclet.2016.10.024
34. Long Q, Li H, Zhang Y, Yao S (2015) Upconversion nanoparticle-based fluorescence resonance energy transfer assay for organophosphorus pesticides. Biosens Bioelectron 68:168–174. https://doi.org/10.1016/j.bios.2014.12.046
35. Hoshi M, Kaneko N, Nishiyabu R, Kubo Y (2018) Formation of emissive nanoparticles from tetraphenylethylene-containing boronate macrocycles: preparation, characterization and functionalization. J Mater Chem C 6:11052–11062. https://doi.org/10.1039/c8tc03748e
36. Jiang Y, McNeill J (2017) Light-harvesting and amplified energy transfer in conjugated polymer nanoparticles. Chem Rev 117:838–859. https://doi.org/10.1021/acs.chemrev.6b00419
37. Xie W, Ju Y, Zhang J, Yang Y, Zeng Y, Wang H, Li L (2022) Highly sensitive and specific determination of imidacloprid pesticide by a novel Fe3O4@ SiO2@ MIPIL fluorescent sensor. Anal Chim Acta 1195:339449
38. Yang Y, Lei X, Liu B, Liu H, Chen J, Fang G, Liu J, Wang S (2023) A ratiometric fluorescent sensor based on metalloenzyme mimics for detecting organophosphorus pesticides. Sensors Actuators B Chem 377:133031
39. Bashir K, Shikha S, Rattu G, Jan K, Krishna P, Pattanayek S (2024) Pesticide residues and their detection techniques in foods using sensors- a review. J Food Sci Technol 62:221. https://doi.org/10.1007/s13197-024-06116-8
40. Chen N, Liu H, Zhang Y, Zhou Z, Fan W, Yu G, Shen Z, Wu A (2018) A colorimetric sensor based on citrate-stabilized AuNPs for rapid pesticide residue detection of terbuthylazine and dimethoate. Sens Actuators B 255:3093–3101. https://doi.org/10.1016/j.snb.2017.09.134
41. Shikha S, Dureja S, Sapra R, Babu J, Haridas V, Pattanayek SK (2021) Interaction of borohydride stabilized silver nanoparticles with sul fur-containing organophosphates. RSC Adv 11(51):32286–32294
42. Ma S, He J, Guo M, Sun X, Zheng M, Wang Y (2018) Ultrasensitive colorimetric detection of triazophos based on the aggregation of silver nanoparticles. Colloids Surfaces A 538:343–349. https://doi.org/10.1016/j.colsurfa.2017.11.030
43. Ghoto A, Khuhawar MY, Janhngir TM (2019) Silver nanoparticles with sodium dodecyl sulfate as colorimetric probe for detection of dithiocarbamate pesticides in environmental samples. Anal Sci 35:631–637. https://doi.org/10.2116/analsci.18P417
44. Nie Y, Teng Y, Li P, Liu W, Shi Q, Zhang Y (2018) Label-free aptamer-based sensor for specific detection of malathion residues by surface-enhanced Raman scattering. Spectrochim Acta, Part A 191:271–276. https://doi.org/10.1016/j.saa.2017.10.030
45. Zhang H, Kang Y, Liu P, Tao X, Pei J, Li H, Du Y (2016) Determination of pesticides by surface-enhanced Raman spectroscopy on gold-nanoparticle-modified polymethacrylate. Anal Lett 49:2268–2278. https://doi.org/10.1080/00032719.2016.1147577

Chapter 3

Advanced Mass Spectrometry Protocols for Food Authentication

Flaviana Coelho Pacheco, Ana Flávia Coelho Pacheco, José Antônio de Queiroz Lafetá Junior, and Paulo Henrique Costa Paiva

Abstract

Food authentication and fraud detection are global challenges that impact food safety, consumer confidence, and market integrity. Traditional analytical methods often do not provide the speed and accuracy needed to identify adulterations and ensure traceability of food origin. In this context, mass spectrometry (MS) has emerged as an advanced and indispensable approach for food authentication. This chapter explores advanced MS protocols applied to the identification of chemical profiles and characteristic markers in different food categories. Innovative techniques such as ambient ionization mass spectrometry (AIMS), ion mobility mass spectrometry, and proton transfer reaction mass spectrometry are analyzed due to their ability to perform fast, accurate, and nondestructive analyses. Furthermore, the combination of MS with multivariate statistical methods and chemometrics has shown great potential to improve food authenticity screening and adulteration detection. The miniaturization of MS instruments is emerging as a promising trend, enabling more efficient field analysis and inspections in the supply chain. The implementation of these advanced technologies, combined with rigorous validation of analytical models, is essential to ensure reliable interpretations of spectral data. Thus, this chapter provides a comprehensive overview of the state of the art and challenges in the application of MS for food authentication, contributing to the development of robust and efficient protocols for global food safety.

Key words Authentication detection, Consumer confidence, Field analysis, Spectral data interpretation, Global food safety

1 Introduction

Globally, practices associated with food fraud represent a major concern, which has driven the continuous advancement of detection and authentication methods [1]. Food authentication refers to a set of scientific and analytical procedures aimed at confirming the veracity of the information declared on the label of a food product [2]. This process seeks to ensure that the attributes reported are correct, covering aspects such as the origin of the food, whether related to the species, geographic location, or genetic profile, as

Tanmay Sarkar (ed.), *Novel Food Analysis: Techniques and Applications*, Methods and Protocols in Food Science,
https://doi.org/10.1007/978-1-0716-4787-5_3,

well as the methods used during its production [3]. These methods may include conventional practices, organic systems, traditional techniques, or free-range farming. In addition, authentication can also involve verifying the technologies used in food processing, such as irradiation, freezing, or microwave heating [2].

The constant search for mechanisms capable of identifying fraud, whether intentional or not, was strongly boosted by initiatives by the European Union, which aims to combat fraudsters through coordinated and global strategies along the food chain [4]. In Europe, origin is one of the main authenticity issues related to food. As a protective measure, the European Union has established three quality labels for high-quality agri-food products: Protected Designation of Origin (PDO), Protected Geographical Indication (PGI), and Traditional Speciality Guaranteed (TSG) [5].

There are several forms of food fraud, these being identified as counterfeiting, adulteration, substitution of ingredients, and intentionally incorrect labeling [1]. These practices are largely motivated by the pursuit of illicit financial gains, which characterizes the concept of adulteration with economic motivation [4]. However, the technologies currently available have proven insufficient to effectively detect fraudulent practices, posing significant risks to both consumers and society [4].

Currently, mass spectrometry (MS) is among the dominant instrumental platforms used for metabolomic analyses, becoming a powerful key to the analysis of various components in food [6]. This technology has the advantage of being coupled with different separation technologies, such as capillary electrophoresis (CE), gas chromatography (GC), and liquid chromatography (LC) [7]. In addition, compared to vibrational methods, it has greater sensitivity and selectivity of the analytical signal, which allows the extraction of more information from the data [1].

Verifying the authenticity of food plays a crucial role in both quality control and ensuring food safety [2]. As consumer demands for transparency and traceability increase, authentication identification methods have become a key criterion for assessing the quality of food and its ingredients [2]. This demand is also directly related to the growing appreciation of regional and traditional products, whose legal protection requires strict proof of origin and specific characteristics.

In view of this scenario, this chapter seeks to examine and synthesize recent advances in MS protocols applied to food authentication. It seeks to offer a critical analysis of emerging techniques, highlighting their capabilities, limitations, and potential to improve accuracy and reliability in the identification of the origin, composition, and integrity of food products.

2 Materials

MS is an advanced and accurate technology used for food authentication. This analytical method detects and quantifies specific chemical compounds, allowing the identification of fraud and ensuring the integrity of food products without the need to destroy the sample. MS operates with high sensitivity and selectivity, enabling detailed analysis of complex matrices, from natural components to contaminants and adulterants. Its application contributes to food safety, protecting consumers and strengthening confidence in the market.

2.1 Materials Required

1. Mass spectrometer.
2. Coupled separation systems (such as LC or GC).
3. Ionization source (responsible for converting the sample molecules into ions in the gas phase).
4. Mass analyzer (where ions are separated according to their respective m/z).
5. Detector (where ions are counted).
6. Food samples to be authenticated.
7. Reagents for sample preparation and extraction.
8. Software for data processing and analysis.

3 Method

3.1 Mass Spectrometry: Fundamentals and Mechanisms

MS is an analytical technique that measures the mass-to-charge ratio (m/z) of ions, allowing the identification and characterization of chemical compounds. The basic foundation of MS involves ionizing the sample, separating the ions based on their m/z ratio, and detecting and analyzing the ions. A typical mass spectrometer includes four main components: sample input, ion source, mass analyzer, and ion detector. The sample is ionized by spraying through a thin capillary under a strong electric field, forming charged droplets. These ions pass through a heated inert gas and are separated by m/z before detection. A high vacuum (10^{-5}–10^{-7} Torr) ensures longer ion paths and minimizes molecular collisions, allowing structural analysis. The results of MS depend mainly on the introduction of the sample and the ionization method, with the mass analyzer having little effect on the observed chemical reactions [8].

3.1.1 Mechanism of Mass Spectrometry in Food Authentication

Chromatography is an effective technique for separating chemically similar compounds from complex food matrices. In food authentication, it faces the challenge of analyzing a wide variety of substances, such as peptides, lipids, carbohydrates, amino acids, fatty acids, phenolic compounds, and additives, which have great chemical diversity and polarity [8]. Authentication is based on comparing the profiles of measured compounds with reference standards, allowing the detection of adulteration in various products [2].

Initially, the molecules of interest are introduced into the ionization source of the spectrometer, where they are converted into positively or negatively charged ions. These ions are then directed to the mass analyzer, which separates them based on the mass-to-charge ratio (m/z). When they reach the detector, the ions generate signals that are captured and processed by a computer system. The result is presented in the form of a mass spectrum, in which the relative abundance of ions is plotted as a function of their respective m/z ratios [9].

1. *Ionization:* The molecules in the sample are transformed into ions, which can be positive or negative, through different ionization methods, such as electron impact ionization (EI) or electrospray ionization (ESI).
2. *Mass Analysis:* The ions generated are separated based on their m/z ratio, which is the ratio of the ion's mass to its electrical charge. This separation can be done by different types of mass analyzers, such as quadrupole, time of flight (TOF), and others.
3. *Detection and Analysis:* The separate ions are detected and quantified, generating a mass spectrum, which is a graph that shows the abundance of each ion as a function of its m/z ratio.

3.1.2 Factors Affecting the Quantitative Capabilities of Mass Spectrometry

The quantitative capability of MS is influenced by several factors, which can be grouped into two main categories: instrument-related factors and sample-related factors.

Instrument-Related Factors

1. Ion source parameter setting.
2. Efficiency in the transmission of ions.
3. Efficiency of the detection system.
4. Internal contamination of the equipment (parts of the spectrometer).

Sample-Related Factors

1. Ionization efficiency of analytes.
2. Analyte concentration.
3. Possible degradation of analytes during the process.
4. Quality of sample preparation.
5. Limitations in signal stability and repeatability.

6. Interference in ionization.
7. Interference with detection.
8. Spectral interferences.

In addition to these, the operator himself can be a source of errors, whether in the configuration, handling, or preparation of samples. However, with technological advances in software, hardware, and protocols, the influence of human error has been significantly reduced.

Among all the factors, sample preparation is considered one of the most critical steps, as inadequate preparation can directly compromise the precision and accuracy of the results. To minimize interference from the matrix and improve quantitative performance, pre-separation techniques such as chromatography are applied.

3.1.3 Modern Mass Spectrometry Protocols

Modern MS protocols use different ionization and separation techniques, depending on the food matrix and target analytes:

1. *Electrospray Ionization (ESI) and Atmospheric Pressure Chemical Ionization (APCI):* employed for polar and semipolar compounds, good for small organic molecules.
2. *Gas Chromatography–MS (GC-MS):* used for volatile organic compounds.
3. *Liquid Chromatography–MS (LC-MS):* used for the targeted quantification of known adulterants.
4. *Isotope Ratio Mass Spectrometry (IRMS):* evaluates stable isotope ratios (e.g., $^{13}C/^{12}C$, $^{15}N/^{14}N$, $^{18}O/^{16}O$) to verify geographic origin and production methods.

3.2 Ionization Techniques

The first step in MS analysis is the ionization of analytes from the liquid or solid phase to the gas phase [10]. There are several ionization techniques, each with specific advantages, and the choice of technique can significantly affect the results, influencing, for example, the sensitivity and the range of quantification [10]. Different ionization methods, such as electrospray ionization (ESI), atmospheric pressure chemical ionization (APCI), desorption electrospray ionization (DESI), or matrix-assisted laser desorption ionization (MALDI), can identify distinct sets of metabolites [11].

3.2.1 Electrospray Ionization (ESI-MS)

In ESI-MS, the liquid sample is introduced into a capillary with high voltage, forming electrically charged droplets that are directed into the atmosphere. An auxiliary gas, such as nitrogen, helps conduct these droplets, which undergo heating to evaporate the solvent. The ions generated in this process enter the mass spectrometer for analysis [12].

Two main models explain the formation of ions in electrospray ionization, ionic evaporation, and residual charge [13]:

1. *Ionic evaporation* describes the reduction of the size of the charged droplets by evaporation of the solvent until the electrostatic repulsion between the charges overcomes the surface tension, releasing the ions.
2. *Residual charge* suggests that the molecule remains in the droplet until the solvent evaporates completely, releasing the ion; this process is more common in large molecules.

ESI is suitable for polar compounds and macromolecules, generating multiple ions that facilitate analysis in spectrometers with a limited m/z range. However, it suffers from ion suppression, caused by interference from nonvolatile compounds or at high concentrations. To minimize this effect, strategies such as sample purification, chromatographic adjustment, ionization alteration, flow rate reduction, and use of isotopically labeled internal patterns are used [14].

3.2.2 Atmospheric Pressure Chemical Ionization (APCI-MS)

The main difference between APCI and other techniques is that ionization occurs outside of a vacuum, at atmospheric pressure. In this process, the volatile analytes carried by the column gas enter the source, where they are ionized by a corona needle. A hole separates this region from the vacuum zone of the mass spectrometer [15].

The mechanism of ionization in APCI is based on ion-molecule reactions in the gas phase. Initially, electrons generated by a corona discharge needle ionize the reactant gas (usually nitrogen), forming ions such as $N_2^{+\cdot}$and $N_4^{+\cdot}$. These, in turn, transfer charge to the analytes with low ionization potential. In the presence of water, the formation of H_3O^+ and its clusters occurs, which promotes ionization by proton transfer, generating $[M + H]^+$ [16].

In negative mode, ions such as $O_2^{-\cdot}$ are formed by electron capture and ionize compounds with high electron affinity, promoting dissociation reactions. The chamber temperature and gas composition directly affect the ionization efficiency in both modes [15].

3.2.3 Desorption Electrospray Ionization (DESI-MS)

The DESI-MS technique stands out in the area of food safety due to its ability to perform direct and fast analyses, eliminating time-consuming sample preparation steps, which is in line with the principles of green analytical chemistry [17], making it a practical and versatile tool. Despite this, quantification with DESI-MS is challenging due to the variability of the signals and the difficulty of experimental control [18].

The operation of the technique is based on the application of charged microdroplets on the surface of the sample by means of a high-speed gas jet. This interaction results in desorption and ionization of the analytes, which are then conducted to the mass spectrometer under atmospheric pressure conditions, similar to the ESI-MS process [18].

The intensity of the signal obtained is influenced by factors such as the way the solute is distributed on the surface and the morphological characteristics of the analyzed material [19]. In addition, the natural variability of surfaces, with different structures, affects the reproducibility and accuracy of the analysis [20]. Parameters such as surface morphology and analyte distribution are critical to ensure consistent results [21].

3.2.4 Matrix-Assisted Laser Desorption/Ionization (MALDI-MS)

The MALDI technique is widely used to generate high-molecular-weight ions, and is especially effective for proteins, peptides, oligonucleotides, glycoconjugates, and polymers. Its main advantage is the production of single-charge ions, which facilitates the analysis of complex mixtures without the need for prior separation by chromatography [22].

Despite its usefulness, MALDI has limitations, such as low reproducibility and low mass resolution for high m/z ratio molecules, as well as the difficulty of ionizing large polysaccharides and glycoproteins due to the lack of appropriate matrices [22].

The ionization mechanism of MALDI is not yet fully understood. The classical model, based on photochemical matrix ionization, does not explain well certain observed effects, such as the suppression of the matrix and the rapid generation of ions. Alternative models, such as cluster ionization and pseudoproton transfer, have been proposed, but there is still no definitive consensus on which mechanism predominates [23].

1. *Traditional photochemical model*: proposes that photoionization of the matrix (after absorption of multiple photons) generates ions that transfer charge to the analyte by collisions in the gas phase. However, it does not explain well-observed phenomena such as the matrix suppression effect, the "sweet spots" of crystallization, and the rapid formation of polymeric ions.
2. *Cluster ionization (CI) model*: suggests that protonated analyte clusters are desorbed from the matrix, and the ions are generated by desolvation, similar to the process in ESI. This model eliminates the need for gas-phase collisions, but has difficulty explaining the predominance of monocharged ions in large biomolecules.
3. *Alternative pseudoproton transfer model*: occurs during crystallization between matrix and analyte, contributing to the generation of ions.

4. *Alternative energy transfer-induced disproportionalization (ETID) model:* explains the simultaneous formation of positive and negative ions from excited analyte dimers.

3.3 Chromatographic Coupling Mass Spectrometry

The combination of chromatographic techniques with MS, such as GC, LC, Ultra-High Performance Liquid Chromatography (UHPLC-MS), and Ion Chromatography (IC), is essential to detect food fraud with high sensitivity and specificity [24]. These techniques have become widely used in modern science for their simplicity, speed, high sensitivity, and accuracy, and are applied in the analysis of metabolites, drugs, foods, environmental samples, and explosive residues [25].

Each technique has advantages depending on the type of target compound and the matrix [24]. For instance:

1. *GC-MS* is more suitable for volatile compounds.
2. *LC-MS/MS* and *UHPLC-MS* are preferred for nonvolatile metabolites and for analysis of proteins, peptides, and small molecules, especially in complex matrices.

Although GC-MS and LC-MS are widely used in chemical analyses, they have significant limitations in the detection of highly polar compounds, unless derivatization is resorted to [26]. However, these derivatization processes tend to be labor-intensive, imprecise, and not always effective for separating substances with similar characteristics [25]. In view of this, IC stands out as a more suitable alternative for this type of analysis, due to its high retention capacity of ionic species and polar compounds, even in complex matrices [27].

In addition to the choice of analytical technique, sample preparation plays a central role in the quality of the results. Methods such as solid-phase extraction (SPE), protein precipitation, and Quick, Easy, Cheap, Effective, Rugged and Safe (QuEChERS) are often used to purify samples and reduce matrix interference, which improves the accuracy and repeatability of analyses. The appropriate choice of solvent, as well as the control of pH and temperature, are also important factors to preserve the analytes and ensure the efficiency of extraction [24].

3. *Protein precipitation* aims to remove proteins from biological matrices that interfere with the analysis. An organic solvent (such as acetonitrile or methanol) is used for the sample. The protein precipitates and is removed by centrifugation, and the supernatant is used for analysis.
4. *QuEChERS Extraction* is efficient in the extraction of multiple residues in food. The sample is mixed with acetonitrile and specific salts. After stirring and centrifugation, the phase containing the analytes is separated. It can be followed by a cleaning step with solid phase dispersion (d-SPE).

5. *Solid-phase extraction* is a rapid purification step after extraction, usually used with QuEChERS. A sorbent (such as $MgSO_4$, PSA, and C18) is added directly to the extract. After stirring and centrifugation, a purified extract is obtained.

3.4 Calibration and Quantification by Mass Spectrometry

To ensure accurate quantitative results, it is essential to carry out a rigorous calibration process, which correlates the response of the instrument with the concentration of the analytes of interest [4]. The calibration procedure involves the preparation of a series of standards containing known concentrations of the target compound. From these standards, a calibration curve is constructed, which relates the signal strength (usually the peak area in the mass spectrum) to the concentration. This curve must present linearity within the work interval, being evaluated by statistical parameters such as the coefficient of determination (R^2) and the analysis of the residuals [28].

The quantification of the samples is performed by comparing the response obtained with the calibration curve. For greater accuracy and compensation for any instrumental or matrix variations, it is common to use internal standards, compounds chemically similar to the analyte, but with different molecular masses, as labeled isotopes [20]. These internal patterns help correct for losses during sample preparation, variations in ionization, and fluctuations in equipment response.

In complex matrices, quantification can be affected by ionization suppression or enhancement effects. Therefore, in addition to external calibration, the standard addition approach is often applied, where known amounts of the analyte are added directly to the sample. This allows you to compensate for the effects of the matrix and increase the accuracy of the results [18].

The choice of the type of quantification (absolute or relative) depends on the objective of the analysis. Absolute quantification aims to determine the exact concentration of a compound in the sample, while relative quantification compares the abundance of different compounds or experimental conditions [29].

3.5 Authentication in Food Using Mass Spectrometry

Mass spectrometry-based methods of food authentication have developed significantly, especially with the use of nontargeted approaches coupled with chromatography and spectral fingerprinting. These methods allow the analysis of complex food matrices, even when it comes to little-known substances, being fast, high-yield tools with wide application in the MS platform [6].

3.5.1 Fatty Acid (FA) and Triglyceride Profiles

Lipids are present in high concentrations in foods such as oilseeds, meats, nuts, liver, milk, and egg yolks. In edible oils and fats, triacylglycerols (TAGs) predominate, formed by three fatty acids linked to glycerol. They can also be present in more complex forms, such as phospholipids and sterol esters. The lipid profiles of foods, which vary according to the composition of TAGs and fatty acids, provide useful information about the origin and quality of products thus being important tools for food authentication [6].

The classification of lipids is challenging due to the variety of their structures. They are grouped based on the chemical entities they have. The main classes include fatty acids, sterols, neutral lipids, glycerophospholipids, and sphingolipids. Glycerophospholipids have a phosphate group attached to glycerol, which can be acylated or alkylated at the sn-1 and sn-2 positions, called "radil." Unsaturations are described as E/Z and, in the absence of radil, are called "smooth" [29].

Due to their high economic value, edible oils are products that are often subject to adulteration [29]. The analysis of TAGs has been shown to be an effective approach to authenticate oils, fats, and other foods, and MS is a fundamental tool in this process. This technique allows the identification of TAGs based on their fragmentation spectra and molecular mass information [30].

Direct MS methods, such as sample analysis [31] and matrix-assisted laser desorption ionization [32], have been widely used in the authentication of TAGs in various foods, especially animal fats and vegetable oils. At the same time, LC is particularly suitable for the analysis of TAGs, since these molecules are sensitive to degradation under the high temperatures of GC [29].

Dairy products, such as milk, have also been investigated through GAT profiling. For this, techniques such as DART-HRMS (real-time MS with direct ionization) associated with statistical methods, such as Principal Component Analysis (PCA) and linear discriminant analysis (LDA), demonstrate good performance in sample authentication and discrimination [33].

In addition, the analysis of fatty acid (FA) composition in edible oils has proven to be an efficient tool, especially with the use of GC-MS operating in the mode of monitoring selected ions. This approach makes it possible to classify oils such as soybean, peanut, sunflower, rapeseed, and sesame through multivariate methods such as PCA, HCA, and random forest (RF) [34].

In situations that require faster analyses, direct electron impact MS, without chromatographic separation, allows the classification and differentiation of oils based on their GA profiles, using statistical strategies such as LDA, PCA, HCA, and class-independent modeling. These profiles encompass both free fatty acids (FFA) and esterified fatty acids [35].

3.5.2 Proteins

Mass spectrometry has been consolidated as one of the main tools for protein analysis, being fundamental in modern proteomics. Initially, its use was focused on the identification of proteins from the detection of specific peptides generated by enzymatic digestion [6]. However, this approach provided only qualitative information, limiting the understanding of variations in protein amounts under different biological conditions. With the advancement of techniques, MS has evolved to also become a quantitative tool, which is essential for systems biology, where subtle changes in protein abundance are biologically relevant and often determinant to understand cellular mechanisms and responses to stimuli [36].

Mass spectrometry coupled to liquid chromatography (LC-MS) and capillary electrophoresis (CE-MS) has been widely used in the analysis of proteins and peptides, both for nutritional studies and for food authentication [6]. Proteins, in addition to providing energy, are essential for physiological functions, such as enzymatic and immunological action. In food analysis, proteins and peptides identified by MS can serve as biomarkers, allowing the differentiation of foods of high and low nutritional value, such as meat, seafood, and dairy products [36].

There are two main strategies in proteomics: bottom-up and top-down. The bottom-up approach is the most common, involving the enzymatic digestion of proteins into peptides, followed by analysis by LC-MS/MS or by combining two-dimensional gel electrophoresis with MALDI-TOF-MS [37]. The top-down approach analyzes intact proteins, offering more complete information about their structure. Both strategies rely on comparing the data obtained with protein databases to ensure correct identification.

For quantification, label-based methods such as iTRAQ and TMT can be used, which use isotopic markers to measure relative and absolute amounts of proteins, offering high accuracy, but with high costs [37]. Alternatively, label-free methods such as MRM and SWATH allow for fast, cost-effective, and efficient quantification, especially suitable for large-volume sample analysis [6].

3.5.3 Volatile Organic Compounds (VOCs)

Volatile organic compounds are responsible for the aromas of food, and therefore, their profiles are widely used in the authentication of food products. The GC-MS technique is one of the main tools for the identification of these compounds, being applied in the analysis of foods such as honey, olive oil, wine, cheese, meat, fruits, vegetables, tea, and coffee. The qualitative identification of VOCs is done by comparing the spectra obtained with standard databases [38].

In addition, due to the great complexity of volatile profiles, more advanced techniques such as GC × GC-MS (two-dimensional gas chromatography) are used to improve the separation of coeluting compounds [38]. Another emerging technique is proton transfer reaction mass spectrometry (PTR-MS), which offers fast and sensitive analysis of VOCs and is especially useful for authentication and characterization of foods with intense aromas. According to studies, in general, the most commonly used pretreatment of volatiles in food is HS-SPME [6].

3.6 Multivariate Statistical Analysis in Food Authentication

Methods that use profiling or fingerprinting for food authentication generate large volumes of data, often complex and multidimensional. Therefore, multivariate statistical analysis techniques are applied, which help to reduce dimensionality, visualization, and classification of data. Models such as the unsupervised PCA and the supervised Partial Least Squares Discriminant Analysis (PLS-DA) are widely used, as demonstrated in several studies [6].

PCA is recommended when seeking to identify subtle variations between samples of the same group and also to detect outliers. The PLS-DA, on the other hand, is more efficient in differentiating between groups that present more significant differences [39]. Both approaches are effective in distinguishing samples from different origins or characteristics, based on their chemical or metabolic profiles.

A promising alternative is the single-class classification model (OCC), which relies only on authentic samples to build the model. This method is advantageous in detecting various types of fraud [40]. However, its efficacy depends on the availability of a robust number of authentic and adulterated samples, which adequately represent the variability of foods.

In addition, selecting the most relevant variables is essential for developing robust models. Omics techniques detect a large number of compounds, many of which are not useful for authentication. Therefore, it is critical to identify the components that truly differentiate legitimate samples from adulterated ones. The Orthogonal Partial Least Squares Discriminant Analysis (OPLS-DA) technique has been successfully applied for this purpose, such as in the detection of linseed oil adulteration with only 5% addition of cheaper oils [41].

3.7 Applications of Mass Spectroscopy in Food Analysis

Mass spectroscopy has excelled in food analysis, making it a powerful tool for complex analyses such as authentication, quality control, fraud detection, and food safety.

In food analysis, MS has been applied both independently and coupled with techniques such as LC-MS and GC-MS, increasing the selectivity and sensitivity of the methods. Its application ranges from the control of volatile compounds in industrial processes, such as fermentation and freshness monitoring, to the authentication of products such as olive oils, cheeses, honey, and beverages, as well as the detection of adulteration and contaminants.

3.7.1 Applications in Food Analysis

In the context of food analysis, MS has proven to be essential in food safety, allowing the detection of a wide range of contaminants such as pesticides, mycotoxins, antibiotic residues, hormones, and environmental contaminants such as PCBs and dioxins. The use of LC-MS/MS and GC-MS/MS allows multi-residue analyses to be performed, detecting hundreds of compounds simultaneously, with high sensitivity and specificity, meeting both regulatory requirements and the demands of the production sector. In addition, MS plays a central role in food authentication and traceability. Techniques such as isotope ratio spectrometry (IR-MS) make it possible to determine the geographical and botanical origin of products, differentiating, for example, real oils from adulterated mixtures, pure honey from honey enriched with syrups, or even characterizing the origin of wines, coffees, and meats. Food fraud detection is also a consolidated application of MS, making it possible to identify substitutions, adulterations, and counterfeits, protecting both consumers and legitimate producers.

1. *Composition:* Characterization of lipids, phenolics, carbohydrates, and volatile compounds in foods such as olive oil, milk, wine, and coffee.
2. *Process Control:* Monitoring of freshness (eggs, meat, and fish), fermentation (beer and wine), and degradation during storage.
3. *Authentication:* Differentiation of geographical origin, purity, and quality of olive oils, meats, cheeses, wines, and honey.
4. *Adulteration Detection:* Identification of fraudulent mixtures in vegetable oils, honey, and dairy products.
5. *Food Safety:* Detection of pesticides, mycotoxins, environmental contaminants, and veterinary residues, even at low levels.

4 Note

MS stands as one of the most powerful, versatile, and sensitive analytical techniques currently available for the authentication and traceability of food products [2]. Its core principle, the precise measurement of m/z of ions enables the detection and quantification of a vast array of chemical compounds, from small metabolites to complex macromolecules, providing unparalleled resolution in compositional fingerprinting [42]. This capability is particularly crucial in a context where food fraud, mislabeling, adulteration, and origin falsification pose serious risks to consumers, industry integrity, and regulatory compliance worldwide.

The advantages of MS lie in its high sensitivity, selectivity, and versatility. When coupled with separation techniques such as LC-MS, GC-MS, ion chromatography (IC-MS), or CE-MS, it allows the resolution of highly complex food matrices into detailed molecular profiles [13]. Moreover, the development of ambient ionization techniques—such as Desorption Electrospray Ionization (DESI), Direct Analysis in Real Time (DART), and Proton Transfer Reaction Mass Spectrometry (PTR-MS)—enables rapid, direct, and often nondestructive analysis of samples with minimal preparation [43]. These approaches are invaluable for real-time monitoring and field-based screening applications in supply chains, ports, and production lines.

Despite its remarkable analytical power, the successful application of MS for food authentication depends heavily on rigorous methodological protocols. Critical parameters that directly influence the reliability and reproducibility of results include the choice of ionization method, sample preparation strategies, calibration protocols, control of matrix effects, and selection of appropriate statistical and chemometric models [42]. Poorly optimized ionization conditions can lead to ion suppression, signal instability, or false negatives, especially in complex food matrices rich in salts, lipids, or polysaccharides.

Sample preparation emerges as one of the most decisive steps. Inadequate extraction, insufficient purification, or failure to account for matrix-specific interferences can drastically compromise data quality [42]. Techniques like QuEChERS, protein precipitation, solid-phase extraction, and filtration are not mere procedural steps but pivotal determinants of the sensitivity, accuracy, and robustness of the analysis [44]. Moreover, the stability of analytes during extraction, storage, and analysis must be carefully considered, especially for thermolabile or reactive compounds.

Calibration is another cornerstone of reliable MS analysis. The use of internal standards, particularly isotopically labeled analogs, is indispensable to correct for variability in ionization efficiency, detector response, and sample handling [45]. External calibration curves must be constructed with rigor, ensuring linearity, reproducibility, and absence of systematic bias. In scenarios where matrix effects are pronounced, the adoption of standard addition protocols becomes essential to accurately compensate for signal suppression or enhancement caused by the sample matrix [45].

In terms of data analysis, the shift toward nontargeted approaches and high-dimensional data generation demands the application of advanced chemometric and machine learning (ML) tools [11]. Models such as PCA, PLS-DA, OPLS-DA, and, more recently, one-class classifiers (OCC) are widely employed [6]. However, these models are only as reliable as the quality, representativeness, and balance of the reference datasets used for training and validation. Overfitting, lack of generalizability, and bias can compromise the interpretability of the results if model validation is not conducted with cross-validation, permutation testing, and external validation datasets [6].

An important limitation often underestimated is the natural variability of food products. Factors such as geographical origin, seasonality, agricultural practices, processing conditions, and storage can introduce significant variability in the chemical profiles of authentic products. Therefore, authentication models must be built on comprehensive and well-curated databases that capture this variability to prevent false positives or negatives in fraud detection.

Another key consideration is the choice of mass analyzer. High-resolution instruments such as Orbitrap, Time-of-Flight (TOF), and Fourier Transform Ion Cyclotron Resonance (FT-ICR) provide accurate mass measurements with ppm-level precision, allowing the identification of unknown compounds and structural elucidation [46]. However, these instruments come with higher operational costs, maintenance complexity, and data processing demands. Conversely, quadrupole-based systems, while more accessible and robust, may have limitations in resolving power, particularly in untargeted or complex analyses [46].

Recent technological advancements are addressing some of these challenges. The development of miniaturized, portable mass spectrometers is expanding the application of MS beyond the laboratory, allowing in situ analyses for rapid authentication at production sites, markets, and import/export facilities [47]. These instruments, combined with simplified ambient ionization sources and cloud-based data analysis platforms, are reshaping the future of food quality control. Nevertheless, portability does not exempt the need for strict standardization of sampling, calibration, and data interpretation protocols [48].

Integration with complementary techniques further amplifies the capabilities of MS. Coupling with stable isotope ratio analysis (IRMS) provides insights into geographical and botanical origin. When combined with spectroscopic methods (NMR, infrared, Raman) or genomic tools (DNA-based assays), MS contributes to a multi-omics approach that enhances reliability, traceability, and authentication robustness [49].

It is equally important to recognize the growing role of bioinformatics and artificial intelligence (AI) in MS-based food authentication [50]. Automated feature extraction, spectral deconvolution, pattern recognition, and predictive modeling are becoming standard practices, necessitating that food scientists acquire or collaborate with expertise in data science. This interdisciplinary convergence is transforming food analysis into a data-driven science capable of tackling increasingly sophisticated fraud schemes.

In conclusion, MS is not merely an analytical tool but a strategic pillar in modern food science [6]. Its application to food authentication goes beyond detecting fraud; it safeguards public health, preserves cultural and regional food heritage, and reinforces consumer trust. However, its successful implementation requires more than access to advanced instrumentation, it demands rigorous methodological execution, robust validation protocols, comprehensive databases, and an interdisciplinary approach that combines analytical chemistry, food science, bioinformatics, and regulatory knowledge [12]. As technological innovation continues to advance, MS will remain indispensable in ensuring the authenticity, safety, and integrity of the global food supply chain [2].

Despite the enormous advances, MS still faces important challenges, such as the complexity of the data generated, which requires the use of bioinformatics and AI tools for its interpretation, in addition to the high cost of high-resolution equipment, which still limits its use in some contexts. Another relevant challenge is the influence of the feed matrix, which can interfere with ionization efficiency and the accuracy of results, requiring robust sample preparation methods and matrix effect correction strategies [6].

The future prospects for MS in food analysis are extremely promising. The trend is that the miniaturization of systems, the development of portable spectrometers, and the automation of analytical processes will allow rapid analyses to be carried out directly in the field, on the production line, or at points of sale. The integration of MS with ML and AI tools will allow for faster, more accurate analyses with less need for human intervention, making quality control more efficient and accessible. In addition, the growth of nontargeted approaches, such as chemical fingerprinting, will play a key role in authentication, traceability, and the development of safer, healthier, and more sustainable foods.

In this way, MS consolidates itself as an indispensable technology in food analysis, capable of meeting society's growing demands for high-quality, safe, authentic products with guaranteed traceability. Its continuous development, combined with technological innovations and integration with other areas of knowledge, ensures that its role will continue to be central in the food sciences in the coming decades.

References

1. Augusti R, Fulgencio ACC, Nogueira HM, Gomes JCL, dos Santos LB, de Macedo AN et al (2024) Enhancing food authentication screening through the integration of chemometrics and ambient ionization mass spectrometry: a comprehensive review. Trends Food Sci Technol 104480
2. Danezis GP, Tsagkaris AS, Camin F, Brusic V, Georgiou CA (2016) Food authentication: techniques, trends & emerging approaches. TrAC Trends Anal Chem 85:123–132
3. Aung MM, Chang YS (2014) Traceability in a food supply chain: safety and quality perspectives. Food Control 39:172–184
4. Esteki M, Regueiro J, Simal-Gándara J (2019) Tackling fraudsters with global strategies to expose fraud in the food chain. Compr Rev Food Sci Food Saf 18(2):425–440
5. Dias C, Mendes L (2018) Protected designation of origin (PDO), protected geographical indication (PGI) and traditional speciality guaranteed (TSG): a bibiliometric analysis. Food Res Int 103:492–508
6. Dou X, Zhang L, Yang R, Wang X, Yu L, Yue X et al (2023) Mass spectrometry in food authentication and origin traceability. Mass Spectrom Rev 42(5):1772–1807
7. Zhang W, Hankemeier T, Ramautar R (2017) Next-generation capillary electrophoresis–mass spectrometry approaches in metabolomics. Curr Opin Biotechnol 43:1–7
8. Cserháti T, Forgács E, Deyl Z, Miksik I (2005) Chromatography in authenticity and traceability tests of vegetable oils and dairy products: a review. Biomed Chromatogr 19(3):183–190
9. Ho CS, Lam CWK, Chan MH, Cheung RCK, Law LK, Lit LCW et al (2003) Electrospray ionisation mass spectrometry: principles and clinical applications. Clin Biochem Rev 24(1):3
10. Buse J, Purves RW, Verrall RE, Badea I, Zhang H, Mulligan CC, Peru KM, Bailey J, Headley JV, El-Aneed A (2014) The

development and assessment of high-throughput mass spectrometry-based methods for the quantification of a nanoparticle drug delivery agent in cellular lysate. J Mass Spectrom 49(11):1171–1180
11. Nordström A, Want E, Northen T, Lehtiö J, Siuzdak G (2008) Multiple ionization mass spectrometry strategy used to reveal the complexity of metabolomics. Anal Chem 80(2): 421–429
12. De Hoffmann E, Stroobant V (2007) Mass spectrometry: principles and applications. Wiley
13. Kraj A, Desiderio DM, Nibbering NM (2008) Mass spectrometry: instrumentation, interpretation, and applications. Wiley
14. Van Eeckhaut A, Lanckmans K, Saarland S, Smolders I, Michotte Y (2009) Validation of bioanalytical LC–MS/MS assays: evaluation of matrix effects. J Chromatogr B 877(23): 2198–2207
15. Fang J, Zhao H, Zhang Y, Lu M, Cai Z (2020) Atmospheric pressure chemical ionization in gas chromatography-mass spectrometry for the analysis of persistent organic pollutants. Trends Environ Anal Chem 25:e00076
16. Horning EC, Horning MG, Carroll DI, Dzidic I, Stillwell RN (1973) New picogram detection system based on a mass spectrometer with an external ionization source at atmospheric pressure. Anal Chem 45(6):936–943
17. Gałuszka A, Migaszewski Z, Namieśnik J (2013) The 12 principles of green analytical chemistry and the SIGNIFICANCE mnemonic of green analytical practices. TrAC Trends Anal Chem 50:78–84
18. Costa AB, Cooks RG (2008) Simulated splashes: elucidating the mechanism of desorption electrospray ionization mass spectrometry. Chem Phys Lett 464(1–3):1–8
19. Kauppila TJ, Wiseman JM, Ketola RA, Kotiaho T, Cooks RG, Kostiainen R (2006) Desorption electrospray ionization mass spectrometry for the analysis of pharmaceuticals and metabolites. Rapid Commun Mass Spectrom 20(3):387–392
20. Mainero Rocca L, Cecca J, L'Episcopo N, Fabrizi G (2017) Ambient mass spectrometry: direct analysis of dimethoate, tebuconazole, and trifloxystrobin on olive and vine leaves by desorption electrospray ionization interface. J Mass Spectrom 52(11):709–719
21. Gurdak E, Green FM, Rakowska PD, Seah MP, Salter TL, Gilmore IS (2014) VAMAS interlaboratory study for desorption electrospray ionization mass spectrometry (DESI MS) intensity repeatability and constancy. Anal Chem 86(19):9603–9611
22. Chang WC, Huang LCL, Wang YS, Peng WP, Chang HC, Hsu NY et al (2007) Matrix-assisted laser desorption/ionization (MALDI) mechanism revisited. Anal Chim Acta 582(1): 1–9
23. Knochenmuss R (2010) MALDI ionization mechanisms: an overview. In: Electrospray and MALDI mass spectrometry, pp 149–183
24. Suman M, Cavanna D, Sammarco G, Lambertini F, Loffi C (2021) Fighting food frauds exploiting chromatography-mass spectrometry technologies: scenario comparison between solutions in scientific literature and real approaches in place in industrial facilities. TrAC Trends Anal Chem 142:116305
25. Li D, Huang W, Huang R (2023) Analysis of environmental pollutants using ion chromatography coupled with mass spectrometry: a review. J Hazard Mater 458:131952
26. Montaseri H, Forbes PB (2018) Analytical techniques for the determination of acetaminophen: a review. TrAC Trends Anal Chem 108: 122–134
27. Bigard A, Cardinael P, Agasse V (2023) Anion exchange chromatography coupled to electrospray-mass spectrometry: an efficient tool for food, environment, and biological analysis. Crit Rev Anal Chem 53(7): 1591–1603
28. Ye H, Yang J, Xiao G, Zhao Y, Li Z, Bai W et al (2023) A comprehensive overview of emerging techniques and chemometrics for authenticity and traceability of animal-derived food. Food Chem 402:134216
29. Wei F, Hu N, Lv X, Dong XY, Chen H (2015) Quantitation of triacylglycerols in edible oils by off-line comprehensive two-dimensional liquid chromatography–atmospheric pressure chemical ionization mass spectrometry using a single column. J Chromatogr A 1404:60–71
30. Xu B, Zhang L, Ma F, Zhang W, Wang X, Zhang Q, Luo D, Ma H, Li P (2018) Determination of free steroidal compounds in vegetable oils by comprehensive two-dimensional gas chromatography coupled to time-of-flight mass spectrometry. Food Chem 245:415–425
31. Fiorino GM, Losito I, De Angelis E, Logrieco AF, Monaci L (2018) Direct analysis in real time coupled to high resolution mass spectrometry as a rapid tool to assess salmon (Salmo salar) freshness. J Mass Spectrom 53(9):781–791
32. Zhang YL, Gong C, Pei XL, Han YL, Huang YY, Xu X (2018) Rapid quantitative determination of triglycerides in edible oils by matrix-

assisted laser desorption/ionisation Fourier transform ion cyclotron resonance mass spectrometry using pencil graphite combined with 2, 5-dihydroxybenzoic acid as matrix. Int J Mass Spectrom 431:56–62

33. Hrbek V, Vaclavik L, Elich O, Hajslova J (2014) Authentication of milk and milk-based foods by direct analysis in real time ionization–high resolution mass spectrometry (DART–HRMS) technique: a critical assessment. Food Control 36(1):138–145
34. Zhang L, Li P, Sun X, Wang X, Xu B, Wang X, Ma F, Zhang Q, Ding X (2014) Classification and adulteration detection of vegetable oils based on fatty acid profiles. J Agric Food Chem 62(34):8745–8751
35. Kenar A, Çiçek B, Arslan FN, Akin G, Elmas ŞN, Yilmaz I (2019) Electron impact–mass spectrometry fingerprinting and chemometrics for rapid assessment of authenticity of edible oils based on fatty acid profiling [Internet]. Food Anal Methods 12:1369–1381
36. Zhang H, Zhang X, Zhao X, Xu J, Lin C, Jing P, Hu L, Zhao S, Wang X, Li B (2019) Discrimination of dried sea cucumber (Apostichopus japonicus) products from different geographical origins by sequential windowed acquisition of all theoretical fragment ion mass spectra (SWATH-MS)-based proteomic analysis and chemometrics. Food Chem 274: 592–602
37. Roberts DS, Loo JA, Tsybin YO, Liu X, Wu S, Chamot-Rooke J et al (2024) Top-down proteomics. Nat Rev Methods Primers 4(1):38
38. Sales C, Portolés T, Johnsen LG, Danielsen M, Beltran J (2019) Olive oil quality classification and measurement of its organoleptic attributes by untargeted GC–MS and multivariate statistical-based approach. Food Chem 271: 488–496
39. Zhang L, Shuai Q, Li P, Zhang QI, Ma F, Zhang W, Ding X (2016) Ion mobility spectrometry fingerprints: a rapid detection technology for adulteration of sesame oil. Food Chem 192:60–66
40. Zhang L, Li P, Sun X, Mao J, Ma F, Ding X, Zhang Q (2015) One-class classification based authentication of peanut oils by fatty acid profiles. RSC Adv 5(103):85046–85051
41. Yuan Z, Zhang L, Wang D, Jiang J, Harrington PD, Mao J, Zhang Q, Li P (2020) Detection of flaxseed oil multiple adulteration by near-infrared spectroscopy and nonlinear one class partial least squares discriminant analysis. LWT 125:109247
42. Pandey A (2025) Mass spectrometry: Techniques and applications. Educohack Press
43. Feider CL, Krieger A, DeHoog RJ, Eberlin LS (2019) Ambient ionization mass spectrometry: recent developments and applications. Anal Chem 91(7):4266–4290
44. Borra PK (2025) Sample preparation techniques. In: Analytical methods for drug development, pp. 38
45. Visconti G, Boccard J, Feinberg M, Rudaz S (2023) From fundamentals in calibration to modern methodologies: A tutorial for small molecules quantification in liquid chromatography-mass spectrometry bioanalysis. Anal Chim Acta 1240(July 2022):340711.
46. Hecht ES, Scigelova M, Eliuk S, Makarov A (2006) Fundamentals and advances of orbitrap mass spectrometry. In: Encyclopedia of analytical chemistry: Applications, Theory and Instrumentation, pp. 1–40
47. Cao J, Jiang T, Xu W (2025) Miniature mass spectrometry for VOCs analysis: Recent developments in instrumentation and applications. Rapid Commun Mass Spectrom 39(18): e10088
48. Wang J, Pursell ME, DeVor A, Awoyemi O, Valentine SJ, Li P (2022) Portable mass spectrometry system: Instrumentation, applications, and path to 'omics analysis. J Proteom 22(23–24):2200112
49. Yadav D (2025) Analytical chemistry foundations. Educohack Press
50. Kamble A, Bajaj S, Dhaulaniya AS, Balan B (2025) Analytical advances in food authentication and origin traceability: from mass spectrometry to AI and IoT-enabled smart sensing systems. J Food Meas Charact 1–29

Chapter 4

Isotopic Ratio Analysis for Tracing Food Origin and Quality

Dipjyoti Sharma, Abhilash Bharadwaj, Pollobi Porasar, Supriya Sahani, Syed Tohshin Ishtiyak, Rashmita Gogoi, Darshana Hazarika, Suman Lamichhane, and Koushik Nandan Dutta

Abstract

This chapter explores the significance of isotopic ratio analysis in verifying the authenticity and quality of food sources. Concerns are escalating regarding the integrity of food products and the potential for manipulation within global supply networks. This discourse highlights the significance of contemporary methodologies such as Liquid Chromatography with tandem mass spectrometry (LC-MS/MS), Gas chromatography-mass spectroscopy (GC-MS), and nuclear magnetic resonance (NMR) in discerning the origins of food, the processes of its production, and its quality assessment. This chapter delves into the realm of stable isotopes, such as carbon, nitrogen, and oxygen, presenting them as intrinsic markers of nature, and the influence of environmental elements like soil composition, climatic conditions, and farming methodologies can be observed in these isotopes. This chapter also explains the application of blockchain technology, interdisciplinary methods like DNA analysis and chemometrics alongside machine learning (ML), to improve data understanding and compliance with regulations. The examination of ethical considerations, fostering trust among consumers, and upholding elevated global standards is likewise discussed. This chapter explores different methodologies, illustrating the application of isotopic analysis in tackling fraud, enhancing transparency, and ensuring food quality within an ever-evolving market landscape.

Key words Isotopic ratio analysis, Food traceability, Stable isotopes, Food adulteration, Consumer protection, Blockchain for food traceability and sustainable food systems

1 Introduction

Authenticity and traceability are the two most important words in the food and beverage industries, which are linked with consumer protection. Food and beverage authenticity is an utmost important issue for the participants participating in the food chain. The word "authentic" means "a sense of actuality and lack of misinterpretation and falsehood" and the word "trace" means "to follow, discover, or ascertain the course of development of something." The word authentic carries a subtext of authoritative certification that an object is what it is claimed to be. Food traceability involves

Tanmay Sarkar (ed.), *Novel Food Analysis: Techniques and Applications*, Methods and Protocols in Food Science, https://doi.org/10.1007/978-1-0716-4787-5_4,

controlling the whole chain of food production and marketing, which helps in tracking the food through each production step, back to its origin. Authentication of food is a major concern, especially for those consumers that have a unique taste for nutrition, culture, and religion [1, 2]. Adequate food quality and value should be ensured by the food producers, and the food authorities should supervise the safety and traceability of the food that is present in the market. Consumers should also have the responsibility to demand honest, safe, and properly labelled foods, along with the promotion of organic products and the implementation of animal-welfare conscious and environmentally friendly production methods [1, 3, 4]. In recent years, food fraud and adulteration are on the rise, which negatively affects the economy and food safety. It is an intentional act of misinterpretation of the food, which includes modification of the food and associated documents for personal gains [5]. Opportunities for food frauds are created due to various factors like international trade, long and complex supply chains, booming e-commerce, and market globalization. Even though several cases of food frauds do not pose any major risk, some frauds have severe consequences on the consumers such as chronic illness and even death [6]. In a world with abundant food frauds and non-authentic food products, authentic foods are a necessity. Food traceability and authenticity are important for building confidence in consumers and help to track the origin of the ingredients and identify the source of any contaminants in food [1].

Consequently, over the past few years, analytical techniques have made significant growth and different analytical methods were developed to evaluate the quality of the food and traceability of the food system. A diverse range of scientific techniques is involved in these methods, which contributes to the comprehensive assessment of food integrity. Stable isotope analysis is one of the most prominent approaches, which provides insight into the geographical origin and environmental history of the food product [7, 8]. Other techniques like liquid chromatography, gas chromatography, spectroscopic techniques, sensor techniques, and DNA-based techniques have their own merits and demerits [8]. Some studies reveal that these methods are used for the authenticity of food from animal origin, while some other studies focused on agro-product traceability using specific analytical techniques [8, 9].

Isotope Ratio Mass Spectrometry (IRMS) is a specialized mass spectrometry technique that is used to determine the relative stable isotopic abundance of the elements (e.g., hydrogen, carbon, oxygen, nitrogen, and sulfur) that comprise a material. Valuable insights are provided by these isotopes into the chemical, biological, and geographical origins of a substance. Even among the chemically identical materials, isotopic compositions can vary due to kinetic and thermodynamic effects, making it a powerful tool for authentication and differentiation. The results of IRMS are reported using Delta (δ), following the international standards. Traditionally, it was expressed in per mil (‰); however, now Urey (Ur) is used. Urey (Ur) aligns with the SI unit conventions and allows flexibility in scientific reporting. This technique helps in various fields like food traceability, environmental studies, and forensic science by providing accurate, precise, and nondestructive isotopic fingerprinting of complex sample [10].

2 Materials

Throughout the world, high-purity solvents, also referred to as analytical grade solvents, are an important part of the scientific research and laboratory analysis [11]. Solvents belonging to the analytical grade adhere to strict purity requirements that reduce hindrance in chemical processes and provide accurate and reproducible experimental results [12]. Water is the first known solvent for chemical analysis. Further research using the theory of "like dissolves like" led to the establishment of various well-known solvents [13]. The tabular form for the general class of organic solvents is given in Table 1.

Reference materials are necessary for evaluating the accuracy and reliability of outcomes from other laboratories as well as for verifying analytical procedures. A reference material that has a metrologically valid technique for certain qualities and an endorsement indicating the property's value, unpredictability, and metrological accountability is known as a Certified Reference Material (CRM) [15]. A diagrammatic representation of worldwide institutions engaged in the production of CRMs is depicted in Fig. 1.

Buffer solutions guard against abrupt changes in the acidity or alkalinity, and the quality of buffer solution is determined by their buffering capacity or resistance to pH changes brought on by strong acids or bases and their power to maintain a constant pH after dilution or the addition of neutral salts [16]. Buffers and their pK values are listed in Table 2.

Table 1
General class of organic solvents

Hydrocarbons	Alcohols	Ethers	Chlorinated solvents	References
n- Pentane	Methanol	Diethyl ether	Methylene chloride	[13, 14]
n- Hexane	Ethanol	D-isopropyl ether	Chloroform	
Toluene	n- Propanol	Dibutyl ether	Carbon tetrachloride	
Ethylbenzene	2- Butanol	Esters	Acetic acid	
Ketones	Ethanediol	Methyl acetate	Aniline	
Acetone	Cyclohexanol	n- Butyl acetate	Nitrobenzene	

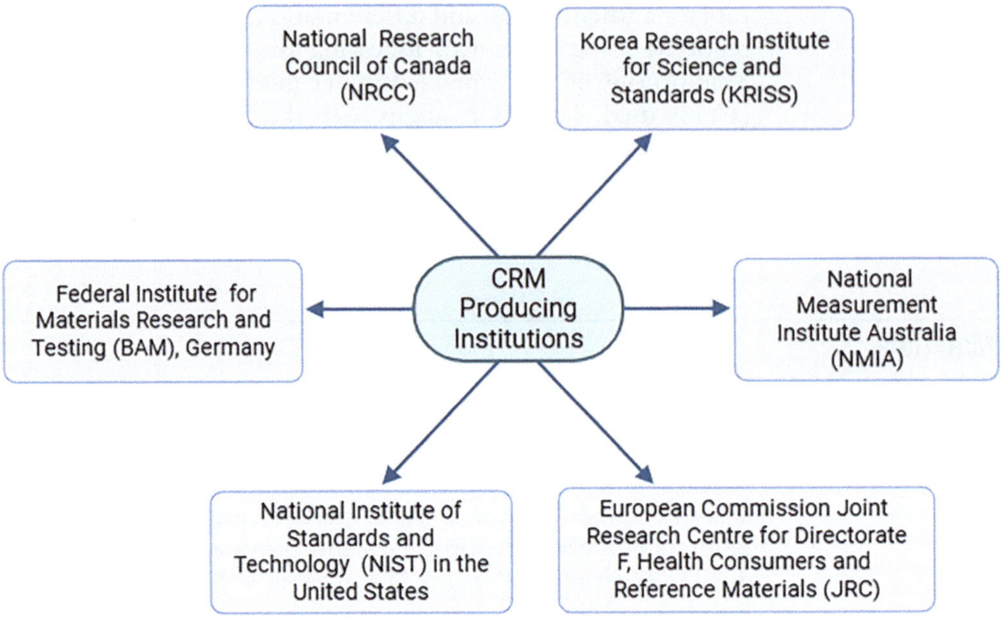

Fig. 1 Worldwide institutions engaged in the production of CRMs

Table 2
Buffers and their pK values

Buffer name	Trivial name	pK_a
2-(N-Morpholino) ethanesulfonic acid	MES	6.10
Dimethylarsinic acid	Cacodylate	6.27
3-(N-Morphino)-2-hydroxypropanesulfonic acid	MOPSO	6.95
3,6-Endomethylene-1,2,3,6 tetrahydrophthalic acid	EMTA	7.23
3-{[Tris(hydroxymethyl)methyl] amino}propanesulfonic acid	TAPS	8.40

The demand for high-quality and diverse food and food supplements is rising as a result of consumers' dedication to lead better lives and eating healthier foods [17]. Consumers' primary concern, which affects their acceptance and decisions, is quality and safety [18]. It is well-known that raw fruits and vegetables are healthier than cooked food. Food is mostly harmed by cooking as it destroys minerals and enzymes, changes the structure of food and digestibility, and also produces toxic byproducts [19]. However, cooking certain foods can increase their digestibility and nutritional bioavailability [20]. Lactose intolerance symptoms are lessened, the immune system is strengthened, digestive tract health is improved, and the risk of various cancers is decreased by the consumption of fermented foods [21]. Food supplements fall within the food-pharmaceutical border and are subjected to certain regulations

[22]. In order to satisfy the needs of the global market, the food business is concentrating on creativity through new technical processes and ways to obtain food and supplements [23, 24]. These products, which focus on dietary plant bioactives, are intended for the people who stand to gain the most from the application of research to real-world situations. However, the successful launch of innovative food items depends on consumers' perceptions of cost advantages and safety [17]. Spectroscopy comprises of the creation, measurement, and analysis of spectra resulting from the interaction of electromagnetic radiation with matter. It is employed for both qualitative and quantitative investigations [25]. The analytical method known as LC-MS blends sensitive mass spectrometric detection with high-resolution chromatographic separation. Due to advancements in technology, enhanced performance of mass spectrometers and High-Performance Liquid Chromatography (HPLC) and the requirement for quick, large-scale analysis of trace mixtures, LC-MS has emerged as a fruitful analytical platform for drug development [26]. Applications of Gas chromatography-mass spectroscopy (GC-MS) are numerous and include particle analysis, pollutant detection, and the study of paper degrading processes [27]. One of the most important techniques for determining and describing chemical structures in infrared spectroscopy, especially Fourier Transform Infra Red (FTIR) spectroscopy. The use of Michelson interferometers reduces energy constraints and facilitates the identification of fiber origin as well as additive consumption in the production of paper [28]. The study of paper degrading processes also benefits greatly from the use of FTIR [29]. Through the use of Raman scattering and infrared absorption, its primary spectroscopies identify substances by their distinctive spectral patterns and provide information on their physical and chemical forms [30]. Biosensors comprise sensing devices that recognize and communicate with an analyte by using a transducer and a bio-recognition element. They then translate changes in the analyte's characteristics into electrical signals. Depending upon the kind of bio-recognition aspects or transducing techniques employed, they can be categorized [31]. The study of working with fluids and materials at micron as well as submicron sizes is known as microfluidics. Its rapid development has been fueled by its promise for enhanced treatments, drug delivery, biosensing, single-cell or molecule assessment, high-throughput drug screening, and point-of-care diagnosis [32]. AI-supported systems have been widely employed in prevention, diagnosis, new medication development, and aftercare, and they provide structure to clinical practice [33]. Regulations governing the accreditation of laboratory equipment are sometimes ambiguous and open to interpretation. Calibration of instruments, equipment, and gauges for recording at appropriate intervals in compliance with a defined program is a requirement of good manufacturing practice. Every

instrument utilized for regulatory submissions must have a defined and documented program that includes procedures for eligibility, maintenance, and failure recording [34].

3 Methods

3.1 Isotopic Fingerprinting

Radioisotopes' analysis of isotope ratios has become a crucial part in food testing where its quality, authenticity, and origin are evaluated. This is done by calculating the relative abundance of stable isotopes (isotopes of hydrogen, carbon, nitrogen, etc.) in food testing samples. This evaluation gives us information about its adulteration, geographical origin, and methods used for its production. Recent advancements in this technology have improved its precision, range of isotopic analysis, which helped in its clearer calibration and interpretation of results. Evaluation of the biological or physicochemical properties is necessary for food safety. Various analytical test is used for this analysis. Usually, these tests are carried out by the techniques such as High-performance liquid chromatography (HPLC), Polymerase chain reaction (PCR), Enzyme-Linked Immunosorbent Assay (ELISA), Nuclear magnetic resonance (NMR), Electron spin resonance (ESR), Mid-Infrared Spectroscopy (MIR), Near-Infrared Spectroscopy (NIR), Gas chromatography-mass spectroscopy (GC-MS) and Laser-Induced Breakdown spectroscopy (LIBS), which are used for the evaluation of biological and physicochemical parameters. But these parameters do not clearly differentiate similar products apart and do not show where the product came from, i.e., the food product's origin. This is because many environmental factors influence the organic part of the food; these factors include the type of soil, weather, and fertility of the field it was grown. So, it is complex to figure out where the food came from. But isotope ratios (such as of hydrogen, nitrogen, lead, strontium, and sulfur) behave like a unique fingerprint that helps in identifying the origin of the food sample [35, 36].

Isotopic fingerprinting shows the unique characters of the food products, which reflect their processing, environmental, and agricultural conditions. Advancements in this technique have extended its utility in the evaluation of the authenticity of the food (e.g., packaged and frozen foods) and its traceability. After implementing modern advanced analytical techniques like different types of mass spectrometry (TIMS, MC-ICP-MS, and DRC-ICP-MS) have been made for enhanced analysis of the chemical and isotopic composition of any food and its origin can be identified. Instruments such as IRMS, which helps in tracing plant or animal food products [35]. Another modern isotopic profiling method like Compound-specific isotope analysis (CSIA) helps to find the origin of the target compounds [37].

Intermixing IRMS with analytical techniques like NMR and DNA analysis enhances the robustness of food authenticity and traceability. By analyzing the stable isotopes, the IRMS talks about the geographical and metabolic origin of the product, whereas the DNA analysis identifies the species and detects any source of adulteration and finally verifies its authenticity. Combining these analytical techniques makes the authentication easier and accurate. For example, in the authentication of Garcinia fruit products, DNA analysis confirms its species and the NMR identifies its metabolites [38].

3.1.1 Importance of Robust Data Interpretation in Harmonizing Analytical Tools for Holistic Food Profiling

Even being its robust potential, poor standardization is a critical problem. Since it leads to variation in reference materials, preparation of sample lowers the range of data reproducibility and comparability. Furthermore, inter-lab harmonization is essential to make the results reliable for food quality and also to comply with the food safety regulatory body [39]. Interpretation of complex isotopic data sets without powerful computational tools may lead to misclarification or overfitting. Incorporating machine learning (ML), chemometrics, or AI into data processing allows classification accuracy and pattern recognition. It ensures that the traceability claims are approved by statistically endorsed evidence, which is important in legal contexts [40]. Holistic food profiling is the new approach for future food traceability, which harmonizes isotopic data with genetic and molecular information. For example, combining IRMS with NMR and dana tools that allow regulators to simultaneously verify origin, composition, and species. The convergence supports the quality assurance, fraud detection, and compliance monitoring in a single unified protocol [41].

3.1.2 Fundamentals of Isotopic Ratio Analysis

Atoms of a single element that differ in their atomic masses due to differing neutron counts in their nuclei are called isotopes. The same binding types are formed by stable isotopes, which do not experience radioactive decay. However, because of their disparate masses, their reaction rates could vary, which might cause isotopic segregation or fractionation in the surrounding environment. The ratio between heavy to light isotopes makes this clear. Our food constituent elements cycle, which results in chemical and biological interactions that may cause isotopic fractionation and the environment's natural isotope partitioning [42]. Light isotopes are the most prevalent, while the ratio among light and heavy isotopes is more balanced, according to research on the isotopic ratios for hydrogen, carbon, nitrogen, oxygen, and sulfur [43]. Isotopic ratios of the common elements are described in Table 3. IRMS is a popular method in many disciplines, including biology, geology, medicine, archaeology, and food authenticity, along with forensic science [44]. Natural isotopic abundance must be measured with a

Table 3
Isotopic ratios of the common elements

Sl no.	Elements	Symbol	Percentage abundance (%)	Abundance ratio of reference standard	Reference
1.1	Hydrogen	^{1}H	99.985	0.00015576	[46]
1.2		^{2}H	0.015		
2.1	Carbon	^{12}C	98.98	0.011237	
2.2		^{13}C	1.11		
3.1	Nitrogen	^{14}N	99.63	0.0036765	
3.2		^{15}N	0.37		
4.1	Oxygen	^{16}O	99.759	0.00200520	
4.2		^{17}O	0.037		
4.3		^{18}O	0.204		
5.1	Sulfur	^{32}S	95.00	0.0450045	
5.2		^{33}S	0.76		
5.3		^{34}S	4.22		
5.4		^{36}S	0.014		

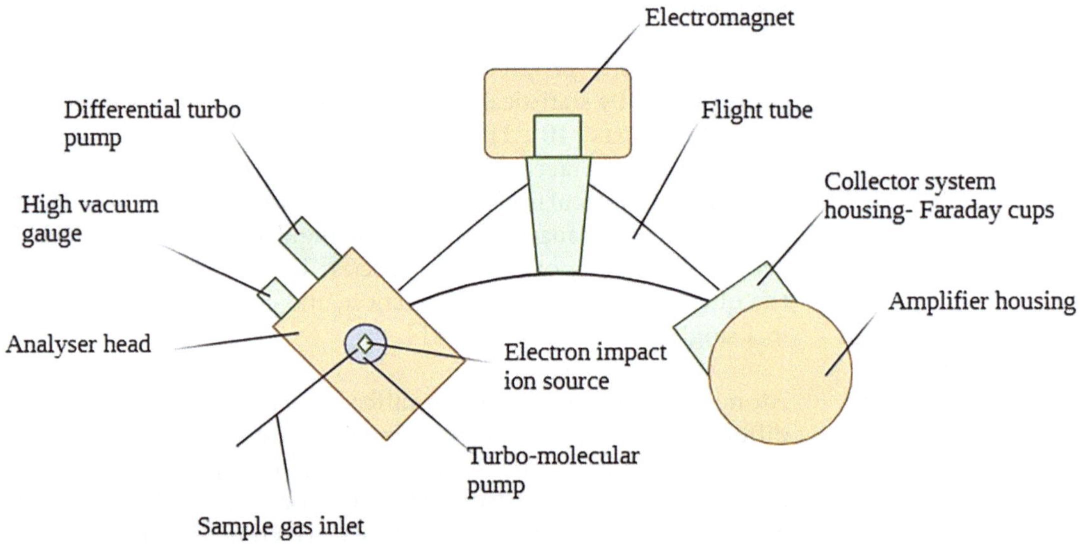

Fig. 2 Diagram showing the main sections of an IRMS instrument

specialized device, such as a multicollector magnetic sector mass spectrometer [45]. Figure 2 shows the diagram of the main sections of an IRMS instrument.

A sample administration system, electron ionization origin, magnetic sector detector, Faraday-collector sensing array, as well as computer-controlled data gathering system are the five primary components of IRMS equipment. Variations in the quantity of light stable isotopes may be measured precisely and accurately with this equipment [44]. Table 4 depicts the common standards and their applications.

Table 4
Common standards and their applications

Element	Standard	Material type	Application
H, O	VSMOW2, SLAP2, GISP	Water	Water isotope analysis (*delta*^{2}*H*)
C	Cellulose, Sucrose, Oil, Calcite	Organic, Inorganic	Carbon isotope analysis (*delta*^{13}*C*)
N	Potassium nitrate, Sodium nitrate, Ammonium sulfate	Organic, Inorganic	Nitrogen isotope analysis (*delta*^{15}*N*)
S	Barium sulfate, silver sulfide	Inorganic	Sulfur isotope analysis (*delta*^{34}*S*)

4 Notes

4.1 Role in Regulatory Approval Processes

Innovative food safety evaluations are conducted by the European Food Safety Authority (EFSA), which has become the sole EU authority since 2018. This regulation requires comprehensive traceability and documentation of novel food products—including lab-grown meat, precision-fermented ingredients, and plant-based alternatives—to ensure safety, quality, and transparency for consumers. A cross-disciplinary, well-prepared application is required to meet EFSA criteria and receive regulatory approval [47]. Authentication techniques such as stable isotope ratio analysis, DNA-based approaches, and metabolomics help traceability by verifying origin, production methods, and ingredient integrity, as well as preventing mislabeling and food fraud. Throughout history, new food types have been introduced into human diets. Rice, potatoes, and coffee were considered novel when imported from various regions of the world hundreds of years ago. The European Commission (EC) defines novel foods as those that were not widely consumed in Europe before May 15, 1997. Before NFs may be marketed, they must undergo premarket authorization for consumer safety [48].

4.2 Some of the International Guidelines

Under the 2018 rule, the EFSA became the sole entity in the European Union capable of conducting risk evaluations of novel foods between 2003 and 2019. The EFSA conducts assessments based on application data that follows established scientific assessment involving materials derived from plants, microbes, fungi, algae, and animals, revealing recurring issues with compositional analysis, manufacturing techniques, and toxicological data. Because novel foods are diverse, EFSA's scientific standards need a comprehensive, interdisciplinary approach [49]. One of the most important aspects in establishing a worldwide food policy is the Codex Alimentarius Commission (CAC), which writes global food standards and works with the legal authority allowed by the World

Trade Organization (WTO). The European Commission (EC) has used its involvement in Codex to boost its role in safeguarding food safety in Europe. Important points are to protect the environment, consider nonscientific factors, and ensure that food modified with biotechnology is known and traceable. Because of the large transatlantic issues at stake, their views have often encountered resistance from the United States. Through Codex, the EC still works to keep safety and transparency rules, even though Codex has some challenges it has resolved [50].

4.3 Ethical Implications of Food Fraud and False Labeling

Food fraud involves deceptive labeling or paperwork around the product. Many people are now interested in buying ethically, and companies are responding. Consumers are increasingly choosing sustainable products, which businesses are responding to. To grow various offerings in different fields and markets all over the world. Today, individuals in the food industry brainstorm about food fraud and how companies' unethical practices may affect their brand experiences. It studies the problem of unintentional errors in labeling foods in the United States. Correlations between brand attachment, ethical judgment, brand commitment, and purchase intention are substantial. Many consumers thought about the brand breach in terms of ethics. Buying and sticking with the brand were influenced a lot, as many felt the behavior of the person was not ethical [51]. In order to meet the rules for food labeling that apply to consumers, a lot of people have negative feelings when it comes to reading food labels. People's opinions about ethical food labeling have not been given a lot of attention [52].

4.4 Scientific Verification in Supply Chains for Transparency

The use of scientific methods, including matching unique chemical key values, DNA analysis, spectroscopic tools, and blockchain technology, allows all parties to ensure the authenticity of the origin, the way products are processed, and what ingredients they contain [53]. The use of scientific methods, including matching unique chemical key values, DNA analysis, spectroscopic tools, and blockchain technology, allows all parties to ensure the authenticity of the origin, the way products are processed, and what ingredients they contain [54].

4.5 Applications in Food Origin and Geographic Traceability

Since the isotopic ratio analysis came into existence, it has paved its way into various different industries as an important tool of analysis. Various application of isotopic ratio analysis is shown in Fig. 3.

The use of analytical methods to ascertain the geographic origin of agricultural goods, especially dairy products like milk and meat, has been highlighted in recent studies. One natural characteristic that may be applied as a natural identifier to determine the provenance of food is its isotopic makeup [55, 56]. In Cyprus, the regional provenance of meat, dairy products, as well as genuine foods like local wine has been evaluated using stable

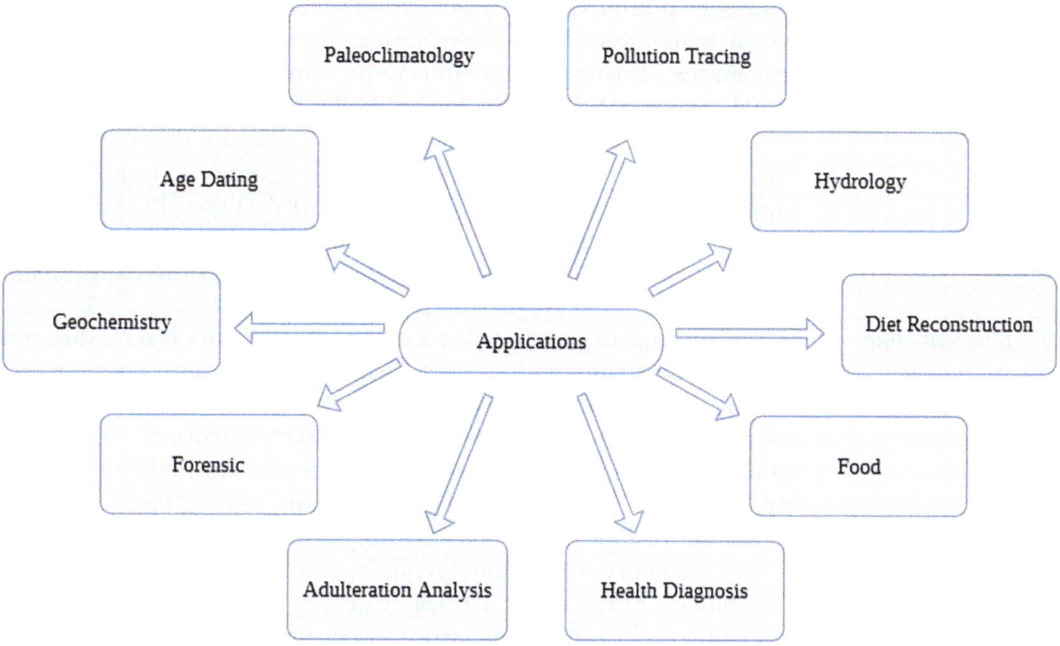

Fig. 3 Applications of Isotopic Ratio Analysis in various fields

isotope analysis. Since adding honey sugar is a frequent way to adulterate honey, studies on honey quality and authenticity are particularly crucial. Research on carbon isotope ratios demonstrates a distinct difference between C4 plants that produce cane or corn sugar and those that are mostly used to make honey (C3 plants) [57].

Since ancient times, people have been drinking alcoholic beverages having an ethanol concentration of 0.5% or more because of their cultural importance and variety of flavors. Issues including adulteration, mislabeling, and counterfeiting have been brought about by the industry's growing internationalization, endangering both consumer protection and market integrity. A variety of technological approaches have been developed to detect adulteration in alcoholic beverages, including spectroscopic techniques, chromatography, along with stable isotopes techniques. Stable isotope methods provide ratios, which make them an effective tool for confirming authenticity by encoding important details about production method and place of origin [58].

4.6 Role in Food Quality and Authenticity

Consumer protection relies heavily on food authenticity, composition, and labeling. Globalization and trade expansion have heightened worries about food fraud and criminal profit-driven deception [59]. Traditionally, authenticity has changed in the field of tourist studies. Mac Cannell popularized the term "staged authenticity" need for authentic encounters. Wang subsequently distinguished four viewpoints, each of which emphasized a distinct aspect ranging from material authenticity to culturally and personally generated meanings: existential, postmodern, constructive, and objective [60]. Support Vector Machine (SVM), Classification and Regression Trees (CART) and Radio frequency (RF) are the sophisticated approach for food verification, while some commonly used approaches include Principal Component Analysis (PCA) and Partial Least-Squares Discriminant Analysis (PLS-DA). Figure 4 shows the role of isotopic analysis in food quality and authenticity [59].

4.7 Differentiating Between Organic Versus Conventional Products

Conventional organic products and organic products have different criteria for quality and safety. Organic products are rich in antioxidants and have lower nitrates; it is necessary to evaluate the authenticity, toxicity, heavy metals, and stable isotopes for better understanding [61]. Organic vegetables contain higher quantities of Vitamin C, phenolic compounds, iron, and nitrates show low pesticide residue. Isotope ratio analysis separates organic origins using markers derived from fertilizer, and authenticity is confirmed. Despite slight variations in sensory quality or total nutritional value, consumers choose organic produce because of the perceived health

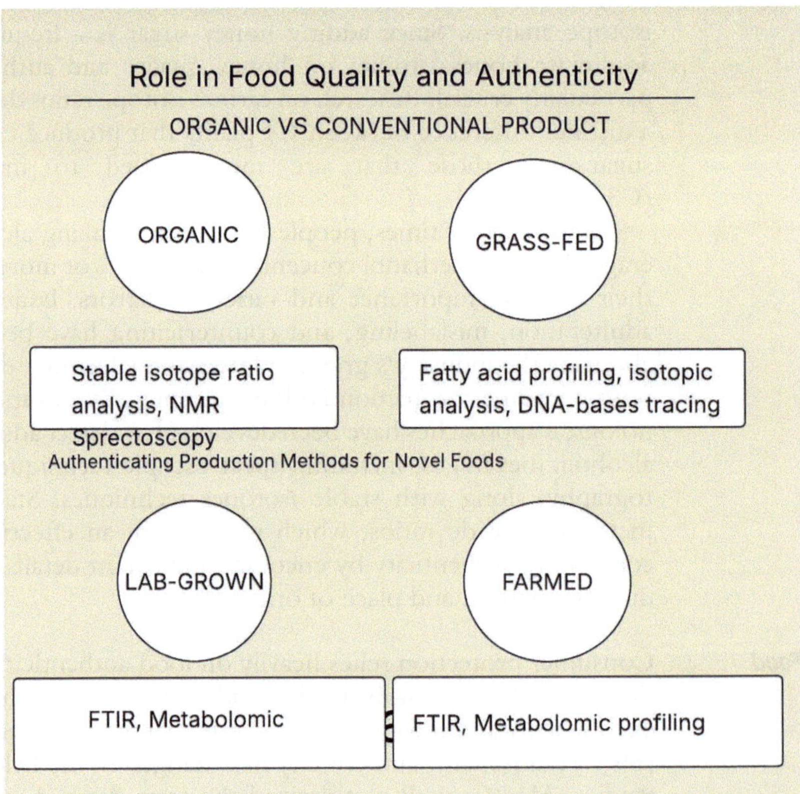

Fig. 4 Role in food quality and authenticity

Table 5
Analytical differences between organic and conventional products

Parameter	Organic products	Conventional products	Analytical methods
Pesticide Residue	Very low to none	It is high because of synthetic pesticide use	LC-MS/MS, GC-MS
Nitrate content	Lower	Higher	UV-Vis spectroscopy, Ion chromatography
Microbial safety	Depending on handling and hygiene, sometimes higher	Low because of chemicals	Plate count, qPCR
Stable isotope Ratio	Distinct due to different types of fertilizer	Less variable	IRMS
Heavy metals	Slightly lower	Lower	HPLC, LC-MS

and environmental benefits, while conventional products have lower nitrate and less heavy metals. Table 5 describes some analytical differences between organic and conventional products [62, 63].

4.8 Feed Sources and Agricultural Practices

Sources such as traceability, quality, storage condition, and ingredient verification of agricultural products and animal feed contribute to the health and well-being of the people and environment. Fed grass and grain fed are the techniques used in animal agriculture, and they assure the quality and authenticity of food. In grass-fed, animals consume grass throughout their entire life. Grass-fed animals are rich in vitamin A, which contributes to yellow fat, along with omega-3 fatty acids and conjugated linoleic acid (CLA). In contrast, grain-fed animals consume grains, growth hormones, and antibiotics to promote faster growth, resulting in paler, off-white fat due to a lack of vitamin A. A reliable distinction between these feeding regimes in meat and dairy products can be made using isotopic signatures, which accurately reflect the animals' nutritional intake [64, 65]. Different parameters on grass-fed and grain-fed are shown in Table 6.

4.9 Authenticating Production Methods for Novel Foods

For novel food manufacturing techniques like lab-grown, meat needs to be authenticated for the quality, safety, and regulatory compliance as it fundamentally differs from typical food. Modern tools like FTIR, NMR, SIRA, PCR, and DNA profiling are used to determine methods and their origin. Authentication is necessary to gain customers, regulatory compliance, and prevention from fraud with proper labeling [66]. Cultured meat is produced in vitro (cell culture) and aspires to be biologically equivalent to traditional meat, and its fat content can be customized with low pathogen

Table 6
Different parameters on grass-fed and grain-fed

Parameters	Grass-fed	Grain-fed
Feed type	Mainly pasture grasses and forage	Cereals, Grains
Fat composition	High Omega-3 Fatty Acids and CLA	High Omega-6 Fatty Acids
Vitamin E	Higher	Lower
Taste profile	Strong, gamey flavor	Mild, buttery taste
Isotopic signature	More ^{13}C-depleted (due to C_3 plants)	More ^{13}C-enriched (due to C_4 grains)
Authenticity methods	Isotope ratio, Fatty acid profiling	Same methods with comparative reference
Customers perception	Perceived as it is healthier and more natural	Perceived as more consistent and tender
Price	Generally higher	More affordable

risk and no antibiotics. SIRA, proteomics, lipidomics, and epigenetics are the authenticating methods used to generate flavors and textures. Animals raised on farms have naturally occurring fats with a higher infection risk, and thus, antibiotics are frequently utilized. SIRA, feed tracking, and labeling are authentication measures.

- Linking blockchain shows real-time isotopic data and inflexible traceability beyond the food supply. Isotopic fingerprints can be transferred onto blockchain nodes, where they are accessible from farm to form. This combination promises enhanced trust, consumer clarity, and liability in the agri-food sector [67]. Moreover, with the advancement of AI and ML, isotopic datasets can be refined using algorithms that recognize slight impressions that are invisible to the human eye. Applications include confirming food purity, predicting geographic origin, and detecting fraud in highly processed products. These tools lower the analysis time and improve decision-making in the regulatory system [40]. In a world more influenced by climate change, stable isotope data can indicate environmental stress and alteration patterns in crops and livestock. This makes IRMS one of the potential tools for monitoring climate flexibility. Plus, its combination into a new digital supply chain program ensures clarity and helps understanding vulnerabilities hence aiding supply chain stability [68].
- In the modern world of food consumption, traceability and authenticity are essential for fostering confidence and thwarting fraud. Isotopic ratio estimation is a potent technique that provides a distinct "fingerprint" of food items, exposing their production methods, geographic origin, and even adulteration. Climate, altitude, soil type, and agricultural methods are some

of the natural factors that cause the predictable variations in the isotopic makeup of elements such as carbon, nitrogen, oxygen, hydrogen, and sulfur across various geographical regions and ecosystems. In order to incorporate the distinctive isotopic ratios into their tissues, plants consume these elements from their environment. Animals and food prepared from them carry these isotopic trace elements. Isotopic ratio analysis uses methods such as IRMS to accurately determine the relative abundance of several isotopes of a certain element inside a food sample. It is feasible to confirm the supposed source of an unidentified food item by comparing its isotopic signature with databases of genuine goods from recognized origins. Possible mislabeling or adulteration may be indicated by a notable divergence. Given that isotopic analysis provides an inherent signature that is hard to tamper with, it also improves food traceability. A strong scientific foundation for confirming the legitimacy of food items and improving traceability across the supply chain is provided by isotopic ratio analysis. It can detect the addition of less expensive sugar to honey or fruit juices and animal feeding for dairy and meat. Isotopic analysis provides quantifiable, objective information that reflects the food products' combined agricultural and environmental history and enhances conventional techniques of quality assurance and authenticity.

The availability of thorough and well-characterized comparison databases is necessary for the effective use of isotopic analysis. This calls for the collection and examination of genuine goods from a variety of geographical locations and manufacturing processes. Understanding analytical methods as well as the agricultural and environmental factors that affect isotopic variances is frequently necessary for interpreting isotopic data. With the creation of more complex analytical methods and the growth of isotopic databases, the area of isotopic ratio analysis is progressing in spite of these obstacles. It is a vital instrument in the continuous attempts to create a more open and reliable food system because of its capacity to improve food authenticity, fortify traceability, and guarantee food quality.

References

1. Kamiloglu S (2019) Authenticity and traceability in beverages. Food Chem 277 (October):12–24
2. Haider A, Iqbal SZ, Bhatti IA, Alim MB, Waseem M, Iqbal M et al (2024) Food authentication, current issues, analytical techniques, and future challenges: a comprehensive review. Compr Rev Food Sci Food Saf 23(3):e13360
3. González-Domínguez R, Sayago A, Fernández-Recamales Á (2022) An overview on the application of chemometrics tools in food

authenticity and traceability. Foods 11(23): 1–13

4. Marozzo V, Vargas-Sánchez A, Abbate T, D'Amico A (2022) Investigating the importance of product trace ability in the relationship between product authenticity and consumer willingness to pay. Sinergie 40(2):21–39
5. Visciano P, Schirone M (2021) Food frauds: global incidents and misleading situations. Trends Food Sci Technol 114(May):424–442
6. Amaral JS (2021) Target and non-target approaches for food authenticity and traceability. Foods. 10(1):172
7. Li C, Kang X, Nie J, Li A, Farag MA, Liu C et al (2023) Recent advances in Chinese food authentication and origin verification using isotope ratio mass spectrometry. Food Chem 398
8. Wadood SA, Boli G, Xiaowen Z, Hussain I, Yimin W (2020) Recent development in the application of analytical techniques for the traceability and authenticity of food of plant origin. Microchem J 152:104295
9. Liu H, Nie J, Liu Y, Wadood SA, Rogers KM, Yuan Y et al (2023) A review of recent compound-specific isotope analysis studies applied to food authentication. Food Chem 415:135791
10. Perini M, Bontempo L (2022) Liquid chromatography coupled to isotope ratio mass spectrometry (LC-IRMS): a review. TrAC Trends Anal Chem 147:116515
11. B R. (2024) A brief review on different chromatographic techniques. Open access. J Pharm Res 8(1)
12. Gautier C, Wund P, Laporte E, Vial J, Thiebaut D, Rivier C, et al (2019) Are analytical standards and reagents really reliable? To cite this version: HAL Id: cea-02384424 Are analytical standards and reagents really reliable?
13. Joshi DR, Adhikari N (2019) An overview on common organic solvents and their toxicity. J Pharm Res Int (June):1–18
14. Grodowska K, Parczewski A (2010) Organic solvents in the pharmaceutical industry. Acta Pol Pharm Drug Res 67(1):3–12
15. Wise SA (2018) What is novel about certified reference materials? Anal Bioanal Chem 410(8):2045–2049
16. Stoll VS, Blanchard JS (2009) Chapter 6 buffers. Principles and practice1. Methods Enzymol 463(C):43–56
17. López-Rodríguez R, Domínguez L, Fernández-Ruiz V, Cámara M (2022) Extracts rich in nutrients as novel food ingredients to be used in food supplements: a proposal classification. Nutrients 14(15)
18. He HJ, Sun DW (2015) Microbial evaluation of raw and processed food products by visible/infrared, Raman and fluorescence spectroscopy. Trends Food Sci Technol [Internet] 46(2): 199–210. https://doi.org/10.1016/j.tifs.2015.10.004
19. Link LB, Potter JD (2004) Raw versus cooked vegetables and cancer risk. Cancer Epidemiol Biomarkers Prev 13(9):1422–1435
20. Salama SM, Mariod AA (2022) Significance of African fermented foods in nutrition and food science. African Fermented Food Prod New Trends:37–44
21. Moslehi-Jenabian S, Pedersen LL, Jespersen L (2010) Beneficial effects of probiotic and food borne yeasts on human health. Nutrients 2(4): 449–473
22. Domínguez Díaz L, Fernández-Ruiz V, Cámara M (2020) The frontier between nutrition and pharma: the international regulatory framework of functional foods, food supplements and nutraceuticals. Crit Rev Food Sci Nutr [Internet] 60(10):1738–1746. https://doi.org/10.1080/10408398.2019.1592107
23. Barrena R, Sánchez M (2013) Neophobia, personal consumer values and novel food acceptance. Food Qual Prefer 27(1):72–84
24. de Roos B, Aura AM, Bronze M, Cassidy A, Conesa MTG, Gibney ER et al (2019) Targeting the delivery of dietary plant bioactives to those who would benefit most: from science to practical applications. Eur J Nutr [Internet] 58 (s2):65–73. https://doi.org/10.1007/s00394-019-02075-5
25. Brandt-Casadevall C, Dimo-Simonin N (2004) Blood analysis. In: Encyclopedia of analytical science, 2nd edn, pp 373–378
26. Lim CK, Lord G (2002) Current developments in LC-MS for pharmaceutical analysis. Biol Pharm Bull 25(5):547–557
27. Al-Rubaye AF, Hameed IH, Kadhim MJ (2017) A review: uses of gas chromatography-mass spectrometry (GC-MS) technique for analysis of bioactive natural compounds of some plants. Int J Toxicol Pharmacol Res 9(01)
28. Hon DNS (1986) Fourier transform IR spectroscopy and electron spectroscopy for chemical analysis. In: Historic textile and paper materials, vol 3, pp 349–361
29. Bitossi G, Giorgi R, Mauro M, Salvadori B, Dei L (2005) Spectroscopic techniques in cultural heritage conservation: a survey. Appl Spectrosc Rev 40(3):187–228
30. Edwards HGM (2005) Modern Raman spectroscopy—a practical approach. Ewen Smith and Geoffrey Dent. Wiley, Chichester,

p 210. ISBN 0 471 49668 5 (cloth, hb); 0 471 49794 0 (pbk) . Vol. 36, Journal of Raman Spectroscopy. 2005. 835–835 p
31. Ding X, Srinivasan B, Tung S (2015) Development and applications of portable biosensors. J Lab Autom 20(4):365–389
32. Yeo LY, Chang HC, Chan PPY, Friend JR (2011) Microfluidic devices for bioapplications. Small 7(1):12–48
33. Grant NV, Bejan A, Kunze C, Burkhardt H (2024) Can AI-supported systems help with aftercare planning? Opportunities and challenges from a clinical perspective. In: ACM Int Conf Proceeding Ser, pp 655–659
34. Sigvardson KW, Manalo JA, Roller RW, Saless F, Wasserman D (2001) Laboratory equipment qualification. Pharm Technol 25(10):102–108
35. Katerinopoulou K, Kontogeorgos A, Salmas CE, Patakas A, Ladavos A (2020) Geographical origin authentication of agri-food products: a review. Foods 9(4):1–16
36. Medina S, Pereira JA, Silva P, Perestrelo R, Câmara JS (2018) Food fingerprints—A valuable tool to monitor food authenticity and safety. Food Chem 2019(278):144–162
37. Dou Q, Canavan A, Fu Y, Xiang L, Wang Y, Wang X et al (2024) Nitrogen stable isotope analysis of sulfonamides by derivatization-gas chromatography-isotope ratio mass spectrometry. Anal Bioanal Chem 416(19):4237–4247
38. Seethapathy GS, Tadesse M, Urumarudappa SKJ, Gunaga SV, Vasudeva R, Malterud KE et al (2018) Authentication of Garcinia fruits and food supplements using DNA barcoding and NMR spectroscopy. Sci Rep 8(1):1–12
39. Camin F, Bontempo L, Perini M, Piasentier E (2016) Stable isotope ratio analysis for assessing the authenticity of food of animal origin. Compr Rev Food Sci Food Saf 15(5):868–877
40. Mahmud M, Kaiser MS, McGinnity TM, Hussain A (2021) Deep learning in mining biological data, Cognitive computation, vol 13. Springer, pp 1–33
41. Sobolev AP, Ingallina C, Spano M, Di Matteo G, Mannina L (2022) NMR-based approaches in the study of foods. Molecules 27
42. O'Brien DM (2015) Stable isotope ratios as biomarkers of diet for health research. Annu Rev Nutr 35(1):565–594
43. Rossmann A (2001) Determination of stable isotope ratios in food analysis. Food Rev Int 17(3):347–381
44. Muccio Z, Jackson GP (2009) Isotope ratio mass spectrometry. Analyst 134(2):213–222
45. Benson S, Lennard C, Maynard P, Roux C (2006) Forensic applications of isotope ratio mass spectrometry – a review. Forensic Sci Int 157(1):1–22
46. Dawson TE, Brooks PD (2001) Fundamentals of stable isotope chemistry and measurement. Springer, pp 1–18
47. Ververis E, Ackerl R, Azzollini D, Colombo PA, de Sesmaisons A, Dumas C et al (2020) Novel foods in the European Union: scientific requirements and challenges of the risk assessment process by the European Food Safety Authority. Food Res Int 137:109515
48. Grimsby S (2021) New novel food regulation and collaboration for innovation. Br Food J 123(1):245–259
49. Poli S (2004) The European Community and the Adoption of International Food Standards within the Codex Alimentarius Commission. Eur Law J 10(5):613–630
50. Masson-Matthee MD (2007) The codex alimentarius commission and its standards
51. Hernández A, Kaeck DL (2019) Ethical judgment of food fraud. Effects on consumer behavior. J Food Prod Mark 25(6):605–627
52. Van der Merwe M, Venter K (2010) A consumer perspective on food labelling: ethical or not? Koers Bull Christ Scholarsh 75(2): 405–428
53. Sodhi MMS, Tang CS (2019) Research opportunities in supply chain transparency. Prod Oper Manag 28(12):2946–2959
54. Manning L, Monaghan J (2019) Integrity in the fresh produce supply chain: solutions and approaches to an emerging issue. J Hortic Sci Biotechnol 94(4):413–421
55. Baroni MV, Podio NS, Badini RG, Inga M, Ostera HA, Cagnoni M et al (2011) How much do soil and water contribute to the composition of meat? A case study: meat from three areas of Argentina. J Agric Food Chem 59(20): 11117–11128
56. Silva AV, Hélie JF, Caxito F d A, Monardes H, Mustafa AF, Stevenson R (2014) Multi-stable isotope analysis as a tool for assessing the geographic provenance of dairy products: a case study using buffalo's milk and cheese samples from the Amazon basin. Brazil Int Dairy J 35(2):107–110
57. Dinca OR, Ionete RE, Costinel D, Popescu R, Miricioiu MG, Stefanescu I et al (2015) Evaluating the origin of honey by stable isotopes ratio (C and N) using multivariate statistical analysis. Smart Energy Sustain Environ 18(1): 73

58. Ma Y, Li Y, Shao F, Lu Y, Meng W, Rogers KM et al (2025) Advancing stable isotope analysis for alcoholic beverages' authenticity: novel approaches in fraud detection and traceability. Foods 14(6):1–21
59. Jiménez-Carvelo AM, González-Casado A, Bagur-González MG, Cuadros-Rodríguez L (2019) Alternative data mining/machine learning methods for the analytical evaluation of food quality and authenticity—a review. Food Res Int 122:25–39
60. Zhang T, Chen J, Hu B (2019) Authenticity, quality, and loyalty: local food and sustainable tourism experience. Sustain 11(12)
61. Rahman SME, Mele MA, Lee YT, Islam MZ (2021) Consumer preference, quality, and safety of organic and conventional fresh fruits, vegetables, and cereals. Foods 10(1):1–17
62. Kuan CH, Rukayadi Y, Ahmad SH, Wan Mohamed Radzi CWJ, Thung TY, Premarathne JMKJK et al (2017) Comparison of the microbiological quality and safety between conventional and organic vegetables sold in Malaysia. Front Microbiol 8(JUL)
63. Novak V, Adler J, Husted S, Fromberg A, Laursen KH (2019) Authenticity testing of organically grown vegetables by stable isotope ratio analysis of oxygen in plant-derived sulphate. Food Chem 291(November 2018): 59–67
64. Chen H, Yang F, Lv L, Fu C, Cai X, Wang S (2020) Interaction among protein, daidzein and surfactants in the WPI-based daidzein self-microemulsifying delivery system. Food Chem 332(June):127461
65. Cordain L. Nutritional differences between grass and grain fed beef. Fort Collins. pp 1–39
66. Fraeye I, Kratka M, Vandenburgh H, Thorrez L (2020) Sensorial and nutritional aspects of cultured meat in comparison to traditional meat: much to be inferred. Front Nutr 7(March)
67. Marchese A, Tomarchio O (2022) A blockchain-based system for agri-food supply chain traceability management. SN Comput Sci 3(4):279
68. Alwi A, Sasongko NA, Suprapto SY, Subagyo H (2024) Blockchain and big data integration design for traceability and carbon footprint management in the fishery supply chain. Egypt Informatics J 26(May):100481

Chapter 5

Lab-on-a-Chip Protocols for Adulterant Detection in Foods

Akshay Sunil, B. S. Shrinidhi, B. K. Jnanesh, Yuvaraj Sivamani, Ragunathan Devendran, and Sumitha Elayaperumal

Abstract

Food adulteration remains a significant threat to global food safety, public health, and consumer trust. The intentional or naïve addition of hazardous materials to consumables inflicts a range of health risks from immediate poisoning to long-term illnesses. Although the classical methods for detection have a great degree of precision, they are frequently limited by exorbitant costs, complicated machinery, skilled staff requirements, and lengthy procedure periods. There has been a need to develop rapid, user-friendly, and compact alternatives.

An emerging promising technology is the Lab-on-a-Chip (LOC) technology, which integrates multiple laboratory procedures into a singular micro device. LOC systems use microfluidic principles to quickly, sensitively, and specifically analyze food samples by manipulating small amounts of fluids. These devices have the ability to detect and identify numerous concerning additives like pesticides, toxins, heavy metals, synthetic dyes, and other unauthorized substances in a wide variety of food matrices such as milk, meat, cereals, oils, fruits, beverages, and spices.

This chapter explores a range of LOC detection methods, including DNA-based verification, surface Plasmon resonance (SPR), fluorescence detection, Dielectrophoresis (DEP), and Clustered Regularly Interspaced Short Palindromic Repeats (CRISPR) systems, in the context of detecting food adulterants. Especially in resource-poor regions, LOC systems' particular features of mobility, cost effectiveness, automation, and lower reagent consumption make them ideal for point-of-care (POC) and in-field testing. Also, advancements in nanomaterials, including gold nanoparticles (AuNPs) and graphene, biosensors, and smartphone technology, have greatly improved the function and availability of LOC systems. Furthermore, these systems owe their high adaptability to the development of new materials such as hybrid polymer and paper-based chips.

Despite the fact that LOC technologies could revolutionize food safety, device standardization, sample matrix interference, and regulatory clearance remain problematic. Nonetheless, their further development and integration into existing food safety systems could alter the entire landscape of monitoring and controlling food fraud worldwide. This assessment, therefore, underlines the promise of LOC technologies as advanced technologies for protecting consumers and verifying the quality of food offered to them.

Key words Lab-on-a-Chip (LOC), Food adulteration, Microfluidics, Point-of-care testing, Food safety, Biosensors, DNA-based detection, Surface Plasmon Resonance (SPR), CRISPR, Dielectrophoresis (DEP), Fluorescence detection, Nanomaterials, Portable diagnostics

Tanmay Sarkar (ed.), *Novel Food Analysis: Techniques and Applications*, Methods and Protocols in Food Science, https://doi.org/10.1007/978-1-0716-4787-5_5,

1 Introduction

Adulteration of food is a significant concern for consumer health and food business, making food safety an issue of international public health concern. Table 1 describes the types of adulterations. Adulteration is defined as the intentional and unintentional addition of inferior or harmful substances to food products, usually done for reasons such as enlarging the bulk, improving the looks of the products, or even reducing the costs of production [1]. These lesser used toxic adulterants not only decrease the food's nutritional value, but also damage vital internal organs, cause food poisoning, and carcinogenic diseases. Because of the fragmented food supply systems, low enforcement of regulatory bodies, and limited social resources to monitor such activities, there is little to no surveillance over food supply systems in many underdeveloped countries [2]. Various techniques, such as high-performance liquid chromatography (HPLC), gas chromatography combined with mass spectrometry (GC-MS), enzyme-linked immunosorbent assays (ELISA), and some spectrophotometric methods, were used for detecting food products' adulterants. While useful, these techniques come at a price; they can be sensitive and specific, but also costly, slow, having a need for skilled staff and advanced laboratory equipment, along with sophisticated infrastructure. More accessible, low-cost options that are focused on accuracy, speed, and portability are in high demand. The need for the new techniques is urgent, especially in areas lacking resources [3]. One of the most recent innovations in science and technology is the Lab-on-a-Chip (LOC) platform. LOC consists of a device that performs a variety of laboratory functions on a single chip that ranges from a few millimeters to a few centimeters in size. Most LOC systems are founded on microfluidic technologies, which manipulate small quantities of liquids through tiny channels engraved in substances like glass, polymers, or even paper. The ability of these systems to perform complex biochemical analyses

Table 1
Types of adulteration

Types of adulterations	Purposes and causes
Intentional [1, 2]	Increase profit margins by reducing quality and cost of production Enhance appearance Increase shelf life Increase quantity and weight
Unintentional [1, 9]	Poor hygiene during processing or storage Contamination due to environmental pollution Improper handling or transportation

in parallel and within a short time period, at low cost and with minimal reagent consumption, has attracted a great deal of attention in the last few years [4].

LOC devices offer numerous advantages with regard to food adulteration. Their portable and user-friendly design allows parts to be tested in situ without needing great skill or bulky equipment. As an example, colorimetric, electrochemical, and fluorescent-based detection systems can be incorporated to detect different food adulterants. Colorimetric indicators on paper-based chips have been used for the rapid field detection of synthetic pigments in turmeric and chilli powder, while ultrasensitive and highly selective melamine immunoassays in electrochemical LOC sensors have been reported. Besides a reduction in turnaround time, these technologies enable real-time decision-making for the surveillance and control of food quality [5, 6]. Recent advancements in materials science and biosensor technology have greatly enhanced the functionalities of LOC systems. Through the implementation of nanomaterials, bio recognition elements such as aptamers and antibodies, as well as smartphone-based readout systems, sensitivity, selectivity, and data accessibility have all been improved [7]. Moreover, LOC systems are now being standardized for commercial food quality testing, making them an attractive alternative to traditional laboratory tests in both developed and developing countries [8].

Regardless of the prospects, LOC technologies have limitations. Problems with sample matrix interference, reagent shelf life, consistency in device manufacturing, and regulatory approvals still pose challenges to their wide implementation. Additionally, integrating these devices within existing food safety systems will require interdisciplinary collaboration and continuity [10]. Admittedly, LOC-based food adulteration detection offers great potential. Research efforts that enhance the performance, scalability, and affordability of devices are likely to make LOC protocols pivotal in democratizing access to food safety monitoring tools X. This study assesses the design and fundamental principles, applications, and recent trends of LOC techniques for food adulterant detection to demonstrate their promise as next-generation tools for ensuring food quality and consumer safety [11].

1.1 Food Adulterants and Their Health Concerns

Adulteration of food is still a critically alarming global concern, which is undermining the safety and quality of food across the globe. An Adulterant is any substance that is added intentionally or unintentionally to any food substance to lower its quality, impose health risks, or falsely deceive consumers regarding the authenticity and nature of the product. Adulterants are most commonly added to food to enhance the appearance, flavor, nutritional value, and shelf life, or even to increase the quantity of food served to amplify the profit margins. This concern is intensified in developing countries because of the low level of consumer education and weak regulations pertaining to food safety [1, 2].

1.1.1 Types of Food Adulterants

There are two types of adulterants, which are incidental and intentional. Intentional adulterants include mixtures like chalk powder in flour, water in milk, metanil yellow in turmeric, argemone oil in mustard oil, and artificial pigments in candies and spices. These adulterants are intentionally added to food. In most cases, these compounds serve as cheaper substitutes or serve as visual enhancers, which pretend to be the real portions. On the other hand, incidental adulterants are those contaminants that get added unintentionally during the handling, packaging, or storage. These types of adulterants include metal pollutants such as lead, insect parts, rodent feces, and pesticide residues. [1, 12]

1.1.2 Common Adulterants and Their Health Impacts

Milk Adulteration

To enhance its appearance and viscosity, milk is one of the most commonly adulterated food commodities, which is diluted with water and adulterated with flour, detergent, or even urea. The kidneys and the gastrointestinal tract can be harmed by consuming detergents and urea, particularly in children and in individuals with compromised immunity [13].

Vegetables and Fruits

There are multiple artificial methods for accelerating fruit ripening, including the use of calcium carbide, which contains acetylene gas. It is important to note, though, that calcium carbide is classified as a carcinogen and can also result in numbness, vertigo, headache, or even seizures when exposed to over long durations [14, 15].

Spices and Sweets

To improve the appearance of spices, metanil yellow and rhodamine B are used to enhance the color. However, these additions are known to be harmful. It is known that metanil yellow is neurotoxic and hepatotoxic in animals and is deemed a likely carcinogen in people [16]. Likewise, confectionery is refined with prohibited colors and silver foils crafted with aluminium or similarly nonpermitted metals, which can cause organ damage and chronic poisoning [17].

Pulses and Cereals

The polishing of pulses may involve the addition of lead chromate for a more appealing yellow color, which subjects consumers to severe health risks such as anemia, brain damage, and chronic kidney failure due to the steady accumulation of lead in the body. Furthermore, grains may be contaminated with ergot or kesari dal (*Lathyrus sativus*), which is known to cause various neurological, paralytic, and devastating disorders [18].

Oils and Fats

Among the most perilous adulterants is argemone oil, which is frequently combined with mustard oil. It possesses toxic alkaloids such as sanguinarine. In its consumption, there has been a correlation with epidemic dropsy, which causes swelling of limbs, cardiac failure, and glaucoma [19, 20].

1.1.3 Chronic and Acute Health Implications

Health issues stemming from the consumption of contaminated or adulterated food is dependent on the type and amount of additive. Also, the duration of consumption and the consumer's health condition are taken into account. Common acute effects of the consumption of such food are vomiting, abdominal pains, and diarrhoea. The more severe consequences of long-term consumption include organ failure, neurodegenerative diseases, castration of endocrine functions, and malignant neoplasm. Children, elderly, or pregnant women tend to be more vulnerable to these impacts [2].

Furthermore, changes in metabolism and immune activity heighten the risk for infections and diseases due to chemical adulterants. Adulterants such as lead, arsenic, and mercury can chronically poison the body through gradual accumulation which results in fatigue, developmental lags in children, reproductive dysfunction, and behavioral modifications [1, 2, 12].

1.1.4 Global Perspective on Food Adulteration

Though the challenge is especially acute when looking at lower middle-income countries, food adulteration remains a worldwide issue. In more developed economies, adulteration takes the form of misbranding, false organic labeling, or the replacement of high-quality constituents with less expensive ones. The 2008 Chinese melamine scandal, where infant formula was fraudulently enhanced with melamine to increase protein content, led to the death of six infants and thousands suffering from kidney damage, culling the need for more robust international monitoring systems for food surveillance [21].

1.1.5 Need for Detection and Regulation

Due to the serious health concerns involved with food adulteration, detection methods alongside regulations must be uncompromising. Though traditional methods of detection can be effective, they are not practical for large-scale field use [22]. As a result, LOC technologies continue to develop because of their ability to identify food adulterants efficiently and accurately. These systems can be extremely beneficial in the prevention of food fraud while simultaneously ensuring public health. [23]

1.2 Techniques Used to Detect Adulterants

Food adulteration poses a grave danger to food safety and public health. In order to detect adulterants in food products, sensitive, reliable, and sometimes rapid methods, which can accommodate a wide range of substances, are required. Over time, a multitude of analytical methods have been developed to detect, quantify, and characterize adulterants in food matrices, including advanced spectroscopic and biosensor-based platforms as well as classical chemical assays. These methods are selected based on the type of food, the nature of the adulterant, the economy, and the level of accuracy needed [24].

1.2.1 Chromatography and Mass Spectrometry

Chromatography is a common analytical technique used in the examination of food adulteration. Exemplary methods like High-Performance Liquid Chromatography (HPLC) [25] and Gas Chromatography (GC) effectively analyze synthetic dyes, preservatives, and pesticide residues. These methods allow the analyst to have a great level of precision and specificity when identifying contaminants, enabling the detection of even meticulous traces of contaminants [26]. In methodical or unfamiliar adulterant analysis, mass spectrometry (MS) is often paired with chromatography to achieve greater precision in detailing the molecular weight and composition of sophisticated unknown adulterants. This is especially applicable in examining the presence of hazardous organic compounds in ready-to-eat foods [27].

1.2.2 Spectroscopic Techniques

Frequently found spectroscopic methods include Nuclear Magnetic Resonance (NMR) spectroscopy [28], Infrared spectroscopy (IR) [29], and Ultraviolet-visible spectroscopy (UV-Vis) [30]. These noninvasive methods tend to be particularly powerful as they gather information concerning both the amount and make-up of a sample using its interaction with light. For example, illegal color additives in spices can be determined through the use of UV-Visible spectroscopy. Due to the speed with which they can analyze materials, near infrared spectroscopy (NIRS) is gaining popularity. It has been applied to the detection of adulterants in dairy products, oils, and cereals [31].

1.2.3 Biosensors and Electrochemical Sensors for On-Site Detection

The rise in popularity of biosensors and electrochemical sensors for on-site detection techniques is primarily because of their portability, low cost, and quick detection capabilities. These sensors function by sensing variations in electric currents when certain impurities are present, and so can be modified for application with milk, fruits, water, and drinks [32]. Unlike these sensors, poisons, allergens, and antibiotics can be readily identified with the use of antibodies in ELISA assay [33]. With highly sensitive and specific immunoassay techniques, ELISA is mostly employed in the detection of contaminants in dairy and meat products [34].

1.2.4 Polymerase Chain Reaction (PCR)

Among various techniques, Polymerase Chain Reaction (PCR) is extensively used for identifying nucleic acids from particular organisms [35]. It is especially useful for determining the species of meat or for verifying the presence of genetically modified organisms (GMOs) or microbial contamination in food [36]. Real-time PCR enables quantification of the level of contamination in a given sample as well [37].

1.2.5 Laser-Induced Breakdown Spectroscopy (LIBS)

Laser-Induced Breakdown Spectroscopy (LIBS) is another advanced analytical technique that is used for detecting heavy metals in food [38]. This method involves focusing a high-intensity laser on a sample to create plasma, and then analyzing the emitted light to determine the sample's elemental composition. It has shown promising results in analyzing vegetables and grains for toxic metal contamination [39].

1.2.6 Electrical Impedance Spectroscopy (EIS)

Electrical Impedance Spectroscopy (EIS) is being developed for the detection of adulteration in liquids such as milk, honey, and juices [40]. EIS has the capability of detecting changes in composition due to adulterants like water, sugar syrup, or additional starch by measuring the sample's electrical current resistance at different frequencies. This method is greatly sensitive to the changes in the physicochemical properties of food that has been altered [41].

Numerous techniques are being adapted into LOC devices and microfluidic systems to satisfy the increasing demand for convenient, accurate, and economical food testing. These small devices can perform complex examinations at the location of interest and perform multiple analytical tasks within one device [42]. Detecting food adulteration is shifting toward simpler and more nanoscale-responsive detector technologies enhanced by smartphone integration, microminiaturization, and advancements in nanotechnology [43].

1.2.7 Introduction to LOC Technology

Biotechnology and analytical fields have been changed because of LOC technology. This technology integrates multicellular, laboratory procedures into a single chip, rendering it a multifunctional research instrument [44]. Various physical and biological activities are performed on LOCs with the help of microfluidics, which is the manipulation of liquids in the nanoliter to picoliter range [45]. Since laboratory procedures are now reproducible on LOC systems, diagnostic tests can be performed rapidly, precisely, and autonomously with minimal human supervision [46].

LOC technology was developed in the latter half of the twentieth century due to the advancements in Integrated Circuit (IC) manufacturing as well as Micro-Electro-Mechanical Systems (MEMS). S.C. Terry created the first miniature gas chromatograph, which is considered the earliest LOC device, in 1979 at Stanford University [47]. Fortissimo's invention of microfluidic chips in the 1990s allowed different industries to commercially exploit LOC systems for environmental monitoring, medical and drug research, food safety, and proteomic and genomic studies [48].

LOC devices are portable, which is one of their primary advantages. This enables easier use in point-of-care (POC) diagnostic applications in remote and resource-poor regions. Their ability to function with very small sample volumes also diminishes the costs and waste associated with reagents [49–51]. The total quality improves when sample handling, chemical reactions, separations, and detection are performed on a single chip, as time to perform the analysis is reduced, human error is minimized, each step is automatable, and all quality metrics are improved [52]. Automated access systems enable real-time analysis, continuous monitoring, and remote access to diabetic results [53].

The decentralization of medical diagnosis has been made possible due to the integration of the LOC systems into healthcare, as they facilitate a quicker response to clinical emergencies, enhance surveillance systems, and monitor disease outbreaks. Also, LOC technology's remarkable speed and efficiency make it attractive across various fields of science and industry [54].

1.3 Benefits of LOC Technology

The myriad benefits of LOC technologies have transformed testing in biochemistry, medicine, and even in environmental science [51]. One of the greatest advantages of integrating laboratories into a chip is being able to perform complex tasks like sample processing, chemical analysis, and detection on one compact unit. This self-contained unit's scope surpasses that of traditional laboratory settings, as it greatly reduces the equipment, time, and space needed. An important advantage of LOC devices is their ability to operate with minimal sample and reagent volumes, which are usually in the nanoliter or even picoliter range. Hence, LOCs are highly cost-effective and green, because the chemical waste and money spent on analysis are greatly reduced. Their small size also increases their response kinetics, which enhances the time it takes to obtain the results. This is particularly useful for on-site testing and for panic and emergency diagnostic situations [54].

Standardization, along with automation, forms the two other major advantages. Pumps, valves, and sensors are used to control fluid metering in LOC systems, which reduces experiment variability and hand-operated errors. This allows for high precision repeatable analyses, which is important in clinical as well as research settings [55].

The portability and ease of use of LOC devices make them suitable for POC diagnostics in remote or low-resource settings due to limited access to sophisticated laboratory equipment. Their high speeds of reaction and ease of use enhance public health surveillance and disease control. Moreover, the integration of LOC devices with computers and smartphones allows for real-time data capture, remote diagnostics, and digital health records, enabling LOC devices to interface with the consumer. This integration supports advancing trends in telemedicine and personalized care [56].

1.4 Current Applications of Lab-on-a-Chip Technology

The compact size and high portability of LOC systems make them very useful for multiple industries, as they provide quick analysis. LOC systems have made an impact in the field of clinical diagnostics by automating disease detection and monitoring. For example, it is used in rapid diagnostic tests for infectious diseases where timely diagnosis is critical, like COVID-19, HIV, and TB. In the course of the COVID-19 pandemic, LOC tools such as lateral flow immunoassays were instrumental for mass testing [57, 58].

LOC systems also help in the early-stage diagnosis, prognosis, and therapy monitoring for cancer by detecting circulating tumor cells (CTCs), tumor markers, and specific genetic changes. Because of their high sensitivity, they are the most suitable for the detection of low-abundance targets in blood or tissue samples [59].

In pharmaceuticals, LOC is used for drug screening, toxicity testing, and even for pharmacokinetics. LOC platforms eliminate the need for animal testing, and the time required for drug development is expedited, as LOC systems permit high-throughput screening of pharmacological compounds on human cells grown in micro chambers [60]. LOC technology is also gaining attention in environmental monitoring. These tools are able to detect pollutants such as microbiological diseases, pesticides, and heavy metals within air, soil, and water. Their quick detection abilities help in mitigating pollution and averting public health crises [51].

In the food industry, LOC systems are used to trace foodborne pathogens, spoilage organisms, and even adulterants. LOCs are significant in proteomic and genomic research, assisting with protein expression studies, sequencing, and even DNA amplification (e.g., PCR) [7, 61]. The ability to carry out complex analyses at an ultra-small scale technologically accelerates research, while concurrently streamlining expenses. For example, Microfluidic chips can automatically detect artificial colors in spices or melamine in milk within minutes. In totality, LOC devices have redefined the paradigm of monitoring and diagnostics, evolving as indispensable tools not only in industrial and scientific research but also in health care and environmental monitoring [11].

2 Materials

2.1 Microfluidic Foundations

Polydimethylsiloxane (PDMS) is the substrate that forms the basis of LOC devices. a common elastomeric substance that is renowned for its ease of melting, flexibility, and transparency. PDMS works well with a variety of bonding methods and is beneficial for quick prototyping. But in some applications, its hydrophobicity and leaching propensity can be problematic.

Polymethyl Methacrylate (PMMA): A thermoplastic substance that provides strength and clarity of vision. PMMA is perfect for scalable manufacturing since it can be attached using adhesive bonding or solvent welding and is appropriate for mass production. Glass and Silicon: Applications needing accurate optical detection techniques benefit from the exceptional chemical resistance and optical transparency of inorganic materials like glass and silicon. Their extensive use may be constrained, nonetheless, by their brittleness and intricate manufacturing procedures.

2.2 Materials for Detection

The sensing mechanisms of LOC devices depend on detection materials, which enable the identification of particular adulterants using a variety of detection techniques (Fig. 1):

AuNPs, or gold nanoparticles: Because of their special optical characteristics, they are used in colorimetric tests. When target analytes are present, AuNPs can agglomerate and produce color changes that are easily noticeable even without specialized equipment.

Because of their high surface area and remarkable electrical conductivity, graphene and graphene oxide are used in LOC devices. The sensitivity of electrochemical sensors for identifying pollutants like mycotoxins can be increased via graphene-based platforms. Carbon Nanotubes (CNTs): Because of their superior electrical characteristics and large surface area, CNTs are used as immunosensors. They make it possible to create sensitive sensors that can identify food samples for infections like Salmonella.

2.3 Reagents and Coatings with Functions

To improve sensitivity and selectivity, functional coatings and reagents are added to sensor surfaces:Aptamers and Antibodies: Sensor surfaces are functionalized with biorecognition

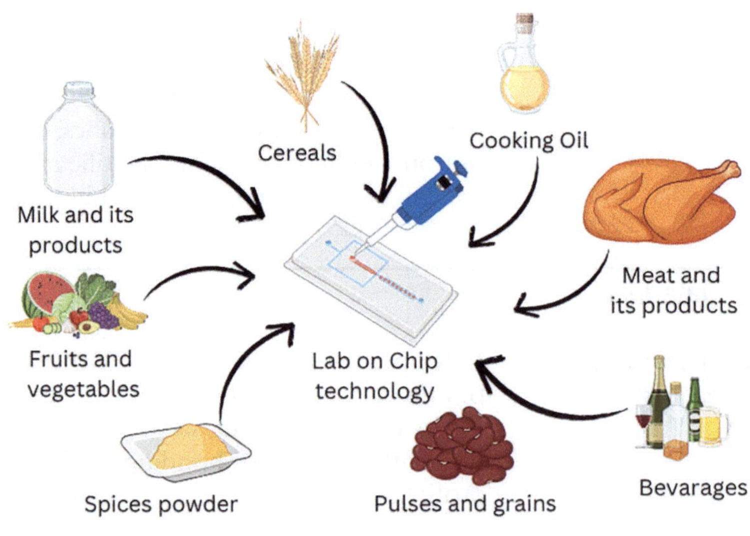

Fig. 1 Application of Lab-on-a-Chip technology for detecting food adulterants in various food matrices, including milk, cereals, oils, meat, beverages, spices, fruits, and pulses

components, such as aptamers and antibodies, to bind target adulterants selectively. Their specificity guarantees precise identification of pollutants such as heavy metals, antibiotics, and pesticides. Conductive Polymers: To improve electrical conductivity, LOC devices use polymers such as polypyrrole and polyaniline. Through the facilitation of electron trans processes, they are employed Hydrogels: Because hydrogels can swell and hold a lot of water, they are used in LOC devices. They are helpful as part of the detection process in biosensors or in applications that call for the controlled release of chemicals.

2.4 Components of Signal Transduction

The following elements of signal transduction are necessary to transform biological signals into quantifiable outputs:

Electrodes: For electrochemical detection, LOC devices have electrodes composed of carbon, platinum, or gold. They make it easier to measure how the presence of adulterants is affecting things now or in the future.

Optical Components: For optical detection techniques like fluorescence or absorbance measurements, LOC devices employ optical fibers and photodetectors. By altering the characteristics of light, these elements make it possible to identify food adulterants. Microvalves and Pumps: These devices, which are frequently made of thermoplastics or PDMS, are used to regulate the fluid flow inside the microchannels. They are necessary for LOC devices to automate sample processing operations.

2.5 Materials for Integration and Packaging

Materials that promote bonding and sealing are necessary for the integration of several components into a coherent LOC device:

Adhesives: UV-curable polymers or epoxy resins are frequently employed to join the device's various layers. These adhesives stop fluid leaks and guarantee structural integrity.

Packaging Materials: The materials used for packaging must guarantee the structural soundness of the gadget and shield delicate parts from the elements. For encapsulation, materials such as thermoplastics are frequently utilized.

2.6 Emerging Materials

New materials have been developed recently to improve LOC devices:

Bioactive Paper: Using enzymes or antibodies, bioactive paper can identify pollutants in food, such as pesticides. These disposable, portable, and reasonably priced paper-based sensors are appropriate for on-site testing.

Regenerated Silk: Regenerated silk can decrease water permeability when used in packaging, increasing the shelf life of perishable foods.

3 Methods

3.1 Lab-on-a-Chip Methods for Beverage Adulteration Detection

3.1.1 DNA-Based Authentication Using Capillary Electrophoresis LOC

Capillary electrophoresis combined with DNA-based authentication, coupled with LOC technologies, is one of the latest methodologies that offers exceptional accuracy for detecting food adulteration [62]. This method encompasses microfluidics and molecular biology to ascertain the originating species of a food sample. In microfluidic devices, capillary electrophoresis (CE) yields fast and precise results owing to the separation of amplified and segmented DNA fragments by size and charge in microchannels. LOC-CE systems are ideal for testing the authenticity of food because they identify even minute levels of adulterants, especially in highly processed or blended food items, through specific DNA probes placed for certain species [63].

With the implementation of the LOC device, DNA is separated, and fluid velocity is altered via microfluidic channels under an electric field. After the specimen's DNA is procured, it is either amplified via Loop-Mediated Isothermal Amplification (LAMP) or PCR [64]. The microcapillary channels within the plastic LOC are laden with the amplified fragments. In the effort to classify the DNA fragments by size, a voltage is applied in order to move them through the channel. A DNA fingerprint identifying the species of the food is generated by detecting the pattern of the fragments through integrated fluorescence or UV sensors [63].

In DNA-based LOC systems, the initial step is sample preparation, which entails the mechanical lysis and purification of DNA from intricate matrices like meats, oils, or liquids. Species-specific primers are used to amplify the skeletal retrieved DNA, like cyt b for mitochondria of mammals and trnL intron of chloroplast for plants. The amplicons obtained are put into the microfluidic CE channels [63].

Applying voltage causes negatively charged DNA molecules to migrate toward the anode. The separative outcome is based on size because smaller pieces diffuse more rapidly than bigger fragments. The chip has a detection module, which captures movement data of DNA in real time, often via laser-induced fluorescence. In the detection process, to determine if adulteration has happened, the banding pattern is analyzed and compared against established databases or predetermined guidelines. With this technique, one can reliably and highly resolutely detect adulterants like apple juice in pomegranate juice or palm oil in olive oil [65]. This approach reduces concerns regarding contamination, accelerates analysis, and minimizes the amount of reagents used, because sample processing and DNA separation are accomplished on-chip. For identification of high-value food products such as pomegranate juice or olive oil, portable and sensitive target devices afford unparalleled specificity: the DNA-CELOC units [66].

3.1.2 Hybrid Paper/Polymer-Based LOC for Pomegranate Juice Authentication

The emerging hybrid paper/polymer-based LOC systems integrate the structural strength of polymers with microchannel precision and the user-friendliness of paper microfluidics. These devices present a low-cost and easily accessible solution for detecting contamination in food authentication systems, particularly in pomegranate juice. Lower quality juices, like apple, grape, and beetroot, are often mixed with pomegranate juice. Hybrid LOC techniques are capable of identifying specific DNA sequences or chemical markers associated with pomegranates to determine such adulteration [65]. The hybrid LOC device integrates an action capture polymer architecture for supporting and sensing frameworks while using paper as a substrate for sample entry and fluid movement through capillary action [65]. Usually, the technique comprises of DNA isolation, LAMP, and visual identification, all being completed on a single chip. Some methods utilize chemical signatures, such as betanin from beetroot, instead of DNA for identification. The pigment that indicates red beetroot adulteration in pomegranate juice is preferentially captured by MIPs. Binding can result in a detectable and quantifiable colorimetric or electrochemical change [67].

The juice sample undergoes concentration and is subsequently placed within the paper channel prior to being sent to the reaction zone. If a DNA analysis is done, the target DNA of pomegranate is retrieved and amplified through LAMP if the processing involves DNA analysis. The additional markers and fluorescent dyes will bind to the DNA in the detection zone, which will show increased brightness. The work focuses on directing the juice sample to a chamber containing MIPs that selectively bind to the pigment in systems that use betanin as a target. This binding can be detected by a change in color or an electrochemical signal. These steps can be accomplished without any laboratory facilities due to the layered hybrid platform that integrates all the technologies described above [68].

These LOC devices provide dependable, affordable, and quick analysis—usually in less than an hour. They are very helpful for field testing in quality control and food safety inspections [69].

3.2 Lab-on-a-Chip Methods for Meat Adulteration Detection

3.2.1 Centrifugal Integrated Purification-CRISPR Array (CIPAM) System

The Centrifugal Integrated Purification-CRISPR Array (CIPAM) system has LOC technology that integrates centrifugal microfluidics and CRISPR molecular detection [70]. This technology is designed for the rapid, sensitive, and specific detection of meat adulteration by analyzing species-specific DNA sequences. CIPAM does everything, from DNA extraction and purification to amplification and CRISPR detection, on a single chip that fully integrates all steps into a centrifugal, valve-controlled system. Its main benefit is the ability to do high-throughput, POC testing without advanced equipment or technical skill [71].

CIPAM uses a disc-shaped microfluidic chip to process the sample and clean the reagents. As the disc spins, centrifugal force moves the fluids through the various modules sequentially. First, DNA extraction is performed using a microneedle device where the chip lyses and purifies the DNA during the process simultaneously. Then, the purified DNA is subjected to an isothermal method suitable for rapid nucleic acid amplification, called Recombinase-Aided Amplification (RAA). For the specific detection of the amplified products, the CRISPR-Cas12a system is utilized. Enzyme Cas12a gets activated after identifying some predefined sites within the DNA strands of interest; for instance, the DNA sequences from pigs, ducks, and or chickens. This cleavage results in the formation of single-strand oligonucleotides, which emit light upon cutting. The level of adulteration is defined by how much light is emitted or the intensity of the signal [70, 72].

To initiate the process, a meat sample is placed in the CIPAM disc's intake. While the integrated chambers dispense the magnetic particles and lysis buffers, the microneedle arrays assist in lysis and capture DNA during the spinning process. These particles that bind to and guide DNA into the purification chamber effectively differentiate the DNA from other impurities. The RAA enzymes within the amplification module repeat the DNA and maintain that temperature (between 37 °C and 42 °C). Subsequently, the preloaded CRISPR-Cas12a complexes are designed to target specific DNA sequences, guiding the amplified DNA to the detection zone. The fluorescent reporter is cleaved by Cas12a, yielding an easily detectable signal that can be read by a portable fluorescence reader [70, 71].

CIPAM is very effective for real-time food safety monitoring since the complete process, from sample input to detection, takes less than 30 min and reaches sensitivity as low as 0.1% adulterant content [70].

3.2.2 Microarray Chip PCR-Directed Microfluidic Lateral Flow Strip (LFS) Device

The device operates on a two-part mechanism. First, the DNA from a meat sample is amplified using multiplex PCR. Each pair of primers targets a distinct species-specific DNA sequence (beef, hog, chicken, or duck). The microfluidic cartridge containing the PCR products is then dispensed into a cartridge with a lateral flow strip (LFS). The amplified DNA fragments flow through the microchannels and bind to pre-immobilized probes or specific antibodies to each meat species that are embedded in the LFS. The binding reaction attesting to the presence of the species in the sample is usually displayed as a colored line, which in most cases employs AuNPs as its chromophore. The presence of gold and other meat species is evaluated qualitatively as additional test lines appear on the strip, enabling four or more meat varieties to be detected simultaneously [73].

The first step in the process involve homogenization of the flesh, followed by homogenate DNA extraction. Then, the DNA collected is divided through multiplex PCR using a primer set designed for differentiating the various origins of meat. Captured amplicons are then placed onto the microarray-based chip downstream integrated with the lateral flow strip. Due to capillary effects, the sample within the LOC device is drawn through microfluidic channels to the LFS. The probes complementary to some of the hybrids capture the products of PCR. When hybridization happens, labeled particles such as biotin-streptavidin conjugates or AuNPs bound to specific test lines identify the test lines and produce signal [74].

The results can be analyzed in 15–20 min without any specialized equipment. This method is quite useful in verifying the non-halal components of processed foods or undeclared meats such as pork masquerading as beef. It can assess complex mixtures and offers remarkable specificity and rapid analysis [73].

3.3 Lab-on-Chip (LOC) Methods for Detecting Adulteration in Milk and Dairy Products

3.3.1 Microfluidic Paper-Based Analytical Devices (μPADs)

Microfluidic paper-based analytical devices (μPADs) for chemical and biological analysis are a simple, low-cost, and mobile technology. Fluids are moved through designed pathways in μPADs without external pumps or power, relying on the capillary action of cellulose paper [75]. These devices are especially useful for food quality monitoring and POC diagnostics in resource-limited settings. Since colorimetric reagents can be incorporated into them, they can be used for the visual detection of adulterants like urea in milk and dairy products [76].

The μPAD employs capillary movement of the milk sample, which is pulled through the patterned paper channels. For example, urea's specific detection reagents were pretreated in specialized detection zones. As the milk sample moves to the detection zone, a sample profile is created by reaction with colorimetric reagents such as diacetyl monoxime or urease-impregnated markers. The reaction yields a color change that is proportional to the concentration of urea present in the sample. To enhance accuracy, measurement of color intensity can also be done visually, or using a portable colorimeter or smartphone app [77].

A biocompatible filter paper mask and a template of human arteries are first constructed using the above techniques, which include photolithography and wax printing, and injecting liquid latex into the hydrophilic filter paper with hydrophobic barriers [78]. On the template, these patterns create response zones and channels. Urease enzyme that catalyzes the decomposition of carbon dioxide and ammonia's precursor, urea, is poured on the paper. Ammonia's interaction with the apex motives paves the way for new paradigm shifts in the domain of perturbation theory [79].

Diphenyl carbonate succinaldehyde can serve as an alternate in a less aggressive milieu, whereby the pink associated with a sugar undergoes alteration. In conducting the test, a diluted sample of milk, with an expected concentration of 20% protein, is placed into a box-like structure with the scanning head mounted on it. Following the channels results in a syringe filled with the enzymes that had previously been exposed to the template and latex. The reagent takes less than 5 min, and, depending on the sticker used in the visible spectrum, achieves the level detectable at the control aiming mark, mark in situ for feedback regulated feedback [80].

For quantitative analysis, color intensity can be either measured digitally or visually compared with a preprepared calibration chart. μPADs, or microfluidic paper-based analytical devices, are useful in the field and for rapid screening due to their disposability, minimal required training, and long shelf life [81].

Research indicates that the low micromolar range for urea detection with μPADs is sufficient for identifying harmful levels of adulteration. As with most modern devices, these come with low cost, portability, and small reagent volumes. Their incorporation into optical or smartphone readers elevates μPADs' analytical functionalities and transforms them into devices capable of data storage for traceability and quality assurance [82].

Detection techniques based on μPAD are particularly useful in economically underdeveloped regions with more basic laboratory facilities. Their application in monitoring the quality of milk allows for the swift identification of urea adulteration and protection of consumers' health [83].

3.3.2 Surface Plasmon Resonance (SPR)-Based LOC Devices

Surface Plasmon Resonance (SPR)-based LOC technology merges optical sensing with microfluidics for rapid, ultrasensitive, and label-free molecular interaction detection. The application of SPR-LOC devices for the detection of melamine adulteration is increasingly common within the context of food safety, especially in monitoring dairy products. Melamine, a nitrogen-rich compound, is added illegally to milk to boost its apparent protein content. This practice is dangerous since it can lead to renal failure. SPR-LOC devices allow for rapid and precise detection of melamine, making them an efficient solution for identifying trace levels of this substance [84].

During SPR, the interface between a metal thin film (commonly gold) and a dielectric medium (usually the sample solution) excites surface plasmons—coherent oscillations of the electrons. The changes in intensity of reflected light at the so-called resonance angle, also known as the minimum angle of reflection, plunge sharply for a given angle of incidence when a beam of polarized light has an angle of incidence near the minimum angle of reflection due to the reorienting of the sensor's refractive index. The intensity of the light reflected off the gold surface is extraordinarily sensitive in proximity to the sensor surface. The SPR sensor measures binding events by tracking melamine molecules in the milk sample that bind to the functionalized surface through specific surface coating with antibodies or molecular receptors using melamine gold immunoassay. Changes in the resonance angle associated with the interactions, which are an increase in response to bound melamine, are proportional to the concentration of melamine [85].

SPR-based LOC devices are formed from a microfluidic channel placed over a sensor surface coated with gold and functionalized with melamine recognition elements like aptamers or monoclonal antibodies. The milk sample flows into the microchannel and glides over the sensor surface. When melamine is present, it attaches to the receptors and alters the refractive index, causing an SPR signal to be generated. Relatively, melamine concentration can be accurately determined through the continuous observation of the interaction without any labelling or amplification [85].

Such systems often include a small unit for processing data, a monitor for observing shifts in resonance, and a light source like a laser. This lightweight compact system enables on-site testing, which is particularly advantageous in dairy farms and processing centers [86].

Due to their extraordinary sensitivity, which is crucial for compliance with global safety standards, SPR-based LOC devices have detection limits in the range of parts-per-billion (ppb). Moreover, the approach is rapid; it typically provides results within minutes and requires minimal preparatory work with the sample [87]. These devices have a proven track record in controlling the quality of foods and diagnosing patients at the site. Particularly, they offer enhanced advantages over custom-built laboratory techniques in speed, automation, and mobility, as they function under complex sample matrices such as milk [87].

3.4 Lab on Chip Methods for Detection of Adulterants in Fruits

3.4.1 Dielectrophoresis (DEP)-Based LOC

The motion of particles within nonuniform electric fields can be utilized in a label-free method called dielectrophoresis (DEP). DEP enables fine-grained separation, concentration, and detection of particles based on dielectric properties in microfluidic LOC devices. In terms of food safety, LOC platforms that utilize DEP offer rapid, precise, and portable detection of adulterants such as formalin, a hazardous substance that is often illicitly used to extend the shelf life of fruits and their derivatives. The toxicity of formalin, a solution of formaldehyde in water, places great demand on detection devices, especially those used in the field [88].

The application of an alternating current electric field to suspended particles of cells, molecules, or compounds induces a dipole moment due to DEP effects. Particles migrate toward (positive DEP) or away (negative DEP) from regions of a stronger electric field gradient, depending on the relative polarizability of the particle in relation to the surrounding medium [89]. This property allows for the manipulation and identification of formalin molecules or reaction products within detection systems of formalin selectively.

For application of DEP in a LOC system, fruit extract or juice is injected into microchannels containing microelectrodes. When alternating current is applied, the particles of the fluid experience DEP forces. To achieve indirect detection of formalin or its derivatives, functionalization using certain chemicals with specific dielectric properties can be employed. These modified molecules exhibit distinct dielectrophoretic responses, allowing for their separation and concentration at specific locations within the chip. Integrated sensors (optical or electrochemical) then detect the presence of formalin based on signal variations, such as impedance shifts or fluorescence changes [90].

Before extracting the sample, such as fruit extract, the DEP-based LOC sample is pretreated using a filter to remove impurities [91]. Microelectrodes are placed around the channels of the microfluidic chip, which is filled with a fluid. An alternating current (AC) electric field is also applied through the microelectrodes at a controlled frequency and strength designed to optimize the dielectric signature of formalin [92].

To enhance selectivity, a formalin capturing reagent that affects the dielectric properties of formalin may be added prior to or during sample loading. The interaction improves DEP-based discrimination between formalin and other naturally occurring fruit product components. After application of an electric field, formalin-associated substances undergo a DEP reaction and migrate to specific places. These places are monitored by the systems for on-chip detection [93].

Detection is possible using impedance spectroscopy, which observes the change in electrical impedance as formalin accumulates at the sensor locations. Another method is reactive formalin fluorescent tagging, which enables optical detection. The whole assay is efficient on reagents and may be executed on-site, calling for real-time monitoring of fruit safety, which makes it rapid, taking only minutes [94]. DEP-based LOC systems have many advantages, including high sensitivity, low required sample volume, fast analysis time, potential for automation, or all of the above for miniaturized devices such as the Altered Integrated Microsystem, which makes them useful in decentralized markets and processing units that need low-level formalin detection [95].

3.4.2 Molecularly Imprinted Polymer (MIP)-Based LOC

Polymers that have recognition sites tailored for specific target molecules are referred to as molecularly imprinted polymers, or MIPs. Incorporating MIPs into LOC systems unlocks incredible and highly focused labs for the identification of even the faintest traces of pollutants [96]. This technique works wonders in food safety cases such as detecting the fungicidal adulterant carbendazim in orange juice and its derivatives. Immediate, precise, and portable detection methods are critical because carbendazim exceeds the danger threshold when ingested, making its health implications severe [97].

The microfluidic channels of MIP LOC devices are equipped with MIP particles or membranes that perform micro- and nanoparticle measurements. As known, MIPs are fabricated in the presence of a template molecule (in this case, carbendazim) that creates a surrounding polymer that can later be extracted [98]. A set of binding sites can be obtained through the template removal, which has bonds that are the right shape and functional groups for the target analyte. During the MIP-LOC operation, sample orange juices are introduced, and molecular recognition leads to permselective attachment of bondable carbendazim molecules on the selective binding sites. In MIP-LOC systems, detection is most frequently done by electrochemical or optical means. On-chip electrochemical sensors may detect current or potential changes induced by the attachment of an analyte [99]. Nevertheless, fluorescence and SPR are some of the optical procedures which may be employed for measuring the analyte concentration. The use of molecular imprinted polymers reduces the chances of nonspecific binding.

The MIP-LOC technique starts with the preparation of orange juice samples by filtration and pH adjustment for optimal binding conditions. The sample is subsequently loaded into the microfluidic device containing MIP-coated microchannels or detection zones. During the flow of the stream, carbendazim selectively binds to the MIP recognition sites. Nonspecific molecules are removed with washing buffers, greatly increasing selectivity [100]. In electrochemical detection, the binding of carbendazim changes the redox reaction at the electrode interface, creating a detectable signal. The signal generated corresponds to the amount of carbendazim contained in the sample. If optical sensors are applied, changes in emission microscopy and refractive index due to the binding of the analyte can also be monitored. This process usually takes less than 30 min and uses very small volumes of reagents [101].

Like other MIP-based methods, LOC has high specificity, low detection limits (i.e., in the nanomolar range), and is particularly robust in complex matrices such as orange juice. It allows on-site, real-time monitoring of food safety without bulky instrumentation. Furthermore, the objections listed serve to only bolster the fact that MIPs are cost-efficient, permanently binding active sites as MIPs make the system economical over repeated usage for quality control purposes [102]. Beyond these immediate implications, this technology can be applied toward the routine surveillance for juice manufacturing, import–export quality assessment, or public health initiatives against adulteration. The versatility in MIP synthesis expands the range of detectable contaminants through simple adjustments to the template molecule used during polymerization [103].

3.5 Lab on Chip Methods to Detect Adulterants in Oil

3.5.1 Magnetic Microfluidic Lab-on-a-Chip (LOC)

Microfluidics and manipulation with magnet systems are integrated in the magnetic microfluidic LOC system, allowing for the precision handling of magnetic beads functionalized with biomolecules. Within a small chip, this technology optimizes sample processing such as target capture, separation, and washing [104]. Magnetic beads can selectively capture analytes such as proteins, infections, or nucleic acids. The external magnetic field enables greatly reduced automation costs and resource expenditure by controlling the movement and placement of the beads within the microchannels. Clinicians using this method reported its advantage in diagnostics, food safety, and environmental monitoring due to higher sensitivity and reduced time for the assay [105].

The sample mixed with magnetic beads enters microchannels. Target molecules bind to beads, which are then immobilized by applying an external magnetic field. Washing buffers flow through to remove unbound materials while beads remain trapped magnetically. Subsequently, labeling agents bind to captured targets. Detection is performed on-chip using optical or electrochemical sensors. Magnetic forces enable bead transport and mixing, improving assay speed and reproducibility. This integration allows rapid, sensitive, and portable detection [106].

Magnetic microfluidic LOC systems enable the careful and selective identification of complex samples using their unique properties with magnetic particles. Their design takes into account all steps in processing the sample, which begins with the introduction of fluid samples, such as blood or food extracts, into the microfluidic chip. In strain form or solution, the sample is preloaded or mixed with magnetic nanoparticles or beads that are functionalized with aptamers, antibodies, or DNA probes [107].

These magnetic beads, interspersed with the sample while in solution, are selectively bound to the sample's target analyte as the sample moves through the microchannel. The external magnetic field (from integrated micro-electromagnets or permanent magnets) enables the precise control of the position of the beads. With the magnetic field able to concentrate or catch the beads at certain positions or spots within the chip, separation of target-bound beads from nontarget components is possible [108].

After capture, washing fluids pass through to remove unbound and contaminating interferences to enhance specificity. The quantity and presence of the analyte–bead complex are evaluated by detection modules integrated downstream. Various methods of detection can be used, which include electrochemical and impedance-based detection as well as optical detection through fluorescence or absorbance [107, 108].

The ability to conduct multiplexing with beads tailored with multiple probes on a single micromagnetic bead, high sensitivity due to the effects of concentration, and rapid handling of samples are a few of the benefits of magnetic microfluidic labs on a chip. Useful in maintaining food security, measuring biomarkers, and in pathogen detection are some of the other applications, particularly in the precise identification of magnetic microfluidic contaminants and adulterants [109].

3.5.2 ELISA on a Chip

ELISA on a chip diminishes the size of the enzyme-linked immunosorbent test (ELISA) by integrating the traditional ELISA onto microfluidic devices. This method permits the rapid and precise antigen detection through the use of immobilized antibodies in the microchannels. The swift reaction kinetics and minimized reagent consumption make it ideal for environmental monitoring, food safety, and point-of-need diagnosis. The microfluidic platform enhances automation and multiplexing capabilities [109, 110].

Capture antibodies are held in place by channel surfaces. While the sample is passing, antigen capture can occur. During washing, unbound particulates are cleared. The secondary antibody, which is linked to an enzyme, binds the antigen sandwiching it in a complex. When a substrate is added, there is an enzymatic reaction which happens that generates an optical signal, the amount of which directly corresponds to the level of the antigen present. The signal can be detected by optical detectors, making quantification faster. Assay speed and sensitivity are improved with microfluidics due to increased diffusion and binding kinetic rates [111].

ELISA on a chip streamlines the process of conducting sensitive and low-volume immunoassays by using a microfluidic apparatus for reagents rather than traditional techniques, unlike the enzyme-linked immunosorbent test, or ELISA. The chip has microchannels or microchambers that have been functionalized by surface-immobilized capture antibodies. As the sample moves through the channels, the target antigens in the sample fluid specifically bind to the antibodies that are affixed to the walls of the channels [112].

In washing buffers post incubation, channels are washed to remove unbound materials or residue, which leads to background noise, and lowers the noise floor. Then, an enzyme-labeled secondary antibody that is specific for the target antigens, and binds covalently, is added and a sandwich complex is formed. Enzymes catalyze a reaction and once the proper substrate is added, notable and optical signals, which can be color, light, or energy are produced. These signals are proportional to the concentration of the analyte [112, 113].

External readers or integrated optical detectors measure the produced signal. The microfluidic format accelerates diffusion and binding kinetics, which leads to shorter test durations and higher sensitivity. In instances where rapid POC testing is needed, proteins, toxins, and allergens can be detected using ELISA on a Chip in clinical diagnostics, environmental monitoring, and food safety applications. Mobility, parallel multiplexing, automation, and reduced use of reagents are some of the benefits. Various materials are used in chip fabrication, including polymers, glass, and PDMS [114].

3.6 Lab on Chip Method for Detection of Adulterants in Spices and Cereals

3.6.1 Paper-Based Lab-on-a-Chip

Cellulose paper allows the self-pumping capabilities of fluids through paper-based LOC devices using capillary action. They are ideal for POC and field testing as they are portable, inexpensive, and easy to use (Table 2). Cellulose-based LOC devices contain hydrophobic barriers, which outline fluidic channels along with reagent-containing test zones. Colorimetric and fluorescent detection is widely used for clinical diagnostics, food safety, and environmental monitoring in resource-limited settings [115, 116].

A sample droplet undergoes wicking through the hydrophilic paper channels, which are bordered by hydrophobic barriers. Absence of merges results in the obvious color change (or fluorescence) to the interacting fluid. Interaction with target analyses leads to colorimetric or fluorescent changes in the fluid's interaction with the embedded dry reagents within the assay regions. To achieve multiplexing, a greater number of channels and detection sites are added. Results are measurable or visually analyzed through portable readers or smartphone cameras. The ease of operation stems from passive movement of fluids without the need for pumps as they are economical [117–119].

Paper-Based LOC, also known as microfluidic paper analytical devices (μPADs), employ fluidic paper technology, which relies on the capillary action of porous cellulose paper to move liquids without the aid of external pumps or electricity [120].

Hydrophilic channels and test zones are formed through the application of hydrophobic barriers (wax, ink) onto the paper substrates, which can be cut with a laser. The sample is driven along predetermined pathways by this design [121].

For the inlet, a liquid sample is introduced, for example, an extract from a spice or cereal. With the aid of capillary forces, the fluid automatically flows through the hydrophilic channels toward reaction zones containing pre-deposited reagents such as enzymes, antibodies, or chemical markers, and also other compounds [75].

A visible signal like a color change occurs when an analyte engages with specific reagents during a chemical or biological interaction. This signal can be captured visually, with portable gauges or smartphone cameras, or through precise measurement [122].

These devices enable multiplexed tests on one chip by creating multiple channels and sensing lines. The devices are very cost-efficient and simple to use, and have no need for complex tools.

Table 2
Lab-on-a-Chip Techniques for Food Adulteration Detection in Various Food Matrices

Sl no	Food material	Type of lab on chip method used for detection	Advantages
01.	Beverage	DNA-Based Authentication Using Capillary Electrophoresis LOC	Identifies even minute levels of adulterants, especially in highly processed or blended food items, through specific DNA probes placed for certain species.
		Polymer-Based LOC	Capable of identifying specific DNA sequences or chemical markers associated with fruits for juices to determine adulteration.
02.	Meat	Centrifugal Integrated Purification-CRISPR Array (CIPAM) System	Rapid, sensitive, and specific detection of meat adulteration by analyzing species-specific DNA sequences
		Microarray Chip PCR-Directed Microfluidic Lateral Flow Strip (LFS) Device	It can simultaneously identify multiple species of meat in a single test, making it incredibly helpful for food safety inspections and authenticity verification in complicated meat products.
03.	Milk and dairy	Microfluidic Paper-Based Analytical Devices	It is a simple, low-cost, and mobile technology. Particularly useful in economically underdeveloped regions with more basic laboratory facilities.
		Surface Plasmon Resonance (SPR)-Based LOC	They offer enhanced advantages over custom-built laboratory techniques in speed, automation, and mobility and function under complex sample matrices such as milk.
04.	Fruits	Dielectrophoretic (DEP)-Based LOC	DEP-based LOC systems have many advantages, including high sensitivity, low required sample volume, fast analysis time, potential for automation.
		Molecularly Imprinted Polymer (MIP)-Based LOC	High specificity, low detection limits (i.e., in the nanomolar range), and is particularly robust in complex matrices. It can be applied toward the routine surveillance for juice manufacturing.
05.	Oil	Magnetic Microfluidic LOC	Useful in maintaining food security, measuring biomarkers, particularly in the precise identification of pathogens
		ELISA on a Chip	Mobility, parallel multiplexing, automation, and reduced use of reagents. Proteins, toxins, and allergens can be detected using ELISA on a Chip in clinical diagnostics, environmental monitoring, and food safety applications.

(continued)

Table 2
(continued)

Sl no	Food material	Type of lab on chip method used for detection	Advantages
06.	Spices and Cereals	Paper-Based LOC	The devices are very cost-efficient and simple to use and have no need for complex tools. They are ideal for in-field examination of contamination, poisons, or adulterants in foods.
		Fluorescence-Based LOC	This method is applicable with better precision in nucleic acid assays, immunoassays, cell and tissue analysis, and other immunoassays. These advantages in combination with rapid detection and the ability to detect multiple targets simultaneously lead to high precision results.

They are ideal for in-field examination of contamination, poisons, or adulterants in foods [83].

Common equipment includes cellulose paper, chemicals that can be printed on the paper, and wax printers. The compact size and low weight of the devices allow for easy transportation and storage. This makes them useful in the field or anywhere care is needed [123].

3.6.2 Fluorescence-Based Lab-on-a-Chip

LOC appearing as an abbreviation of "lab on a chip" fluorescence system combines highly specific and sensitive detection of biomolecules, more than traditional methods of detection using microscope-based scanning and imaging methods. Target analytes are labeled with fluorescence tags that are later measured quantitatively. This method is applicable with better precision in nucleic acid assays, immunoassays, cell and tissue analysis, and other immunoassays. These advantages, in combination with rapid detection and the ability to detect multiple targets simultaneously, lead to high precision results [91, 124].

The excitation light from integrated sources stimulates fluorophores, which causes emission of fluorescence. Samples containing the fluorescently labeled targets are streamed through microchannels where they interact with capture probes fixed on surfaces or beads [125]. Microfluidics improves the kinetics of reactions, lowers volumes, and enhances sensitivity, enabling real-time monitoring. Photodetectors or cameras capture emission signals whose intensity correlates with analytes. Multiplexed detection is achieved via spatial or spectral separation [124, 125].

Fluorescence-based LOC devices integrate the sensitivity and specificity of fluorescence detection with microfluidics for the

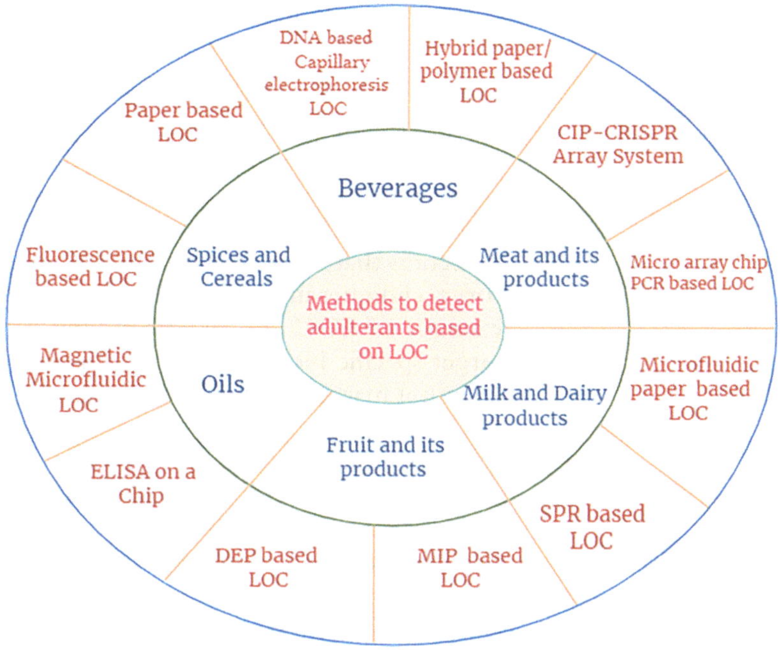

Fig. 2 Different methods to detect Adulterants in different categories of food products

analysis of a wide range of analytes such as proteins, toxins, and nucleic acids occurring in minute volumes. In the system, the sample is injected into microchannels where it can meet fluorescent probes or dyes designed to specifically bind or interact with the target molecule. When binding occurs, the fluorophores are excited by an internal or external light source (LEDs or lasers) and emit light at specific wavelengths, which yields an intense emission. These signals are automatically captured by built-in optical detectors (photodetectors, CCD sensors) or external ones. Changes in the intensity or the wavelength shifts of the emitted light yield in correlation to the analytes concentration [126].

Microfluidics improves the speed of chemical reactions within the sample as well as fluid movement for mixing, leading to the possibility of performing rapid assays with minute sample volumes alongside minimal reagent usage [127]. The presence of fluorophores with different emission spectra enables the use of multiple distinct probes, leading to the possibility of multiplexing. Its ability to perform real-time monitoring along with high sensitivity makes this method popular in food safety, clinical diagnostics, and environmental monitoring [128] (Fig. 2).

4 Notes

LOC systems have evolved as efficacious tools for detecting food fraud, contamination, and authenticity in numerous food categories like beverages, meat, dairy, fruits, oils, spices, and cereals. These miniature devices that integrate numerous laboratory processes onto a single chip enable rapid, sensitive, and ultra-specific detection that, more often than not, exceeds conventional laboratory techniques in mobility, rapidity, and cost effectiveness [23]. An example of LOC devices' capabilities is their ability to accurately target specific biomolecules such as proteins, DNA sequences, or chemical markers and aid in precise species or contaminant detection. In the case of DNA-based authentication, processes such as capillary electrophoresis and CRISPR array systems focus on identifying species-specific DNA sequences, which allow for successful identification of adulterants in processed food and beverages. This degree of accuracy is vital, especially for complex or blended foods where traditional methods might be used to identify adulteration [129].

Magnetic microfluidic systems and paper-based analytical devices are examples of microfluidic LOC devices that demonstrate effortless use, economic value, and portability. They are particularly beneficial in the field or in resource-poor environments that lack operational laboratories. While magnetic microfluidic chips enhance pathogen detection and food security monitoring in oils and other food matrices, paper-based LOCs are ideal for rapid on-site screening of contamination or adulteration in spices and cereals [130].

Fluorescence-based detection methods and SPR-based LOCs provide higher sensitivity and automation. Fluorescence-based LOCs offer highly accurate, multiplexed tests for proteins and nucleic acids, while SPR devices excel at analyzing complex matrices such as milk, where multiple components can hinder detection [131]. Additionally, the capability provided by ELISA-on-a-chip technology in detecting critical clinical and food safety diagnostics extends to proteins, toxins, and allergens [132]. With regards to fruit juices, the specific class of MIPS LOCs exhibits unparalleled durability and selectivity, which aids in overcoming complex matrices. The ability to detect biological moieties as low as nanomolar concentrations elevates the reliability of scrutiny monitoring and enforces routine observation in juice production and quality control [133].

Keeping LOCs together, these technologies offer multifaceted strategies for food safety challenges such as rapid processes, small sample sizes, low reagent consumption, automation, and multiplexing capabilities. With these technologies, food industries and authorities are equipped with means to validate the authenticity and scientific safety compliance while ensuring consumer health safety [134].

In LOC technologies, the integration vis-à-vis miniaturization of multiple detection devices into a singular multifunctional LOC platform is anticipated to dominate as a primary direction for advancement. The capability to conduct rapid and multiple evaluations of contaminants or adulterants will be possible through the combination of DNA-based assays, immunoassays, and sensor technologies, significantly improving throughput and reducing time to results [23].

Future advancements in nanotechnology biosensors will enhance LOC devices' sensitivity and specificity. The inclusion of metallic nanoparticles, graphene, and quantum dots nanomaterials allows for the amplification of signals, which makes it feasible to detect allergens or pollutants at trace levels. In combination with CRISPR-based gene editing techniques, LOC systems stand to revolutionize diagnostic precision for food safety [71].

An equally important area of development is wireless communication and Internet of Things (IoT) connectivity. In the future, LOC devices will have the capability to communicate diagnostic data and monitor data in real-time via mobile devices, enabling rapid response throughout the food supply chain. The ability to remotely monitor LOC devices will enhance transparency, traceability, and efficiency in food safety management by regulators, producers, and distributors [135].

The widespread implementation of rapid, portable LOC diagnostics will be encouraged by the increased confidence that customers and industry stakeholders will experience.

Additionally, before buying or consuming food products, consumers may be able to independently test them for contamination, freshness, or authenticity using consumer-facing LOC devices. Democratizing food safety testing could reduce food fraud and alter consumer behavior [136]. Lastly, LOC platforms might be increasingly used for environmental monitoring associated with food production, such as detecting soil and water contaminants that affect the well-being of crops and cattle. This will promote more ecologically friendly farming practices and reduce environmental risks that compromise food safety [51].

References

1. Anagaw YK et al (2024) Food adulteration: causes, risks, and detection techniques—review. SAGE Open Med 12. https://doi.org/10.1177/20503121241250184
2. Momtaz M, Bubli SY, Khan MS (2023) Mechanisms and health aspects of food adulteration: a comprehensive review. Foods 12(1):199. https://doi.org/10.3390/foods12010199
3. Artavia G, Cortés-Herrera C, Granados-Chinchilla F (2021) Selected instrumental techniques applied in food and feed: quality, safety and adulteration analysis. Foods 10(5): 1081. https://doi.org/10.3390/foods10051081
4. Neužil P, Giselbrecht S, Länge K, Huang TJ, Manz A (2012) Revisiting lab-on-a-chip technology for drug discovery. Nat Rev Drug

Discov 11(8):620–632. https://doi.org/10.1038/nrd3799

5. Liang M, Zhang G, Song J, Tan M, Su W (2023) Paper-based microfluidic chips for food Hazard factor detection: fabrication, modification, and application. Foods 12(22): 4107. https://doi.org/10.3390/foods12224107
6. Wang A et al (2024) Magnetic-gold nanoparticle-mediated paper-based biosensor for highly sensitive colorimetric detection of food adulteration. J Hazard Mater 475: 134849. https://doi.org/10.1016/j.jhazmat.2024.134849
7. Zolti O, Suganthan B, Ramasamy RP (2023) Lab-on-a-Chip electrochemical biosensors for foodborne pathogen detection: a review of common standards and recent Progress. Biosensors (Basel) 13(2):215. https://doi.org/10.3390/bios13020215
8. Mohammadi Z, Jafari SM (2020) Detection of food spoilage and adulteration by novel nanomaterial-based sensors. Adv Colloid Interf Sci 286:102297. https://doi.org/10.1016/j.cis.2020.102297
9. Lebelo K, Malebo N, Mochane MJ, Masinde M (2021) Chemical contamination pathways and the food safety implications along the various stages of food production: a review. Int J Environ Res Public Health 18(11):5795. https://doi.org/10.3390/ijerph18115795
10. Sridhar A, Kapoor A, Kumar PS, Ponnuchamy M, Sivasamy B, Vo D-VN (2022) Lab-on-a-chip technologies for food safety, processing, and packaging applications: a review. Environ Chem Lett 20(1):901–927. https://doi.org/10.1007/s10311-021-01342-4
11. Ray R et al (2022) Paper-based microfluidic devices for food adulterants: cost-effective technological monitoring systems. Food Chem 390:133173. https://doi.org/10.1016/j.foodchem.2022.133173
12. Choudhary A, Gupta N, Hameed F, Choton S (2020) An overview of food adulteration: concept, sources, impact, challenges and detection. Int J Chem Stud 8(1): 2564–2573. https://doi.org/10.22271/chemi.2020.v8.i1am.8655
13. Choudhary M, Sharma N (2024) Milk adulterants: serious impact on human health. J Med Evid 5(2):124–128. https://doi.org/10.4103/JME.JME_145_23
14. Per H, Kurtoğlu S, Yağmur F, Gümüş H, Kumandaş S, Poyrazoğlu MH (2007) Calcium carbide poisoning via food in childhood. J Emerg Med 32(2):179–180. https://doi.org/10.1016/j.jemermed.2006.05.049
15. Tai Y-T et al (2022) The preferential accumulation of cadmium ions among various tissues in mice. Toxicol Rep 9:111–119. https://doi.org/10.1016/j.toxrep.2022.01.002
16. Kourani K, Kapoor N, Badiye A, Shukla RK (2020) Detection of synthetic food color 'Metanil Yellow' in sweets: a systematic approach. JPC—J Planar Chromatogr—Mod TLC 33(4):413–418. https://doi.org/10.1007/s00764-020-00046-9
17. DAS L (2023) Determination of toxic elements in silver leaf coated sugar confectionaries by inductively coupled plasma—mass spectrometry and their health risk assessment. https://doi.org/10.2139/ssrn.4587063
18. Wang J, Liang L, Wei S, Zhang S, Hu Y, Lv Y (2022) Histone 2-Hydroxyisobutyryltransferase encoded by Afngg1 is involved in pathogenicity and aflatoxin biosynthesis in Aspergillus flavus. Toxins (Basel) 15(1):7. https://doi.org/10.3390/toxins15010007
19. Sood NN, Sachdev MS, Mohan M, Gupta SK, Sachdev HPS (1985) Epidemic dropsy following transcutaneous absorption of Argemone mexicana oil. Trans R Soc Trop Med Hyg 79(4):510–512. https://doi.org/10.1016/0035-9203(85)90079-3
20. Sharma BD, Malhotra S, Bhatia V, Rathee M (1999) Epidemic dropsy in India. Postgrad Med J 75(889):657–661. https://doi.org/10.1136/pgmj.75.889.657
21. Pei X, Tandon A, Alldrick A, Giorgi L, Huang W, Yang R (2011) The China melamine milk scandal and its implications for food safety regulation. Food Policy 36(3): 412–420. https://doi.org/10.1016/j.foodpol.2011.03.008
22. Chen W, Li H, Wang Y, De Silva P, Adhikari B, Wang B (2020) Advances in technologies used in the detection of food adulteration. In: Biotechnological approaches in food adulterants, 1st edn. CRC Press/Taylor & Francis Group, Boca Raton, pp 49–78. https://doi.org/10.1201/9780429354557-3
23. Yoon J-Y, Kim B (2012) Lab-on-a-chip pathogen sensors for food safety. Sensors 12(8): 10713–10741. https://doi.org/10.3390/s120810713
24. Song D et al (2024) Research advances in detection of food adulteration and application of MALDI-TOF MS: a review. Food Chem 456:140070. https://doi.org/10.1016/j.foodchem.2024.140070
25. Macrae R (1981) Recent applications of high pressure liquid chromatography to food analysis. Int J Food Sci Technol 16(1):1–11.

https://doi.org/10.1111/j.1365-2621.1981.tb00990.x

26. de Andrade FI et al (2014) Determination of synthetic food dyes in commercial soft drinks by TLC and ion-pair HPLC. Food Chem 157:193–198. https://doi.org/10.1016/j.foodchem.2014.01.100
27. Chen Z et al (2025) Recent progress in the application of chromatography-coupled mass-spectrometry in the analysis of contaminants in food products. Food Chem X 27:102397. https://doi.org/10.1016/j.fochx.2025.102397
28. Qu Q, Jin L (2022) Application of nuclear magnetic resonance in food analysis. Food Sci Technol 42. https://doi.org/10.1590/fst.43622
29. Yegon D, Ojijo NK, Tybussek T, Owino W (2023) Application of portable near-infrared spectroscopy for rapid detection and quantification of adulterants in baobab fruit pulp. Int J Food Sci Technol 58(3):1465–1473. https://doi.org/10.1111/ijfs.16313
30. Gräb P, Geidel E (2019) Spectroscopic studies of food colorings. World J Chem Educ 7(2):136–144. https://doi.org/10.12691/wjce-7-2-13
31. Di Anibal CV, Odena M, Ruisánchez I, Callao MP (2009) Determining the adulteration of spices with Sudan I-II-II-IV dyes by UV–visible spectroscopy and multivariate classification techniques. Talanta 79(3):887–892. https://doi.org/10.1016/j.talanta.2009.05.023
32. Mohan B, Singh G, Chauhan A, Pombeiro AJL, Ren P (2023) Metal-organic frameworks (MOFs) based luminescent and electrochemical sensors for food contaminant detection. J Hazard Mater 453:131324. https://doi.org/10.1016/j.jhazmat.2023.131324
33. Amine A, Mohammadi H, Bourais I, Palleschi G (2006) Enzyme inhibition-based biosensors for food safety and environmental monitoring. Biosens Bioelectron 21(8):1405–1423. https://doi.org/10.1016/j.bios.2005.07.012
34. Inês A, Cosme F (2025) Biosensors for detecting food contaminants—an overview. PRO 13(2):380. https://doi.org/10.3390/pr13020380
35. Salihah NT, Hossain MM, Lubis H, Ahmed MU (2016) Trends and advances in food analysis by real-time polymerase chain reaction. J Food Sci Technol 53(5):2196–2209. https://doi.org/10.1007/s13197-016-2205-0
36. Liu Y, Cao Y, Wang T, Dong Q, Li J, Niu C (2019) Detection of 12 common food-borne bacterial pathogens by TaqMan real-time PCR using a single set of reaction conditions. Front Microbiol 10. https://doi.org/10.3389/fmicb.2019.00222
37. Karabasanavar N et al (2022) Polyclonal hen egg yolk antibodies could confer passive protection against *Salmonella* serotypes in broiler chicks. J Food Saf 42(4). https://doi.org/10.1111/jfs.12987
38. Stefas D et al (2021) A laser-based method for the detection of honey adulteration. Appl Sci 11(14):6435. https://doi.org/10.3390/app11146435
39. Yang Z, Ren J, Du M, Zhao Y, Yu K (2022) Enhanced laser-induced breakdown spectroscopy for heavy metal detection in agriculture: a review. Sensors 22(15):5679. https://doi.org/10.3390/s22155679
40. Tsopelas F (2025) Electroanalytical approaches to combatting food adulteration: advances in non-enzymatic techniques for ensuring quality and authenticity. Molecules 30(4):876. https://doi.org/10.3390/molecules30040876
41. Cheng CJ, Jones OG (2019) Effect of drying temperature and extent of particle dispersion on composite films of methylcellulose and zein nanoparticles. J Food Eng 250:26–32. https://doi.org/10.1016/j.jfoodeng.2019.01.012
42. Khalaf EM et al (2023) Smartphone-assisted microfluidic sensor as an intelligent device for on-site determination of food contaminants: developments and applications. Microchem J 190:108692. https://doi.org/10.1016/j.microc.2023.108692
43. Janakiraman K, Ramesh A, Sethuraman V, Shanmugasundaram Prema S (2024) Prospects of nanotechnology advances in food adulterant detection, spoilage detection, packaging and preservation applications. Int J Food Eng 20(7):483–494. https://doi.org/10.1515/ijfe-2023-0304
44. Bale AS et al (2021) Advancements of lab on chip in reducing human intervention: a study. In: 2021 3rd International Conference on Advances in Computing, Communication Control and Networking (ICAC3N). IEEE, pp 38–42. https://doi.org/10.1109/ICAC3N53548.2021.9725466
45. Dutse SW, Yusof NA (2011) Microfluidics-based lab-on-chip systems in DNA-based biosensing: an overview. Sensors 11(6):5754–5768. https://doi.org/10.3390/s110605754
46. Wang C, Liu M, Wang Z, Li S, Deng Y, He N (2021) Point-of-care diagnostics for

infectious diseases: from methods to devices. Nano Today 37:101092. https://doi.org/10.1016/j.nantod.2021.101092

47. Abgrall P, Gué A-M (2007) Lab-on-chip technologies: making a microfluidic network and coupling it into a complete microsystem—a review. J Micromech Microeng 17(5):R15–R49. https://doi.org/10.1088/0960-1317/17/5/R01
48. Figeys D, Pinto D (2000) Lab-on-a-chip: a revolution in biological and medical sciences. Anal Chem 72(9):330 A–335 A. https://doi.org/10.1021/ac002800y
49. Gómez-Arconada L, Díaz-Fernández A, Ferapontova EE (2022) Ultrasensitive disposable apatasensor for reagentless electrocatalytic detection of thrombin: an O2-dependent hemin-G4-aptamer assay on gold screen-printed electrodes. Talanta 245:123456. https://doi.org/10.1016/j.talanta.2022.123456
50. Arshavsky-Graham S, Segal E (2020) Lab-on-a-chip devices for point-of-care medical diagnostics, pp 247–265. https://doi.org/10.1007/10_2020_127
51. Aryal P, Hefner C, Martinez B, Henry CS (2024) Microfluidics in environmental analysis: advancements, challenges, and future prospects for rapid and efficient monitoring. Lab Chip 24(5):1175–1206. https://doi.org/10.1039/D3LC00871A
52. Ma X et al (2023) Advances in integration, wearable applications, and artificial intelligence of biomedical microfluidics systems. Micromachines (Basel) 14(5):972. https://doi.org/10.3390/mi14050972
53. Cappon G, Vettoretti M, Sparacino G, Facchinetti A (2019) Continuous glucose monitoring sensors for diabetes management: a review of technologies and applications. Diabetes Metab J 43(4):383. https://doi.org/10.4093/dmj.2019.0121
54. Gavina K, Franco LC, Khan H, Lavik J-P, Relich RF (2023) Molecular point-of-care devices for the diagnosis of infectious diseases in resource-limited settings—a review of the current landscape, technical challenges, and clinical impact. J Clin Virol 169:105613. https://doi.org/10.1016/j.jcv.2023.105613
55. Reyes DR et al (2021) Accelerating innovation and commercialization through standardization of microfluidic-based medical devices. Lab Chip 21(1):9–21. https://doi.org/10.1039/D0LC00963F
56. Mejía-Salazar JR, Rodrigues Cruz K, Materón Vásques EM, Novais de Oliveira O Jr (2020) Microfluidic point-of-care devices: new trends and future prospects for eHealth diagnostics. Sensors 20(7):1951. https://doi.org/10.3390/s20071951
57. Cabibbe AM et al (2015) Lab-on-Chip-based platform for fast molecular diagnosis of multidrug-resistant tuberculosis. J Clin Microbiol 53(12):3876–3880. https://doi.org/10.1128/JCM.01824-15
58. Tharakan S, Faqah O, Asghar W, Ilyas A (2022) Microfluidic devices for HIV diagnosis and monitoring at point-of-care (POC) settings. Biosensors (Basel) 12(11):949. https://doi.org/10.3390/bios12110949
59. Vaidyanathan R, Soon RH, Zhang P, Jiang K, Lim CT (2018) Cancer diagnosis: from tumor to liquid biopsy and beyond. Lab Chip. https://doi.org/10.1039/C8LC00684A
60. Cong Y et al (2020) Drug toxicity evaluation based on organ-on-a-chip technology: a review. Micromachines (Basel) 11(4):381. https://doi.org/10.3390/mi11040381
61. Zhang J, Ma C, Du Y, Huang J, Xue L (2025) Microfluidic biosensors for rapid detection of foodborne pathogenic bacteria: recent advances and future perspectives. Front Chem 13. https://doi.org/10.3389/fchem.2025.1536928
62. Spaniolas S, Bazakos C, Awad M, Kalaitzis P (2008) Exploitation of the chloroplast *trn* L (UAA) intron polymorphisms for the authentication of plant oils by means of a lab-on-a-chip capillary electrophoresis system. J Agric Food Chem 56(16):6886–6891. https://doi.org/10.1021/jf8008926
63. Lee I, Kim H-Y (2024) Lab-on-a-chip devices for nucleic acid analysis in food safety. Micromachines (Basel) 15(12):1524. https://doi.org/10.3390/mi15121524
64. Zhang H, Xu Y, Fohlerova Z, Chang H, Iliescu C, Neuzil P (2019) LAMP-on-a-chip: revising microfluidic platforms for loop-mediated DNA amplification. TrAC Trends Anal Chem 113:44–53. https://doi.org/10.1016/j.trac.2019.01.015
65. Hu Y, Lu X (2020) Rapid pomegranate juice authentication using a simple sample-to-answer hybrid paper/polymer-based lab-on-a-Chip device. ACS Sens 5(7):2168–2176. https://doi.org/10.1021/acssensors.0c00786
66. Xu Z, Morris R, Bencsik M, Newton M (2014) Detection of virgin olive oil adulteration using low field unilateral NMR. Sensors 14(2):2028–2035. https://doi.org/10.3390/s140202028
67. Ghasempour Z, Alizadeh Khaled-Abad M, Vardast MR, Rezazad Bari M, Moghaddas

Kia E (2019) Fabrication of betanin imprinted polymer for rapid detection of red beet adulteration in pomegranate juice. Polym Bull 76(4):1793–1805. https://doi.org/10.1007/s00289-018-2444-5
68. Zheng Y, Li J, Zhou B, Ian H, Shao H (2021) Advanced sensitivity amplification strategies for voltammetric immunosensors of tumor marker: state of the art. Biosens Bioelectron 178:113021. https://doi.org/10.1016/j.bios.2021.113021
69. Hurtado-López EF, Vásquez-Garibay EM, Trujillo X, Larrosa-Haro A (2017) Body composition predicts growth in infants and toddlers with chronic liver disease. J Pediatr Gastroenterol Nutr 65(6):e117–e119. https://doi.org/10.1097/MPG.0000000000001746
70. Xiang X et al (2024) High-throughput identification of meat ingredients in adulterated foods based on centrifugal integrated purification-CRISPR array. Food Chem 443:138507. https://doi.org/10.1016/j.foodchem.2024.138507
71. Sun Y, Wen T, Zhang P, Wang M, Xu Y (2024) Recent advances in the CRISPR/Cas-based nucleic acid biosensor for food analysis: a review. Foods 13(20):3222. https://doi.org/10.3390/foods13203222
72. Wang X et al (2023) CRISPR/Cas12a-mediated enzymatic recombinase amplification for rapid visual quantitative authentication of halal food. Anal Chim Acta 1255:341144. https://doi.org/10.1016/j.aca.2023.341144
73. Wang H et al (2023) Accurate molecular identification of different meat adulterations without carryover contaminations on a microarray chip PCR-directed microfluidic lateral flow strip device. Food Chem 7:100180. https://doi.org/10.1016/j.fochms.2023.100180
74. Zhao L et al (2021) Development of a polymerase chain reaction lateral flow immunoassay for rapid authentication of venison in food products. ACS Food Sci Technol 1(1):12–16. https://doi.org/10.1021/acsfoodscitech.0c00085
75. Yetisen AK, Akram MS, Lowe CR (2013) Paper-based microfluidic point-of-care diagnostic devices. Lab Chip 13(12):2210. https://doi.org/10.1039/c3lc50169h
76. Tian T, Bi Y, Xu X, Zhu Z, Yang C (2018) Integrated paper-based microfluidic devices for point-of-care testing. Anal Methods 10(29):3567–3581. https://doi.org/10.1039/C8AY00864G
77. Shalileh F, Sabahi H, Golbashy M, Dadmehr M, Hosseini M (2023) A simple smartphone-assisted paper-based colorimetric biosensor for the detection of urea adulteration in milk based on an environment-friendly pH-sensitive nanocomposite. Anal Chim Acta 1284:341935. https://doi.org/10.1016/j.aca.2023.341935
78. Altundemir S, Uguz AK, Ulgen K (2017) A review on wax printed microfluidic paper-based devices for international health. Biomicrofluidics 11(4):041501. https://doi.org/10.1063/1.4991504
79. Ferreira FTSM, Rangel AOSS, Mesquita RBR (2025) A microfluidic paper-based device for monitoring urease activity in saliva. Biosensors (Basel) 15(1):48. https://doi.org/10.3390/bios15010048
80. Nongonierma AB, FitzGerald RJ (2014) Susceptibility of milk protein-derived peptides to dipeptidyl peptidase IV (DPP-IV) hydrolysis. Food Chem 145:845–852. https://doi.org/10.1016/j.foodchem.2013.08.097
81. Noviana E et al (2021) Microfluidic paper-based analytical devices: from design to applications. Chem Rev 121(19):11835–11885. https://doi.org/10.1021/acs.chemrev.0c01335
82. Suresh V, Qunya O, Kanta BL, Yuh LY, Chong KSL (2018) Non-invasive paper-based microfluidic device for ultra-low detection of urea through enzyme catalysis. R Soc Open Sci 5(3):171980. https://doi.org/10.1098/rsos.171980
83. Boobphahom S et al (2020) Recent advances in microfluidic paper-based analytical devices toward high-throughput screening. Molecules 25(13):2970. https://doi.org/10.3390/molecules25132970
84. Lu Y, Xia Y, Pan M, Wang X, Wang S (2014) Development of a surface plasmon resonance immunosensor for detecting melamine in milk products and pet foods. J Agric Food Chem 62(51):12471–12476. https://doi.org/10.1021/jf504055g
85. Wu H, Li H, Chua FZH, Li SFY (2013) Rapid detection of melamine based on immunoassay using portable surface plasmon resonance biosensor. Sens Actuators B Chem 178:541–546. https://doi.org/10.1016/j.snb.2012.12.089
86. Sharma S, Paliwal A, Bassi M, Tomar M, Gupta V, Gulati S (2021) Investigation of adulteration in milk using surface plasmon resonance. ECS J Solid State Sci Technol 10(9):091004. https://doi.org/10.1149/2162-8777/ac1f71

87. Ravindran N et al (2023) Recent advances in surface Plasmon resonance (SPR) biosensors for food analysis: a review. Crit Rev Food Sci Nutr 63(8):1055–1077. https://doi.org/10.1080/10408398.2021.1958745
88. Pethig R (2017) Dielectrophoresis. Wiley. https://doi.org/10.1002/9781118671443
89. Pethig R (2010) Review article—Dielectrophoresis: status of the theory, technology, and applications. Biomicrofluidics 4(2). https://doi.org/10.1063/1.3456626
90. Martín A, Vilela D, Escarpa A (2012) Food analysis on microchip electrophoresis: an updated review. Electrophoresis 33(15): 2212–2227. https://doi.org/10.1002/elps.201200049
91. Ryu G et al (2011) Highly sensitive fluorescence detection system for microfluidic lab-on-a-chip. Lab Chip 11(9):1664. https://doi.org/10.1039/c0lc00586j
92. Cai D, Xiao M, Xu P, Xu Y-C, Du W (2014) An integrated microfluidic device utilizing dielectrophoresis and multiplex array PCR for point-of-care detection of pathogens. Lab Chip 14(20):3917–3924. https://doi.org/10.1039/C4LC00669K
93. Mohd Maidin NN, Buyong MR, Rahim RA, Mohamed MA (2021) Dielectrophoresis applications in biomedical field and future perspectives in biomedical technology. Electrophoresis 42(20):2033–2059. https://doi.org/10.1002/elps.202100043
94. Sherif S, Ghallab YH, Ismail Y (2025) Comprehensive characterization of a microfluidic platform for DEP manipulation and bio-impedance detection using multi-sized polystyrene microbeads. Microfluid Nanofluidics 29(2):11. https://doi.org/10.1007/s10404-024-02785-1
95. Nikolic-Jaric M et al (2012) Electronic detection of dielectrophoretic forces exerted on particles flowing over interdigitated electrodes. Biomicrofluidics 6(2):24117. https://doi.org/10.1063/1.4709387
96. Lazarević-Pašti T, Tasić T, Milanković V, Potkonjak N (2023) Molecularly imprinted plasmonic-based sensors for environmental contaminants—current state and future perspectives. Chem 11(1):35. https://doi.org/10.3390/chemosensors11010035
97. Saylan Y, Akgönüllü S, Yavuz H, Ünal S, Denizli A (2019) Molecularly imprinted polymer based sensors for medical applications. Sensors 19(6):1279. https://doi.org/10.3390/s19061279
98. Saylan Y, Kılıç S, Denizli A (2024) Biosensing applications of molecularly imprinted-polymer-based nanomaterials. PRO 12(1): 177. https://doi.org/10.3390/pr12010177
99. Fauzi DI, Saputri FA (2019) Molecularly imprinted polymer nanoparticles (MIP-NPs) applications in electrochemical sensors. Int J Appl Pharma:1–6. https://doi.org/10.22159/ijap.2019v11i6.35088
100. Vodova M et al (2022) Detection of pesticides in food products using paper-based devices by UV-induced fluorescence spectroscopy combined with molecularly imprinted polymers. Food Chem 380:132141. https://doi.org/10.1016/j.foodchem.2022.132141
101. Wisnuwardhani HA, Ibrahim S, Mukti RR, Damayanti S (2022) Molecularly-imprinted-SERS: a potential method for bioanalysis. Sci Pharm 90(3):54. https://doi.org/10.3390/scipharm90030054
102. Oliveira S et al (2025) Advances in optical devices for the detection of contaminants in food and water. TrAC Trends Anal Chem 184:118139. https://doi.org/10.1016/j.trac.2025.118139
103. Ali A et al (2023) Ensuring food safety with molecularly imprinted polymers: innovative methods for the detection of aflatoxins in food and feed samples. Front Sustain Food Syst 7. https://doi.org/10.3389/fsufs.2023.1210104
104. Gijs MAM, Lacharme F, Lehmann U (2010) Microfluidic applications of magnetic particles for biological analysis and catalysis. Chem Rev 110(3):1518–1563. https://doi.org/10.1021/cr9001929
105. Zhang Y, Nguyen N-T (2017) Magnetic digital microfluidics—a review. Lab Chip 17(6): 994–1008. https://doi.org/10.1039/C7LC00025A
106. Cai G et al (2023) Magnetic bead manipulation in microfluidic chips for biological application. Cyborg Bionic Syst 4. https://doi.org/10.34133/cbsystems.0023
107. Afonicheva PK, Esikova NA, Tupik AN, Evstrapov AA (2019) Magnetic beads-based nucleic acids extraction in microfluidic chip. J Phys Conf Ser 1400(3):033012. https://doi.org/10.1088/1742-6596/1400/3/033012
108. Liu F, Ni L, Zhe J (2018) Lab-on-a-chip electrical multiplexing techniques for cellular and molecular biomarker detection. Biomicrofluidics 12(2):021501. https://doi.org/10.1063/1.5022168
109. Fang X, Yang Y, Wang H, Xu H (2024) Bead-based microfluidic platforms for multiplex and ultrasensitive immunoassays in clinical diagnosis and treatment. Mechanobiol Med

2(2):100063. https://doi.org/10.1016/j.mbm.2024.100063

110. Zhao Y, Jiang S, Cai G, Wang L, Zhao J, Feng S (2025) A programmable magnetic digital microfluidic platform integrated with electrochemical detection system. Microsyst Nanoeng 11(1):82. https://doi.org/10.1038/s41378-025-00914-6
111. Hou F et al (2012) Development of a microplate reader compatible microfluidic chip for ELISA. Biomed Microdevices 14(4):729–737. https://doi.org/10.1007/s10544-012-9654-7
112. Yoshikawa H, Yoshinaga M, Tamiya E (2018) An optical pickup enzyme-linked immunosorbent assay (ELISA) with a microfluidic disk. RSC Adv 8(26):14510–14514. https://doi.org/10.1039/C8RA01149D
113. Tan X, Khaing Oo MK, Gong Y, Li Y, Zhu H, Fan X (2017) Glass capillary based microfluidic ELISA for rapid diagnostics. Analyst 142(13):2378–2385. https://doi.org/10.1039/C7AN00523G
114. Wu K et al (2022) Recent progress of microfluidic chips in immunoassay. Front Bioeng Biotechnol 10. https://doi.org/10.3389/fbioe.2022.1112327
115. Shen L, Zhang G, Etzold BJM (2020) Paper-based microfluidics for electrochemical applications. ChemElectroChem 7(1):10–30. https://doi.org/10.1002/celc.201901495
116. Fu L-M, Wang Y-N (2018) Detection methods and applications of microfluidic paper-based analytical devices. TrAC Trends Anal Chem 107:196–211. https://doi.org/10.1016/j.trac.2018.08.018
117. Chauhan A, Toley BJ (2021) Barrier-free microfluidic paper analytical devices for multiplex colorimetric detection of analytes. Anal Chem 93(25):8954–8961. https://doi.org/10.1021/acs.analchem.1c01477
118. Verma S, Toley BJ (2024) Saturation equation: an analytical expression for partial saturation during wicking flow in paper microfluidic channels. Langmuir 40(22):11419–11427. https://doi.org/10.1021/acs.langmuir.4c00245
119. Yuan M et al (2024) A portable multi-channel fluorescent paper-based microfluidic chip based on smartphone imaging for simultaneous detection of four heavy metals. Talanta 266:125112. https://doi.org/10.1016/j.talanta.2023.125112
120. Martinez AW, Phillips ST, Whitesides GM, Carrilho E (2010) Diagnostics for the developing world: microfluidic paper-based analytical devices. Anal Chem 82(1):3–10. https://doi.org/10.1021/ac9013989
121. Das S, Gagandeep, Bhatia R (2022) Paper-based microfluidic devices: Fabrication, detection, and significant applications in various fields. Rv Anal Chem 41(1):112–136. https://doi.org/10.1515/revac-2022-0037
122. Mutlu AY, Kılıç V, Özdemir GK, Bayram A, Horzum N, Solmaz ME (2017) Smartphone-based colorimetric detection via machine learning. Analyst 142(13):2434–2441. https://doi.org/10.1039/C7AN00741H
123. Ruiz RA, Gonzalez JL, Vazquez-Alvarado M, Martinez NW, Martinez AW (2022) Beyond wax printing: fabrication of paper-based microfluidic devices using a thermal transfer printer. Anal Chem 94(25):8833–8837. https://doi.org/10.1021/acs.analchem.2c01534
124. Novak L, Neuzil P, Pipper J, Zhang Y, Lee S (2007) An integrated fluorescence detection system for lab-on-a-chip applications. Lab Chip 7(1):27–29. https://doi.org/10.1039/B611745G
125. Sugino H, Ozaki K, Shirasaki Y, Arakawa T, Shoji S, Funatsu T (2009) On-chip microfluidic sorting with fluorescence spectrum detection and multiway separation. Lab Chip 9(9):1254. https://doi.org/10.1039/b815765k
126. Jiang D et al (2018) A novel screen-printed mast cell-based electrochemical sensor for detecting spoilage bacterial quorum signaling molecules (N-acyl-homoserine-lactones) in freshwater fish. Biosens Bioelectron 102:396–402. https://doi.org/10.1016/j.bios.2017.11.040
127. Écija-Arenas Á, Zafra-Poyato A, Fernández-Romero JM (2023) Integrated microfluidic systems for fluorescence monitoring rapid kinetic reactions in bioanalysis. Microchim Acta 190(6):209. https://doi.org/10.1007/s00604-023-05786-z
128. Chen F et al (2023) Multiplex detection of infectious diseases on microfluidic platforms. Biosensors (Basel) 13(3):410. https://doi.org/10.3390/bios13030410
129. Deng R et al (2024) CRISPR-based nucleic acid assays for food authentication. Trends Food Sci Technol 145:104351. https://doi.org/10.1016/j.tifs.2024.104351
130. Wang M et al (2022) Microfluidic paper-based analytical devices for the determination of food contaminants: developments and applications. J Agric Food Chem 70(27):8188–8206. https://doi.org/10.1021/acs.jafc.2c02366

131. Dursun AD, Borsa BA, Bayramoglu G, Arica MY, Ozalp VC (2022) Surface plasmon resonance aptasensor for Brucella detection in milk. Talanta 239:123074. https://doi.org/10.1016/j.talanta.2021.123074
132. Ashley J et al (2017) An SPR based sensor for allergens detection. Biosens Bioelectron 88: 109–113. https://doi.org/10.1016/j.bios.2016.07.101
133. Ashley J et al (2018) Development of a β-Lactoglobulin sensor based on SPR for Milk allergens detection. Biosensors (Basel) 8(2):32. https://doi.org/10.3390/bios8020032
134. Bimolata W et al (2013) Analysis of nucleotide diversity among alleles of the major bacterial blight resistance gene Xa27 in cultivars of rice (Oryza sativa) and its wild relatives. Planta 238(2):293–305. https://doi.org/10.1007/s00425-013-1891-3
135. Yu Z et al (2022) Smart traceability for food safety. Crit Rev Food Sci Nutr 62(4): 905–916. https://doi.org/10.1080/10408398.2020.1830262
136. Bartold K et al (2022) Molecularly imprinted polymer-based extended-gate field-effect transistor (EG-FET) chemosensor for selective determination of matrix metalloproteinase-1 (MMP-1) protein. Biosens Bioelectron 208:114203. https://doi.org/10.1016/j.bios.2022.114203

Chapter 6

NMR-Based Metabolomic Profiling of Functional Foods

Javeria Shabbir, Nizwa Itrat, Beenish Israr, Walid Elfalleh, and Muhammad Rizwan Majeed

Abstract

Nuclear Magnetic Resonance (NMR)-based metabolomic profiling is a powerful analytical technique for characterizing the biochemical composition of functional foods. This chapter explores the fundamental principles of NMR spectroscopy in metabolomics, focusing on its role in identifying metabolites, bioactive compounds, and food authentication by nutritional quality assessment. NMR metabolomics examines essential nutrients, including macro- and micronutrients, and secondary metabolites that influence metabolic pathways, highlighting their relevance in functional food science. The materials required for metabolomic profiling include sample selection protocols, deuterated solvents, buffer solutions, and spectrometer specifications, all of which are followed to ensure precision and accuracy in metabolite extraction and spectral handling, in accordance with GMP standards. Furthermore, the first step of NMR spectroscopy is sample preparation, solvent extraction protocols, and then advanced NMR applications such as 2D spectroscopy and High-Resolution Magic Angle Spinning (HRMAS) NMR. The selection of extraction or NMR methods depends totally on the desired outcomes. In addition, post-processing and statistical-computational strategies enhance the accuracy of metabolite classification. NMR-based metabolomic profiling acts as a transformative tool for functional food studies. It also contributes to personalized nutrition, ensures food safety, and helps to build therapeutic dietary applications.

Key words NMR spectroscopy, Metabolomics, Functional food, GMP, Transformative tool

1 Introduction

Functional foods have gained significant attention in modern nutrition science due to their ability to provide health benefits beyond basic nourishment [1]. Unlike conventional foods, functional foods contain secondary metabolites that act as bioactive compounds, which boost immunity, prevent disease by performing pathophysiological functions, and contribute to overall well-being. Such foods include polyphenols, amino acids, terpenoids, flavonoids, phenolic acids, omega-3 or 6 fatty acids, essential vitamins or minerals, and probiotics that play a crucial role in modulating physiological metabolism at the cellular level [2].

Tanmay Sarkar (ed.), *Novel Food Analysis: Techniques and Applications*, Methods and Protocols in Food Science,
https://doi.org/10.1007/978-1-0716-4787-5_6,

To understand the biochemical composition of functional foods, advanced analytical techniques are required that can accurately identify and quantify the essential nutrients and metabolites present in food [3]. Metabolomics is an emerging discipline in food science that allows researchers to make a profile of the molecular makeup of functional foods and assess their nutritional quality, authenticity by food quality, and potential for their pathophysiological functions that provide therapeutic benefits. The scientists can explore metabolic pathways and discover how food components interrelate with human biological systems through metabolomics [4].

Among various metabolomic techniques, Nuclear Magnetic Resonance (NMR) spectroscopy has proven to be a powerful tool for profiling complex mixtures of metabolites [5]. Unlike other analytical methods that require extensive sample preparation or separation procedures, NMR offers a nondestructive, highly reproducible, and robust approach to analyzing food samples with minimal interference. Its ability to simultaneously detect multiple metabolites makes it particularly suitable for studying the intricate biochemical composition of functional foods [6].

1.1 Metabolomics in Functional Food Research

Metabolomics is an emerging term that serves as a fundamental link between food science, nutrition, and health outcomes. By applying metabolomic approaches to functional foods, researchers can identify bioactive metabolites, trace metabolic shifts, and evaluate food quality, fortification, and processing effects. This holistic method enables scientists to gain deeper insights into how nutritional components contribute to human health and disease prevention [7].

Functional foods undergo numerous biochemical transformations during processing, storage, and digestion. These alterations can significantly impact their nutrient profile and bioactivity [8]. Metabolomic studies help quantify changes in metabolite composition, ensuring that functional foods retain their intended health-promoting attributes. Furthermore, metabolomics facilitates the authentication and standardization of food products, preventing mislabeling and ensuring consumer safety [9].

In clinical applications, metabolomic profiling of functional foods aids in understanding their impact on gut microbiota, metabolism, and disease management. For example, metabolomic analysis of fermented foods like yogurt and kombucha has revealed the presence of beneficial probiotic metabolites that improve gut health and immune function. Similarly, polyphenol-rich foods such as green tea and berries have been extensively studied for their antioxidant properties and potential neuroprotective effects [10].

1.2 Role of NMR Spectroscopy in Metabolomics

NMR spectroscopy has emerged as a preferred analytical tool for metabolomic studies due to its unique ability to characterize a wide range of metabolites efficiently. It operates on the principle of nuclear spin interactions, where atoms in a molecule resonate in response to a magnetic field, producing spectral data that provides valuable chemical information [11].

One of the key advantages of NMR spectroscopy is its nondestructive nature, allowing repeated analyses of the same sample without degradation. Additionally, NMR exhibits high specificity and accuracy, enabling the identification of complex metabolic profiles in functional foods without requiring extensive chromatographic separation. Its ability to detect metabolites in their natural state without chemical modification makes it superior in capturing a food's true biochemical fingerprint [12].

Applications of NMR spectroscopy in functional food metabolomics include quantitative analysis of bioactive compounds, structural elucidation of novel metabolites, and comparative profiling of food samples based on their metabolic signatures. The technique provides crucial insights into food composition, dietary effects, and metabolic interactions, thereby aiding in the development of nutritional guidelines, quality control standards, and therapeutic applications in personalized nutrition [13].

1.3 Nutrients Present in Functional Foods

Macronutrients, micronutrients, probiotics, prebiotics, and secondary metabolites are the basic nutrients that are present in functional foods (Table 1). Macronutrients are the fundamental building blocks of food that influence metabolism, energy production, and cellular functions. Their metabolomic profiling provides insight into their bioavailability, metabolic interactions, and nutritional impact. Carbohydrates in functional foods include simple sugars, complex polysaccharides, and fiber. Monosaccharides include glucose, fructose, and galactose, and they are found in fruits and honey, essential for quick energy metabolism. Disaccharides like sucrose, lactose, and maltose are present in dairy and cereals, influencing digestion rates. Polysaccharides such as starch, cellulose, and pectin are found in grains and vegetables, impacting gut microbiota and insulin regulation. Dietary fibers like β-glucan, inulin, and lignin act as prebiotics, nurturing beneficial microbial growth and improving gut health [14, 15].

Proteins in functional foods are analyzed for their amino acid composition, digestibility, and bioactive potential. Essential amino acids like lysine, valine, and leucine are required for muscle repair, immune function, and metabolic regulation. Functional peptides having antihypertensive and immunomodulatory properties are derived from dairy, soy, and pulses, which contribute to therapeutic effects. Enzymatic proteins such as bromelain and papain are present in fruits like pineapple and papaya, aiding digestion and anti-inflammatory responses [18, 19]. Lipids serve as structural

Table 1
Key nutrients in NMR-based metabolomic profiling of functional foods [2, 16, 17]

Nutrient category	Examples	Sources	Metabolic significance
Carbohydrates	Glucose, β-Glucans, Fructose	Whole Grains, Fruits, Legumes	Basic source of energy, and support gut microbiota
Proteins	Peptides, Essential Amino Acids	Soy, Dairy, Lentils,	Muscle repair, enhances enzymatic activity, boosts immunity
Lipids	Phospholipids, Omega-3	Fish, Nuts, Seeds	Cellular signaling, neuro and cardio protective
Vitamins	Vitamin B-complex, C, E	Leafy Greens, Citrus, Nuts and Seeds	Immunomodulation, neurological health and regulate metabolism
Minerals	Magnesium, Iron, Zinc	Legumes, Dairy, Seafood	Improves bone health, aids in oxygen transport and enzymatic reactions
Polyphenols & Flavonoids	Anthocyanins, Catechins, Quercetin	Fruits, Vegetables, Herbs, Spices	Antioxidant and anti-inflammatory potential
Probiotics & Prebiotics	Lactobacillus, Fructo-oligosaccharides, Beta-Glucans	Fermented Foods, Yogurt, Oats	Boosts immunity and regulates gut health
Terpenes & Alkaloids	Curcumin, Caffeine, Capsaicin	Chili Peppers, Turmeric, Coffee, Herbs	Anti-inflammatory, enhances metabolic functions

components, energy reserves, and signaling molecules, playing critical roles in functional food applications. Omega-3 and 6 fatty acids support cognitive, cardiovascular functions, and they are found in seafood (fish) and nuts (flaxseeds, walnuts, etc.). Medium-Chain Triglycerides (MCTs) aid in energy utilization, improve metabolism, and are enriched in coconut oil. Sterols and Phospholipids are core components of cell membranes; NMR analysis helps to assess these fats and their functional benefits in food products [20, 21].

Micronutrients play a crucial role in antioxidant defense, enzymatic reactions, and supporting immune function. NMR-based metabolomics precisely identifies them in food products and ensures their optimal nutrient profiling and standardization. Vitamin A (retinol, carotenoids) regulates the immune system, supports eye health, and its sources are leafy greens, carrots, liver, etc. Vitamin C (ascorbic acid) aids in collagen production, acts as a strong antioxidant, and is found in citrus and berries [22–24]. Magnesium acts as a key component in over 300 enzymatic reactions, supports bone health, improves muscle functioning, and can be sourced from leafy vegetables, nuts, and seeds. Iron plays a critical role in the oxygen transport system, acts as a core component in hemoglobin synthesis, and can be obtained from lentils, red meat, and iron-fortified foods. Zinc aids in cellular growth, which fastens the wound healing, boosts immunity, and can be found in seafood, nuts, seeds, pumpkin seeds, and dairy products [16, 25, 26].

Therapeutic efficacy of functional foods is enhanced by secondary metabolites that act as bioactive compounds; they perform powerful antioxidant activity against oxidative stress, which lowers the inflammatory effects and ultimately prevents disease. Polyphenols and flavonoids include quercetin, anthocyanins, or catechins that can be obtained from herbs, fruits, or spices, support the immune system, reduce oxidative stress, and have anti-cancerous and cardio-protective properties [27]. Terpenes and alkaloids include caffeine, curcumin, and capsaicinoids that can be sourced from coffee, herbs, chili peppers, or spices, support metabolism, enhance cognitive function, anti-inflammatory and anti-cancerous potential [17]. Beta-glucans and fructo-oligosaccharides (FOS) are fibers that are fermented by the gut microflora. These fermented fibers enhance immune responses, regulate cholesterol levels, and support the growth of beneficial bacteria like *Lactobacillus* and *Bifidobacterium* that help to regulate and modulate microbial balance and improve overall gut health [28, 29].

1.4 Functional Nutrients in Disease Prevention

Functional foods provide targeted therapeutic benefits, making their metabolomic profiling crucial for disease prevention strategies. Omega-3 fatty acids, sterols, and polyphenols regulate heart arrhythmia, which promotes cardiovascular health. Resistant starches, flavonoids, and soluble fibers from chia seeds or basil seeds help to improve insulin sensitivity, which acts as a big aid in diabetes management. Curcumin, anthocyanins, and phospholipids regulate neuronal integrity and enhance cognitive function, which directly regulates neuro health. Antioxidants, anthocyanidins, gingerol, alkaloids, and flavonoids promote phagocytosis of ROS species that reduces oxidative damage and also promotes apoptosis of malignant cells [17, 30].

1.5 Objectives

The main objective of this chapter is to investigate the metabolic profiles of functional foods using NMR spectroscopy, which helps to identify and characterize core nutrients and bioactive compounds/metabolites. By leveraging NMR-based metabolomic profiling, authors aim to establish a comprehensive molecular database that enhances nutritional analysis and food quality assessment.

2 Materials

2.1 Functional Food Samples

A diverse range of functional food samples is considered to be necessary for accurate profiling of metabolomics. These food samples should reflect a variety of food matrices, the composition of their nutrients, and bioactive compounds to ensure comprehensive analysis.

1. *Plant-Based Functional Foods*: Fruits like blueberries, grapes, strawberries, cherries, and vegetables like leafy greens and tomatoes are rich in bioactive compounds like polyphenols, alkaloids, and flavonoids. Whole grains like quinoa, sorghum, and legumes like lentils, beans, and chickpeas provide essential amino acids and fermentable fibers [31].
2. *Fermented Functional Foods*: Dairy products like yogurt, cheese, and kefir contain probiotic strains including *Bifidobacteria, Lactobacillus*, and *Saccharomyces*. Plant-based fermented food products including kimchi, sauerkraut, and kombucha are well known for their potential to maintain gut microflora balance [32].
3. *Beverages with Bioactive Compounds*: Green tea, black tea, and coffee are known for their high catechin and caffeine content. Functional herbal infusions with medicinal plant extracts (e.g., turmeric tea) of herbs or spices also have beneficial effects on gut health on consumption in moderation [33].
4. *Fortified or Enriched Foods*: Includes nutrient-enhanced or nutrient-fortified bakery products, dairy alternatives, and protein-enriched snacks [34].

Each sample must be standardized, ensuring consistency in sourcing, processing, and storage conditions before analysis.

2.2 Reagents and Solvents

2.2.1 Deuterated Solvents for NMR Analysis

Deuterated solvents are essential in NMR spectroscopy because they reduce interference from hydrogen (^1H) signals and provide clearer spectral resolution [35, 36]. Each solvent is chosen based on its compatibility with different metabolite classes:

1. *D2O (Deuterium Oxide):* Used for water-soluble metabolites, including sugars, amino acids, and certain organic acids. Also minimizes solvent peaks, ensures consistent sample preparation, and is often mixed with buffers to maintain physiological pH levels.
2. *CDCl3 (Deuterated Chloroform):* Preferred for lipid and non-polar metabolite extraction and is commonly used in analyzing sterols, triglycerides, and phospholipids. It also ensures minimal solubility interactions with aqueous metabolites.

3. *Methanol-d4 and Acetonitrile-d3:* Applied in organic solvent extraction for semipolar compounds and are suitable for flavonoids, polyphenols, and secondary plant metabolites. It also provides clean spectral backgrounds without overlapping peaks.

2.2.2 Buffer Solutions for Metabolite Stability

Maintaining stability during NMR analysis requires precise pH control and ionic strength adjustments [37].

1. *Phosphate-Buffered Saline (PBS)* has 7.4 pH that preserves biological metabolites, including proteins and peptides. Enhances the solubility of hydrophilic compounds in aqueous media.
2. *Acetate and Citrate Buffers* are utilized for weak acid/base metabolite stabilization and prevent degradation of organic acids and phenolic compounds.

2.2.3 Standard Reference Compounds for NMR Calibration

Internal standards improve quantification accuracy and spectral reproducibility [36, 38].

1. *Trimethylsilyl Propionate (TSP)* is used in ^1H NMR as a universal chemical shift reference at 0 ppm, and it does not interfere with most food metabolites.
2. *2,2-dimethyl-2-silapentane-5-sulfonate (DSS)* acts as a stable internal standard for aqueous metabolite profiling that ensures reproducibility across biological samples.

2.3 NMR Spectroscopy Equipment

2.3.1 Core NMR Instruments

Different NMR spectrometer frequencies influence resolution, sensitivity, and metabolite detection capability [39].

1. *NMR Spectrometers* have 400 MHz–600 MHz frequency, which is a standard resolution for food metabolomics, and it is effective for routine profiling of macronutrients and micronutrients.
2. *Ultra-High-Field NMR Spectrometers* had a frequency of >700 MHz, which enhanced resolution for complex metabolite identification. Useful for detailed structural elucidation of novel bioactive compounds.

2.3.2 Key Accessories for NMR Sample Handling

Optimized sample handling reduces spectral variability and enhances data accuracy [40].

1. *NMR Tubes* of 5 mm are considered standard size for liquid-state metabolomic analysis and act as precision-grade tubes that prevent sample contamination.
2. *Cryoprobes* help to improve sensitivity for low-abundance metabolite detection and minimize sample degradation over prolonged data acquisition.

3. *Temperature-controlled sample Chambers* help to regulate and maintain spectral consistency under variable conditions that ensure reproducibility for temperature-sensitive food metabolites.

2.3.3 Advanced NMR Techniques for Functional Food Analysis

Modern NMR setups provide multidimensional metabolomic analysis capabilities [9, 41].

1. *2D NMR Techniques* include COSY, TOCSY, and HSQC, enabling correlation-based structural analysis of food compounds.
2. *High-Resolution Magic Angle Spinning (HRMAS)* is used for solid-state metabolomics, including whole food matrix studies.

2.4 Sample Preparation Tools

2.4.1 Homogenization and Extraction Equipment

Proper sample extraction ensures a representative metabolic profile.

1. *Tissue Homogenizers are* used for grinding food samples into fine suspensions for NMR analysis, and prevent degradation and oxidation of bioactive compounds.
2. *Ultrasonic Processors aid in cell disruption that* facilitates the release of intracellular metabolites from plant and microbial sources. Also ensures high recovery rates of water-soluble and lipid-soluble compounds.

2.4.2 Filtration and Centrifugation for Purification

Removing impurities enhances spectral clarity and minimizes noise in metabolomic analysis.

1. *Centrifuges at 10,000–15,000 RPM could* separate bioactive compounds based on molecular weight and solubility. Used for protein precipitation and lipid fraction isolation.
2. *Microfiltration and syringe filters help to* remove particulates from liquid food extracts and prevent debris or any contaminants from interfering with NMR spectral resolution.

2.4.3 Precision Handling and Storage

Ensuring sample integrity during preparation and storage minimizes degradation.

1. *Cryogenic freezers have a* -80 *°C temperature and* preserve functional food metabolites over extended storage periods by preventing enzymatic breakdown of unstable bioactives.
2. *Pipettes are used for micro-volume precision* to get precise reagent and solvent handling in NMR tube preparation [38, 40, 42, 43].

2.5 Computational Tools for Data Processing

Advanced software tools are required for spectral analysis and metabolite identification. Spectral processing software such as MestReNova, TopSpin, and Chenomx helps to analyze and interpret NMR spectral data. Metabolite Databases like Human Metabolome Database (HMDB) and Biological Magnetic Resonance

Bank (BMRB) support metabolite matching. Multivariate statistical tools like SIMCA and MetaboAnalyst support principal component analysis (PCA) and clustering. These tools ensure accurate data interpretation and integration with metabolic pathways [44, 45].

3 Methods

This section outlines sample preparation techniques, NMR applications, data analysis strategies, and integrative approaches for comprehensive profiling (Fig. 1).

3.1 Sample Preparation Techniques

In metabolomic profiling, proper sample preparation and handling techniques following the Good Manufacturing Practices (GMP) standards ensure the accuracy and reproducibility of results. Food matrices contain diverse bioactive compounds or metabolites, so it is required to optimize a proper extraction method tailored to obtain different compound classes.

3.1.1 Homogenization and Extraction

1. Grinding can be done by cryogenic grinders, and homogenization by mechanical homogenizers ensures the sample uniformity and consistency. These processes help to preserve biochemical stability and prevent metabolic degradation.
2. Aqueous extraction is used to extract hydrophilic metabolites (sugars or amino acids), organic solvent extraction is preferred to extract lipophilic metabolites (fatty acids or sterols), while methanol or ethanol extraction ensures the optimization of flavonoid and polyphenol retrieval.

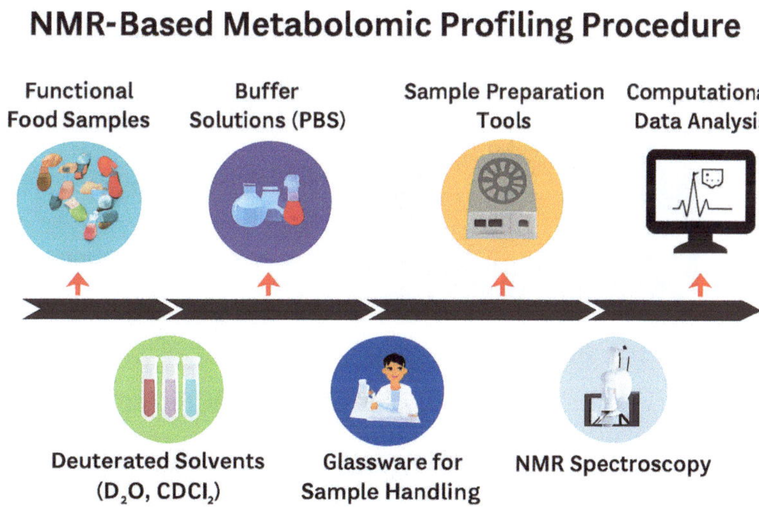

Fig. 1 Materials and Methods for NMR-based metabolomic profiling

3.1.2 Filtration and Centrifugation

1. Membrane Filtration removes particulate matter, preventing spectral noise in NMR.
2. High-speed centrifugation separates compounds based on density, isolating metabolite-rich fractions.
3. Freeze-drying, also known as Lyophilization, ensures long-term stability of sensitive metabolites before analysis. Reduces moisture interference in spectral resolution [40, 42].

3.2 NMR Spectroscopy Applications

NMR spectroscopy enables qualitative and quantitative metabolite profiling in functional foods. Different analytical approaches enhance precision in detecting metabolic variations (Fig. 2).

Qualitative Analysis Referred as metabolic fingerprinting and identifies major metabolites contributing to bioactivity. Also useful for comparing food sources, detecting adulteration, and authenticity testing.

Quantitative Analysis Provides absolute metabolite concentration and measures concentration levels of functional metabolites such as polyphenols, alkaloids, and peptides. Also enhances food quality assessment and nutritional standardization.

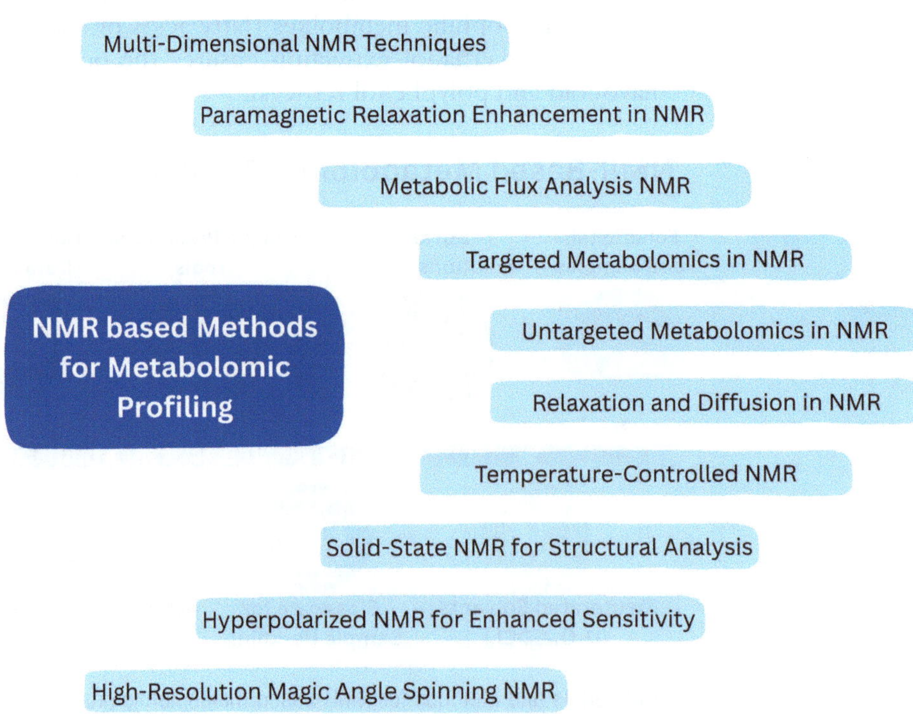

Fig. 2 NMR-based methods that are used for targeted metabolomic profiling

Multi-dimensional NMR Spectroscopy 2D Correlation Spectroscopy (COSY), Total Correlation Spectroscopy (TOCSY), and Heteronuclear Single Quantum Coherence (HSQC) provide structural confirmation and metabolite interactions. HRMAS NMR is used for semisolid or intact food matrix analysis. Diffusion-Ordered Spectroscopy (DOSY) helps in differentiating metabolite size and molecular dynamics [46–48].

High-Resolution Magic Angle Spinning (HRMAS) NMR HRMAS NMR provides a breakthrough in the analysis of semisolid food matrices, allowing direct metabolomic profiling of intact food samples without requiring complex extraction procedures. Unlike liquid-state NMR, which necessitates metabolite isolation, HRMAS preserves the natural biochemical state of foods such as grains, seeds, plant tissues, and dairy products. This method improves structural elucidation of lipophilic compounds, fiber-associated metabolites, and protein-bound bioactives, making it ideal for studying nutrient bioavailability in fortified foods, organic produce, and protein-enriched functional products [49, 50].

Targeted vs. Untargeted Metabolomics in NMR Metabolomic profiling can be divided into targeted and untargeted approaches, depending on research objectives [51, 52].

1. Targeted Metabolomics *focuses on specific metabolites with predefined biological significance, such as omega-3 fatty acids, polyphenols, or antioxidants. This approach provides precise quantification using reference standards.*
2. Untargeted Metabolomics *seeks to identify all detectable metabolites in a sample, uncovering novel bioactive compounds and biochemical interactions. This method is crucial for food authentication, quality control, and the discovery of unknown nutritional biomarkers.*

Both strategies leverage NMR spectral databases and multivariate statistical tools to classify metabolites effectively, enhancing food composition analysis.

Temperature-Controlled NMR for Food Stability Studies - Temperature fluctuations influence the stability and metabolic composition of functional foods, requiring Variable Temperature (VT) NMR to assess heat-induced changes. This method allows researchers to analyze oxidative degradation, microbial growth, enzymatic activity, and nutrient preservation under different storage conditions. For example, VT NMR helps evaluate polyphenol stability in herbal teas exposed to prolonged heating, ensuring retention of antioxidant properties. By integrating temperature-controlled NMR, food scientists can develop shelf-life optimization strategies, reducing product wastage while maintaining nutritional potency [47, 53].

Solid-State NMR (SSNMR) for Structural Analysis Solid-State NMR (SSNMR) is essential for analyzing semi-solid and crystalline food matrices, such as whole grains, seeds, and fiber-rich foods, which are difficult to dissolve in conventional solvents. Unlike liquid-state NMR, SSNMR can directly examine food texture attributes, protein folding mechanisms, and molecular interactions within polysaccharide networks. This method is widely used in studying fortified bakery products, encapsulated nutrients, and plant-based proteins, where ingredient interactions influence structural integrity and nutritional value. SSNMR also helps characterize starch modifications in functional flours, ensuring optimal digestibility and glycemic responses in specialized diets [54, 55].

3.3 Advanced Data Analysis Methods

After acquiring NMR spectral data, several computational techniques refine and interpret the results. Preprocessing methods, such as baseline correction, peak alignment, and normalization, ensure spectral clarity and consistency. Once optimized, statistical tools like Principal Component Analysis (PCA) and Partial Least Squares Discriminant Analysis (PLS-DA) classify metabolites, distinguishing between different food samples based on their biochemical profiles. Machine learning enhances metabolite identification by automating classification and biomarker discovery, making it particularly useful for food authentication studies. These artificial intelligence (AI)-driven models improve precision, allowing researchers to detect bioactive compounds efficiently. By integrating NMR spectroscopy with AI and multivariate statistical analysis, researchers can conduct large-scale metabolomic profiling with high accuracy, leading to deeper insights into food composition, nutritional quality, and health applications [56–58].

4 Notes

1. Foods that contain bioactive compounds beyond basic nutrients and provide additional health benefits are known as functional foods.
2. To understand the whole biochemical composition of such foods is termed as metabolomics.
3. NMR spectroscopy has a highly reproducible nature and ability to analyze several bioactive compounds or metabolites simultaneously without destroying them, yet a powerful technique in metabolomics.

4. Macronutrients are the major nutrients that aid in energy and body buildup, while micronutrients are minor nutrients that are essential for the body to perform physiological functions. On the other hand, bioactive compounds are those that aid in disease prevention or treatment beyond the basic nutrition and are found in functional foods.
5. Fruits, vegetables, prebiotics, probiotics, herbs, spices, and fermented foods are all examples of functional foods.
6. Multidimensional NMR techniques help to resolve the overlapping of spectral peaks that clarify the identification of bioactive compounds or metabolites.
7. Targeted metabolomics only focuses on known bioactive compounds like polyphenols, terpenoids, etc., while untargeted metabolomics has the ability to explore the unknown compounds present in food.
8. MFA combined with the isotope tracing technique allows the identification of the actual track of nutrient or bioactive compound bioavailability in the body, widely applied in research related to precision nutrition and gut microbiome.

References

1. Banwo K, Olojede AO, Adesulu-Dahunsi AT, Verma DK, Thakur M, Tripathy S et al (2021) Functional importance of bioactive compounds of foods with Potential Health Benefits: a review on recent trends. Food Biosci 43: 101320
2. Abdul Hakim BN, Xuan NJ, Oslan SNH (2023) A comprehensive review of bioactive compounds from lactic acid bacteria: potential functions as functional food in dietetics and the food industry. Foods 12(15):2850
3. Martirosyan D, Lampert T, Lee M (2022) A comprehensive review on the role of food bioactive compounds in functional food science. Funct Food Sci 2(3):64–78
4. Cevallos-Cevallos JM, Reyes-De-Corcuera JI (2012) Metabolomics in food science. Adv Food Nutr Res 67:1–24
5. Chowdhury CR, Kavitake D, Jaiswal KK, Jaiswal KS, Reddy GB, Agarwal V et al (2023) NMR-based metabolomics as a significant tool for human nutritional research and health applications. Food Biosci 53:102538
6. Fernández-Ochoa Á, Leyva-Jiménez FJ, De la Luz Cádiz-Gurrea M, Pimentel-Moral S, Segura-Carretero A (2021) The role of high-resolution analytical techniques in the development of functional foods. Int J Mol Sci 22(6): 3220
7. Fraga-Corral M, Carpena M, Garcia-Oliveira P, Pereira AG, Prieto M, Simal-Gandara J (2022) Analytical metabolomics and applications in health, environmental and food science. Crit Rev Anal Chem 52(4):712–734
8. Gupta E, Shyam H, Devi S, Maurya NK (2020) Nutrigenomics in functional foods and personalized nutrition. Gedrag Organisatie Учредители: Uitgeverij Lemma Bv 33(2): 670–686
9. Li S, Tian Y, Jiang P, Lin Y, Liu X, Yang H (2021) Recent advances in the application of metabolomics for food safety control and food quality analyses. Crit Rev Food Sci Nutr 61(9): 1448–1469
10. Mondal S, Soumya NPP, Mini S, Sivan SK (2021) Bioactive compounds in functional food and their role as therapeutics. Bioact Compd Health Dis 4(3):24–39
11. Valentino G, Graziani V, D'Abrosca B, Pacifico S, Fiorentino A, Scognamiglio M (2020) NMR-based plant metabolomics in nutraceutical research: an overview. Molecules 25(6):1444
12. Moco S (2022) Studying metabolism by NMR-based metabolomics. Front Mol Biosci 9:882487
13. Cao R, Liu X, Liu Y, Zhai X, Cao T, Wang A et al (2021) Applications of nuclear magnetic

resonance spectroscopy to the evaluation of complex food constituents. Food Chem 342: 128258

14. Kiely LJ, Hickey RM (2021) Characterization and analysis of food-sourced carbohydrates. In: Glycosylation: Methods protocols, pp 67–95
15. Neri-Numa IA, Arruda HS, Geraldi MV, Júnior MRM, Pastore GM (2020) Natural prebiotic carbohydrates, carotenoids and flavonoids as ingredients in food systems. Curr Opin Food Sci 33:98–107
16. Akram M, Munir N, Daniyal M, Egbuna C, Găman M-A, Onyekere PF et al (2020) Vitamins and Minerals: types, sources and their functions. In: Functional foods nutraceuticals: bioactive components, formulations innovations, pp 149–172
17. Essa MM, Bishir M, Bhat A, Chidambaram SB, Al-Balushi B, Hamdan H et al (2023) Functional foods and their impact on health. J Food Sci Technol:1–15
18. Grossmann L, Weiss J (2021) Alternative protein sources as technofunctional food ingredients. Ann Rev Food Sci Technol 12(1):93–117
19. Tan M, Nawaz MA, Buckow R (2023) Functional and food application of plant proteins–a review. Food Rev Intl 39(5):2428–2456
20. Katiyar R, Arora A (2020) Health promoting functional lipids from microalgae pool: a review. Algal Res 46:101800
21. Tietel Z, Hammann S, Meckelmann SW, Ziv C, Pauling JK, Wölk M et al (2023) An overview of food lipids toward food lipidomics. Compr Rev Food Sci Food Safe 22(6):4302–4354
22. Handu S, Prakash B, Jan S, Habib M, Jan K (2024) Vitamins as functional foods and nutraceuticals. In: Functional foods and nutraceuticals: chemistry, health benefits and the way. Springer, Forward, pp 113–144
23. Mitra S, Paul S, Roy S, Sutradhar H, Bin Emran T, Nainu F et al (2022) Exploring the immune-boosting functions of vitamins and minerals as nutritional food bioactive compounds: a comprehensive review. Molecules 27(2):555
24. Xiao S, Li J, editors (2020) Study on functional components of functional food based on food vitamins. In: Journal of Physics: Conference Series. IOP Publishing
25. Awuchi CG, Igwe VS, Amagwula I, Echeta CK (2020) Health benefits of micronutrients (vitamins and minerals) and their associated deficiency diseases: a systematic review. Int J Food Sci 3(1):1–32
26. Weyh C, Krüger K, Peeling P, Castell L (2022) The role of minerals in the optimal functioning of the immune system. Nutrients 14(3):644
27. Maqsood S, Adiamo O, Ahmad M, Mudgil P (2020) Bioactive compounds from date fruit and seed as potential nutraceutical and functional food ingredients. Food Chem 308: 125522
28. Chugh B, Kamal-Eldin A (2020) Bioactive compounds produced by probiotics in food products. Curr Opin Food Sci 32:76–82
29. Ranjha MMAN, Shafique B, Batool M, Kowalczewski PŁ, Shehzad Q, Usman M et al (2021) Nutritional and health potential of probiotics: a review. Appl Sci 11(23):11204
30. Simó C, García-Cañas V (2020) Dietary bioactive ingredients to modulate the gut microbiota-derived metabolite TMAO. New opportunities for functional food development. Food. Science 11(8): 6745–6776
31. Marchev AS, Vasileva LV, Amirova KM, Savova MS, Balcheva-Sivenova ZP, Georgiev MI (2021) Metabolomics and health: from nutritional crops and plant-based pharmaceuticals to profiling of human biofluids. Cell Mol Life Sci 78(19–20):6487–6503
32. Brigante F, Solovyev P, Bontempo L (2024) Nuclear magnetic resonance applications in fermented foods and plant-based beverages: challenges and opportunities. Food Rev Intl 40(10):3370–3397
33. Wang K, Xu Z, Liao X (2022) Bioactive compounds, health benefits and functional food products of sea buckthorn: a review. Crit Rev Food Sci Nutr 62(24):6761–6782
34. Vishwakarma S, Dalbhagat CG, Mandliya S, Mishra HN (2022) Investigation of natural food fortificants for improving various properties of fortified foods: a review. Food Res Int 156:111186
35. Reif B (2021) Deuteration for high-resolution detection of protons in protein magic angle spinning (MAS) solid-state NMR. Chem Rev 122(10):10019–10035
36. Silva RA, Pereira TC, Souza AR, Ribeiro PR (2020) 1H NMR-based metabolite profiling for biomarker identification. Clin Chim Acta 502:269–279
37. Brudar S, Hribar-Lee B (2021) Effect of buffer on protein stability in aqueous solutions: a simple protein aggregation model. J Phys Chem B 125(10):2504–2512
38. Richards SA, Hollerton JC (2023) Essential practical NMR for organic chemistry. John Wiley & Sons
39. Sobolev AP, Ingallina C, Spano M, Di Matteo G, Mannina L (2022) NMR-based approaches in the study of foods. Molecules 27(22):7906

40. Ben-Tal Y, Boaler PJ, Dale HJ, Dooley RE, Fohn NA, Gao Y et al (2022) Mechanistic analysis by NMR spectroscopy: a users guide. Prog Nucl Magn Reson Spectrosc 129:28–106
41. Zhang J, Sun M, Elmaidomy AH, Youssif KA, Zaki AM, Kamal HH et al (2023) Emerging trends and applications of metabolomics in food science and nutrition. Funct Food Sci 14(20):9050–9082
42. Hurkul MM, Cetinkaya A, Yayla S, OZKAN SA. (2024) Advanced sample preparation and chromatographic techniques for analyzing plant-based bioactive chemicals in nutraceuticals. J Chromatogr Open 5:100131
43. Salem MA, Perez de Souza L, Serag A, Fernie AR, Farag MA, Ezzat SM et al (2020) Metabolomics in the context of plant natural products research: from sample preparation to metabolite analysis. Metabolites. 10(1):37
44. Kirwan JA, Gika H, Beger RD, Bearden D, Dunn WB, Goodacre R et al (2022) Quality assurance and quality control reporting in untargeted metabolic phenotyping: mQACC recommendations for analytical quality management. Metabolomics 18(9):70
45. Kim HC, Yim D-G, Kim JW, Lee D, Jo C (2021) Nuclear magnetic resonance (NMR)-based quantification on flavor-active and bioactive compounds and application for distinguishment of chicken breeds. Food Sci Anim Resour 41(2):312
46. Adeeyo AO, Amao IO, Ngandjui YA, Alabi MA, Msagati TA (2024) Metabolic profiling of plant and other natural products: platforms, applications and quality evaluation. Microchem J 111779:111779
47. Günther H (2013) NMR spectroscopy: basic principles, concepts and applications in chemistry. John Wiley & Sons
48. Nagana Gowda G, Raftery D (2021) NMR-based metabolomics. In: Cancer metabolomics: methods applications, pp 19–37
49. Singh N, Joshi R (2024) Metabolomics applications in food science, technology and nutrition. Oxford University Press, pp 4166–4168
50. Tampieri A, Szabó M, Medina F, Gulyás H (2021) A brief introduction to the basics of NMR spectroscopy and selected examples of its applications to materials characterization. Phys Sci Rev 6(1):20190086
51. Garcia-Perez I, Posma JM, Serrano-Contreras JI, Boulangé CL, Chan Q, Frost G et al (2020) Identifying unknown metabolites using NMR-based metabolic profiling techniques. Nat Protoc:1–30
52. Anand A, Sharma A, Kaur Saini H, Sharma S, Sharma R, Thakur C et al (2022) Profiling of plant derived natural constituents by using magnetic resonance techniques. Concepts Magn Reson Part A 2022(1):5705637
53. Galvan D, Tanamati AAC, Casanova F, Danieli E, Bona E, Killner MHM (2021) Compact low-field NMR spectroscopy and chemometrics applied to the analysis of edible oils. Food Chem 365:130476
54. Demarque DP, Dusi RG, de Sousa FD, Grossi SM, Silvério MR, Lopes NP et al (2020) Mass spectrometry-based metabolomics approach in the isolation of bioactive natural products. Sci Rep 10(1):1051
55. Macomber RS (1997) A complete introduction to modern NMR spectroscopy. John Wiley & Sons
56. Schorn C, Taylor BF (2004) NMR spectroscopy: data acquisition. John Wiley & Sons
57. Gupta P, Verma A, Rai N, Singh AK, Singh SK, Kumar B et al (2021) Mass spectrometry-based technology and workflows for studying the chemistry of fungal endophyte derived bioactive compounds. ACS Chem Biol 16(11): 2068–2086
58. Solovyev PA, Fauhl-Hassek C, Riedl J, Esslinger S, Bontempo L, Camin F (2021) NMR spectroscopy in wine authentication: an official control perspective. Compr Rev Food Sci food Safe 20(2):2040–2062

Chapter 7

Biosensor-Based Methods for Mycotoxin and Allergen Detection: Advances, Applications, and Future Prospects

Robin Kaura, Harmeet Singh Dhillon, Nishchal Thakur, Mohit Singla, Rajpreet Kaur Goraya, and Chitranayak Sinha

Abstract

Ensuring food safety amid the growing prevalence of naturally occurring contaminants such as mycotoxins and food allergens remains a critical global imperative. These xenobiotic agents, originating from fungal metabolism and allergenic proteins, respectively, pose substantial threats to public health, necessitating the development of rapid, reliable, and field-deployable detection technologies. This chapter delineates the recent paradigm shift toward biosensor-based diagnostics, offering a nuanced synthesis of their operational principles, biomolecular recognition elements, and physicochemical transduction mechanisms. Emphasis is placed on the integration of nanostructured materials, molecular imprinting, and aptameric systems to augment sensor sensitivity, selectivity, and operational stability. Moreover, this chapter explores advanced modalities including electrochemical impedance spectroscopy, surface plasmon resonance (SPR), and gravimetric assays, highlighting their utility in detecting trace-level contaminants across complex food matrices. The translational potential of these systems is further enhanced through their convergence with Internet of Things (IoT) frameworks and artificial intelligence (AI) for predictive analytics and real-time supply chain surveillance. Finally, regulatory harmonization, standardization of reference materials, and integration into HACCP-based critical control strategies are discussed as essential enablers for the widespread commercial adoption of biosensor technologies. Collectively, these advancements underscore the pivotal role of biosensors in fortifying global food safety systems through intelligent, multiplexed, and decentralized analytical platforms.

Key words Biosensors, Electrochemical and Optical transduction, Nanostructure, Smart Food Diagnostics, Surface Plasmon Resonance (SPR)

1 Introduction

Food safety remains a critical global challenge due to the persistent risk posed by a variety of biological and chemical contaminants. Among these, mycotoxins and food allergens represent two major classes of naturally occurring hazards that can compromise food quality, safety, and public health. Mycotoxins, toxic secondary metabolites produced primarily by filamentous fungi, such as

Tanmay Sarkar (ed.), *Novel Food Analysis: Techniques and Applications*, Methods and Protocols in Food Science, https://doi.org/10.1007/978-1-0716-4787-5_7,

Aspergillus, *Fusarium*, and *Penicillium*, are known to contaminate a broad spectrum of dietary commodities, including cereals, grains, nuts, dried fruits, spices, and beverages [1]. Their production is influenced by environmental factors, such as temperature, humidity, and water activity, with contamination possible across the entire food production chain, from preharvest to storage and processing. Despite regulatory efforts, mycotoxins like aflatoxins (AFs), ochratoxins (OTs), fumonisins (FUM), zearalenone (ZEA), and trichothecenes (THs) persist as formidable public health hazards, owing to their nephrotoxic, immunosuppressive, carcinogenic, and endocrine-disrupting properties [2].

Conversely, food allergens, typically proteins or glycoproteins found in foods, such as milk, eggs, nuts, wheat, soy, and shellfish, can provoke immunoglobulin E (IgE)-mediated responses in sensitized individuals [3]. Even trace-level contamination can lead to severe reactions, including anaphylaxis, thereby necessitating strict monitoring and regulation. The growing incidence of food allergies has prompted more stringent regulatory frameworks, increasing the demand for accurate, sensitive, and rapid detection methods.

Traditional detection techniques, including enzyme-linked immunosorbent assay (ELISA), thin-layer chromatography (TLC), high-performance liquid chromatography (HPLC), gas chromatography (GC), and various mass spectrometry (MS) techniques, have played an instrumental role in ensuring food safety compliance [4, 5]. However, these methods are often time-consuming, costly, and reliant on specialized expertise, limiting their application in real-time or field-based environments.

To overcome these limitations, recent years have witnessed significant advances in biosensor technologies, synergistically coupling biological recognition elements with transducers to translate specific biomolecular interactions into quantifiable analytical signals [6]. These platforms, especially when enhanced by nanotechnology, microfluidics, and machine learning (ML) algorithms, offer unprecedented sensitivity, specificity, and portability. Notably, multiplex biosensors now allow for the simultaneous detection of multiple mycotoxins and allergens, representing a major leap forward in comprehensive food safety assessment.

Crucially, the reliability of any detection system depends not only on the analytical platform but also on the quality of extraction procedures and the availability of certified reference materials (CRMs). These components are essential for ensuring reproducibility, accuracy, and cross-platform comparability, particularly when testing across complex food matrices.

This chapter provides a comprehensive review of the occurrence of mycotoxins and food allergens, the recent innovations in their detection, and the pivotal role of reference materials and validation protocols. Special focus is placed on biosensor-based technologies, exploring their mechanisms, advantages, current limitations, and potential for integration into smart food safety systems. In light of climate variability, evolving trade dynamics, and stringent food regulations, these technological advancements offer a promising avenue to enhance food safety, protect public health, and fortify the resilience of global food systems.

2 Materials

2.1 Common Mycotoxins in Food

Mycotoxins are toxic secondary metabolites produced by filamentous fungi, primarily from the genera *Aspergillus*, *Fusarium*, and *Penicillium* (Table 1). These fungi can contaminate a wide range of agricultural commodities, including cereals, nuts, oilseeds, dried fruits, spices, and milk. Mycotoxin contamination can occur at any stage of the food production chain, thereby constituting a major threat to both human and animal health [7]. Aflatoxins, produced by *Aspergillus flavus* and *A. parasiticus*, are commonly found in maize, groundnuts, cottonseed, spices, and tree nuts. Furthermore, aflatoxin M1, a hydroxylated metabolite of aflatoxin B1 (AFB1), can be released into the milk of animals that have ingested contaminated feedstuffs [8]. AFB1 is a potent carcinogen, classified as a Group 1 human carcinogen by the International Agency for Research on Cancer (IARC). It causes hepatocellular carcinoma, particularly in individuals with hepatitis B virus infection [9]. To mitigate health risks, regulatory bodies have established stringent maximum residue limits (Table 2); the European Union Regulation (EC) No. 1881/2006 stipulates permissible levels of 2 μg/kg for AFB1 and 4 μg/kg for total aflatoxins in cereals intended for direct human consumption [10]. In the United

Table 1
Common mycotoxins in food

Mycotoxin	Fungal source	Primary food sources	Health effects
Aflatoxins	*Aspergillus flavus*, *A. parasiticus*	Maize, peanuts, tree nuts, milk (AFM1)	Hepatotoxicity, carcinogenicity (AFB1—Group 1 carcinogen)
Ochratoxin A	*Aspergillus*, *Penicillium*	Cereals, coffee, dried fruits, wine	Nephrotoxicity, immunotoxicity, possible human carcinogen (Group 2B)
Fumonisins	*Fusarium verticillioides*	Maize and maize products	Neurotoxicity, esophageal cancer, neural tube defects
Deoxynivalenol (DON)	*Fusarium graminearum*	Wheat, maize, barley	Emesis, growth retardation, immune modulation
Zearalenone	*Fusarium* spp.	Maize, wheat, barley	Estrogenic effects, reproductive toxicity
T-2/HT-2 toxins	*Fusarium sporotrichioides*	Cereal grains	Immunosuppression, gastrointestinal damage, hematotoxicity

Table 2
Regulatory thresholds of mycotoxins in foods

Mycotoxin	EU limit (μg/kg)	FDA limit (μg/kg)
Aflatoxin B1	2 (ready-to-eat), 20 (feeds)	20 (total aflatoxins for human food)
Aflatoxin M1	0.05 (milk)	0.5 (milk)
Ochratoxin A	3 (cereals), 5 (dried fruits)	No specific FDA limit
DON	750 (bread), 1250 (raw wheat)	1000 (finished wheat products)
Fumonisins	2000–4000 (maize products)	2000–4000 (maize-based products)
Zearalenone	100 (cereals), 350 (maize)	No specific FDA limit

States, the Food and Drug Administration (FDA) sets an action level of 20 μg/kg for most food products and 0.5 μg/kg for aflatoxin M1 in milk [11].

Ochratoxins, produced by *Aspergillus ochraceus* and *Penicillium verrucosum*, are often found in stored cereals, coffee, wine, dried fruits, and meat products [12]. Ochratoxin A (OTA) is nephrotoxic, immunosuppressive, and teratogenic, and is classified as a possible human carcinogen (Group 2B, IARC) [9]. In the European Union, permissible levels of OTA range from 3 μg/kg in cereals to 10 μg/kg in dried vine fruits [10]. The FDA monitors OTA under import alerts and surveillance programs [11].

FUM, produced by *Fusarium verticillioides* and *F. proliferatum*, are primarily associated with maize [13]. Fumonisin B1 (FB1) disrupts sphingolipid metabolism, leading to leukoencephalomalacia in equines, pulmonary edema in swine, and esophageal cancer in humans. The European Union has established a cumulative maximum residue limit of 1000 μg/kg for FB1 and fumonisin B2 (FB2) in maize-based products designated for direct human consumption, as part of its regulatory framework to safeguard public health [10]. The FDA provides guidance levels ranging from 2000 to 4000 μg/kg in human food [11].

Deoxynivalenol (DON), produced by *Fusarium* species, is prevalent in temperate cereals such as wheat, barley, and oats [14]. DON inhibits protein synthesis, causing gastrointestinal distress, immunotoxicity, and emetic responses. In the European Union, regulatory limits for DON are set at 750 μg/kg in flour and 200 μg/kg in cereal-based infant foods [10]. The FDA sets an advisory level of 1000 μg/kg in finished wheat products [11].

ZEA, produced by *Fusarium* species, is commonly found in maize and related products [15]. ZEA exhibits estrogenic activity, leading to reproductive disorders in swine and hyperestrogenism in other livestock. The European Union sets limits of 100 μg/kg in refined maize oil and 20 μg/kg in baby food [10]. The FDA monitors ZEA under import alerts and surveillance programs [11].

2.2 Common Food Allergens in Food

Food allergens are proteins that induce hypersensitive immune responses in genetically predisposed individuals (Table 3). The major food allergens are responsible for over 90% of food-induced allergic reactions globally. Milk is one of the most common allergens, particularly in children, with casein and whey proteins (-α-lactalbumin and β-lactoglobulin) being the primary immunogenic components. Although some children outgrow this allergy, it can be severe and persistent in others. Peanuts are a leading cause of fatal anaphylaxis, with major allergenic proteins including Ara h1, Ara h2, and Ara h3. Peanut allergy is typically lifelong and highly sensitive. Tree nuts, such as almonds, walnuts, cashews, and pecans, are also common allergens. Cross-reactivity among different tree nuts is common, but not universal, and the risk of severe systemic reactions warrants strict avoidance [16].

Wheat contains IgE-binding proteins like ω-5 gliadin, which provoke allergic responses, particularly in exercise-induced anaphylaxis and baker's asthma. Soy contains Gly m proteins (Gly m5, Gly m6), which are known allergens. Although soy allergy is often milder than peanut allergy, it can be persistent due to the ubiquity of soy in processed foods [16].

Eggs are another common allergen, with ovalbumin and ovomucoid in egg white being potent allergens, especially in pediatric populations. Although heat treatment reduces allergenicity, some proteins remain stable, complicating dietary management. Fish, particularly species like salmon or tuna, contain parvalbumins, which are heat-stable allergens. Reactions to fish often persist throughout life and can be severe even after consuming minute quantities. Shellfish, including shrimp, lobster, and crab, contain the major allergen tropomyosin, which is highly resistant to heat and digestive enzymes, rendering shellfish allergy notably severe [16].

Table 3
Common allergens in food

Allergen source	Major allergenic proteins	Heat stability	Prevalence (approx.)	Risk of anaphylaxis
Milk	Casein, β-lactoglobulin	Medium	High in children	Moderate
Peanut	Ara h 1, 2, 3	High	Moderate to high	High
Tree nuts	Jug r 1, Ana o 1	High	Moderate	High
Wheat	ω-5 gliadin, HMW glutenin	Medium	Moderate	Low to moderate
Soy	Gly m 5, Gly m 6	Medium	Moderate in children	Low
Egg	Ovomucoid, Ovalbumin	Variable	High in children	Moderate
Fish	Parvalbumins	High	Low to moderate	High
Shellfish	Tropomyosin	High	Moderate	High

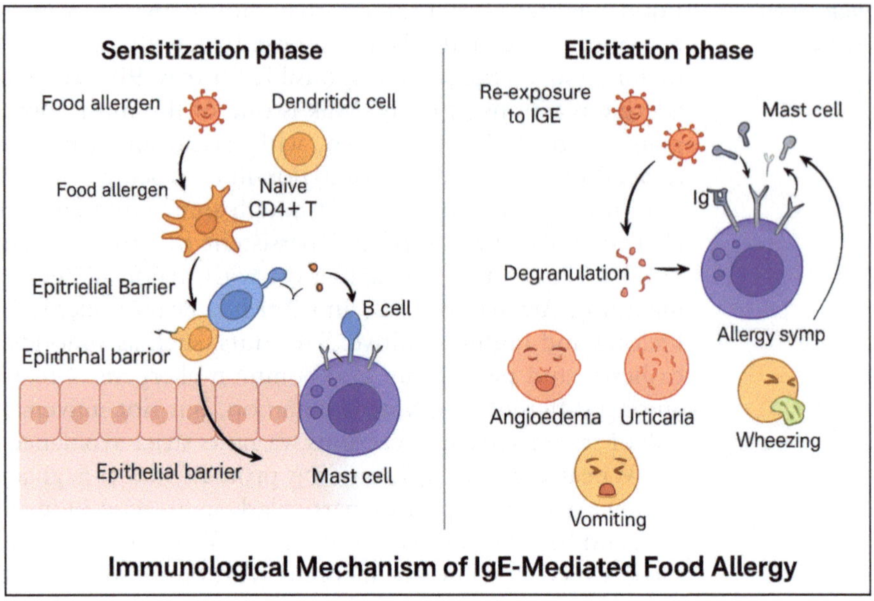

Fig. 1 Immunological mechanisms of IgE-mediated food allergy

Food allergies are primarily Type I hypersensitivity reactions (Fig. 1) mediated by IgE. The immune system responds to allergens by releasing specific antibodies during sensitization. Upon subsequent exposure, IgE antibodies trigger the release of histamine and other chemicals, leading to allergic symptoms ranging from mild (e.g., hives and itching) to severe (e.g., anaphylaxis) [16]. Effective labelling and detection are crucial for managing food allergens. Regulatory bodies mandate the declaration of major allergens on food packaging. For example, the EU Regulation No. 1169/2011 requires clear labelling of 14 allergens, including milk, peanuts, tree nuts, soy, wheat, eggs, fish, and shellfish [17]. The FDA's Food Allergen Labeling and Consumer Protection Act (FALCPA) requires explicit labelling of eight major allergens [11].

2.3 Fundamentals of Biosensor Technology

Biosensors are advanced analytical tools designed to detect biological or chemical interactions by transducing these responses into quantifiable signals that correlate with the concentration of specific analytes or mycotoxins [18]. Structurally, biosensors are composed of two fundamental components: a biorecognition element (bioreceptor) and a signal transduction unit (transducer). The bioreceptor serves as the selective recognition interface, typically involving biological entities such as enzymes, nucleic acids, aptamers (synthetic oligonucleotides), whole cells, or antibodies, which interact specifically with target mycotoxins [6, 19]. These elements have increasingly been integrated into detection techniques, particularly immunochemical and electrochemical methods to enhance analytical sensitivity, specificity, and response time [20].

Biosensors are classified based on the underlying signal transduction mechanism employed, which includes electrochemical, optical, and thermal modalities. These primary categories can be further subdivided into specialized types such as amperometric, piezoelectric, chromogenic, and optical fiber-based sensors [18, 21, 22]. In addition to the core sensing elements, biosensors are integrated with electronic modules such as signal amplifiers, data processors, and digital displays. These components collectively function to amplify, process, and visually represent the transduced signal, enabling accurate interpretation via computer interfaces [23].

2.3.1 Types of bioreceptors

The initial and critical phase in rapid mycotoxin detection involves the specific recognition of target analytes, which fundamentally determines the analytical method's specificity and selectivity. This section provides an overview of various recognition elements employed for mycotoxin detection, including antibodies, aptamers, and molecularly imprinted polymers (MIPs).

Antibodies

Antibodies are immunoglobulin molecules synthesized by the immune system with the capability to specifically bind to their corresponding antigens. Depending on their production methods, antibodies are categorized into polyclonal, monoclonal, and recombinant types. Owing to their exceptional specificity and high binding affinity, antibody–antigen interactions are considered the gold standard in molecular recognition. Antibodies and antigens are extensively employed as recognition and capture elements in immunoassays and biosensors for diverse applications, including clinical diagnostics, therapeutic monitoring, environmental surveillance, and food safety assessment [24, 25].

Despite their widespread use, antibodies exhibit several limitations. The generation of high-affinity antibodies involves complex immunization and purification protocols, which are both time-intensive and costly. Additionally, antibodies are susceptible to degradation under extreme pH and temperature conditions, restricting their utility in certain practical settings. Furthermore, their recognition is confined to immunogenic and immunoreactive molecules.

To address these constraints, alternative antibody formats such as heavy-chain variable domain antibodies (nanobodies) and phage-displayed mimotope peptides have been proposed [26]. Nanobodies, which naturally lack light chains, possess high thermal and chemical stability and a small molecular size, making them amenable to storage and integration into analytical platforms. Meanwhile, phage-displayed mimotope peptides can mimic the epitopes of target analytes, eliminating the need for hapten-carrier conjugates and enabling the development of eco-friendly, mycotoxin-free sensing systems [27, 28].

Aptamers

Aptamers are short single-stranded oligonucleotides, composed of either DNA or RNA, typically comprising 10–100 nucleotides. These sequences attain high-affinity binding to specific targets by folding into unique three-dimensional conformations, enabling the identification of sequences capable of binding a wide array of targets, from small ions to entire cells.

Due to their favorable physicochemical properties, aptamers are regarded as promising alternatives to antibodies for molecular recognition. In contrast to antibodies, aptamers offer several advantages: they are cost-effective to synthesize, exhibit high thermal and chemical stability, are readily reversible in their binding interactions, and are not constrained by the immunogenicity of the target. Additionally, they exhibit high bioavailability and can be easily chemically modified to enhance functionality. These attributes make aptamers well-suited for applications in biosensing, diagnostics, and targeted drug delivery systems [29, 30].

Nonetheless, the selection of aptamers with high affinity and specificity, particularly for small-molecule targets, remains a challenge, with relatively low efficiency in traditional SELEX approaches. In recent years, advancements in aptamer selection methodologies have significantly enhanced the efficiency and performance of aptamer discovery. Techniques have been developed to improve binding affinity, specificity, and reduce overall selection time and cost [31]. Despite the growing body of research and promising results reported in the literature, commercially available aptamer-based detection kits for mycotoxins have yet to be realized [32].

Molecularly Imprinted Polymers (MIPs)

Molecular imprinting is a biomimetic technique inspired by the specific recognition mechanisms observed in natural biological systems, such as enzyme–substrate interactions and antigen–antibody binding. MIPs are synthetic materials engineered through the polymerization of functional monomers and cross-linkers around a specific target compound, known as the template. After the polymerization process, the template is extracted, leaving behind uniquely shaped and chemically compatible cavities designed to rebind the original molecule.

Due to their high stability and specificity, MIPs are often referred to as "artificial antibodies." They exhibit excellent resistance to harsh environmental conditions, including high temperatures, extreme pH, and pressure, and are not easily degraded by biological processes [33]. These properties make MIPs ideal candidates for diverse applications, including their use in optical and electrochemical biosensors, quartz crystal microbalance sensors, biomimetic catalysis, membrane separations, and solid-phase extraction (SPE) methods [34].

Despite their advantages, the practical application of MIPs presents several challenges. One major limitation is the occurrence of nonspecific binding, particularly in the detection of small molecules. To account for this, non-imprinted polymers (NIPs) are often employed as controls to evaluate and correct for nonspecific interactions [35]. Furthermore, the selectivity and binding affinity of MIPs must be further optimized to enable reliable detection of target analytes in complex sample matrices [36].

2.3.2 Types of Biosensors

Electrochemical Biosensors

Electrochemical biosensors (ECBs) have gained prominence as analytical tools for the detection of mycotoxins, offering rapid, sensitive, and cost-efficient analysis [37]. These biosensors function by converting the biochemical interaction between a specific bioreceptor and the target mycotoxin into a measurable electrical signal, which is quantitatively correlated with the analyte concentration.

ECBs encompass various types of electrochemical transducers, including amperometric sensors, which detect current variations resulting from redox reactions; potentiometric and voltammetric sensors, which monitor changes in electric potential and applied voltage, respectively; and conductometric sensors, which measure changes in electrical conductivity [22, 38].

Numerous studies have demonstrated the utility of ECBs in mycotoxin detection. For example, scientists successfully quantified deoxynivalenol (DON) and FB1 in maize using an ECB platform, achieving limits of detection (LOD) of 8.66 ppm and 4.2 ppm, respectively [39]. In another study, researchers compared the performance of ECBs and conventional ELISA for the detection of T-2 toxin in wheat [40]. The ECB exhibited a significantly lower LOD of 0.10 μg/kg compared to 10.0 μg/kg for ELISA, highlighting its superior sensitivity. Moreover, ECBs demonstrated faster assay times relative to ELISA.

Additional applications include the detection of OTA in fruit-derived beverages such as wine and beer, where the study reported up to 105% recovery using ECBs, notably without requiring sample pretreatment [41]. The incorporation of electrochemical sensing into portable devices has also been made easier by recent technical developments. For example, a voltammetric aptasensor that is connected to a smartphone was developed for the detection of patulin (PAT) in fruit juices. This device operated by measuring shifts in redox probe signals and achieved 0.47 pg/mL LOD in diluted apple juice, with recovery rates reaching 93% [42].

Optical Biosensors

Optical biosensors (OBs) have emerged as highly promising tools in food safety, as well as in clinical diagnostics and environmental monitoring, due to their high sensitivity, rapid response, and broad applicability. Unlike electrochemical biosensors (ECBs), which rely on the generation of an electrical signal upon biorecognition events, OBs utilize optical transduction mechanisms. These systems convert biorecognition events into quantifiable optical outputs, such as variations in light intensity, wavelength, or fluorescence, upon interaction between the bioreceptor and the target analyte [43].

OBs can be broadly categorized into label-free and label-based formats. Label-free OBs detect analytes directly without the need for signal-enhancing tags, whereas label-based OBs rely on optical labels such as fluorescent dyes, chemiluminescent agents, or quantum dots (QDs) that emit detectable signals upon target binding [44]. Common transduction mechanisms in OBs include surface plasmon resonance (SPR), optical fiber-based systems, chromogenic, luminogenic, and chemiluminescent methods [45].

Label-based OBs have been extensively investigated for mycotoxin detection. One such approach involved a fluorescence-based biosensor for the detection of AFB1, employing graphene QDs (GQDs) whose fluorescence was quenched in the presence of AFB1 [46]. This system enabled detection within a concentration range of 5–800 ng/mL, achieving a 0.158 ng/mL LOD and a recovery rate of up to 106% in spiked samples. Similarly, a group of scientists introduced a portable, optical fiber-based aptasensor incorporating chemiluminescence for multiplexed detection of mycotoxins, including OTA, AFB1, FB1, and ZEN [47]. The biosensor demonstrated a wide detection range and remarkably low LODs ranging from 0.015 to 0.043 pg/mL.

Label-free optical biosensors, particularly those based on SPR, are widely utilized in medical diagnostics, food analysis, and pharmaceutical quality control. SPR offers real-time, label-free detection by monitoring shifts in the refractive index at a metal sensor surface, thereby preserving the native interaction dynamics without introducing external labeling that may interfere with target recognition [48]. In a study, SPR-based label-free detection of ZEN in maize was performed using both direct and competitive assay approaches [49]. Optimization of covalent immobilization in the competitive format led to an exceptionally low detection limit of 0.002 pg/mL and a dynamic range of 0.01–1 pg/mL. Additionally, the method demonstrated high selectivity, effectively distinguishing ZEN from its structural analogues such as α-ZEN and unrelated mycotoxins.

Gravimetric, Acoustic, and Thermal Biosensors

Gravimetric biosensors (GMBs) function by detecting minute changes in mass on the sensor surface, typically through shifts in resonance frequency. These changes are transduced using acoustic wave-based technologies such as piezoelectric and magnetoelastic systems, which are highly sensitive to variations in surface-bound mass [38]. Common platforms employed in GMBs include Quartz Crystal Microbalance (QCM) and Surface Acoustic Wave (SAW) devices [50].

Numerous studies have demonstrated the effectiveness of gravimetric biosensors in mycotoxin detection. For instance, researchers reported the use of a piezoelectric QCM sensor for the detection of AFB1, achieving a limit of detection (LOD) of 0.01 ng/mL [51]. Similarly, a study highlighted the application of QCM in identifying multiple mycotoxins, including ZEN, T-2 toxin, and FB1, with a detection limit of 0.37 ng/mL and a working range extending up to 24.4 ng/mL [52]. Additionally, investigators described the use of SAW sensors for deoxynivalenol (DON) detection, reporting reproducibility between 4.03% and 16.48%, indicating consistent performance [18].

Thermal biosensors, on the other hand, measure the thermal energy released during biochemical reactions. One such innovation involved a photothermal probe designed to convert heat signals into multidimensional optical outputs [53]. This sensor platform was capable of detecting ZEN with an extremely low LOD of 1.06×10^{-7} ng/mL and a linear dynamic range from 10^{-7} to 10^{-1} ng/mL. These developments highlight the promise of thermal biosensors for portable, real-time testing, offering high sensitivity and user-friendly features ideal for field or personal use.

3 Methods

3.1 Electrochemical Biosensors for Mycotoxin and Allergen Detection

Electrochemical biosensors function by converting biochemical interactions between the target analyte and a specific substrate into measurable electrical signals. These signals can be analyzed using various techniques, each offering distinct analytical advantages [54]. Amperometric sensing measures the current at a constant potential, providing direct correlation with analyte concentration. Voltammetric methods evaluate the current response as the potential is varied over time, offering detailed insights into redox behavior. Impedimetric techniques measure the impedance, including both resistance and reactance components, allowing for sensitive detection of binding events and changes at the sensor interface.

Amperometric and voltammetric biosensors operate by measuring the electric current generated during redox reactions at the electrode interface. In amperometric systems, the current is recorded at a fixed potential, whereas voltammetric techniques involve varying the applied potential. A wide array of electrode materials is utilized in these biosensors, including noble metals, carbon-based materials, screen-printed electrodes, metal oxide nanoparticles, boron-doped diamond, and various composite materials. Biorecognition elements, such as enzymes, antibodies, or synthetic peptides, are immobilized on the transducer surface, enabling specific molecular detection.

Second-generation biosensors address the limitations of oxygen-dependent electron transfer by incorporating redox mediators that act as artificial electron acceptors. In typical catalytic oxidation reactions, the enzyme reduces the mediator, which is then reoxidized at the electrode, producing an electron flow measurable as an electrical current. The use of mediators allows for operation at lower working potentials, thereby enhancing selectivity by reducing interference from nontarget electroactive species. The intensity of the current generated correlates directly with the concentration of the analyte being measured [55].

An example of this approach can be seen in the voltammetric detection of AFB1 using a biosensor comprising aflatoxin oxidase as the biorecognition element immobilized within a silica-based sol–gel matrix. This enzyme-containing matrix was integrated onto a platinum electrode modified with multi-walled carbon nanotubes (MWCNTs). The nanostructured modification significantly enhanced the electrode's surface area, electrocatalytic properties, and overall sensitivity. The biosensor exhibited a linear response range from 3.2 to 721 nmol/L, with a detection limit of 1.6 nmol/L. The silica sol–gel matrix not only stabilized enzymatic activity but also prevented enzyme leaching and provided a biocompatible microenvironment conducive to efficient catalysis [56].

In related work, experimenters described amperometric biosensors for mycotoxin detection employing electrodes fabricated from inert metals (platinum [Pt] and Au), carbon materials (graphite and glassy carbon), and screen-printed platforms [57]. For the detection of OTA, an advanced microfabricated amperometric biosensor was developed using the synthetic peptide NFO4 as the recognition element. This peptide mimicked the binding characteristics of an OTA-specific antibody and demonstrated high binding affinity for OTA.

The analytical protocol involved competitive binding between free OTA and horseradish peroxidase (HRP)-conjugated OTA, followed by incubation on the peptide-modified transducer surface. Amperometric measurements were recorded after 5 min of enzymatic reaction in the presence of hydrogen peroxide and luminol. Concurrently, chemiluminescent signals were also measured, allowing comparative assessment of amperometric and luminescent detection modes. The biosensor demonstrated the capability to detect OTA concentrations up to 10 μg/L, with the amperometric modality offering a wider dynamic range compared to the luminescent method [58].

Impedimetric biosensors operate by measuring the impedance of an electrochemical system, which is influenced by specific biorecognition events occurring at the electrode–electrolyte interface [59]. In electrochemical impedance biosensors, a small-amplitude alternating current (AC) potential, typically sinusoidal in nature, is applied to the system. The resulting current response is analyzed to determine changes in impedance, which correlate with the occurrence of molecular interactions at the sensor surface. Biorecognition elements, such as antibodies, enzymes, or aptamers, are immobilized on the electrode to confer specificity [60].

Electrochemical Impedance Spectroscopy (EIS) is a powerful analytical technique used to probe interfacial phenomena associated with biological recognition processes, including antigen–antibody binding, enzyme–substrate reactions, and whole-cell interactions [61]. EIS measurements are conducted by applying a sinusoidal voltage or current over a wide frequency range, typically from below 1 mHz to above 1 MHz, while monitoring the system's sinusoidal response. The resulting impedance is expressed in ohms and reflects both resistive and capacitive components of the electrochemical interface. Because of its label-free, nondestructive nature and high sensitivity, EIS is widely used for quantitative and qualitative biosensing applications. For instance, EIS has demonstrated a LOD of 0.5 ng/mL for the mycotoxin OTA [62]. In a specific application, EIS was utilized for OTA detection using biosensors functionalized with monoclonal antibodies against OTA [63].

3.1.1 Applications of Electrochemical Biosensors for Mycotoxin and Allergen Detection

Trace levels of deoxynivalenol (DON) in maize flour have been successfully detected using electrochemical (EC) aptamers [64]. One voltammetric aptasensor employed $NiFe_2O_4$ nanotubes, deposited gold (Au) to increase the electrode surface area, enabling sensitive DON detection in corn through differential pulse voltammetry (DPV) [65]. Similarly, for ZEN detection, a screen-printed carbon electrode (SPCE) directly immobilized with a bovine serum albumin–zearalenone (BSA-ZEN) conjugate was developed, achieving 0.25 ng/mL detection limit while reducing mismatches and false positives [66]. Advanced nanomaterials have further enhanced sensitivity to the femtogram per milliliter (fg/mL) level for ZEN detection in maize matrices. Notable examples include a thionine (Thi)-labeled 3D DNA-PtNi@Co-metal–organic framework network on $CoSe_2$/Au, which amplified the DPV signal and achieved an LOD of 1.37 fg/mL [67], and a GO-AptZEN/BSA/cDNA/Au/CS@acetylene black-based aptasensor, which attained an LOD of 3.64 fg/mL in corn oil and corn flour [68].

A Cyclic voltammetry (CV) sensor was developed for AFB1 detection in maize samples using an AbAFB1/GQD-Au/ITO configuration [69]. The incorporation of gold nanoparticle (AuNP)-modified GQD-Au) significantly improved the electrode‘s conductivity and biocompatibility, facilitating antibody immobilization and enhancing sensitivity (LOD: 0.008 ng/mL).

Electrophoretic deposition allowed precise control over GQD-Au film thickness and morphology, further improving detection performance. Aptamer-based sensors also show strong promise for detecting aflatoxin M1 (AFM1). For instance, aptasensors using

AuNPs [70] and polyaniline (PANI) [71] modified electrodes achieved low LODs in the nanogram-per-liter range. A dual-layer electrode incorporating graphene oxide–chitosan (GO-CS) and ordered cerium oxide–chitosan (CeO_2-CS) on SPCE, evaluated via CV and DPV, demonstrated high sensitivity in milk samples, with an LOD of 0.009 μg/L [72].

For FB1 detection in wheat, Thi-graphene nanocomposites were integrated into an AptFB1/cDNA/Au/glassy carbon electrode (GCE) platform. This sensor exhibited excellent performance, attributed to the nanocomposite's high electrical conductivity, structural stability, and large specific surface area [73]. Additionally, a linear sweep voltammetry (LSV)-based sensor using α-cyclodextrin–AFM1 hydrogen bonding as the recognition element demonstrated potential for AFM1 detection in milk [74].

Electrochemical impedance spectroscopy (EIS) was used to develop a disposable AptFB1/Au/polydimethylsiloxane (PDMS)/SPCE aptasensor for FB1 detection in corn [75]. The sensor leveraged a 3D AuNP-modified hollow PDMS film for thiolated aptamer immobilization, achieving a superior LOD of 3.4 pg/mL and a broader detection range (10–50,000 pg/mL) [76].

Furthermore, a study reported an ultrasensitive integrated immunosensor for AFB1 detection in maize utilizing both optical and EIS readouts [77]. The sensor employed indium tin oxide (ITO) substrates modified with AuNPs and gold nanobipyramids-enhanced sensitivity, yielding an LOD of 0.1 nM. Scientists developed an amperometric immunosensor using pAbZEN/Au/multi-walled carbon nanotube (MWCNT)/polyethylenimine (PEI)/SPCE architecture for ZEN detection in maize. This system enabled simplified sample preparation and required minimal sample volumes (20 μL), improving applicability for field diagnostics [78].

3.2 Optical Biosensors for Mycotoxin and Allergen Detection

Optical biosensors have emerged as a transformative analytical platform in food safety applications, particularly for the detection of hazardous contaminants such as mycotoxins and food allergens [79]. These biosensors leverage changes in optical properties induced by the specific binding of analytes to biorecognition elements. The ability to perform real-time, noninvasive, and highly sensitive analyses makes optical biosensors particularly attractive for on-site monitoring of food safety hazards [80]. Depending on the detection strategy, optical biosensors are broadly classified into two major categories: label-based and label-free biosensors. Each of these approaches has unique advantages, limitations, and suitable applications in detecting diverse analytes in complex food matrices [81].

3.2.1 Label-Based Optical Biosensors

Label-based optical biosensors rely on the use of tags or reporters, such as fluorescent dyes, QDs, enzymes, or metallic nanoparticles, to generate an optical signal upon interaction with the target analyte. These tags either enhance or modulate a measurable property (e.g., fluorescence, absorbance, and scattering), making the analyte–bioreceptor interaction more detectable [82].

Fluorescence-Based Biosensors

Fluorescent biosensors are widely employed for allergen and mycotoxin detection because of their superior sensitivity, ability to perform multiplexed analyses, and suitability for miniaturization. In these systems, fluorophore-labeled antibodies or aptamers are used to detect the presence of target compounds. Upon binding, the proximity or conformational changes induce fluorescence changes, either in intensity or wavelength shift [83]. For example, aptamer-based fluorescent sensors have been developed for detecting OTA and aflatoxins, achieving LOD values in the low nanomolar or even picomolar range [84]. A study reviewed carbon dot-based systems where fluorescence resonance energy transfer (FRET) between donor–acceptor pairs allows for precise quantification of OTA in food products [85].

Colorimetric Biosensors

Colorimetric biosensors exploit the optical properties of metallic nanoparticles such as AuNPs and silver nanoparticles (AgNPs), which exhibit localized surface plasmon resonance (LSPR). When nanoparticles aggregate or disperse in the presence of a target analyte, there is a visible color change due to shifts in LSPR bands. This enables visual detection without the need for sophisticated instrumentation. These biosensors are particularly suitable for field testing [86]. For instance, aggregation of aptamer-functionalized AuNPs in the presence of OTA or peanut allergens results in a red-to-blue shift visible to the naked eye. Such systems have proven to be cost-effective and user-friendly, albeit sometimes less sensitive compared to fluorescence-based platforms [87].

Surface-Enhanced Raman Scattering (SERS)

SERS biosensors utilize roughened metal surfaces or nanoparticles (typically silver or gold) to amplify Raman scattering signals from labelled molecules. These sensors are capable of detecting target compounds at femtomolar or even attomolar concentrations, which is particularly advantageous for identifying low-abundance mycotoxins in grains, spices, and beverages [88]. In recent work, SERS-based immunoassays were successfully implemented for AFB 1 and ZEN detection in cereals, with extremely low LODs and high specificity, even in complex matrices [89].

3.2.2 Label-Free Optical Biosensors

These types of biosensors eliminate the need for external tags or reporters. Instead, they detect intrinsic physical or chemical changes (e.g., refractive index, optical density, polarization, or resonance frequency) that occur upon analyte binding. These changes are often measured in real-time and offer simpler assay formats, enhanced quantitative accuracy, and the possibility of continuous monitoring [90].

Surface Plasmon Resonance (SPR)

SPR is one of the most established label-free techniques in biosensing. It measures changes in the refractive index at the interface between a thin metal film (commonly gold) and the sample medium. Binding of analytes to immobilized ligands on the sensor surface induces a measurable shift of the resonance angle, allowing for quantitative and kinetic analyses [91]. SPR-based sensors have been widely employed for OTA, AFB1, gluten, and milk allergen detection. The combination of SPR with nanomaterials and aptamers has significantly improved sensitivity, enabling detection in the low ng/mL to pg/mL range [92, 93]. Additionally, imaging SPR (iSPR) allows multiplexing and array-based analysis, making it suitable for high-throughput food testing.

Localized Surface Plasmon Resonance (LSPR)

Unlike traditional SPR, which uses a continuous metal film, LSPR biosensors exploit metallic nanostructures or nanoparticles that exhibit resonance at localized regions. The presence of an analyte causes shifts in the LSPR peak, which can be detected via UV-vis absorption or scattering spectroscopy [94]. The use of gold nanoislands and nano-rods has allowed LSPR-based sensors to achieve picomolar LODs for OTA and other mycotoxins [95]. In a notable study, it was reviewed that combining LSPR with total internal reflection ellipsometry (TIRE) further enhances sensitivity, reaching detection limits as low as 0.01 ppb [96].

Total Internal Reflection Ellipsometry (TIRE)

TIRE is a label-free optical method that merges ellipsometric phase detection with Kretschmann-configured SPR. It measures minute changes in optical phase upon molecular binding, offering higher sensitivity than traditional SPR. When TIRE is combined with LSPR (e.g., using annealed gold nanostructures), the refractive index sensitivity is significantly enhanced [96]. This hybrid system has demonstrated ultrasensitive detection of AFs, ZEA, and OTA at sub-ppt concentrations, making it suitable for regulatory compliance monitoring in food and feed [97].

Optical Waveguides and Polarization Interferometry

Emerging technologies like planar optical waveguides and polarization interferometry (PI) are gaining attention for label-free biosensing. These platforms guide light through engineered waveguides, where binding events induce phase shifts or interference patterns. Such devices are compact, integrate easily with microfluidics, and are well-suited for multiplexed and portable systems. In PI-based biosensors, differential phase shifts between reference and sample arms are used to detect analyte binding [98]. These systems have demonstrated ppt-level LODs for mycotoxins in cereals and beverages, showcasing their utility in high-performance food monitoring.

3.2.3 Comparative Analysis and Integration Strategies

A juxtaposition of label-based and label-free approaches elucidates their respective strengths and operational niches. While label-based biosensors often offer higher signal amplification and multiplexing, their reliance on secondary reagents and susceptibility to photobleaching may impede real-time applications. Conversely, label-free systems facilitate kinetic monitoring and are inherently simpler in assay design, although they may occasionally suffer from lower signal-to-noise ratios. A comparative overview is presented in Table 4. Hybrid biosensing systems, incorporating both label-based amplification and label-free detection, are increasingly explored. These include SERS-enhanced SPR, fluorescence-coupled waveguides, and LSPR-integrated interferometry, which combine the strengths of both platforms to achieve superior analytical performance. Moreover, the convergence of biosensing with microfluidic automation, smartphone-based readouts, and artificial intelligence (AI) for data processing heralds the next frontier in point-of-care diagnostics for food safety.

3.2.4 Applications of Electrochemical Biosensors for Mycotoxin and Allergen Detection

Numerous optical biosensing platforms have been developed in recent years for the detection of mycotoxins and food allergens, employing both label-based and label-free technologies. Table 5 summarizes a selection of these systems, highlighting their analytical performance, working principles, and application contexts. Fluorescence-based aptasensors have proven particularly effective in detecting OTA, as demonstrated by [79]. In their study, a QD-based platform using CdTe QDs and a FRET mechanism achieved 1.2 ng/mL LOD, demonstrating its effectiveness in OTA analysis in cereal and wine matrices. In another example, a colorimetric assay was developed that utilized AuNPs-enabled visual detection of AFB1, with 5 ng/mL LOD in corn flour [50]. This method operates on the principle of nanoparticle aggregation, leading to a visible color shift, and is suitable for rapid, instrument-free screening. The same group also reported a SERS immunosensor for ZEN detection in maize extract, which provided 0.1 ng/mL LOD [50]. This system used AgNP-coated substrates and leveraged the high sensitivity of SERS to detect low analyte concentrations.

Among label-free approaches, SPR biosensors have shown high potential. For instance, a sensor reported in 2024 detected gluten proteins at 0.3 ng/mL in bakery products using a gold SPR chip and polyclonal antibodies [99]. This technique enables real-time and direct monitoring of allergenic proteins. Researchers designed an LSPR biosensor using gold nano-islands functionalized with truncated aptamers to detect AFB1 at 0.04 ng/mL in grains and milk [96]. The use of nanostructures increased the local field sensitivity, allowing label-free detection at very low concentrations. Their work also introduced TIRE technique for the detection of FUM in maize extract, achieving a remarkable LOD of 0.01 ppb

Table 4
Comparative analysis of label-based and label-free optical biosensors

Parameter	Label-based optical biosensors	Label-free optical biosensors
Detection principle	Signal generation via external tags (fluorophores, nanoparticles, dyes) attached to biorecognition elements.	Direct monitoring of intrinsic changes (e.g., refractive index, phase shift, and mass loading) upon analyte binding.
Common techniques	Fluorescence, Colorimetry, Surface-Enhanced Raman Scattering (SERS), Quantum Dot-based assays.	Surface Plasmon Resonance (SPR), Localized Surface Plasmon Resonance (LSPR), Total Internal Reflection Ellipsometry (TIRE), Optical Waveguides, Polarization Interferometry.
Signal generation	Requires secondary labeling; signal depends on label proximity, photophysical properties.	Label-free signal arises directly from biorecognition events at the sensor interface.
Sensitivity	Very high (femtomolar to attomolar limits achievable, especially with SERS).	High to ultra-high (low picomolar to sub-ppt achievable with SPR-LSPR hybrid platforms).
Real-time monitoring	Limited; mostly end-point detection due to labeling kinetics.	Excellent; real-time kinetic binding and dissociation curves can be recorded.
Multiplexing capability	High; different labels (fluorophores, QDs) allow simultaneous multi-analyte detection.	Moderate; achieved via sensor arrays or imaging platforms (iSPR, waveguide arrays).
Assay time	Longer; additional labeling and washing steps are often required.	Shorter; minimal sample preparation, direct detection without labels.
Instrumentation requirement	May require complex optical setups (fluorescence readers, Raman spectrometers).	SPR instruments, ellipsometers, or compact optical setups; increasingly portable.
Assay complexity	Higher due to multiple reagent additions and potential label instability.	Lower; simplified assay formats suitable for automation.
Cost implications	Moderate to high due to label synthesis, conjugation steps.	Moderate; cost primarily driven by chip fabrication and sensor surface functionalization.
Application suitability	Best for ultrasensitive endpoint detection and multiplexed assays.	Best for real-time quantification, kinetic analysis, and portable field devices.
Examples	Fluorescent aptasensor for OTA; SERS immunosensor for AFB1 detection; AuNP-based colorimetric assays for peanut allergens.	SPR aptasensor for milk allergens; LSPR-based OTA sensor using gold nano-islands; TIRE biosensors for zearalenone and fumonisins.

[96]. TIRE combines polarization-based phase shift detection with SPR to deliver enhanced refractive index sensitivity. A related high-sensitivity method, polarization interferometry (PI), was also used by experimenters to detect OTA in beverages like coffee and wine,

Table 5
Applications of Optical Biosensors for Mycotoxin and Allergen Detection

Biosensor type	Target analyte	Detection limit	Transduction principle	Biorecognition element	Food matrix	Key materials	References
Fluorescence aptasensor	Ochratoxin A (OTA)	1.2 ng/mL	Fluorescence intensity change	DNA aptamer	Cereals, wine	Quantum dots (CdTe QDs), FRET-based	[79]
Colorimetric assay	Aflatoxin B1 (AFB1)	5 ng/mL	LSPR-induced visible color change	Antibody	Corn flour	AuNP aggregation, solution-phase assay	[50]
SERS immunosensor	Zearalenone (ZEN)	0.1 ng/mL	Surface-enhanced Raman scattering	Monoclonal antibody	Maize extract	Silver nanoparticle-coated SERS substrate	[50]
SPR immunosensor	Gluten proteins	0.3 ng/mL	Refractive index shift	Polyclonal antibody	Bakery products	SPR gold chip, prism-based platform	[99]
LSPR aptasensor	Aflatoxin B1 (AFB1)	0.04 ng/mL	LSPR wavelength shift	Truncated aptamer	Grains, milk	Au nano-islands on glass, label-free	[96]
TIRE sensor	Fumonisins (FUM)	0.01 ppb	Polarization phase shift (ellipsometry)	Aptamer fragment	Maize extract	Gold thin films, total internal reflection	[96]
Optical waveguide sensor	Peanut allergens	0.5 ng/mL	Phase change in guided light	Allergen-specific aptamer	Processed snacks	Integrated waveguide biosensor	[100]
PI-based sensor	Ochratoxin A (OTA)	<0.001 ppb (sub-ppt)	Polarization interferometry	Anti-OTA antibody	Coffee, wine	Planar optical chip, photonic interferometer	[96]
Fluorescent immunoassay	β-Lactoglobulin	0.75 ng/mL	Fluorescence increases via labeled antibody	IgG	Milk-based products	Alexa Fluor-labeled secondary antibody	[100]
QD-aptasensor	Aflatoxin M1	0.15 ng/mL	Quenching of quantum dot emission	DNA aptamer	Milk	CdSe/ZnS QDs, microplate assay	[79]

achieving sub-ppt detection levels. The PI sensor utilizes phase shifts caused by analyte binding on an optical planar waveguide, offering exceptional precision for trace-level analysis. In allergen detection, optical waveguide sensors incorporating allergen-specific aptamers detected peanut proteins in processed snacks at a limit of 0.5 ng/mL [100]. These waveguide platforms are suitable for integration into lab-on-a-chip devices. The same group reported a fluorescent immunoassay for β-lactoglobulin in milk products, employing labeled IgG antibodies and achieving a 0.75 ng/mL detection limit. This demonstrates the versatility of fluorescence-based formats for protein allergen detection. Lastly, a QD-based aptasensor for aflatoxin M1 detection in milk was highlighted with a detection limit of 0.15 ng/mL, showcasing the quenching behavior of QDs in a microplate assay format [79]. This method is particularly suitable for high-throughput screening in dairy safety monitoring.

3.3 Commercialization and Field Applications

3.3.1 Quality Control Points in Production

The commercialization of biosensor-based methods for mycotoxin and allergen detection necessitates stringent quality control measures to ensure the reliability and accuracy of these devices. The following measures are essential for preserving the reliability of biosensors and ensuring consistent performance across different environments and applications:

Selection of High-Purity Materials

One of the primary quality control points in the production of biosensors is the selection of high-purity materials. For instance, the use of AuNPs is common in biosensors due to their excellent electrical conductivity, stability, and biocompatibility. However, the purity of these materials is paramount. High-purity AuNPs ensure consistent performance and reduce the likelihood of interference or degradation, which could compromise the biosensor's accuracy [101]. Impurities in the materials can lead to variability in the biosensor's response, affecting its sensitivity and selectivity.

Precise Manufacturing Processes

The manufacturing process of biosensors must be precise and consistent. This includes maintaining uniformity in the size and shape of nanoparticles, which are often used as key components in biosensor construction. The size and shape of nanoparticles directly influence the surface area available for binding and the overall performance of the biosensor. For example, co-reduced molybdenum disulfide and AuNPs have shown excellent electron transfer rates and large surface areas, enabling sensitive and selective detection of mycotoxins [102]. Variations in the size or shape of these nanoparticles can lead to inconsistencies in the biosensor's performance, making it essential to have tight control over the manufacturing process.

Rigorous Calibration Procedures

Calibration is another critical quality control point in the production of biosensors. Proper calibration ensures that the devices provide accurate readings across varying concentration levels. This is especially critical when detecting low levels of mycotoxins and allergens, where even slight inaccuracies can have significant implications for food safety [103]. Calibration procedures must be standardized and validated to ensure that each biosensor performs within specified limits. This involves using well-characterized standards and reference materials to adjust the biosensor's response, ensuring that it provides reliable and reproducible results.

Quality Assurance and Testing

In addition to material selection, manufacturing precision, and calibration, comprehensive quality assurance and testing protocols are essential. These protocols should include both in-process and final product testing to verify that each biosensor meets the required performance specifications. In-process testing can identify and correct issues early in the production cycle, reducing waste and improving overall quality. Final product testing ensures that each biosensor performs as expected before it is released for commercial use.

Environmental and Stability Testing

Environmental and stability testing are also important aspects of quality control. Biosensors must be tested under various environmental conditions, such as different temperatures, humidity levels, and exposure to potential contaminants. This ensures that the biosensors remain stable and reliable in real-world applications. Stability testing can also help identify any degradation or drift in performance over time, allowing for appropriate adjustments or recalibration.

3.3.2 Real-time Monitoring in Supply Chains

Biosensor-based techniques play a crucial role in real-time supply chain monitoring (Fig. 2), offering transformative capabilities for food safety and quality assurance. These biosensors can detect mycotoxins and allergens at various stages, from raw material procurement to final product distribution, ensuring that contaminants are identified and mitigated early in the process.

Early Detection at the Farm Level

The early identification of mycotoxins in crops is one of the most important uses for biosensors. Portable biosensors can be deployed directly on farms to monitor crops for contamination before harvesting. This proactive approach ensures that contaminated produce does not enter the supply chain, thereby reducing the risk of mycotoxin exposure to consumers [104]. For example, biosensors capable of detecting aflatoxins in corn can be used in the field to identify contaminated crops, allowing farmers to take immediate action to prevent the spread of contamination. This not only enhances food safety but also minimizes economic losses for farmers by enabling timely intervention.

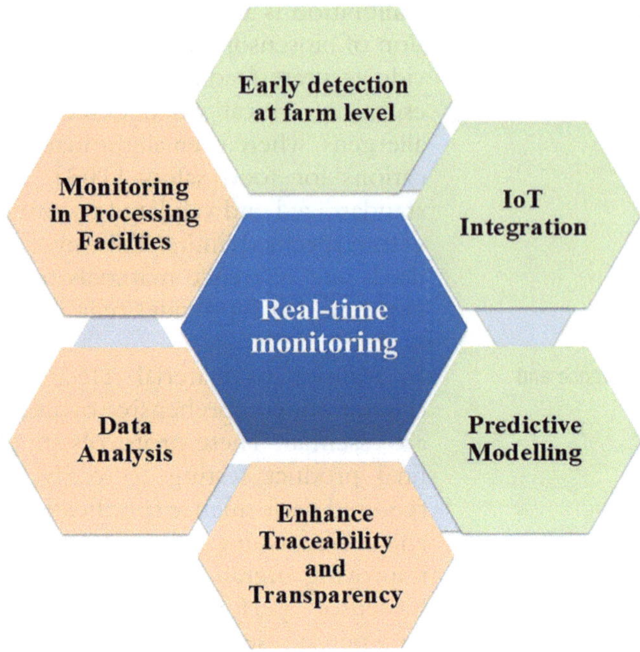

Fig. 2 Realtime monitoring using bio-sensors

Monitoring in Food Processing Facilities

In food processing facilities, real-time biosensors play a crucial role in monitoring allergen levels during production. These biosensors can detect trace amounts of allergens, ensuring that cross-contamination is prevented and that products comply with food safety regulations [105]. For instance, in a facility that processes both gluten-containing and gluten-free products, biosensors can be used to monitor surfaces and equipment for gluten residues, ensuring that gluten-free products remain safe for consumers with celiac disease or gluten sensitivity. This real-time monitoring allows for immediate corrective actions, such as cleaning or sanitizing, to be taken if contamination is detected, thereby maintaining the integrity of the production process.

Integration with IoT Technology

Combining biosensors with Internet of Things (IoT) technology further enhances real-time monitoring capabilities. IoT-enabled biosensors can transmit data continuously to centralized systems, facilitating instant analysis and informed decision-making [103]. This synergy supports the development of an extensive monitoring system covering every stage of the supply chain—ranging from farms and processing plants to distribution hubs and retail stores. For example, biosensors placed in storage facilities can monitor for the presence of mycotoxins in stored grains, alerting managers to potential contamination issues before they become widespread. Similarly, biosensors in transportation vehicles can monitor allergen levels during transit, ensuring that products remain safe throughout the supply chain.

Data Analytics and Predictive Modeling

The continuous data collected by biosensors can be analyzed using advanced analytics and predictive modeling techniques. This allows for the identification of patterns and trends that can inform proactive measures to prevent contamination. For instance, data analytics can help identify specific conditions or practices that are associated with higher levels of mycotoxin contamination, enabling farmers and processors to implement targeted interventions. Predictive modeling can also forecast potential contamination risks, allowing for preemptive actions to be taken before contamination occurs.

Enhancing Traceability and Transparency

Real-time monitoring through biosensors also enhances traceability and transparency in the supply chain. By providing up-to-date information on the presence of mycotoxins and allergens, biosensors enable stakeholders to track the safety and quality of products throughout their journey. This transparency is crucial for building consumer trust and ensuring that food safety standards are consistently met. This level of transparency not only empowers consumers to make informed choices but also holds supply chain participants accountable for maintaining high standards of food safety.

3.3.3 HACCP Integration

The integration of biosensor-based methods into Hazard Analysis and Critical Control Points (HACCP) systems is essential for comprehensive food safety management. HACCP systems are designed to identify critical control points (CCPs) where contamination risks are highest and implement measures to mitigate these risks. By strategically placing biosensors at these CCPs, food processing facilities can achieve rapid and accurate detection of mycotoxins and allergens, thereby enhancing overall food safety [106].

Identifying Critical Control Points

HACCP systems use a structured method to recognize possible hazards in food production and implement controls to prevent or mitigate these hazards. CCPs are defined stages in the process where interventions can effectively reduce or eliminate contamination threats. For instance, in yogurt manufacturing, CCPs may include the raw material stage, where milk is received and stored, and the processing stage, where ingredients are mixed and pasteurized. By identifying these critical points, HACCP systems ensure that potential hazards are addressed at the most effective stages of the production process.

Strategic Placement of Biosensors

Biosensors can be strategically placed at these CCPs to enable continuous supervision and contaminant detection. For instance, in a yogurt processing plant, biosensors can be used to monitor for allergens at the raw material stage, ensuring that milk and other ingredients are free from allergenic contaminants before they enter the production process [106]. During processing, biosensors can continuously monitor for allergens, providing immediate alerts if contamination is detected. This allows for prompt corrective actions, such as cleaning or sanitizing, to be taken before the final product is packaged and distributed.

Enhancing Food Safety and Efficiency

The integration of biosensors into HACCP systems not only enhances food safety but also boosts the system's overall efficiency. By delivering concurrent data, biosensors enable prompt decision-making, facilitating proactive interventions instead of reactive responses. For example, if a biosensor detects the presence of an allergen during processing, the production line can be halted immediately to address the issue, preventing the release of contaminated products into the market. This real-time data also enables more effective risk management, as potential hazards can be identified and mitigated before they become significant issues.

Data Integration and Analysis

The data collected by biosensors can be integrated into existing HACCP management systems, providing a comprehensive view of food safety across the entire production process. This integration allows for continuous monitoring and analysis, enabling food processors to identify trends and patterns that may indicate potential hazards. For example, if a particular batch of raw materials consistently shows higher levels of allergens, this information can be used to identify contamination sources and enable the implementation of preventive strategies. Additionally, combining biosensor data with HACCP systems also facilitates compliance with food safety regulations, as it provides detailed records of monitoring and corrective actions taken.

Training and Implementation

To ensure the successful integration of biosensors into HACCP systems, proper training and implementation strategies are crucial. Food processing staff must be trained to use and maintain biosensors effectively, ensuring that they are calibrated correctly and functioning optimally. Additionally, protocols for responding to biosensor alerts must be established and communicated to all relevant personnel. This ensures that corrective actions are taken promptly and effectively, minimizing the risk of contamination.

3.4 Future Directions and Research Opportunities

3.4.1 Advancements in Biosensor Technology

Future research in biosensor technology should prioritize enhancing sensitivity, selectivity, and stability—key factors for the reliable detection of trace-level contaminants such as mycotoxins and allergens in food products. Timely and precise detection of foodborne hazards is essential for maintaining food safety. One promising avenue involves the use of cutting-edge materials—especially nanomaterials and aptamers—which have demonstrated considerable potential in enhancing the sensitivity and reliability of biosensors [102]. For example, coreduced molybdenum disulfide combined with AuNPs offers excellent electron transfer capabilities and a high surface area, enabling highly sensitive and selective detection of mycotoxins [102]. These materials also contribute to improved detection limits, reliability, and reproducibility.

In addition, integrating biosensors with microfluidics and lab-on-a-chip technologies opens the door to high-throughput, automated analysis. Such systems offer several advantages, including miniaturization, portability, reduced sample and reagent consumption, and faster turnaround times—attributes that are vital for real-time food safety monitoring [105]. Future research should aim to optimize these integrated platforms to enhance sensitivity and specificity further, while also ensuring they remain user-friendly and cost-effective for widespread application.

3.4.2 Multi-target Detection

An important area of ongoing research is the development of biosensors capable of detecting multiple targets simultaneously. Most current biosensors are designed for single-analyte detection, which limits their effectiveness in complex food matrices where multiple contaminants—such as mycotoxins, allergens, and pathogens—may coexist [102]. Advancements in multitarget detection can significantly enhance the practical utility of biosensors in food safety monitoring. For instance, a dual-mode aptasensor has been effectively engineered to simultaneously detect ZEN and FB1, demonstrating high sensitivity and a well-defined linear response range [102]. This kind of approach enables comprehensive screening, reducing the risk of undetected contaminants.

Future efforts should focus on broadening this capability to include a wider range of analytes. The development of multiplex biosensors that can identify several contaminants in a single assay would improve efficiency, reduce costs, and provide a more complete safety profile. Achieving this requires innovative sensor designs incorporating multifunctional nanomaterials and advanced signal processing to ensure accurate, reliable, and simultaneous detection of multiple targets.

3.4.3 Field-Deployable Biosensors

Field-deployable biosensors are vital for real-time monitoring across various points in the food supply chain, from farms to processing and distribution centers. For effective deployment in such settings, these biosensors must be portable, user-friendly, and capable of delivering rapid results without reliance on laboratory infrastructure [104]. Smartphone-based biosensors integrated with lateral flow immunoassays have demonstrated potential for on-site, multiplex detection of mycotoxins in cereals [107]. These systems take advantage of smartphones' widespread availability and processing power to facilitate immediate detection and data analysis, making them highly practical for field use.

Future research should aim to enhance the robustness, accuracy, and reliability of these biosensors under varying environmental conditions. Key challenges include minimizing environmental interference, simplifying sample preparation, and improving data interpretation. Additionally, incorporating user-friendly interfaces and automated data transmission could significantly boost their effectiveness, enabling real-time decision-making and wider adoption in food safety monitoring.

3.4.4 Data Integration and Analytics

The integration of biosensor data with advanced analytics and data management systems is another important area of future research. Integrating biosensor-generated real-time data with additional inputs—such as environmental conditions, production logs, and past contamination incidents—can offer a holistic perspective on food safety. Advanced analytical tools, including ML and AI, can process this combined data to uncover patterns and anticipate potential contamination threats [103]. This proactive approach allows for early intervention and mitigation strategies, enhancing overall food safety management.

3.4.5 Collaborative Research, Regulatory, and Standardization Efforts

Ensuring regulatory compliance and standardization is also a critical aspect of future research. For biosensors to gain broad acceptance in food safety applications, they must comply with the rigorous guidelines established by regulatory authorities like the FDA and the European Food Safety Authority (EFSA). This requires rigorous validation and calibration procedures to ensure that biosensors provide accurate and reproducible results. Collaborative efforts between academia, industry, and regulatory bodies are essential for driving innovation and ensuring the practical application of biosensor technology. Academic research can provide the foundational knowledge and innovative solutions, while industry partners can contribute to the development of scalable and cost-effective biosensor systems. By fostering a collaborative environment, the development and implementation of advanced biosensor technologies can be accelerated, ultimately enhancing food safety and consumer protection.

4 Note

Despite substantial technological advances, the translation of biosensors from laboratory prototypes to commercial, field-deployable devices remains fraught with challenges. Issues such as surface fouling, bioreceptor instability, matrix interference, and regulatory harmonization require concerted efforts across disciplines. However, ongoing innovation in nanofabrication, materials engineering, and integrated photonics indicates that biosensors are poised to become central tools for real-time monitoring of food safety [108]. Upcoming advancements will likely involve the development of disposable, eco-friendly biosensor substrates, multiplexed point-of-need diagnostics, and automated platforms for seamless integration into food production and supply chains. The imperative of global food security necessitates not merely the innovation of analytical tools but their democratization and accessibility across diverse socioeconomic landscapes. Targeted research in these crucial domains can greatly improve the effectiveness and application of biosensors in managing food safety, helping to safeguard consumer health and uphold the quality and security of the food supply [109].

References

1. Maphaisa TC, Akinmoladun OF, Adelusi OA, Mwanza M, Fon F, Tangni E, Njobeh PB (2025) Advances in mycotoxin detection techniques and the crucial role of reference material in ensuring food safety. A review. Food Chem Toxicol 200:115387. https://doi.org/10.1016/j.fct.2025.115387
2. Okechukwu VO, Adelusi OA, Kappo AP, Njobeh PB, Mamo MA (2024) Aflatoxins: occurrence, biosynthesis, mechanism of action and effects, conventional/emerging detection techniques. Food Chem 436: 137775. https://doi.org/10.1016/j.foodchem.2023.137775
3. Ortolani C, Pastorello EA (2005) Food allergies and food intolerances. Best Pract Res Clin Gastroenterol 20(3):467–483. https://doi.org/10.1016/j.bpg.2005.11.010
4. Nji QN, Babalola OO, Ekwomadu TI, Nleya N, Mwanza M (2022) Six main contributing factors to high levels of mycotoxin contamination in African foods. Toxins 14(5):318. https://doi.org/10.3390/toxins14050318
5. Manzano-Sánchez L, Díaz-Galiano FJ, Martínez JA, Ferrer C, Fernández-Alba AR (2024) Improved efficiency of ion trapping time-of-flight mass spectrometry for the analysis of pesticide residues and mycotoxins at trace levels in baby food. Talanta 278: 126448. https://doi.org/10.1016/j.talanta.2024.126448
6. Szelenberger R, Cichoń N, Zajaczkowski W, Bijak M (2024) Application of biosensors for the detection of mycotoxins for the improvement of food safety. Toxins 16(6):249. https://doi.org/10.3390/toxins16060249
7. Bhatnagar D, Bhatnagar M, Ehrlich KC (2010) Mycotoxins and their impact on human health. J Appl Microbiol 109(5): 1421–1431
8. Prandini A, Tansini G, Sigolo S, Filippi L, Laporta M, Piva G (2009) On the occurrence of aflatoxin M1 in milk and dairy products. Food Chem Toxicol 47(5):984–991. https://doi.org/10.1016/j.fct.2007.10.005
9. IARC (2012) IARC monographs on the evaluation of carcinogenic risks to humans. International Agency for Research on Cancer. Retrieved from: https://monographs.iarc.fr/
10. Regulation (EC) No. 1881/2006 (2006) Setting maximum levels for certain contaminants in foodstuffs. Official Journal of the European Union. Retrieved from: https://eur-lex.europa.eu/legal-content/EN/TXT/?uri=CELEX:32006R1881
11. FDA (2023) Food Allergen Labeling and Consumer Protection Act of 2004. U.S. Food and Drug Administration. Retrieved from https://www.fda.gov/food/food-labeling-nutrition/food-allergen-labeling-and-consumer-protection-act-2004-falcpa
12. Samuel MS, Jeyaram K, Datta S, Chandrasekar N, Balaji R, Selvarajan E (2021) Detection, contamination, toxicity, and prevention methods of ochratoxins: an update review. J Agric Food Chem 69(46): 13974–13989. https://doi.org/10.1021/acs.jafc.1c05994
13. Yang D, Ye Y, Sun J, Wang J, Huang C, Sun X (2022) Occurrence, transformation, and toxicity of fumonisins and their covert products during food processing. Crit Rev Food Sci Nutr 64(12):3660–3673. https://doi.org/10.1080/10408398.2022.2134290
14. Zhang Y, Ouyang B, Zhang W, Guang C, Xu W, Mu W (2023) Deoxynivalenol: occurrence, toxicity, and degradation. Food Control 155:110027. https://doi.org/10.1016/j.foodcont.2023.110027
15. Caglayan MO, Şahin S, Üstündağ Z (2020) Detection strategies of zearalenone for food safety: a review. Crit Rev Anal Chem 52(2): 294–313. https://doi.org/10.1080/10408347.2020.1797468

16. Sampson HA (2004) Update on food allergy. J Allergy Clin Immunol 113(5):805–819. https://doi.org/10.1016/j.jaci.2004.03.014
17. Regulation (EU) No. 1169/2011 (2011) Food information to consumers. Official Journal of the European Union. Retrieved from https://eur-lex.europa.eu/legal-content/EN/TXT/?uri=CELEX:32011R1169
18. Pisoschi AM, Iordache F, Stanca L, Mitranescu E, Bader Stoica L, Geicu OI, Bilteanu L, Serban AI (2024) Biosensors for food mycotoxin determination: a comparative and critical review. Chem 12(6):92. https://doi.org/10.3390/chemosensors12060092
19. Zhang M, Guo X, Wang J (2023) Advanced biosensors for mycotoxin detection incorporating miniaturized meters. Biosens Bioelectron 224:115077. https://doi.org/10.1016/j.bios.2023.115077
20. Fan Y, Li J, Amin K, Yu H, Yang H, Guo Z, Liu J (2023) Advances in aptamers, and application of mycotoxins detection: a review. Food Res Int 170:113022. https://doi.org/10.1016/j.foodres.2023.113022
21. Li R, Wen Y, Wang F, He P (2021) Recent advances in immunoassays and biosensors for mycotoxins detection in feedstuffs and foods. J Anim Sci Biotechnol 12(1):108. https://doi.org/10.1186/s40104-021-00629-4
22. Hassan RA, Abu Hanifah S, Heng LY, Al-badaii F, Ulianas A (2024) Systematic review on biosensors for the early detection of mycotoxins as endocrine disruptors. Food Control 157:110195. https://doi.org/10.1016/j.foodcont.2023.110195
23. Ramesh M, Janani R, Deepa C, Rajeshkumar L (2022) Nanotechnology-enabled biosensors: a review of fundamentals, design principles, materials, and applications. Biosensors 13(1):40. https://doi.org/10.3390/bios13010040
24. Alhamoud Y, Yang D, Fiati Kenston SS, Liu G, Liu L, Zhou H, Ahmed F, Zhao J (2019) Advances in biosensors for the detection of ochratoxin A: bio-receptors, nanomaterials, and their applications. Biosens Bioelectron 141:111418. https://doi.org/10.1016/j.bios.2019.111418
25. Ding L, Chen X, He L, Yu F, Yu S, Wang J, Tian Y, Wang Y, Wu Y, Liu L, Qu L (2020) Fluorometric immunoassay for the simultaneous determination of the tumor markers carcinoembryonic antigen and cytokeratin 19 fragment using two kinds of CdSe/ZnS quantum dot nanobeads and magnetic beads. Microchim Acta 187(3):171. https://doi.org/10.1007/s00604-019-3914-7
26. Bueno D, Istamboulie G, Muñoz R, Marty JL (2015) Determination of mycotoxins in food: a review of bioanalytical to analytical methods. Appl Spectrosc Rev 50(9):728–774. https://doi.org/10.1080/05704928.2015.1072092
27. Hou S, Ma Z, Meng H, Xu Y, He Q (2019) Ultrasensitive and green electrochemical immunosensor for mycotoxin ochratoxin A based on phage displayed mimotope peptide. Talanta 194:919–924. https://doi.org/10.1016/j.talanta.2018.10.081
28. Ren W, Xu Y, Huang Z, Li Y, Tu Z, Zou L, He Q, Fu J, Liu S, Hammock BD (2020) Single-chain variable fragment antibody-based immunochromatographic strip for rapid detection of fumonisin B1 in maize samples. Food Chem 319:126546. https://doi.org/10.1016/j.foodchem.2020.126546
29. He F, Wen N, Xiao D, Yan J, Xiong H, Cai S, Liu Z, Liu Y (2018) AptaMer-based targeted drug delivery systems: current potential and challenges. Curr Med Chem 27(13):2189–2219. https://doi.org/10.2174/0929867325666181008142831
30. Sinha K, Mukhopadhyay CD (2020) Quantitative detection of neurotransmitter using aptamer: from diagnosis to therapeutics. J Biosci 45(1). https://doi.org/10.1007/s12038-020-0017-x
31. Darmostuk M, Rimpelova S, Gbelcova H, Ruml T (2015) Current approaches in SELEX: an update to aptamer selection technology. Biotechnol Adv 33(6):1141–1161. https://doi.org/10.1016/j.biotechadv.2015.02.008
32. Chen A, Yang S (2015) Replacing antibodies with aptamers in lateral flow immunoassay. Biosens Bioelectron 71:230–242. https://doi.org/10.1016/j.bios.2015.04.041
33. Gaudin V (2017) Advances in biosensor development for the screening of antibiotic residues in food products of animal origin—a comprehensive review. Biosens Bioelectron 90:363–377. https://doi.org/10.1016/j.bios.2016.12.005
34. Bezdekova J, Zemankova K, Hutarova J, Kociova S, Smerkova K, Adam V, Vaculovicova M (2020) Magnetic molecularly imprinted polymers used for selective isolation and detection of Staphylococcus aureus. Food Chem 321:126673. https://doi.org/10.1016/j.foodchem.2020.126673
35. Ahmad OS, Bedwell TS, Esen C, Garcia-Cruz A, Piletsky SA (2019) Molecularly imprinted polymers in electrochemical and optical sensors. Trends Biotechnol 37(3):294–309

36. Fu H, Xu W, Wang H, Liao S, Chen G (2020) Preparation of magnetic molecularly imprinted polymers for the identification of zearalenone in grains. Anal Bioanal Chem 412(19):4725–4737. https://doi.org/10.1007/s00216-020-02729-y
37. Kaur M, Gaba J, Singh K, Bhatia Y, Singh A, Singh N (2023) Recent advances in recognition receptors for electrochemical biosensing of mycotoxins—a review. Biosensors 13(3): 391. https://doi.org/10.3390/bios13030391
38. Agriopoulou S, Stamatelopoulou E, Varzakas T (2020) Advances in analysis and detection of major mycotoxins in foods. Foods 9(4): 518. https://doi.org/10.3390/foods9040518
39. Lu L, Seenivasan R, Wang Y, Yu J, Gunasekaran S (2016) An electrochemical immunosensor for rapid and sensitive detection of mycotoxins fumonisin B1 and deoxynivalenol. Electrochim Acta 213:89–97. https://doi.org/10.1016/j.electacta.2016.07.096
40. Fernandez Solis LN, Silva Junior GJ, Bertotti M, Angnes L, Pereira SV, Fernández-Baldo MA, Regiart M (2024) Electrochemical microfluidic immunosensor with graphene-decorated gold nanoporous for T-2 mycotoxin detection. Talanta 273: 125971. https://doi.org/10.1016/j.talanta.2024.125971
41. Pacheco JG, Castro M, Machado S, Barroso MF, Nouws HP, Delerue-Matos C (2015) Molecularly imprinted electrochemical sensor for ochratoxin A detection in food samples. Sensors Actuators B Chem 215:107–112. https://doi.org/10.1016/j.snb.2015.03.046
42. Erdem A, Senturk H (2024) Smartphone-controlled aptasensor for voltammetric detection of patulin in apple juice. Sensors 24(3): 754. https://doi.org/10.3390/s24030754
43. Wei L, Wang Z, Chen Y (2022) Optical biosensor for ochratoxin a detection in grains using an Enzyme-Mediated click reaction and polystyrene nanoparticles. J Agric Food Chem 70(46):14798–14804. https://doi.org/10.1021/acs.jafc.2c05137
44. Singh AK, Mittal S, Das M, Saharia A, Tiwari M (2023) Optical biosensors: a decade in review. Alex Eng J 67:673–691. https://doi.org/10.1016/j.aej.2022.12.040
45. Rasheed S, Kanwal T, Ahmad N, Fatima B, Najam-ul-Haq M, Hussain D (2024) Advances and challenges in portable optical biosensors for onsite detection and point-of-care diagnostics. TrAC Trends Anal Chem 173:117640. https://doi.org/10.1016/j.trac.2024.117640
46. Singh AK, Sri S, Garimella LBVS, Dhiman TK, Sen S, Solanki PR (2022) Graphene Quantum Dot-Based Optical Sensing platform for aflatoxin B1 detection via the resonance Energy Transfer phenomenon. ACS Appl Bio Mater 5(3):1179–1186. https://doi.org/10.1021/acsabm.1c01224
47. Jia Y, Zhao S, Li D, Yang J, Yang L (2023) Portable chemiluminescence optical fiber aptamer-based biosensors for analysis of multiple mycotoxins. Food Control 144:109361. https://doi.org/10.1016/j.foodcont.2022.109361
48. Pandey PS, Raghuwanshi SK, Shadab A, Ansari MTI, Tiwari UK, Kumar S (2022) SPR based biosensing chip for COVID-19 diagnosis—a review. IEEE Sensors J 22(14): 13800–13810. https://doi.org/10.1109/JSEN.2022.3181423
49. Székács I, Adányi N, Szendrő I, Székács A (2020) Direct and competitive optical grating immunosensors for determination of fusarium mycotoxin zearalenone. Toxins 13(1): 43. https://doi.org/10.3390/toxins13010043
50. Majer-Baranyi K, Adányi N, Székács A (2023) Current trends in mycotoxin detection with various types of biosensors. Toxins 15(11): 645. https://doi.org/10.3390/toxins15110645
51. Jubeen F, Batool A, Naz I, Sehar S, Sadia H, Hayat A, Kazi M (2024) Mycotoxins detection in food using advanced, sensitive and robust electrochemical platform of sensors: a review. Sensors Actuators A Phys 367: 115045. https://doi.org/10.1016/j.sna.2024.115045
52. Majer-Baranyi K, Adányi N, Székács A (2021) Biosensors for deoxynivalenol and zearalenone determination in feed quality control. Toxins 13(7):499. https://doi.org/10.3390/toxins13070499
53. Chen Y, Zhang S, Huang Y, Lv L, Dai H, Lin Y (2020) A bio-bar-code photothermal probe triggered multi-signal readout sensing system for nontoxic detection of mycotoxins. Biosens Bioelectron 167:112501. https://doi.org/10.1016/j.bios.2020.112501
54. Grieshaber D, MacKenzie R, Vörös J, Reimhult E (2008) Electrochemical biosensors – sensor principles and architectures. Sensors 8(3):1400–1458. https://doi.org/10.3390/s80314000
55. Nguyen HH, Lee SH, Lee UJ, Fermin CD, Kim M (2018) Immobilized enzymes in

biosensor applications. Materials 12(1):121. https://doi.org/10.3390/ma12010121

56. Li SC, Chen JH, Cao H, Yao DS, Liu DL (2010) Amperometric biosensor for aflatoxin B1 based on aflatoxin-oxidase immobilized on multiwalled carbon nanotubes. Food Control 22(1):43–49. https://doi.org/10.1016/j.foodcont.2010.05.005
57. Vidal JC, Bonel L, Ezquerra A, Hernández S, Bertolín JR, Cubel C, Castillo JR (2013) Electrochemical affinity biosensors for detection of mycotoxins: a review. Biosens Bioelectron 49:146–158. https://doi.org/10.1016/j.bios.2013.05.008
58. Tria SA, Lopez-Ferber D, Gonzalez C, Bazin I, Guiseppi-Elie A (2016) Microfabricated biosensor for the simultaneous amperometric and luminescence detection and monitoring of Ochratoxin A. Biosens Bioelectron 79:835–842. https://doi.org/10.1016/j.bios.2016.01.018
59. Bahadır EB, Sezgintürk MK (2014) A review on impedimetric biosensors. Artif Cells Nanomed Biotechnol 44(1):248–262. https://doi.org/10.3109/21691401.2014.942456
60. Magar HS, Hassan RY, Mulchandani A (2020) Electrochemical impedance spectroscopy (EIS): principles, construction, and biosensing applications. Sensors 21(19):6578. https://doi.org/10.3390/s21196578
61. Lazanas AC, Prodromidis MI (2023) Electrochemical impedance spectroscopy—a tutorial. ACS Measur Sci Au 3(3):162–193. https://doi.org/10.1021/acsmeasuresciau.2c00070
62. Radi A, Muñoz-Berbel X, Lates V, Marty J (2009) Label-free impedimetric immunosensor for sensitive detection of ochratoxin A. Biosens Bioelectron 24(7):1888–1892. https://doi.org/10.1016/j.bios.2008.09.021
63. Malvano F, Albanese D, Crescitelli A, Pilloton R, Esposito E (2016) Impedimetric label-free immunosensor on disposable modified screen-printed electrodes for ochratoxin A. Biosensors 6(3):33. https://doi.org/10.3390/bios6030033
64. Subak H, Selvolini G, Macchiagodena M, Ozkan-Ariksoysal D, Pagliai M, Procacci P, Marrazza G (2021) Mycotoxins aptasensing: from molecular docking to electrochemical detection of deoxynivalenol. Bioelectrochemistry 138:107691. https://doi.org/10.1016/j.bioelechem.2020.107691
65. He B, Wang K (2021) A "signal off" aptasensor based on NiFe2O4 NTs and Au@Pt NRs for the detection of deoxynivalenol via voltammetry. Microchim Acta 188(1):23. https://doi.org/10.1007/s00604-020-04666-0
66. Yugender Goud K, Sunil Kumar V, Hayat A, Vengatajalabathy Gobi K, Song H, Kim K, Marty JL (2018) A highly sensitive electrochemical immunosensor for zearalenone using screen-printed disposable electrodes. J Electroanal Chem 832:336–342. https://doi.org/10.1016/j.jelechem.2018.10.058
67. He B, Yan X (2020) Ultrasensitive electrochemical aptasensor based on CoSe2/AuNRs and 3D structured DNA-PtNi@Co-MOF networks for the detection of zearalenone. Sensors Actuators B Chem 306:127558. https://doi.org/10.1016/j.snb.2019.127558
68. Mu Z, Ma L, Wang J, Zhou J, Yuan Y, Bai L (2021) A target-induced amperometic aptasensor for sensitive zearalenone detection by CS@AB-MWCNTs nanocomposite as enhancers. Food Chem 340:128128. https://doi.org/10.1016/j.foodchem.2020.128128
69. Bhardwaj H, Pandey MK, Rajesh N, Sumana G (2019) Electrochemical Aflatoxin B1 immunosensor based on the use of graphene quantum dots and gold nanoparticles. Microchim Acta 186(8):592. https://doi.org/10.1007/s00604-019-3701-5
70. Jalalian SH, Ramezani M, Danesh NM, Alibolandi M, Abnous K, Taghdisi SM (2018) A novel electrochemical aptasensor for detection of aflatoxin M1 based on target-induced immobilization of gold nanoparticles on the surface of electrode. Biosens Bioelectron 117:487–492. https://doi.org/10.1016/j.bios.2018.06.055
71. Kulikova TN, Porfireva AV, Evtugyn GA, Hianik T (2019) Electrochemical aptasensor with layer-by-layer deposited polyaniline for aflatoxin M1 voltammetric determination. Electroanalysis 31(10):1913–1924. https://doi.org/10.1002/elan.201900274
72. An X, Shi X, Zhang H, Yao Y, Wang G, Yang Q, Xia L, Sun X (2019) An electrochemical immunosensor based on a combined amplification strategy with the GO–CS/CeO2–CS nanocomposite for the detection of aflatoxin M1. New J Chem 44(4):1362–1370. https://doi.org/10.1039/c9nj04804a
73. Shi Y, Zheng T, Zhang B, He H, Wu D, Zhang B (2015) DNA electrochemical aptasensor for detecting fumonisins B1 based on graphene and thionine nanocomposite. Electroanalysis 27(5):1097–1103. https://doi.org/10.1002/elan.201400504

74. Shadjou R, Hasanzadeh M, Shadjou N (2018) Electrochemical monitoring of aflatoxin M1 in milk samples using silver nanoparticles dispersed on α-cyclodextrin-GQDs nanocomposite. J Mol Recognit 31(6): e2699. https://doi.org/10.1002/jmr.2699
75. Ren C, Li H, Lu X, Qian J, Zhu M, Chen W, Liu Q, Hao N, Li H, Wang K (2017) A disposable aptasensing device for label-free detection of fumonisin B1 by integrating PDMS film-based micro-cell and screen-printed carbon electrode. Sensors Actuators B Chem 251:192–199. https://doi.org/10.1016/j.snb.2017.05.035
76. Shu M, Xu Y, Liu X, Li Y, He Q, Tu Z, Fu J, Gee SJ, Hammock BD (2016) Anti-idiotypic nanobody-alkaline phosphatase fusion proteins: development of a one-step competitive enzyme immunoassay for fumonisin B1 detection in cereal. Anal Chim Acta 924:53–59. https://doi.org/10.1016/j.aca:2016.03.053
77. Bhardwaj H, Sumana G, Marquette CA (2021) Gold nanobipyramids integrated ultrasensitive optical and electrochemical biosensor for Aflatoxin B1 detection. Talanta 222:121578. https://doi.org/10.1016/j.talanta.2020.121578
78. Riberi WI, Tarditto LV, Zon MA, Arévalo FJ, Fernández H (2017) Development of an electrochemical immunosensor to determine zearalenone in maize using carbon screen printed electrodes modified with multi-walled carbon nanotubes/polyethyleneimine dispersions. Sensors Actuators B Chem 254:1271–1277. https://doi.org/10.1016/j.snb.2017.07.113
79. Meira DI, Barbosa AI, Borges J, Reis RL, Correlo VM, Vaz F (2023) Recent advances in nanomaterial-based optical biosensors for food safety applications: ochratoxin-A detection, as case study. Crit Rev Food Sci Nutr 64(18):6318–6360. https://doi.org/10.1080/10408398.2023.2168248
80. Sagar Shrikrishna N, Sharma R, Sahoo J, Kaushik A, Gandhi S (2024) Navigating the landscape of optical biosensors. Chem Eng J 490:151661. https://doi.org/10.1016/j.cej.2024.151661
81. Samuel VR, Rao K (2022) A review on label free biosensors. Biosens Bioelectron 11: 100216. https://doi.org/10.1016/j.biosx.2022.100216
82. Sadani K, Nag P, Thian XY, Mukherji S (2022) Enzymatic optical biosensors for healthcare applications. Biosens Bioelectron 12:100278. https://doi.org/10.1016/j.biosx.2022.100278
83. Zhao Q, Tao J, Feng W, Uppal JS, Peng H, Le XC (2020) Aptamer binding assays and molecular interaction studies using fluorescence anisotropy – a review. Anal Chim Acta 1125:267–278. https://doi.org/10.1016/j.aca.2020.05.061
84. Chen X, Gao D, Sun F, Li Z, Wang Y, Qiu C, He K, Wang J (2022) Nanomaterial-based aptamer biosensors for ochratoxin A detection: a review. Anal Bioanal Chem 414(9): 2953–2969. https://doi.org/10.1007/s00216-022-03960-5
85. Gopal AR, Joy F, Dutta V, Devasia J, Dateer R, Nizam A (2023) Carbon Dot-Based Fluorescence Resonance Energy Transfer (FRET) systems for biomedical, sensing, and imaging applications. Part Part Syst Charact 41(1). https://doi.org/10.1002/ppsc.202300072
86. Wang W, You Y, Gunasekaran S (2021) LSPR-based colorimetric biosensing for food quality and safety. Compr Rev Food Sci Food Saf 20(6):5829–5855. https://doi.org/10.1111/1541-4337.12843
87. Sun R, Li Y, Du T, Qi Y (2023) Recent advances in integrated dual-mode optical sensors for food safety detection. Trends Food Sci Technol 135:14–31. https://doi.org/10.1016/j.tifs.2023.03.013
88. Langer J, De Aberasturi DJ, Aizpurua J, Alvarez-Puebla RA, Auguié B, Baumberg JJ, Bazan GC, Bell SEJ, Boisen A, Brolo AG, Choo J, Cialla-May D, Deckert V, Fabris L, Faulds K, De Abajo FJG, Goodacre R, Graham D, Haes AJ et al (2019) Present and future of Surface-Enhanced Raman Scattering. ACS Nano 14(1):28–117. https://doi.org/10.1021/acsnano.9b04224
89. Zhai W, You T, Ouyang X, Wang M (2021) Recent progress in mycotoxins detection based on surface-enhanced Raman spectroscopy. Compr Rev Food Sci Food Saf 20(2): 1887–1909. https://doi.org/10.1111/1541-4337.12686
90. Zanchetta G, Lanfranco R, Giavazzi F, Bellini T, Buscaglia M (2017) Emerging applications of label-free optical biosensors. Nano 6(4):627–645. https://doi.org/10.1515/nanoph-2016-0158
91. Ravindran N, Kumar S, Yashini M, Rajeshwari S, Mamathi CA, Nirmal Thirunavookarasu S, Sunil CK (2021) Recent advances in Surface Plasmon Resonance (SPR) biosensors for food analysis: a review. Crit Rev Food Sci Nutr 63(8):1055–1077. https://doi.org/10.1080/10408398.2021.1958745

92. Shrivastava A, Sharma RK (2021) Biosensors for the detection of mycotoxins. Toxin Rev 41(2):618–638. https://doi.org/10.1080/15569543.2021.1894175
93. Evtugyn G, Porfireva A, Kulikova T, Hianik T (2021) Recent achievements in electrochemical and surface plasmon resonance aptasensors for mycotoxins detection. Chem 9(7):180. https://doi.org/10.3390/chemosensors9070180
94. Kim DM, Park JS, Jung S, Yeom J, Yoo SM (2020) Biosensing applications using nanostructure-based localized surface plasmon resonance sensors. Sensors 21(9):3191. https://doi.org/10.3390/s21093191
95. Pellas V, Hu D, Mazouzi Y, Mimoun Y, Blanchard J, Guibert C, Salmain M, Boujday S (2020) Gold nanorods for LSPR biosensing: synthesis, coating by silica, and bioanalytical applications. Biosensors 10(10):146. https://doi.org/10.3390/bios10100146
96. Nabok A, Al-Rubaye A, Al-Jawdah A, Tsargorodska A, Marty J, Catanante G, Szekacs A, Takacs E (2018) [INVITED] Novel optical biosensing technologies for detection of mycotoxins. Opt Laser Technol 109:212–221. https://doi.org/10.1016/j.optlastec.2018.07.076
97. Miklós G, Angeli C, Ambrus Á, Nagy A, Kardos V, Zentai A, Kerekes K, Farkas Z, Józwiak Á, Bartók T (2021) Corrigendum: detection of aflatoxins in different matrices and Food-Chain positions. Front Microbiol 12. https://doi.org/10.3389/fmicb.2021.669714
98. Chalyan T, Ottevaere H, Pasquardini L (2023) Optical biosensors: from working principles to detection methods of label-free devices. In: Biophotonics and Biosensing, pp 15–48. https://doi.org/10.1016/B978-0-44-318840-4.00009-7
99. Zhu D, Fu S, Zhang X, Zhao Q, Yang X, Man C, Jiang Y, Guo L, Zhang X (2024) Recent progresses on emerging biosensing technologies and portable analytical devices for detection of food allergens. Trends Food Sci Technol 148:104485. https://doi.org/10.1016/j.tifs.2024.104485
100. Nazaret L, Carmen MD, Cruz M (2018) Biosensing based on nanoparticles for food allergens detection. Sensors 18(4):1087. https://doi.org/10.3390/s18041087
101. Aquino A, Adam C (2020) A systematic review of food allergy: nanobiosensor and food allergen detection. Biosensors 10(12):194. https://doi.org/10.3390/bios10120194
102. Han et al (2023) Study on co-reduced molybdenum disulfide and AuNPs for biosensor applications. J Anal Chem. https://doi.org/10.1016/j.trac.2023.116001
103. Chen Y, Wang Y, Zhang Y, Wang X, Zhang C, Cheng N (2023) Intelligent biosensors promise smarter solutions in food safety 4.0. Foods 13(2):235. https://doi.org/10.3390/foods13020235
104. Nnachi RC, Sui N, Ke B, Luo Z, Bhalla N, He D, Yang Z (2022) Biosensors for rapid detection of bacterial pathogens in water, food and environment. Environ Int 166:107357. https://doi.org/10.1016/j.envint.2022.107357
105. Zhou Q, Tang D (2020) Recent advances in photoelectrochemical biosensors for analysis of mycotoxins in food. TrAC Trends Anal Chem 124:115814. https://doi.org/10.1016/j.trac.2020.115814
106. Mureşan CC, Marc RA, Jimborean M, Rusu I, Mureşan A, Nistor A, Cozma A, Suharoschi R (2019) Food safety system (HACCP) as quality checkpoints in a spin-off small-scale yogurt processing plant. Sustainability 12(22):9472. https://doi.org/10.3390/su12229472
107. Liu Z, Hua Q, Wang J, Liang Z, Li J, Wu J, Shen X, Lei H, Li X (2020) A smartphone-based dual detection mode device integrated with two lateral flow immunoassays for multiplex mycotoxins in cereals. Biosens Bioelectron 158:112178. https://doi.org/10.1016/j.bios.2020.112178
108. Ateia M, Wei H, Andreescu S (2024) Sensors for Emerging water Contaminants: Overcoming Roadblocks to innovation. Environ Sci Technol 58(6):2636–2651. https://doi.org/10.1021/acs.est.3c09889
109. Sobhan A, Hossain A, Wei L, Muthukumarappan K, Ahmed M (2025) IoT-Enabled biosensors in food Packaging: A breakthrough in food safety for monitoring risks in real time. Foods 14(8):1403. https://doi.org/10.3390/foods14081403

Chapter 8

DNA Barcoding and Next-Generation Sequencing for Food Authentication

Amit Kumar Das, Akramul Ansary, Siddhartha Sankar Das, Koushik Nandan Dutta, Suman Lamichhane, Jitendra Pandey, and Darshana Hazarika

Abstract

Food authentication is a critical process in verifying the species, origin, and production methods of food products to ensure safety, regulatory compliance, and consumer trust. Traditional methods, such as protein-based assays, often fail in processed or mixed food samples due to protein denaturation. DNA-based techniques, especially DNA barcoding and next-generation sequencing (NGS), offer superior specificity, sensitivity, and robustness. DNA barcoding targets short, standardized genetic regions (e.g., COI, rbcL, ITS) for species identification, while NGS enables high-throughput analysis of complex or multispecies products. These techniques are instrumental in detecting food fraud such as adulteration, mislabeling, and species substitution across a wide range of matrices, including meat, seafood, herbs, spices, and processed foods. This chapter also explores technical challenges like degraded DNA, database reliability, and the need for regulatory validation, as well as emerging solutions such as portable sequencing and AI-driven bioinformatics. Overall, DNA-based authentication is a powerful and evolving tool for ensuring food integrity in modern supply chains.

Key words DNA barcoding, Food authentication, Species identification, Next-generation sequencing (NGS), Food fraud, Adulteration detection

1 Introduction

Food authentication verifies that a food product matches its label claims regarding species, origin, or production method [1, 2]. In complex global supply chains, verifying authenticity protects consumers, ensures regulatory compliance, and maintains market confidence. Mislabeling can lead to economic losses, public health risks (e.g., allergen exposure, toxic species substitution), and reputational damage for honest producers [3, 4].

Tanmay Sarkar (ed.), *Novel Food Analysis: Techniques and Applications*, Methods and Protocols in Food Science, https://doi.org/10.1007/978-1-0716-4787-5_8,

1.1 Definition of Food Authentication

Food authentication encompasses analytical techniques to confirm a product's declared identity, such as species (e.g., Atlantic cod), varietal identity (e.g., Pinot Noir), geographical origin (e.g., Tuscany), or production method (e.g., organic). Techniques compare sample characteristics—chemical, molecular, isotopic, or sensory—to reference standards. Traditional assays (protein-based enzyme-linked immunosorbent assay [ELISA], chromatography, spectroscopy) often fail in highly processed or mixed samples, as processing can denature marker proteins or alter chemical profiles [5, 6].

1.2 Importance of Food Safety, Labeling Compliance, and Consumer Protection

1. *Food Safety*: Substituting fish species (e.g., escolar for tuna) may cause toxin ingestion; undeclared allergens risk severe reactions. Authenticity checks avert these hazards [7–10].
2. *Labeling Compliance*: Regulatory bodies (e.g., FDA and EFSA) mandate accurate labeling. Noncompliance can trigger costly recalls, fines, or legal action. Certification schemes (e.g., USDA Organic and MSC seafood) rely on verification to maintain accreditation [11, 12].
3. *Consumer Protection*: Consumers pay premiums for attributes such as "grass-fed" meat or "PDO" cheese. Fraudulent labeling erodes trust, penalizes genuine producers, and distorts pricing [13, 14].

1.3 Challenges in Food Fraud

- *Adulteration*: Adding unauthorized substances (e.g., water in milk, melamine in protein analyses, cheaper oils in olive oil) to increase volume or mask inferior quality. Detecting low-level adulterants against complex matrices demands highly sensitive methods [15, 16].
- *Mislabeling*: Incorrectly declaring geographic origin (e.g., claiming "Italian" olive oil when sourced elsewhere) or production method (e.g., labeling intensively farmed fish as "wild-caught"). Similar terroir profiles often complicate detection [15, 16].
- *Species Substitution*: Swapping high-value species (e.g., cod) with lower value alternatives (e.g., pollock), or mixing meat types (e.g., pork in beef products). Processing (filleting, mincing, and cooking) eliminates morphological clues, and protein assays may fail if proteins are denatured [17].

1.4 Role of DNA-Based Techniques

DNA-based methods leverage the stability and specificity of genetic material to overcome traditional assay limitations:

- *Species-Specificity*: Short mitochondrial or chloroplast regions survive processing, enabling identification even when proteins are denatured [18].
- *High Sensitivity*: PCR amplification detects trace amounts of adulterant DNA (down to 0.1–1%) [19, 20].

Table 1
Comparison of Traditional Protein-Based Assays and DNA-Based Techniques

Criterion	Protein-Based (e.g., ELISA)	DNA-Based (e.g., PCR, Barcoding, NGS)
Analyte Stability	Proteins denature under heat/ processing	Short DNA fragments (<200 bp) remain intact after processing
Specificity	Antibody cross-reactivity may cause false positives	High specificity via species-specific primers and sequences
Sensitivity	Moderate (detects ~1–5% adulterant)	High (detects 0.1–1% adulterant)
Throughput	Moderate; one or few targets per assay	Moderate (qPCR) to very high (NGS, multiplexing dozens of targets)
Cost per Test	Low to moderate per sample (kits required)	Moderate to high for NGS; costs decreasing over time
Multiplexing Capability	Limited	High (multiple targets per run with NGS or multiplex PCR)
Database/ Reference Need	Relies on validated antibodies	Requires curated sequence databases (e.g., BOLD and GenBank)
Limitations	False negatives if proteins denature; limited targets	PCR inhibitors, initial equipment/setup costs, database gaps

- *Broad Applicability*: A single workflow can detect multiple species in complex mixtures (e.g., multispecies sausage and honey blends) [21].
- *Matrix Versatility*: DNA extraction protocols adapt to diverse matrices (high-fat and high-phenolic foods), ensuring robust recovery [22].

Limitations include potential PCR inhibitors in certain foods, and reliance on comprehensive reference databases to avoid misidentification. Table 1 provides a brief comparison of Traditional Protein-Based Assays and DNA-Based Techniques.

1.5 DNA Barcoding for Species Identification

1.5.1 Principles and Markers

DNA barcoding targets a standardized, short genetic fragment that is conserved within species but divergent between species. Animal barcoding relies on mitochondrial cytochrome c oxidase I (COI, ~650 bp); plants use rbcL (~600 bp) and matK (~800 bp); and fungi utilize the ITS region (~500–700 bp). Sequence similarity ≥98–99% against reference libraries (Barcode of Life Data Systems [BOLD], GenBank) enables species assignment [23]. Figure 1 shows DNA barcoding workflow for food authentication.

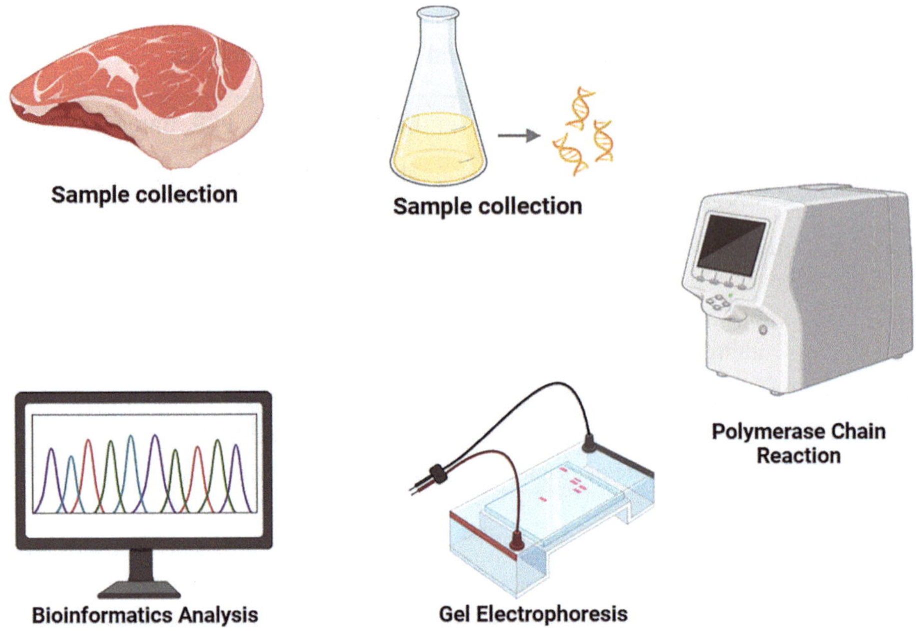

Fig. 1 DNA Barcoding Workflow for Food Authentication

1.6 Key Steps

1. *Sample Collection and Preparation.*
   (a) Obtain representative subsamples (e.g., multiple fillet areas) to address sample heterogeneity.
   (b) Use sterile tools to avoid cross-contamination; assign unique IDs [23].
2. *DNA Extraction.*
   (a) Select protocols adapted to matrix: CTAB/phenol-chloroform for high-fat or phenolic samples (e.g., olive oil, red wine).
   (b) Assess yield and purity (A_{260}/A_{280} ratios via spectrophotometry or fluorometry). Evaluate DNA fragmentation on agarose gel [24].
3. *PCR Amplification.*
   (a) Choose universal or species-specific primers for barcode regions. For suspected adulterants (e.g., pork in beef), employ targeted primers.
   (b) Optimize annealing temperature, Mg^{2+} concentration, and cycle number. Include internal positive controls to monitor inhibition.
   (c) Confirm amplicon size via agarose gel electrophoresis (e.g., 1.5% gel) [24].
4. *Sanger Sequencing (if single-species).*
   (a) Purify PCR products (e.g., ExoSAP).
   (b) Perform bidirectional sequencing; assemble forward and reverse reads into a consensus.

   (c) Use software (e.g., Geneious) for trimming and alignment [25].

5. *Bioinformatics Analysis.*
   (a) Query consensus against reference databases (BOLD and GenBank).
   (b) Construct phylogenetic trees (neighbor-joining or maximum likelihood) to verify clustering with reference sequences.
   (c) Prepare a result table listing sample IDs, barcode region, top BLAST hits, similarity percentages, and identified species [26].

1.7 Applications

- *Seafood*: Detection of substitution (e.g., tilapia sold as snapper).
- *Meat Products*: Identification of undeclared species (e.g., horse meat in beef sausages).
- *Herbal Supplements and Spices*: Uncovering fillers (e.g., corn flour in saffron).
- *Dairy*: Differentiating cow, sheep, and goat milk in cheese [27].

1.8 Limitations

- Best suited for single-species or minimally processed samples; mixed or heavily degraded samples require mini-barcodes (~100–200 bp) at some resolution cost.
- Reliance on complete, accurate reference databases; in-house libraries may be necessary for local or rare species [26].

1.9 Next-Generation Sequencing (NGS) for High-Throughput Authentication

1.9.1 Rationale

NGS platforms (Illumina MiSeq, Ion Torrent, and Oxford Nanopore) generate millions of reads per run. They excel at analyzing complex mixtures by sequencing multiple barcode regions or entire genomes in parallel (Table 2). NGS overcomes Sanger limitations in multiplex detection and processed food analysis [28, 29]. Figure 2 provides a workflow for sequencing processes.

- *Ensure Regulatory Compliance*: Rapidly verify "wild-caught" versus "farmed" fish, safeguarding eco-certification claims.

Table 2
Comparison of DNA Barcoding and NGS in Common Food Authentication Applications

Common Applications	Seafood labeling, single-species meat products	Processed food, herbal supplements, meat mixtures
Sequencing Method	Sanger	Illumina and Nanopore
Bioinformatics Need	Minimal	Essential

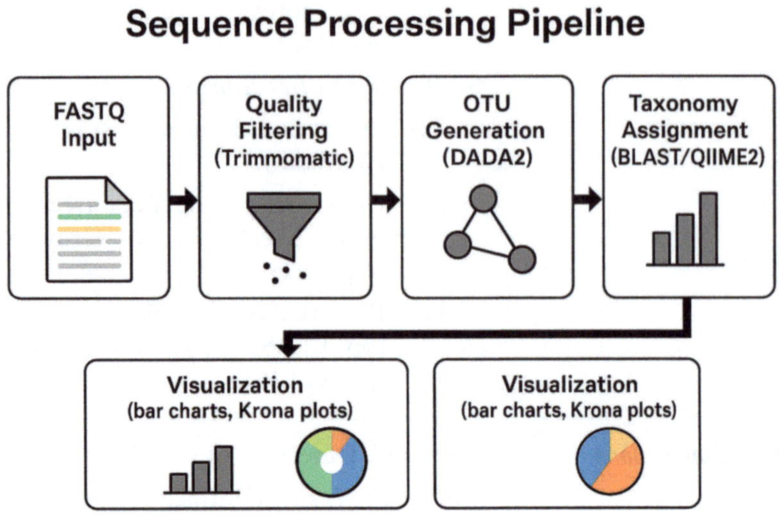

Fig. 2 Workflow for sequencing processes

- *Build Digital Trace Records*: Link DNA-authenticated data to lot numbers in blockchain systems, enabling transparent, immutable provenance tracking.
- *Deter Fraud*: As testing becomes cheaper and more accessible, fraudsters face increased risk of detection, reducing fraudulent activity.

2 Materials

2.1 DNA Barcoding

2.1.1 Sample Types and Collection

DNA barcoding can be applied to a wide range of food products to verify authenticity, detect adulteration, and ensure regulatory compliance. The choice of sample depends on the target organism and the level of processing [28].

(a) *Meat and Poultry*
   - *Fresh/Frozen Meat*: Muscle tissue is ideal for DNA extraction due to high cellularity.
   - *Processed Products* (sausages and canned meats): May require optimized protocols due to DNA degradation from heat/pressure.
   - *Challenges*: Fat and collagen can inhibit PCR; defatting with organic solvents may be needed [30].

(b) *Seafood*
   - *Whole Fish/Filets*: Skin or muscle tissue provides high-quality DNA.

- *Processed Seafood* (surimi and fish balls): Often contains mixed species, requiring multiplex PCR.
- *Adulteration Cases*: Substitution of high-value species (e.g., Atlantic cod with cheaper pollock).

(c) *Herbs and Spices*

- *Dried Leaves/Powders* (e.g., oregano and saffron): High risk of adulteration with fillers (e.g., olive leaves in oregano).
- *Challenges*: Secondary metabolites (e.g., polyphenols) can inhibit PCR; polyvinylpolypyrrolidone (PVPP) is added during extraction to bind inhibitors [31].

(d) *Processed and Composite Foods*

- *Snack Foods and Sauces*: DNA may be fragmented; short barcodes (mini-barcodes) are preferred.
- *Vegetarian Products*: Detection of undeclared animal DNA (e.g., cow milk in "vegan" cheese) [32].

2.1.2 DNA Extraction Methods

The quality of DNA is critical for successful barcoding. Extraction methods must balance yield, purity, and compatibility with downstream PCR [33].

(a) *CTAB-Based Extraction*

- *Protocol*
  - Homogenize sample in CTAB buffer (2% CTAB, 1.4 M NaCl, 20 mM EDTA).
  - Incubate at 65 °C (30 min) to lyse cells.
  - Add chloroform:isoamyl alcohol (24:1) to separate proteins.
  - Precipitate DNA with isopropanol, wash with ethanol.
- *Advantages*: Low cost, effective for plants/fungi.
- *Disadvantages*: Time-consuming; not ideal for processed foods.

(b) *Commercial Kits*

- *Silica-Column Kits* (e.g., DNeasy Blood & Tissue Kit):
  - Bind DNA to silica membranes, wash inhibitors away.
  - Elute in 50–100 μL buffer.
- *Magnetic Bead Kits* (e.g., NucleoSpin Food):
  - Suitable for high-throughput labs.
- *Key Considerations*
  - *Inhibitor Removal*: Additives like PTB (patented inhibitor-removal technology) for challenging matrices.
  - *Yield*: 5–20 μg DNA/g tissue (fresh samples); <1 μg for processed foods.

2.1.3 PCR Amplification

Barcode regions are amplified using species-specific primers. The choice of primers depends on the taxonomic group [34].

(a) *Primer Selection: Table 3 shows the primer selection for PCR amplification.*

(b) *PCR Setup*

- *Reaction Mix*
  - 10X PCR buffer (2.5 μL).
  - dNTPs (0.2 mM each).
  - Primers (0.5 μM each).
  - Taq polymerase (1 U).
  - Template DNA (50–100 ng).
- *Cycling Conditions*
  - Initial denaturation: 95 °C, 5 min.
  - 35 cycles of: 95 °C (30 s), 52 °C (40 s), 72 °C (1 min).
  - Final extension: 72 °C, 5 min.

(c) *Troubleshooting*

- *No Amplification*
  - Check DNA quality (A260/A280 ratio ~ 1.8).
  - Increase $MgCl_2$ concentration (up to 3 mM).
- *Nonspecific Bands*
  - Optimize annealing temperature (gradient PCR).
  - Use touchdown PCR.

2.1.4 Sequencing and Data Analysis

(a) *Sanger Sequencing*

- *Purification*: ExoSAP-IT treatment to remove primers/dNTPs.
- *Sequencing Reaction*: BigDye Terminator v3.1 (Thermo Fisher).
- *Capillary Electrophoresis*: Applied Biosystems 3500xl.

Table 3
Primer selection for PCR amplification

Taxonomic Group	Barcode Gene	Primer Pairs	Amplicon Size
Animal	*COI*	FishF2/FishR2	650 bp
Plants	*rbcL*	rbcLa-F/rbcLa-R	550 bp
Fungi	*ITS*	ITS1/ITS4	300–700 bp

(b) *Bioinformatics*
- *Sequence Quality Control*
  - Trim low-quality bases (Phred score < 20).
  - Assemble forward/reverse reads (e.g., Geneious, Codon Code).
- *BLAST Search*
  - Compare against *BOLD Systems* (https://boldsystems.org) or *NCBI GenBank*.
  - Threshold: ≥98% similarity for species-level ID.

2.2 Next-Generation Sequencing (NGS)

NGS enables high-throughput sequencing of complex food matrices. The choice of platform depends on read length, accuracy, and cost [35].

2.2.1 NGS Platforms

(a) *Illumina (Short-Read Sequencing)*
- *MiSeq*
  - Read length: 2 × 300 bp.
  - Output: 15 GB/run.
  - Best for: 16S/ITS metabarcoding.
- *NovaSeq*
  - Read length: 2 × 150 bp.
  - Output: 6 TB/run.
  - Best for: Whole-genome sequencing.

(b) *Oxford Nanopore (Long-Read Sequencing)*
- *MinION*
  - Read length: Up to 2 Mb.
  - Portable (USB-sized).
  - Best for: Real-time field testing.

2.2.2 Library Preparation

(a) *Metabarcoding (Targeted Amplicon Sequencing)*
- *Primer Design*
  - 16S rRNA (V3–V4 region) for bacteria.
  - *ITS2* for fungi.
- *Dual Indexing*
  - Nextera XT Index Kit (Illumina) for multiplexing.

(b) *Whole-Genome Shotgun Sequencing*
- *DNA Fragmentation*: Covaris ultrasonicator (300 bp fragments).
- *Adapter Ligation*: Illumina TruSeq DNA PCR-Free Kit [36].

2.2.3 Bioinformatics Analysis

(a) *Quality Control*

- *FastQC*: Check per-base sequence quality.
- *Trimmomatic*: Remove adapters/low-quality reads [37].

(b) *Taxonomic Classification*

- *QIIME2*
  - DADA2 for denoising.
  - Feature classifier (sklearn) for OTU assignment.
- *Kraken2*
  - Ultrafast metagenomics classification.

(c) Functional Analysis (Shotgun Data)

- *MetaPhlAn*: Strain-level profiling.
- *HUMAnN3*: Pathway abundance prediction.

2.2.4 Applications in Food Authentication

(a) *Multispecies Detection*

- *Herbal Teas*: Identification of >50 plant species in a single sample.
- *Meat Products*: Detection of 1% pork adulteration in halal beef.

(b) *Allergen Screening*

- *Gluten in "Gluten-Free" Foods*: NGS detects trace contamination.

3 Method

3.1 DNA Barcoding Workflow

DNA barcoding relies on the sequencing of a short, standardized region of DNA that is sufficiently variable to distinguish between species. In animals, the mitochondrial cytochrome c oxidase subunit I (COI) gene is commonly used, while in plants and fungi, rbcL, matK, and ITS regions are preferred. This method is especially useful for the identification of single-species food products and is based on Sanger sequencing technology [38].

3.1.1 Sample Preparation

Homogenization and DNA Extraction

The first step in the DNA barcoding workflow involves collecting and preparing a representative sample of the food product. For consistent DNA extraction, the sample is homogenized through mechanical disruption methods such as grinding, bead beating, or liquid nitrogen pulverization. The goal is to break down cell walls and membranes to release DNA.

DNA is then extracted using the CTAB method or commercial kits such as the Qiagen DNeasy or NucleoSpin kits. The method used is determined by the sample matrix and the level of processing.

The purity and amount of DNA are checked by spectrophotometry (such as NanoDrop), fluorometry (such as Qubit), or by running an agarose gel.

Challenges The DNA extracted from processed foods is generally in a degraded state, as it has been exposed to high temperatures, changes in pH, or preservatives. Recovery of DNA from such conditions should be done with suitable and/or modified protocols [34, 36, 38–41].

3.1.2 PCR Amplification

Target Barcode Region (e. g., COI)

The DNA extracted is a template in a Polymerase Chain Reaction (PCR) to amplify the barcode region. Primers such as LCO1490 and HCO2198 are the most common ones for the COI gene of animal species. The PCR reaction mixture consists of primers, dNTPs, a buffer of the right pH, and Taq DNA polymerase.

Thermal cycling conditions that are optimized are necessary to guarantee the specific amplification and reduction of the nonspecific products. Through agarose gel electrophoresis, one can verify if the PCR products have the expected size (for COI about 650 bp) [38, 39, 42].

3.1.3 Gel Electrophoresis

Confirm PCR Success

The success of the PCR is verified by running the amplified products on a 1–2% agarose gel stained with ethidium bromide or a safer alternative like SYBR Safe. Successful amplification is indicated by a distinct band at the expected size. Failed amplifications may require re-optimization of PCR conditions or use of alternative primers [34].

3.1.4 Sanger Sequencing

Purify and Sequence PCR Products

Amplified products are purified using enzymatic cleanup methods (e.g., ExoSAP-IT) or spin columns. Purified PCR products are then subjected to Sanger sequencing, which provides high-quality, single-read sequences. Both forward and reverse primers are typically used to ensure complete and accurate sequence coverage.

Sequencing results are visualized using chromatogram software such as FinchTV or Geneious. A consensus sequence is generated by aligning forward and reverse reads [34, 42].

3.1.5 Data Analysis

Compare Sequences to Reference Databases

The resulting sequences are compared to reference databases using Basic Local Alignment Search Tool (BLAST) against NCBI GenBank or Barcode of Life Data Systems (BOLD). A species match is typically confirmed when sequence similarity is equal to or greater than 98%.

Limitations DNA barcoding is effective for identifying dominant species in a sample but is limited in detecting multiple species or minor components in mixed samples [43].

3.2 Next-Generation Sequencing (NGS) Workflow

NGS technologies enable the simultaneous sequencing of millions of DNA fragments, making them ideal for analyzing complex or mixed food products. NGS offers high sensitivity, the ability to detect low-abundance species, and is useful in examining processed and degraded food items [35].

3.2.1 DNA Extraction

High-Quality DNA for Degraded Samples

As with DNA barcoding, DNA extraction is the first and crucial step in the NGS workflow. However, NGS often requires higher DNA quality and quantity. Specialized extraction methods may be required for processed foods, which typically yield fragmented or chemically modified DNA.

Kits designed for environmental or forensic samples (e.g., PowerSoil, QIAamp DNA Investigator) often include inhibitor removal steps and yield cleaner DNA suitable for downstream applications. DNA quality is assessed using spectrophotometry and fluorometry, and fragment size can be evaluated via gel electrophoresis or Bioanalyzer [44].

3.2.2 Library Preparation

Fragmentation, Adapter Ligation, and Amplification

Library preparation involves several steps:

- *Fragmentation*: DNA is fragmented to a specific size range (typically 200–500 bp) using enzymatic digestion or sonication.
- *End Repair and Adapter Ligation*: Fragmented DNA ends are polished and ligated to platform-specific sequencing adapters. Unique index sequences (barcodes) are added for sample multiplexing.
- *PCR Amplification*: The adapter-ligated fragments are amplified using PCR to enrich the library.
- *Size Selection and Cleanup*: Libraries are purified using magnetic bead-based methods (e.g., AMPure XP).

  The final libraries are quantified using Qubit or real-time PCR to ensure consistent loading on the sequencing platform [36].

3.2.3 Sequencing

Illumina/Nanopore Platforms

Prepared libraries are sequenced using high-throughput platforms such as:

- *Illumina MiSeq/NextSeq*: Known for high accuracy and short read lengths (75–300 bp).
- *Oxford Nanopore Technologies (ONT)*: Provides longer read lengths and real-time data acquisition, but with higher error rates.

  The choice of platform depends on the study objectives, desired read length, cost, and turnaround time. For most food authentication studies, Illumina's short-read technology is preferred due to its accuracy [45].

3.2.4 Bioinformatics Analysis

Read Trimming, OTU Clustering, Taxonomic Assignment

NGS data requires extensive bioinformatics processing. The main steps include:

- *Quality Control*: Remove low-quality reads, adapter sequences, and chimeras using tools like Trimmomatic, Cutadapt, and FastQC.
- *Operational Taxonomic Unit (OTU) Clustering*: Reads are clustered based on sequence similarity (e.g., 97%) using tools like USEARCH, VSEARCH, or DADA2 (for amplicon sequence variants).
- *Taxonomic Assignment*: Representative sequences from each OTU are compared to curated databases (e.g., SILVA, BOLD, and UNITE) to assign taxonomy.
- *Visualization and Reporting*: Relative abundances of identified species are plotted using statistical and visualization tools such as QIIME2, Krona, or MEGAN.

Strengths NGS can identify all species in a complex sample, including minor or unexpected components (Table 4). It is particularly valuable in detecting food fraud, substitution, contamination, or allergenic ingredients [37].

Table 4
Comparative Analysis: DNA Barcoding vs. NGS

Feature	DNA Barcoding (Sanger)	Next-Generation Sequencing (NGS)
Species Detection	Single species	Multiple species
Sensitivity	Moderate	High
Throughput	Low	High
DNA Quality Requirement	Moderate	High (can be optimized for degraded DNA)
Cost per Sample	Moderate	Lower with multiplexing
Turnaround Time	Short (2–3 days)	Moderate (3–5 days)
Complexity	Low (simple lab and software needs)	High (requires bioinformatics expertise)
Equipment Required	Basic PCR and sequencer	Advanced NGS sequencer
Best Use Case	Simple or single-species products	Mixed or processed food authentication

4 Notes

Primer Selection: Ensuring Specificity and Avoiding Cross-Species Amplification

The selection of appropriate primers is foundational to the success of any DNA barcoding analysis. Primers must bind specifically to the target DNA region without amplifying non-target species. Poorly chosen primers can result in cross-species amplification, leading to ambiguous or incorrect species identification. This is particularly crucial in processed food matrices, where multiple species may be present.

Primers should be designed based on the genetic variability of the region targeted for barcoding, such as cytochrome c oxidase I (COI) for animals or rbcL and matK for plants. Universal primers can be useful for amplifying a wide range of taxa, but they may lack the specificity needed for closely related species. In such cases, species-specific primers are essential. Tools such as Primer-BLAST and in silico PCR simulations are invaluable for assessing primer specificity before laboratory validation [46].

Database Limitations: Incomplete Reference Libraries and False Identifications

Species identification through DNA barcoding can only be as accurate as the reference databases like GenBank, BOLD, and others that are used. The queried sequence will remain unrecognized or will be assigned incorrectly if it does not fit any of the reference sequences in the database, even if it is a closely related species.

In these databases, incomplete or incorrect entries can result in false identification, especially when dealing with taxa that are poorly represented or newly described. Hence, it is important for those who curate the databases to be very careful in their work and to include voucher specimens, so that the results of barcoding can be relied upon. Scientists and industry workers should think about giving these databases verified sequences so that they can be of greater help [47].

Degraded DNA: Challenges in Processed Foods and Optimization of Extraction Protocols

Processed food is often given a hard time by heat and chemicals, or it can be mechanically transformed resulting in damaged DNA that is difficult to extract and amplify. The fragmentation of DNA inevitably results in shorter sequences that may not cover the whole barcode region.

It is very important to use optimized DNA extraction protocols that take into account processed food matrices specifically if we want to safely detect DNA in processed food. Some of the techniques that can be used to achieve that are, for instance, the elimination of impurities by the help of the silica column or magnetic beads, etc. Besides that, the targeting of shorter barcode regions (mini-barcodes), which are more likely to be intact in degraded samples, can increase the chances of success.

Standard protocol for quality control is to use several methods all at once (spectrophotometry, fluorometry, gel electrophoresis) for determining the concentration and integrity of DNA. The selection of the enzyme for the DNA polymerase and the setting of PCR must be compatible with the condition of the degraded DNA, too [48].

Regulatory Compliance: Validation According to ISO, FDA, or EU Standards

DNA barcoding methods that are going to be used in regulatory and legal contexts have to be validated in accordance with the internationally recognized standards. Examples of such standards are those that the ISO, FDA, or EU issue. These validation parameters cover a wide range of factors including specificity, sensitivity, repeatability, reproducibility, and the limit of detection. Accreditation according to the standard ISO 17025 means that the laboratories have been found technically competent and that their results are valid. The FDA and EU have also issued guidance documents on DNA-based methods for food authentication, which should be consulted during method development.

It should be noted that going through proficiency testing and inter-laboratory comparisons is a big step in solidifying the reliability of DNA barcoding methods. Tracing and documentation of procedures and results are of utmost importance for compliance at regulatory and audit requirements [49].

Applications of DNA Barcoding in Food Authentication

Seafood Authentication: Uncovering Species Substitution and Mislabeling

Seafood is undoubtedly the most misinterpreted food products that are recognized as a category globally. It has been found that most of the species which are of high value are usually replaced with those of lower value without the knowledge of consumers. In the world of economics, high-value species are often replaced with cheaper substitutes of lower value. If we take one example, the "red snapper" that can be changed with the various fish species that are indistinguishable only visually but of lesser quality.

DNA barcoding is a method that can be very effective to ensure the authenticity of seafood as it is the similarity in the morphology of species and the fact that the products are usually filleted or processed that makes the absence of the identifying features. DNA barcoding by focusing on mitochondrial COI genes can not only detect fish species in their various forms of processing but also can do so with high precision.

Government authorities and nongovernmental organizations (NGOs) have extensively performed seafood monitoring with the help of DNA barcoding, which has resulted in the identification of a very large number of mislabelings in various food service establishments and retail outlets. The statistics published seem to be the reason for the coming of the new regulations that call for more rigorous labeling and the increased habit of the consumer in the matter of labeling. The method of DNA barcoding, for example, enables the traces of fishing to be followed and assists in the enforcement of regulations about the total catch of each species, thus ensuring that the sources of the seafood are legal and sustainable [50].

Herbs and Spices: Detecting Adulteration and Ensuring Authenticity

The high market value and powdered form of many herbs and spices make them vulnerable to adulteration. Common adulterants include cheaper plant materials that mimic the texture or color of the authentic product. For example, oregano is frequently adulterated with olive leaves, myrtle, or other green leaves. DNA barcoding provides a reliable method for detecting such adulteration. Chloroplast genes like rbcL and matK are often targeted in plant barcoding. However, due to the complex matrices and potential for DNA degradation in dried and ground samples, mini-barcoding and high-sensitivity PCR techniques are often employed.

Authenticating herbal supplements and spices not only protects consumers from economic fraud but also from potential health risks associated with consuming unidentified or allergenic plant material. DNA barcoding also supports regulatory enforcement against counterfeit products in the global spice trade [51].

Meat Products: Species Identification in Sausages and Ready-to-Eat Foods

Meat products, especially those that are minced or processed, are difficult for species identification if one relies on visual inspection only. Such a situation, caused by visual inspection loopholes in the meat market, creates opportunities for fraudulent practices in which, for example, the substitution of flavorful beef or lamb with inexpensive meats like pork or poultry occurs.

Examples of such are pork, which is often sold disguised as beef and chicken meat. DNA barcoding can spot such cases with accuracy that is near to perfect. The mitochondrial COI gene, together with nuclear genes such as 12S and 16S rRNA, is used most often for animal species identification. When it comes to complicated products such as sausages, multiplex PCR, and quantitative PCR (qPCR) methods greatly facilitate the work by allowing the researchers to detect and quantify several species concomitantly in the same assay.

Recently, many incidents in the food industry, examples being the horse meat scandal which happened in Europe, strongly attest that the institutions responsible for the verification of the presence or absence of food fakes are truly necessary. Barcoding DNA is a great method to confirm not only the economic deception but it also indicates they are cooperating with the religion and diet restrictions (e.g., Halal or Kosher) set forth by the authorities [28].

Future Directions in DNA Barcoding for Food Authentication

As DNA barcoding matures and its adoption becomes more widespread, several technological and analytical advancements are poised to enhance its capabilities. These innovations aim to make barcoding faster, more portable, and more accurate, thereby broadening its applicability and increasing its value in real-time and field-based contexts [52].

Portable Sequencing Technologies: Real-Time Authentication Using Nanopore Devices

One of the most promising developments in DNA barcoding is the coming of portable sequencing technologies like Oxford Nanopore Technologies (ONT). They open up many new possibilities for real-time DNA sequencing and species identification in nontraditional lab settings.

Nanopore sequencing also enables food inspectors, customs officers, and quality control personnel to perform species verification at points of entry, processing facilities, and retail outlets. The portability feature of these devices, together with cloud-based analysis platforms, enables nearly instant results.

While nanopore sequencing still has its limitations, such as high error rates and the necessity of good sample preparation protocols, it brings into play speed and flexibility. For instance, the constant advancements in base-calling algorithms, sequencing chemistry, and sample handling serve to alleviate these problems swiftly [53].

AI-Assisted Bioinformatics: Rapid and Accurate Species Identification

Artificial intelligence (AI) and machine learning (ML) are among the most significant advancements in bioinformatics, which have already drastically changed the way DNA barcode data is dealt with. Traditional sequence alignment methods can be very slow and require a lot of computational power, especially when large datasets or mixed-species samples are being handled.

AI algorithms can very quickly identify unknown sequences by drawing examples from curated training datasets. Such systems become better as they gain experience and can even be modified to detect the unexpected or the exceedingly rare sequences that, by definition, are not going to be in the existing databases.

With the help of AI, portable sequencing, it becomes feasible to realize the dream of a complete automated pipeline for food authentication. This would facilitate the nonspecialists in actually performing DNA barcoding analyses accurately and with minimum training, thus not only democratizing the technology but also increasing its reach [54].

Metabarcoding and Multi-locus Approaches

Scientists anticipate that future implementations will almost certainly shift from single-locus barcoding aimed to metabarcoding, and multi-locus identification methods. Such techniques identify several DNA regions at the same time thus giving them higher resolving power and being more robust, especially for complicated food mixtures.

Metabarcoding, which is most of the time accomplished by NGS, makes it possible not only to detect but also to quantify different species' presence in a single sample. Hence, this may be of great assistance when one needs to find out what is there as a contaminant or to corroborate the list of ingredients in the products that contain several species, such as pet food, processed meats, and herbal teas.

On the other hand, the simultaneous utilization of several barcode loci can consolidate the diagnosis and eliminate the mistakes, which are caused by the biases that are specific to the loci. Besides the fact that this tendency goes hand in hand with the movement toward comprehensive and transparent food labeling, it is also an indicator of the trend in the food industry that is aimed at such a direction [55].

Blockchain Integration and Digital Traceability

DNA barcoding data can be integrated with blockchain technology to create immutable, transparent records of food provenance. When a food product is authenticated using DNA barcoding, that verification can be logged on a blockchain ledger accessible throughout the supply chain.

Such integration enhances consumer trust and simplifies audits, recalls, and regulatory inspections. Blockchain also facilitates real-time traceability from farm to fork, supporting sustainability and ethical sourcing initiatives [56].

References

1. Wadood SA et al (2025) Authentication of edible herbal materials and food products using mass spectrometry based metabolites and inorganic constituents. Food Chem 463 (Pt 4):141424. https://doi.org/10.1016/j.foodchem.2024.141424
2. Tinacci L, Stratev D, Strateva M, Zhelyazkov G, Kyuchukova R, Armani A (2023) An authentication survey on retail seafood products sold on the Bulgarian market underlines the need for upgrading the

traceability system. Foods 12(5). https://doi.org/10.3390/foods12051070

3. Chaudhary V, Kajla P, Dewan A, Pandiselvam R, Socol CT, Maerescu CM (2022) Spectroscopic techniques for authentication of animal origin foods. Front Nutr 9:979205. https://doi.org/10.3389/fnut.2022.979205
4. Nehal N, Choudhary B, Nagpure A, Gupta RK (2021) DNA barcoding: a modern age tool for detection of adulteration in food. Crit Rev Biotechnol 41(5):767–791. https://doi.org/10.1080/07388551.2021.1874279
5. Darnal A, Poggesi S, Longo E, Arbore A, Boselli E (2023) Decoding the identity of pinot gris and pinot noir wines: a comprehensive chemometric fusion of sensory (from dual panel) and chemical analysis. Foods 13(1). https://doi.org/10.3390/foods13010018
6. Vignani R, Lio P, Scali M (2019) How to integrate wet lab and bioinformatics procedures for wine DNA admixture analysis and compositional profiling: case studies and perspectives. PLoS One 14(2):e0211962. https://doi.org/10.1371/journal.pone.0211962
7. Do AB, Khuda SE, Sharma GM (2018) Undeclared food allergens and gluten in commercial food products analyzed by ELISA. J AOAC Int 101(1):23–35. https://doi.org/10.5740/jaoacint.17-0384
8. Blickem ER, Bell JW, Baumgartel DM, DeBEER J (2022) Review and analysis of tuna recalls in the United States, 2002 through 2020. J Food Prot 85(1):60–72. https://doi.org/10.4315/JFP-21-254
9. Aksun Tumerkan ET (2022) Detection of Parvalbumin fish allergen in canned tuna by real-time PCR driven by tuna species and can-filling medium. Molecules 27(17). https://doi.org/10.3390/molecules27175674
10. Blanco-Fernandez C et al (2021) Fraud in highly appreciated fish detected from DNA in Europe may undermine the development goal of sustainable fishing in Africa. Sci Rep 11(1):11423. https://doi.org/10.1038/s41598-021-91020-w
11. Maddux PT, Farrell MB, Ewing JA, Tilkemeier PL (2018) Improved compliance with reporting standards: a retrospective analysis of Intersocietal accreditation commission nuclear cardiology laboratories. J Nucl Cardiol 25(3):986–994. https://doi.org/10.1007/s12350-016-0713-y
12. Food and Drug Administration, HHS (2016) Amendments to accreditation of third-party certification bodies to conduct food safety audits and to issue certifications to provide for the user fee program. Final rule. Fed Regist 81(240):90186–90194
13. Menozzi D, Yeh C-H, Cozzi E, Arfini F (2022) Consumer preferences for cheese products with quality labels: the case of Parmigiano Reggiano and Comte. Animals (Basel) 12(10). https://doi.org/10.3390/ani12101299
14. Wang J et al (2025) Food labeling and Chinese consumer preference for naturalness: a new way to differentiate grass-fed dairy products. J Dairy Sci 108(3):2340–2353. https://doi.org/10.3168/jds.2024-25576
15. Wei X, Shao X, Wei Y, Cheong L, Pan L, Tu K (2018) Rapid detection of adulterated peony seed oil by electronic nose. J Food Sci Technol 55(6):2152–2159. https://doi.org/10.1007/s13197-018-3132-z
16. Dou X et al (2018) Multispecies adulteration detection of camellia oil by chemical markers. Molecules 23(2). https://doi.org/10.3390/molecules23020241
17. Garcia JL et al (2024) Fishy business in Seattle: Salmon mislabeling fraud in sushi restaurants vs grocery stores. PLoS One 19(11):e0311522. https://doi.org/10.1371/journal.pone.0311522
18. Chrzanowska-Lightowlers ZM, Lightowlers RN (2024) Mitochondrial RNA maturation. RNA Biol 21(1):28–39. https://doi.org/10.1080/15476286.2024.2414157
19. Xie B, Chen J, Wang Z, Yin Q, Dai Z-M (2024) Sweet enhancers of polymerase chain reaction. PLoS One 19(10):e0311939. https://doi.org/10.1371/journal.pone.0311939
20. Xu M-R et al (2023) Development of a specific and sensitive diagnostic PCR for rapid molecular authentication of the medicinal plant Portulaca oleracea. Mol Cell Probes 67:101890. https://doi.org/10.1016/j.mcp.2022.101890
21. Rice ME, Zou Y, Millar JG, Hanks LM (2020) Complex blends of synthetic pheromones are effective multi-species attractants for Longhorned beetles (Coleoptera: Cerambycidae). J Econ Entomol 113(5):2269–2275. https://doi.org/10.1093/jee/toaa157
22. Yin Z, Yuan B, Lyu W, Huang Q, Simon JE, Wu Q (2022) Method development and validation for analysis of phenolic compounds in fatty complex matrices using enhanced matrix removal (EMR) lipid cleanup and UHPLC-QqQ-MS/MS. Food Chem 373 (Pt A):131096. https://doi.org/10.1016/j.foodchem.2021.131096
23. Madeira S, Duarte A, Boinas F, Costa Osorio H (2021) A DNA barcode reference library of Portuguese mosquitoes. Zoonoses Public Health 68(8):926–936. https://doi.org/10.1111/zph.12885

24. Dieki R, Eyang-Assengone E-R, Makouloutou-Nzassi P, Bangueboussa F, Nsi Emvo E, Akue JP (2022) Comparison of six methods for Loa loa genomic DNA extraction. PLoS One 17(3):e0265582. https://doi.org/10.1371/journal.pone.0265582
25. Maki A, Rissanen AJ, Tiirola M (2016) A practical method for barcoding and size-trimming PCR templates for amplicon sequencing. BioTechniques 60(2):88–90. https://doi.org/10.2144/000114380
26. Damaso N, Elwick KE, Robertson JM (2024) Guidelines for the analysis of DNA barcoding/metabarcoding sequencing data and interpretation of publicly available databases. Methods Mol Biol 2744:391–402. https://doi.org/10.1007/978-1-0716-3581-0_25
27. Zhou H-Q et al (2024) Identification of Cervi Cornu (Cervus elaphus) and its formula granules based on PCR-RFLP. Zhongguo Zhong Yao Za Zhi 49(6):1517–1525. https://doi.org/10.19540/j.cnki.cjcmm.20231213.102
28. Mualim M, Latif H, Pisestyani H, Rahayu P (2024) Analysis of species adulteration in beef sausage using real-time polymerase chain reaction in Makassar, Indonesia. Vet World 17(10):2355–2364. https://doi.org/10.14202/vetworld.2024.2355-2364
29. Kourkouli A, Thomaidis N, Dasenaki M, Markou A (2024) Novel and sensitive touchdown polymerase chain reaction assays for the detection of goat and sheep milk adulteration with cow milk. Molecules 29(8). https://doi.org/10.3390/molecules29081820
30. Ghiasvand R, Marvast LD, Shariatpanahi SP, Pourmasoumi M, Clark CCT, Haeri F (2020) The association between animal flesh foods consumption and semen parameters among infertile Iranian men: a cross-sectional study. Nutr J 19(1):113. https://doi.org/10.1186/s12937-020-00633-w
31. Yao L, Wang S, Xue H, Fei Y, Xiong X, Xiong X (2024) Fish species mislabelling in China: evidence from a survey of frozen, dried, and surimi-based fish products marketed as cod-related species. Food Addit Contam Part A Chem Anal Control Expo Risk Assess 41(2):134–142. https://doi.org/10.1080/19440049.2024.2302962
32. Prager L, Simon JC, Treudler R (2023) Food allergy - new risks through vegan diet? Overview of new allergen sources and current data on the potential risk of anaphylaxis. J Dtsch Dermatol Ges 21(11):1308–1313. https://doi.org/10.1111/ddg.15157
33. Schiebelhut LM, Abboud SS, Gomez Daglio LE, Swift HF, Dawson MN (2017) A comparison of DNA extraction methods for high-throughput DNA analyses. Mol Ecol Resour 17(4):721–729. https://doi.org/10.1111/1755-0998.12620
34. Green MR, Sambrook J (2019) Polymerase Chain Reaction (PCR) amplification of GC-Rich templates. Cold Spring Harb Protoc 2019(2). https://doi.org/10.1101/pdb.prot095141
35. Chen J et al (2024) Solution to a case involving the interpretation of trace degraded DNA mixtures. Int J Legal Med 138(6):2325–2330. https://doi.org/10.1007/s00414-024-03302-3
36. Rozenberg A, Leese F, Weiss LC, Tollrian R (2016) Digital gene expression analysis with sample multiplexing and PCR duplicate detection: a straightforward protocol. BioTechniques 61(1):26–32. https://doi.org/10.2144/000114434
37. Fitzpatrick AH, Rupnik A, O'Shea H, Crispie F, Keaveney S, Cotter PD (2023) Benchmarking bioinformatic tools for amplicon-based sequencing of norovirus. Appl Environ Microbiol 89(1):e0152222. https://doi.org/10.1128/aem.01522-22
38. Antil S et al (2023) DNA barcoding, an effective tool for species identification: a review. Mol Biol Rep 50(1):761–775. https://doi.org/10.1007/s11033-022-08015-7
39. Shinozuka H et al (2015) A simple method for semi-random DNA amplicon fragmentation using the methylation-dependent restriction enzyme MspJI. BMC Biotechnol 15:25. https://doi.org/10.1186/s12896-015-0139-7
40. Green MR, Sambrook J (2019) Agarose gel electrophoresis. Cold Spring Harb Protoc 2019(1). https://doi.org/10.1101/pdb.prot100404
41. Edwin NR, Fitzpatrick AH, Brennan F, Abram F, O'Sullivan O (2024) An in-depth evaluation of metagenomic classifiers for soil microbiomes. Environ Microbiome 19(1):19. https://doi.org/10.1186/s40793-024-00561-w
42. Wang Y et al (2021) High-throughput identification of allergens in a food system via hybridization probe cluster-targeted next-generation sequencing. J Agric Food Chem 69(40):11992–12001. https://doi.org/10.1021/acs.jafc.1c03595
43. Versmessen N et al (2024) Comparison of DeNovix, NanoDrop and Qubit for DNA quantification and impurity detection of bacterial DNA extracts. PLoS One 19(6):e0305650. https://doi.org/10.1371/journal.pone.0305650

44. Viljoen R, Reid KM, Mole CG, Rangwaga M, Heathfield LJ (2022) Towards molecular autopsies: development of a FFPE tissue DNA extraction workflow. Sci Justice 62(2): 137–144. https://doi.org/10.1016/j.scijus.2021.12.005
45. Chang JJM, Ip YCA, Neo WL, Mowe MAD, Jaafar Z, Huang D (2024) Primed and ready: nanopore metabarcoding can now recover highly accurate consensus barcodes that are generally indel-free. BMC Genomics 25(1): 842. https://doi.org/10.1186/s12864-024-10767-4
46. Kurata S, Mano S, Nakahama N, Hirota SK, Suyama Y, Ito M (2024) Development of mitochondrial DNA cytochrome c oxidase subunit I primer sets to construct DNA barcoding library using next-generation sequencing. Biodivers Data J 12:e117014. https://doi.org/10.3897/BDJ.12.e117014
47. Kjaerandsen J (2022) Current state of DNA barcoding of sciaroidea (Diptera)-highlighting the need to build the reference library. Insects 13(2). https://doi.org/10.3390/insects13020147
48. Shokralla S, Hellberg RS, Handy SM, King I, Hajibabaei M (2015) A DNA mini-barcoding system for authentication of processed fish products. Sci Rep 5:15894. https://doi.org/10.1038/srep15894
49. Okezue MA, Adeyeye MC, Byrn SJ, Abiola VO, Clase KL (2020) Impact of ISO/IEC 17025 laboratory accreditation in sub-Saharan Africa: a case study. BMC Health Serv Res 20(1):1065. https://doi.org/10.1186/s12913-020-05934-8
50. Willette DA et al (2017) Using DNA barcoding to track seafood mislabeling in Los Angeles restaurants. Conserv Biol 31(5):1076–1085. https://doi.org/10.1111/cobi.12888
51. Parvathy VA, Swetha VP, Sheeja TE, Sasikumar B (2015) Detection of plant-based adulterants in turmeric powder using DNA barcoding. Pharm Biol 53(12):1774–1779. https://doi.org/10.3109/13880209.2015.1005756
52. Akintan OA, Gebremedhin KG, Uyeh DD (2025) Linking animal feed formulation to milk quantity, quality, and animal health through data-driven decision-making. Animals (Basel) 15(2). https://doi.org/10.3390/ani15020162
53. Oliva M, Milicchio F, King K, Benson G, Boucher C, Prosperi M (2020) Portable nanopore analytics: are we there yet? Bioinformatics 36(16):4399–4405. https://doi.org/10.1093/bioinformatics/btaa237
54. Howard C, Lockie-Williams C, Slater A (2020) Applied barcoding: the practicalities of DNA testing for herbals. Plants (Basel) 9(9). https://doi.org/10.3390/plants9091150
55. Yu J, Wu X, Liu C, Newmaster S, Ragupathy S, Kress WJ (2021) Progress in the use of DNA barcodes in the identification and classification of medicinal plants. Ecotoxicol Environ Saf 208:111691. https://doi.org/10.1016/j.ecoenv.2020.111691
56. Conter M (2024) Recent advancements in meat traceability, authenticity verification, and voluntary certification systems. Ital J Food Saf 14(1). https://doi.org/10.4081/ijfs.2024.12971

Chapter 9

Spectroscopic Techniques (NIR, FTIR, and Raman) for Nutritional Profiling

Aminu Abdullahi, Tarangini Korumilli, Saranyadevi Subburaj, and Sufiyanu Jiga Abubakar

Abstract

Nutritional profiling is critical for evaluating food safety, quality, and regulatory compliance. Spectroscopic techniques such as Near Infrared (NIR), Fourier Transform Infrared (FTIR), and Raman spectroscopy have emerged as reliable, rapid, and nondestructive techniques for analyzing nutritional components in food. These approaches provide insights into macromolecules and micronutrients with minimal sample preparation. NIR spectroscopy is favored for the analysis of moisture, fat, and protein due to its swift results and user-friendly nature. FTIR spectroscopy enables the precise identification of molecular signatures, supporting the analysis of functional groups and structural changes in food compositions. Raman spectroscopy complements these techniques and offers specific vibrational data, particularly in aqueous media. This chapter examines the principles and advancements of NIR, FTIR, and Raman spectroscopy in nutritional profiling, emphasizing their superiority over conventional methods. Moreover, it explores the methods for data interpretation, chemometric strategies, and prospective advancements in the realm of spectroscopic food analysis, thereby illustrating their capacity for instantaneous quality assurance and nutritional evaluation within the food sector.

Key words NIR spectroscopy, FTIR spectroscopy, Raman spectroscopy, Nutritional profiling, Food analysis

1 Introduction

Spectroscopic techniques such as Near-Infrared (NIR), Fourier Transform Infrared (FTIR), and Raman have become pivotal in the nutritional profiling of food products [1]. These methods offer rapid, nondestructive, and cost-effective means to analyze the chemical composition of foods, providing detailed insights into their nutritional content. By leveraging the unique properties of molecules, these techniques can identify and quantify various nutritional components, making them invaluable tools in food quality and safety assessments. Moreover, spectroscopic methods can provide metabolomic fingerprinting, allowing for the detection

Tanmay Sarkar (ed.), *Novel Food Analysis: Techniques and Applications*, Methods and Protocols in Food Science,
https://doi.org/10.1007/978-1-0716-4787-5_9,

and quantification of key nutritional constituents such as aromatic amino acids, peptides, carotenoids, and flavonoids [2]. Researchers and food industries has frequently utilized these techniques to achieve high-throughput, cost-effective, and rapid assessments of food quality and safety, enhancing understanding of nutritional content in various food matrices.

Among the spectroscopic techniques, NIR spectroscopy has gained wide acceptance in the field of food analysis. This has been mainly due to its ease of application and the ready availability of chemometric programs for the development of models. NIR spectroscopy can be used to measure the absorption of near-infrared light by molecular bonds, particularly those involving hydrogen, carbon, and oxygen [3]. This technique is widely used for both qualitative and quantitative analysis of food nutrients. It is effective in determining the content of macronutrients such as proteins, fats, and carbohydrates in food products. NIR has been successfully applied to classify cereal foods based on their nutritional components using principal component analysis (PCA) and partial least squares (PLS) regression [4]. The major advantages of NIR spectroscopy is rapid and requires minimal sample preparation, making it suitable for at-line and online food analysis [5]. The use of NIR monochromatic light excitation reduces the interference from fluorescence, which is often experienced with its use in the analysis of pigmented materials. As a Fourier-transform method, it has the advantage of excellent frequency precision. These attributes, plus the ease of sample preparation, make it amenable for consideration as a routine method of analysis in food technology.

FTIR was among the renowned and reliable spectroscopic techniques, which is capable of high-throughput, cost-effective, easy, and green analyses with minimum or no sample preparation. In other words, it is possible to obtain nutritional fingerprints for a wide variety of foods without any reagent or preprocessing. Additionally, FTIR has the spectacular advantages of rapid, effective, and robust detection in herbal products when compared to traditional analytical methods. It involves measuring the absorption of mid-infrared light, which causes molecular vibrations to obtain a molecular fingerprint of food samples [5]. FTIR can detect and quantify various food constituents, including vitamins, amino acids, and other metabolites [6]. It has been used to accurately measure vitamin C content in food products with high prediction correlation [7]. FTIR can also be combined with multivariate statistical techniques to enhance data interpretation.

Moreover, Raman spectroscopy also emerges as an essential tool for nutritional profiling, which can be used to study molecular vibrations and offer comprehensive spectral information. It makes it possible to analyze important nutrients, bioactive substances, and even dietary toxins all at once. It can provide information about molecular structure through measurement of the energy of molecular vibrational-state transitions. This technique is particularly appealing for food quality control, product development, and regulatory compliance due to its nondestructive nature, low sample preparation requirements, and real-time analysis.

Raman spectroscopy, especially in the form of Surface-Enhanced Raman Spectroscopy (SERS) and Raman imaging, will continue to improve the accuracy and scope of food analysis as technology develops, with great promise for both commercial and research applications. However, Raman spectroscopy has been, until recently, a less accessible and more instrumentally demanding technique in the field of food analysis. Due to the fact that Raman scattering effect is based on polarizability of bonds and not their dipoles, it is relatively insensitive to the presence of water.

This chapter provides an overview on the applications of FTIR, NIR, and Raman spectroscopy in the determination of authenticity, quality, safety, and other essential parameters in foods. It also delves into working principles, mode of operation, data retrieval, and interpretation while using each of these techniques.

1.1 Near Infrared (NIR) Spectroscopy in Nutritional Profiling

NIR spectroscopy is an essential analytical technique that utilizes near-infrared spectrum to examine molecular structure through light absorption. It enables noninvasive, real-time measurements, serving as a significant asset for research and practical use [8]. Its nondestructive qualities facilitate quick and economic analysis, positioning it as a viable substitute for conventional laboratory techniques. The NIR section of the electromagnetic spectrum shows remarkable properties in the wavelength range 700–2500 nm that allow for effective material characterization. This portion of the spectrum provides an ideal amalgamation of characteristics for the examination of a diverse array of samples, encompassing solids, slurries, and liquids. Light interacts with various bonds (such as OH, NH, and CH), with specific wavelengths linked to each type [9]. Proteins, for example, are defined by the presence of NH bonds in amino acids. NH bonds exhibit absorption of NIR radiation across multiple segments of the NIR spectrum at different intensities. Increased protein concentration correlates with enhanced energy absorption in specific regions, resulting in diminished reflected energy intensity in those areas.

1.2 Fourier-Transform Infrared (FTIR) Spectroscopy

FTIR Spectroscopy is a sophisticated analytical method used in nutritional profiling to detect and measure different food ingredients. FTIR produces a molecular fingerprint of a sample by measuring the absorption of infrared light at various wavelengths, indicating the presence of important nutrients such as proteins, lipids, carbohydrates, and moisture [10]. The technique is preferred in food analysis because it is quick, accurate, and requires little to no sample preparation, which makes it perfect for regular quality control and studies.

Among the common FTIR techniques, Attenuated Total Reflectance (ATR-FTIR) is widely used. Because it is simple to use and can be applied to a variety of sample types, including liquids, solids, and semisolids [11]. It makes it possible to analyze food samples directly without requiring complicated preparation procedures. However, transmission FTIR requires the sample to be in a liquid or thin film condition to generate high-resolution data. When applied to powdered or heterogeneous food samples, the Diffuse Reflectance Infrared Fourier Transform (DRIFT) provides information about intricate food matrices like flour and grains.

Accurate nutrient estimation in food products can be accomplished by calibrating these spectral areas using reference standards [12]. FTIR is employed not only for nutritional profiling but also for food adulteration detection, shelf-life monitoring, and product authenticity verification. It can detect undesirable or replacement ingredients in food, including artificial additives in honey or diluted milk. Moreover, FTIR helps monitor chemical alterations like protein breakdown or lipid oxidation that occur during storage. It is a useful technique in contemporary food science and nutritional assessment because of its capacity to deliver thorough, quick, and nondestructive analysis [5]. To accommodate various food matrices and analytical requirements, several FTIR techniques have been created and improved. Every one of these methods has unique benefits and uses in nutritional analysis [4]. The instrumentation setup of FTIR is diagrammatically represented in Fig. 1.

The FTIR spectrometer uses an instrument known as an interferometer to simultaneously gather data at all wavelengths. It

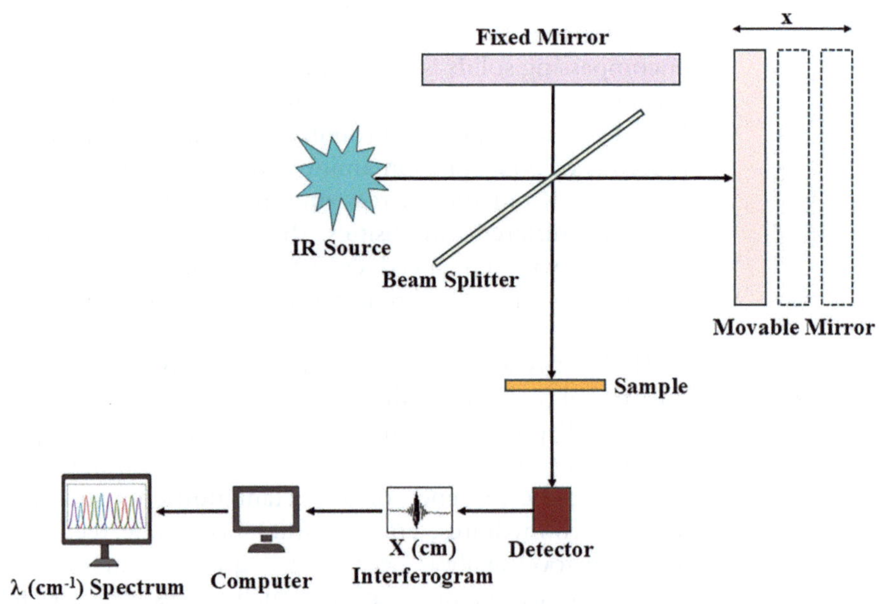

Fig. 1 Instrumentation setup of Fourier Transform Infrared Spectroscopy

records an interferogram, which contains all the spectrum data stored in the time domain, as opposed to recording absorption one wavelength at a time (as in traditional dispersive IR) [13]. A conventional infrared spectrum (intensity vs. frequency or wavenumber) is then generated by mathematically transforming this interferogram using the Fourier Transform technique. Comparing this method to traditional infrared (IR) spectroscopy, data can be acquired more quickly and accurately.

Based on their distinctive absorption bands, FTIR aids in the identification and quantification of important macronutrients such as proteins, lipids, and carbohydrates in the context of nutritional profile. For instance, the carbohydrate region usually appears between 1000 and 1200 cm^{-1}, the C–H stretching vibrations between 2850 and 2950 cm^{-1} belong to lipids, and the amide I and II bands at about 1650 cm^{-1} and 1540 cm^{-1} are linked to protein structures [14]. The amounts of these nutrients in a sample can be precisely ascertained by examining the position and intensity of these bands and applying calibration models with established standards. As a result, FTIR is an effective tool in food science that offers quick, nondestructive, and reagent-free nutritional content analysis of a variety of food matrices. Figure 1 depicts the instrumentation profile of FTIR.

1.3 Raman Spectroscopy

By examining the inelastic scattering of light, Raman spectroscopy is a nondestructive analytical method that yields comprehensive details about the molecular makeup of a material. Raman spectroscopy is an essential method for nutritional profiling due to its ability to produce intricate molecular fingerprints that provide information about the structure and chemical makeup of food such as proteins, fats, carbohydrates, vitamins, and other bioactive substances [15]. It is highly sensitive, nondestructive in nature, and requires very simple sample preparations. Raman spectroscopy offers additional information to methods like FTIR spectroscopy, making it useful for a wide range of food analysis.

Raman spectroscopy's basic idea depends on how monochromatic light interacts with the molecules in the sample. The molecules in the sample are excited when the light strikes them, and the information about the vibrational energy levels of the molecules may be found in the scattered light [16]. The Raman shift, which is the energy difference between incident and dispersed light, creates a distinct spectrum that represents the vibrations of molecules.

Stokes Shift This is the most frequently seen Raman scattering. It causes a red shift in the spectrum because the scattered light has less energy than the original light. This happens when a portion of the photon's energy is transferred to the vibrational modes of the sample [17].

Anti-Stokes Shift This less frequent phenomenon is caused by a blueshift when the energy of the scattered light exceeds that of the incident light. It takes place when the molecules in the sample are already in a vibrational state of excitement [18].

The main Raman spectroscopy types used in nutritional profiling are: Traditional Raman Spectroscopy, SERS, Raman Imaging (Chemical Mapping), and Fiber-Optic Raman Spectroscopy.

In traditional Raman Spectroscopy, the sample is illuminated with a laser in traditional Raman spectroscopy, and the scattered light is gathered and examined to create a spectrum. Because of its great versatility, the approach can be used with solids, liquids, and semisolid foods. Through the detection of vibrations linked to important chemical bonds, including C–H, N–H, C=O, and O–H, it offers direct insights into the composition of food [19]. The sensitivity and detection limits of Raman signals are greatly increased in SERS, a sophisticated variant of conventional Raman spectroscopy. To do this, the Raman signal is enhanced by using nanoparticles, usually made of gold or silver, which, when exposed to laser light, produce localized surface plasmon resonance (Fig. 2) [20]. SERS is very helpful for identifying trace nutrients or bioactive substances in food samples because of the enhancement, which makes it possible to detect lower analyte concentrations, as shown in Fig. 2.

In addition, using Raman Imaging (Chemical Mapping) makes it possible to see how various chemical components are distributed spatially within a sample. Raman imaging may provide intricate maps of the chemical makeup of food by gathering Raman spectra at various locations throughout the sample [21]. This method is very helpful for researching how various nutrients are distributed throughout intricate food matrices. In fiber-optic Raman Spectroscopy, the scattered light from the specimen is collected using a fiber-optic probe [22]. By inserting the probe into food containers or manufacturing processes, this approach enables remote sampling, which analyzes food samples without requiring direct handling, as exemplified in Fig. 3.

To ensure uniformity in nutritional content and identify any deviations from required standards, fiber-optic Raman

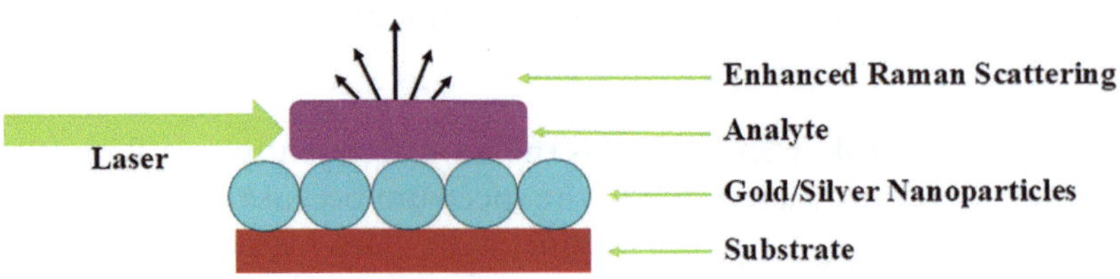

Fig. 2 Schematization of the SERS mechanism

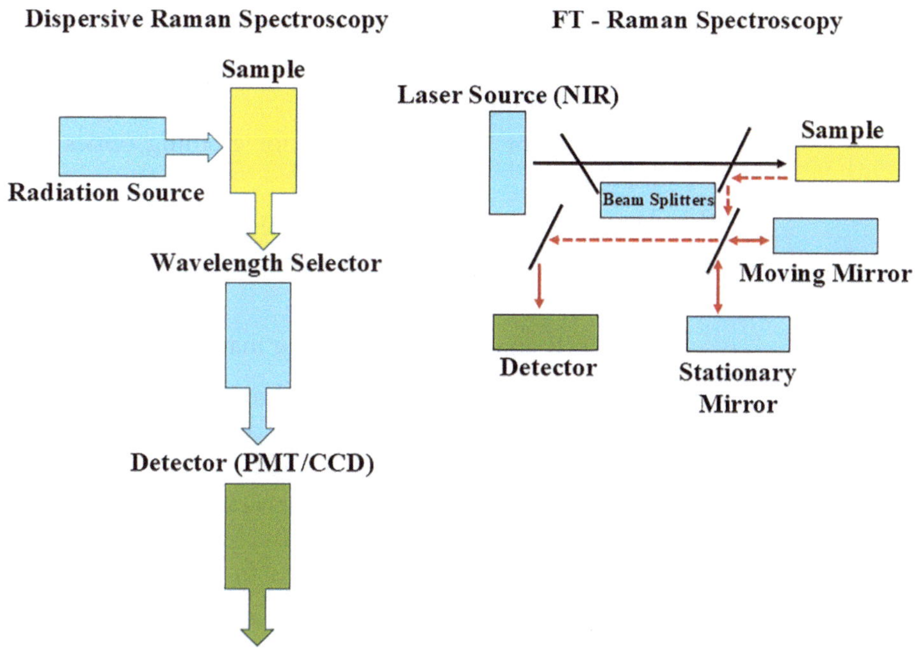

Fig. 3 Schematic representation of Dispersive and FT Raman Spectroscopy

spectroscopy is utilized for real-time monitoring of food production operations. In the food production industry, it can also be used to ensure the quality of final goods or raw components.

2 Materials

For NIR spectroscopy following materials are needed.

1. NIR Spectrometer with suitable range, for example, 780–2500 nm.
2. Computer or Data Processor with chemometric software (e.g., Unscrambler, OPUS, SIMCA, and MATLAB).
3. Sample Holders: Reflectance cups or quartz cuvettes (depending on solid or liquid samples); glass vials or Petri dishes (for powders or pastes).
4. Grinder or Mill: Needed to homogenize solid samples like grains and cereals.
5. Analytical weighing balance: for accurate weighing of samples.
6. Moisture Analyzer: Although optional, may be needed to determine or standardize sample moisture content.
7. Reference standards for calibration and validation: Samples with known nutritional composition (protein, fat, etc.).

8. Desiccants for storage of dry samples (e.g., silica gel).
9. Deionized water or filtration media to be employed for preparing or cleaning liquid samples.
10. Kimwipes or Lint-free tissues for cleaning optics and sample holders.
11. Hand gloves to prevent sample contamination.
12. Cleaning alcohol like 70% ethanol or isopropanol to be used for cleaning sample cups.

For FTIR analysis, the following materials are required

1. FTIR Spectrometer: Attenuated total reflectance (ATR) accessory, Mid-IR range (4000–400 cm^{-1}), and high-sensitivity detector (e.g., DTGS or MCT).
2. NIR spectrometer: Diffuse reflectance or transmittance mode with a wavelength range of (700–2500 nm), fiber optic probes (for bulk samples).
3. Raman Spectrometer: Laser source (e.g., 785 nm or 1064 nm to avoid fluorescence).
4. Software: Spectroscopic data processing software (e.g., OPUS, unscrambler, MATLAB).
5. Chemometric tools: Example OPUS and unscrambler, WinISI for spectral analysis and model development.
6. Sample presentation Accessories: Cup or holder for solid samples, Quartz cuvettes for liquid samples, and rotating sample cup for homogeneous scanning (for powdered samples).
7. Reference method equipment: For calibration validation including Kjeldahl apparatus (protein), Soxhlet extraction (fat), oven drying (moisture), and high-performance liquid chromatography/gas chromatography (HPLC/GC) (for specific nutrients).
8. Data processing tools: Example Savitzky–Golay smoothing and multiplicative scatter correction. Origin 8.1 (OriginLab, Northampton) and Pirouette v.4.5 (Infometrix, Inc. Bothell-WA) softwares.

The primary Materials of Raman spectroscopy are:

1. Laser Source: The sample's molecules are excited by monochromatic light produced by the laser. Because they correspond to areas where the Raman signal is strong and the chance of fluorescence (interfering with the Raman signal) is low, lasers with wavelengths between 532 nm (green), 633 nm (red), or 785 nm (near-infrared) are typically utilized [23].

2. Sample Holder: The food sample is placed in the sample holder so that it may be measured. Samples used in Raman spectroscopy can be solids, liquids, or semisolids, and the holder should be made to fit the type of sample being used. When it comes to solid foods like cereals, the sample is frequently positioned either directly underneath the microscope objective or on a stage that has a positioning device. A fiber-optic probe or a tiny cuvette can be used to analyze liquid samples, such as juices or oils. A similar method is applied to semisolid meals (such as dough, yogurt, or spreads), or a thin layer of the sample can be applied to a glass slide.
3. Optical System: Is needed to concentrate the laser beam onto the sample and gather the dispersed Raman signal. The optical system is made up of lenses and mirrors. After that, a spectrometer receives the gathered scattered light. In a standard configuration, an objective lens is used to concentrate the light on the sample, and either the same or a separate lens is used to collect the dispersed light. Then, a spectrometer receives this light to perform additional analysis.
4. Spectrometer: The Raman scattered light is split up into its wavelengths using a spectrometer. Usually, a prism or grating that divides light into its wavelengths is used for this. The Raman shift, which represents the energy differences between the incident and dispersed light, may be precisely measured thanks to the spectrometer. This energy differential gives the sample a molecular fingerprint that aids in the identification and measurement of food ingredients.
5. Detector: The scattered Raman light is captured by the detector to be transformed into an electrical signal. Photomultiplier tubes (PMT) or charge-coupled devices (CCD) are often used as detectors in Raman spectroscopy. CCD detectors are frequently chosen because they can quickly acquire the Raman spectrum by simultaneously capturing a large variety of wavelengths.
6. Computer and Software: Used for the operation of the Raman spectrometer and also processes the data. The program can help with peak identification and nutrient quantification in the sample, evaluate the spectral data, and provide visual displays of the Raman spectra. To account for baseline changes, fluorescence interference, and other elements that could compromise the spectrum's accuracy, sophisticated algorithms are used. To correlate Raman spectral features with the concentrations of various nutrients, this software can also integrate multivariate analysis techniques.

3 Methods

3.1 NIR

1. Sample Preparation.
   - (a) Grind solid samples (such as grains and powders) uniformly to ensure homogeneity and reduce particle size variation.
   - (b) Liquid samples like juices and milk should be filtered if necessary to remove particulates and degas to eliminate bubbles.
   - (c) Note that the moisture content should be consistent or corrected during analysis, as NIR is sensitive to water.
2. Instrument Calibration.
   - (a) Warm up the NIR spectrometer as per the manufacturer's guidelines.
   - (b) Perform a background scan (reference scan) using a standard material (e.g., a ceramic or Teflon disk).
   - (c) Validate the calibration using a known standard sample to ensure instrument accuracy.
3. Sample Loading.
   - (a) Place the sample into the appropriate holder or sample cup (e.g., cuvette for liquids, reflectance cell for solids).
   - (b) Ensure the sample is evenly packed and covers the window completely to avoid scattering artifacts.
4. Spectral Data Acquisition.
   - (a) Choose an appropriate wavelength range, typically 780–2500 nm, for food profiling.
   - (b) Select the mode of measurement (for example, transmission, reflectance, or transflectance, depending on sample type).
   - (c) Acquire the NIR spectrum of the sample.
5. Data Preprocessing.
   - (a) Apply necessary spectral preprocessing techniques to improve data quality: Baseline correction, smoothing (using Savitzky–Golay filter option), derivatives (first or second) to resolve overlapping peaks, and normalization to correct for scattering.
6. Nutritional Profiling.
   - (a) Input new sample spectra into the model to predict nutrient content.
   - (b) Obtain quantitative results for parameters such as protein, moisture, fat, carbohydrates, ash, and fiber.

7. Model Validation and Verification.
   (a) Compare NIR results with reference methods (e.g., Kjeldahl for protein, Soxhlet for fat) to validate model accuracy.
   (b) Perform cross-validation or external validation to assess model robustness.
   (c) Record and interpret results.
   (d) Document the nutritional profile.
   (e) Interpret results with respect to nutritional standards or product specifications.

3.2 FTIR

A detailed methodology for utilizing FTIR to perform nutritional profiling of food is as follows:

3.2.1 Step 1: Sample Preparation

1. Solid samples (e.g., grains, meats, dairy): A tiny amount of the food sample is put straight onto the ATR crystal for ATR-FTIR. There is no need for additional preparation. To create a pellet or thin film for transmission FTIR, solid specimens are ground into a fine powder and combined with an IR-transparent substance, like potassium bromide (KBr).
2. Liquid Samples (e.g., oils, juices, milk): A drop of the liquid is put straight onto the ATR crystal for ATR-FTIR. The liquid is put in a liquid cell with a transparent glass, like quartz, for transmission FTIR.
3. Semisolid Samples (e.g., yogurt, cheese, spreads): By simply placing the sample on the ATR crystal and exerting light pressure to establish proper contact, these can be examined using ATR-FTIR. To avoid contamination and guarantee precise measurements, the sample must be handled carefully. To preserve the sample's integrity, make sure it is homogeneous and, if required, pretreat or store it.

3.2.2 Step 2: Calibration of the FTIR Instrument

The FTIR instrument must be calibrated before starting the analysis.

1. *Background Measurement*: To remove environmental interference (such as CO_2 and H_2O vapor), do a background scan using a clean crystal or an empty sample container.
2. *Calibration with standards*: The FTIR spectrometer is calibrated using recognized standards with nutrient contents (such as pure fat, protein, or carbohydrate standards) to guarantee the correctness of the data. To link the absorption bands with concentration, calibration curves are created for every nutrient (fat, protein, etc.).

3. *Data Collection*: Depending on the type of sample, the prepared sample is loaded on the FTIR instrument (either on the ATR crystal, in the transmission cell, or the DRIFT accessory). The interferometer inside the FTIR spectrometer splits and recombines the IR beam to create an interferogram. The FTIR system is configured to scan the infrared spectrum, which is normally in the range of 4000–400 cm^{-1} and includes the vibrational frequencies of the majority of organic molecules found in food. The Fourier Transform technique converts this data into a spectrum, producing a graph of absorption strength vs wavenumber (cm^{-1}). The FTIR records the distinctive absorption peaks of functional groups linked to macronutrients throughout this procedure, including:
   (a) Proteins: Amide I and Amide II bands (~1650 cm^{-1} and ~1540 cm^{-1}).
   (b) Lipids: C-H stretching vibrations (~2850–2950 cm^{-1}).
   (c) Carbohydrates: C–O and C–O–C stretching (~1000–1200 cm^{-1}).
   (d) Water: O-H stretching (~3200–3400 cm^{-1}) or bending (~1600 cm^{-1}).

3.2.3 Step 3: Data Analysis and Spectrum Interpretation

Key absorption peaks that correspond to the chemical bonds in nutrients are identified by analyzing the resulting FTIR spectra. The absorption bands are frequently identified and quantified using software tools. The concentration of various macronutrients is ascertained by measuring the intensity of these peaks.

1. The C=O stretching vibration in proteins is represented by the amide I band, which is located at about 1650 cm^{-1} and provides information on protein concentration.
2. The intensity of the CH stretching band, which measures lipid content, is around 2925 cm^{-1}.
3. The amount of carbohydrates is measured using the C-O-C stretching region (1000–1200 cm^{-1}). A calibration model, which is based on previous calibration data or a standard reference database, is used to quantify each nutrient. The relationship between peak intensity and nutrient concentration is given by this model.

3.2.4 Step 4: Nutritional Profiling and Reporting

FTIR spectra offer a comprehensive "fingerprint" of the food's molecular makeup with distinct absorption bands at wavelengths for each nutrient.

1. Chemical bonds like C–H, N–H, O–H, C=O, and others vibrate to produce the absorption bands. These bonds are common in macronutrients (proteins, fats, and carbs) and occasionally in micronutrients or bioactive substances.

2. Proteins are characterized by the amide I band, which is approximately 1650 cm^{-1} and is caused by C=O stretching vibrations, and the amide II band, which is around 1540 cm^{-1} and is linked to N–H bending. Usually, these bands are used to evaluate the amount of protein in the diet.
3. Because of C–H stretching vibrations, lipids show strong absorption bands, especially about 2850–2950 cm^{-1}. Lipid or fat content in food samples is measured in this area.
4. The amount of starch, cellulose, and simple sugars can be obtained from the stretching vibrations of C–O–C bonds, which are located between 1000 and 1200 cm^{-1}.
5. The O–H stretching band at around 3400 cm^{-1} and the bending vibration at about 1600 cm^{-1} can both be used by FTIR to measure moisture content.

3.2.5 Step 5: Validation and Quality Control

For validation purposes, FTIR data can be cross-checked with those from other analytical techniques, such as chromatography (e.g., HPLC or GC) or conventional wet chemistry techniques. The FTIR instrument needs to be calibrated and maintained regularly to guarantee accuracy and consistency.

3.3 Raman Spectroscopy

A step-by-step guide on using Raman spectroscopy for nutritional profiling and nutrient identification is provided here.

3.3.1 Sample Collection and Preparation

For an accurate analysis, make sure the food samples are representative and consistent.

- Gathering: Take food samples from the batch under examination. The food item should be represented by the sample.
- Preparation: Make the sample from the food (solid, liquid, or semisolid). To improve the laser light's ability to penetrate solid meals, break or crush the sample into smaller bits. To make measurement easier, make sure liquid or semisolid items are in a sterile container, like a petri dish or cuvette.
- Consistency: Make sure the sample is uniform and devoid of impurities or other materials that could skew the Raman signal.

3.3.2 Procedure

For precise measurements, set up the Raman spectrometer.

Step 1: Instrument Setup and Calibration

Procedure

1. *Laser Selection*: Depending on the type of food and the desired nutrient, select a suitable laser wavelength (532 nm, 633 nm, or 785 nm). Various wavelengths can reduce fluorescence interference and increase the signal of specific components.

2. *Optical Setup*: Focus the laser on the sample by adjusting the fiber-optic probe, sample holder, or microscope. For the best signal collection, the laser should be accurately focused on the sample surface.
3. *Calibration*: To make sure the Raman spectrometer is correctly oriented and functioning within its ideal spectral range, use a conventional calibration technique with recognized materials (such as silicon or polystyrene).

Step 2: Raman Spectrum Acquisition

Gathering the food sample's Raman spectra.

Procedure

1. *Gathering Spectra*: Collect the Raman scattered light at different wavelengths after directing the laser beam onto the sample surface. Make sure the sample is exposed for a long enough period to produce a high-quality spectrum without causing damage.
2. *Data Acquisition*: The detector, usually a CCD or PMT, captures the Raman shifts while the spectrometer analyzes the scattered light. A plot of intensity against Raman shift (cm^{-1}) is used to depict the Raman spectrum.
3. *Multiple scans*: If required, do multiple scans, particularly for diverse samples. To increase the signal-to-noise ratio and get a more precise picture of the food's composition, average the spectra.

Step 3: Data Processing and Preprocessing

The raw Raman spectra should be cleaned and ready for analysis.

Procedure

1. *Background Subtraction*: Eliminate any background signals that could obstruct the Raman bands of interest, such as fluorescence. To make sure that only the Raman signal is examined, this step is essential.
2. *Baseline Correction*: Adjust for baseline shifts brought on by instrument drift or irregular sample characteristics that might have distorted peak intensity.
3. *Noise Reduction*: To improve the clarity of the peaks in the spectra, apply smoothing techniques like Savitzky–Golay filters.

Step 4: Spectral Interpretation

Determine and analyze the molecular components of the food sample.

Procedure

1. Peak Identification: Determine which major peaks in the Raman spectrum correspond to certain chemical vibrations. For instance, bands of amide I (1650 cm^{-1}) and amide II (1540 cm^{-1}) on proteins.

2. Lipids: Look for C–H stretching bands (2800–3000 cm^{-1}) and C–H bending bands (1440 cm^{-1}). Identify the peaks for carbohydrates at approximately 1000 cm^{-1} (C–C stretching) and 1100 cm^{-1} (C–O stretching).
3. Water: One important measure of water content is the O–H stretch, which is around 3200–3400 cm^{-1}.
4. Fingerprinting: To determine the precise nutrients included in the food sample, compare the acquired spectra with reference spectra of various dietary components.

Step 5: Quantification of Nutrients

Determine how much of each nutrient is present in the sample.

Procedure

1. Calibration curves: Create calibration curves for nutrients if quantification is necessary. These curves are produced by measuring the intensity of matched Raman peaks in food samples that have known quantities of the target nutrient.
2. Intensity Analysis: Determine the concentration of the related nutrient in the sample by comparing the intensity of Raman peaks (for proteins, lipids, or carbohydrates, for example). Certain peaks become more intense in proportion to the nutrient's concentration.
3. Multivariate Analysis: Complex food sample spectra can be analyzed using sophisticated methods like PCA or PLSR to simultaneously provide quantitative information about several nutrients. These techniques are especially helpful for examining food items with intricate compositions.

Step 6: Data Interpretation and Nutritional Profiling

Create a nutritional profile of the food sample by compiling the data.

Procedure

1. Nutrient profile creation: Compile a thorough nutritional profile of the food sample using the Raman spectra and quantification findings. Values for important macronutrients like protein, fat, carbohydrate, and moisture content, as well as micronutrients like vitamins and antioxidants, may be included in this profile.
2. Multivariate data analysis: This technique can be used to evaluate data from numerous Raman bands in situations involving complicated food matrices or mixes, allowing for the development of a more precise and trustworthy nutritional profile.
3. Comparison with Standards: Verify that the food satisfies certain nutrient criteria or claims by comparing the nutritional profile with regulatory standards or nutritional guidelines

(such as those issued by the Food and Drug Administration [FDA] or the European Food Safety Authority [EFSA]).

Step 7: Reporting and Decision-Making

Present the findings for additional consideration or action.

Procedure

1. Reporting: Write a report outlining the sample's nutritional makeup, noting any noteworthy discoveries about nutrient levels, food quality, and any contaminants or additions found.
2. Applications: Make use of the information for several objectives, including Quality Control: Verify that food items fulfil the requirements for quality and nutritional value.
3. Labeling and Compliance: For food labeling or regulatory compliance, provide accurate nutritional information.
4. Product Development: Make sure new food products satisfy consumer expectations and health requirements by optimizing their nutritional composition using Raman spectroscopy. Research and Development: Examine the nutritional bioavailability and composition of food under varied processing and storage circumstances.

4 Notes

NIR, FTIR, and Raman spectroscopy offer unique benefits and limitations depending on the application. These vibrational spectroscopic methods are highly valuable for nondestructive chemical analysis and molecular characterization that are widely employed in food safety analysis, pharmaceutical quality assurance, and microbial detection [24]. They have transformed nutritional profiling through swift, noninvasive, and efficient analytical approaches in food science and industry. These techniques address critical challenges in food safety, quality control, and regulatory compliance, offering distinct yet complementary advantages. These spectroscopic techniques demonstrate superior performance in the bulk assessment of macronutrients with minimal preparatory requirements, rendering them optimal for the real-time surveillance of food safety. Table 1 provides the comparative analysis of NIR, FTIR, and Raman spectroscopy along with their principle and sampling types. Despite prospects, their limitations must be addressed through method optimization and complementary techniques. Future advancements in portable sensors, artificial intelligence (AI)-driven analysis, and hybrid systems could also help enhance their reliability in food and nutrition science.

Table 1
Comparative study of NIR, FTIR, and Raman Spectroscopy

Feature	Near Infrared (NIR) Spectroscopy	Fourier-Transform Infrared (FTIR) Spectroscopy	Raman Spectroscopy
Principle	Measures the absorption of light in the near-infrared region (750–2500 nm), related to overtones and combinations of molecular vibrations.	Measures absorption of infrared light by the sample, corresponding to molecular vibrations (4000–400 cm^{-1}).	Measures inelastic scattering (Raman scattering) of monochromatic light, providing vibrational and rotational information.
Sample Type	Solid, liquid, or powder samples; especially used for bulk analysis like grains, powders, and liquids.	Solids, liquids, powders, and thin films; typically used for food, pharmaceuticals, and chemicals.	Solids, liquids, gels, and some powders; particularly suitable for biological samples and surface analysis.
Wavelength Range	750–2500 nm (Near-Infrared region).	4000–400 cm^{-1} (Mid-Infrared region).	Typically 100–4000 cm^{-1} (Depends on laser wavelength).
Sensitivity to Water	High sensitivity to water content.	Moderate sensitivity to water, with water bands visible.	Moderate to low sensitivity to water (water bands often visible).
Analysis Type	Primarily used for quantitative analysis of moisture, fats, proteins, and carbohydrates.	Qualitative and quantitative analysis of organic and inorganic compounds, including macronutrients and contaminants.	Qualitative and quantitative analysis of organic compounds, protein, lipids, and other food molecules.
Sample Preparation	Minimal sample preparation, can analyze bulk samples directly.	Minimal preparation, but often requires sample placement in a holder (e.g., ATR crystal).	Minimal preparation, can analyze the sample in a variety of forms (e.g., liquid, solid, or gel).
Measurement Mode	Diffuse reflectance, transmission, or interaction with the sample (typically nondestructive).	Transmission, reflection, or Attenuated Total Reflectance (ATR) mode.	Backscattering, forward scattering, or transmission modes.
Resolution	Lower resolution (typically 10–20 nm).	Higher resolution (1–4 cm^{-1}).	Medium resolution (1–10 cm^{-1}).
Quantitative Capability	High, especially for moisture, protein, fat, and carbohydrate content.	High, especially with proper calibration (for food ingredients and additives).	High, especially with proper calibration and multivariate analysis.

(continued)

Table 1 (continued)

Feature	Near Infrared (NIR) Spectroscopy	Fourier-Transform Infrared (FTIR) Spectroscopy	Raman Spectroscopy
Fluorescence Interference	Susceptible to fluorescence, which can interfere with analysis.	Fluorescence can interfere, especially in biological samples.	Less susceptible to fluorescence compared to NIR and FTIR.
Speed of Analysis	Fast, can analyze multiple samples in a short time.	Moderate, requires sample preparation and analysis time.	Moderate to slow, depending on the sample and analysis method.
Cost	Moderate to low (cheaper instruments available).	Moderate to high (expensive instruments with high precision).	Moderate to high, particularly for high-end systems.
Complexity of Interpretation	Relatively simple, especially with chemometric methods for quantitative analysis.	Requires some experience and advanced interpretation for complex spectra.	Requires expertise, especially for complex spectra and mixtures.
Advantages	Fast, nondestructive, minimal sample preparation, and effective for bulk material analysis.	Provides detailed molecular information, versatile in sample types, and widely used in food and pharmaceutical industries.	Nondestructive, capable of analyzing solids and surfaces, minimal sample preparation, and provides detailed molecular vibrational information.
Disadvantages	Lower resolution, susceptibility to interference from water and other substances.	Sensitive to sample thickness, background interference, and requires high-quality calibration.	Low sensitivity for concentration components and susceptibility to interference from background fluorescence or noise

References

1. Cebi N, Bekiroglu H, Erarslan A (2023) Nondestructive metabolomic fingerprinting: FTIR, NIR and Raman spectroscopy in food screening. Molecules 28:7933. https://doi.org/10.3390/molecules28237933
2. Workman J (2024) A review of the latest spectroscopic research in food and beverage analysis. In: Spectrosc. Online. https://www.spectroscopyonline.com/view/latest-spectroscopic-research-in-food-and-beverage-analysis. Accessed 8 Apr 2025
3. Le Pevelen DD, Tranter GE (2017) FT-IR and Raman spectroscopies, polymorphism applications. In: Lindon JC, Tranter GE, Koppenaal DW (eds) Encyclopedia of spectroscopy and spectrometry, 3rd edn. Academic Press, Oxford, pp 750–761
4. Wang Z, Wu Q, Kamruzzaman M (2022) Portable NIR spectroscopy and PLS based variable

selection for adulteration detection in quinoa flour. Food Control 138:108970. https://doi.org/10.1016/j.foodcont.2022.108970

5. Kassem A, Abbas L, Coutinho O et al (2023) Applications of Fourier Transform-Infrared spectroscopy in microbial cell biology and environmental microbiology: advances, challenges, and future perspectives. Front Microbiol 14:1304081. https://doi.org/10.3389/fmicb.2023.1304081
6. Martins MS, Nascimento MH, Barbosa LL et al (2022) Detection and quantification using ATR-FTIR spectroscopy of whey protein concentrate adulteration with wheat flour. LWT 172:114161. https://doi.org/10.1016/j.lwt.2022.114161
7. Borba KR, Spricigo PC, Aykas DP et al (2021) Non-invasive quantification of vitamin C, citric acid, and sugar in 'Valência' oranges using infrared spectroscopies. J Food Sci Technol 58:731–738. https://doi.org/10.1007/s13197-020-04589-x
8. Beć KB, Grabska J, Huck CW (2020) Near-infrared spectroscopy in bio-applications. Molecules 25:2948. https://doi.org/10.3390/molecules25122948
9. Reich G (2005) Near-infrared spectroscopy and imaging: basic principles and pharmaceutical applications. Adv Drug Deliv Rev 57:1109–1143. https://doi.org/10.1016/j.addr.2005.01.020
10. Al-Kelani M, Buthelezi N (2024) Advancements in medical research: exploring Fourier Transform Infrared (FTIR) spectroscopy for tissue, cell, and hair sample analysis. Skin Res Technol 30:e13733. https://doi.org/10.1111/srt.13733
11. Canteri MHG, Renard CMGC, Le Bourvellec C, Bureau S (2019) ATR-FTIR spectroscopy to determine cell wall composition: application on a large diversity of fruits and vegetables. Carbohydr Polym 212:186–196. https://doi.org/10.1016/j.carbpol.2019.02.021
12. Subramanian A, Rodriguez-Saona L (2009) Chapter 7 - Fourier Transform Infrared (FTIR) spectroscopy. In: Sun D-W (ed) Infrared spectroscopy for food quality analysis and control. Academic Press, San Diego, pp 145–178
13. Pirutin SK, Jia S, Yusipovich AI et al (2023) Vibrational spectroscopy as a tool for bioanalytical and biomonitoring studies. Int J Mol Sci 24:6947. https://doi.org/10.3390/ijms24086947
14. Pasieczna-Patkowska S, Cichy M, Flieger J (2025) Application of Fourier Transform Infrared (FTIR) spectroscopy in characterization of green synthesized nanoparticles. Molecules 30:684. https://doi.org/10.3390/molecules30030684
15. Hong C, Shi M, Wang S et al (2025) Novel analysis based on Raman spectroscopy in nutrition science. Anal Methods 17:1977–1996. https://doi.org/10.1039/D4AY02129K
16. Brewer PG, Kirkwood WJ (2013) 16 - Raman spectroscopy for subsea applications. In: Watson J, Zielinski O (eds) Subsea optics and imaging. Woodhead Publishing, pp 409–433
17. Jones RR, Hooper DC, Zhang L et al (2019) Raman techniques: fundamentals and frontiers. Nanoscale Res Lett 14:231. https://doi.org/10.1186/s11671-019-3039-2
18. Langer J, Jimenez de Aberasturi D, Aizpurua J et al (2020) Present and future of surface-enhanced Raman scattering. ACS Nano 14:28–117. https://doi.org/10.1021/acsnano.9b04224
19. Fodor M, Matkovits A, Benes EL, Jókai Z (2024) The role of near-infrared spectroscopy in food quality assurance: a review of the past two decades. Foods 13:3501. https://doi.org/10.3390/foods13213501
20. Cialla-May D, Bonifacio A, Markin A et al (2024) Recent advances of surface enhanced Raman spectroscopy (SERS) in optical biosensing. TrAC Trends Anal Chem 181:117990. https://doi.org/10.1016/j.trac.2024.117990
21. Petersen M, Yu Z, Lu X (2021) Application of Raman spectroscopic methods in food safety: a review. Biosensors 11:187. https://doi.org/10.3390/bios11060187
22. Liu J, Zhang B, Abdullah A et al (2024) A fiber optics based surface enhanced Raman spectroscopy sensor for chemical and biological sensing. Sens Bio-Sens Res 46:100686. https://doi.org/10.1016/j.sbsr.2024.100686
23. Li Y-S, Church JS (2014) Raman spectroscopy in the analysis of food and pharmaceutical nanomaterials. J Food Drug Anal 22:29–48. https://doi.org/10.1016/j.jfda.2014.01.003
24. Castro RC, Ribeiro DSM, Santos JLM, Páscoa RNMJ (2021) Comparison of near infrared spectroscopy and Raman spectroscopy for the identification and quantification through MCR-ALS and PLS of peanut oil adulterants. Talanta 230:122373. https://doi.org/10.1016/j.talanta.2021.122373

Chapter 10

Electrochemical Techniques for Food Contaminant Detection

Ann Justin Kappen, Nikhilesh A, V. C. Anitha Krishanan, H. P. Vasantha Rupasinghe, and Jobil J. Arackal

Abstract

Food safety remains a critical global concern as contamination incidents continue to pose significant risks to public health and economic stability. Traditional analytical methods, while reliable, often require sophisticated laboratory infrastructure, extensive sample preparation, and considerable time for analysis. Electrochemical techniques have emerged as powerful alternatives, offering rapid, sensitive, and cost-effective solutions for detecting various food contaminants. The core electrochemical techniques employed in food contaminant detection encompass voltammetry, amperometry, potentiometry, and impedance spectroscopy. Voltammetric techniques excel in trace metal detection through their ability to measure current responses as a function of applied potential. Amperometric sensors offer excellent sensitivity for real-time monitoring applications, while potentiometric methods provide selective ion detection capabilities. Impedance spectroscopy contributes valuable information about interfacial processes and binding events, particularly useful in biosensor applications. Modern electrochemical sensors incorporate advanced electrode materials and surface modifications to enhance performance characteristics. Carbon-based electrodes, noble metal electrodes, and screen-printed electrodes serve as foundational platforms, often modified with nanomaterials to improve sensitivity and selectivity. Nanomaterial integration, including carbon nanotubes, graphene, gold nanoparticles (AuNPs), and quantum dots (QDs), has revolutionized sensor capabilities by increasing surface area and enhancing electron transfer kinetics. Biosensors and immunosensors represent sophisticated developments that combine biological recognition elements with electrochemical transduction, enabling highly specific detection of target analytes.

Key words Electrochemical sensors, Food safety, Contaminant detection, Voltammetry, Amperometry, Biosensors, Nanomaterials, Heavy metals, Pesticides, Mycotoxins, Pathogen detection, Portable analysis, Multiplexed detection, Food quality control

1 Introduction to Electrochemical Detection in Food Safety

The need to guarantee food safety and security has become an emerging concern in the context of public health, food security, economic development, and environmental sustainability. Guaranteeing a safe and dependable food supply is crucial for the well-

Tanmay Sarkar (ed.), *Novel Food Analysis: Techniques and Applications*, Methods and Protocols in Food Science,
https://doi.org/10.1007/978-1-0716-4787-5_10,

being of society, influencing individual health as well as economic viability and international trade dynamics [1]. "Food safety" pertains to the conditions and practices that maintain the quality of food, preventing contamination and illness caused by food. It is an issue that impacts everyone, irrespective of location, culture, or type of cuisine. The need to ensure the safety of the food we eat is more crucial than ever as the global population continues to grow and become more diverse [2]. Reports indicate that around 600 million instances of foodborne illness occur globally each year, leading to approximately 420,000 fatalities, which accounts for roughly 7.5% of the total deaths worldwide [3]. Food contaminants have consistently posed a significant threat to human health over extended periods. Therefore, it is crucial to create methods that are simple, efficient, rapid, and cost-effective for detecting food contaminants [4]. The pressing demand for food safety has been brought to the attention of analytical and testing researchers. The electrochemical technique has garnered interest among researchers working in food safety detection and sensor technologies. Electrochemical detection operates based on the principle that certain analytes in the sample could cause measurable changes in electrical characteristics, either in current, voltage, or even in impedance. Such changes are directly proportional to the concentration of the analyte and provide a reliable means of detecting and quantifying contaminants and quality markers in food samples. The high adaptability of electrochemical sensors to different detection situations is achieved, thanks to a variety of sensing approaches, developed for different kinds of analytes and analytical needs. Methods for detecting food contaminants through electrochemistry can be divided into four categories: amperometric, voltametric, potentiometric, and impedance spectroscopy. This chapter discusses the use of these four techniques in detecting food contaminants.

2 Materials

The methodological approach for electrochemical detection of food contaminants involves the systematic implementation of various electrochemical techniques tailored to specific analytical requirements. Standard protocols begin with proper electrode preparation, including surface cleaning and modification procedures to ensure optimal performance [1]. The methodology encompasses multiple electrochemical techniques (Fig. 1), each offering distinct advantages for different types of contaminants and food matrices. *Voltammetric Methods:* Cyclic voltammetry (CV) is employed as the primary characterization technique to evaluate electrode behavior and redox properties of target analytes. Linear sweep voltammetry (LSV) and differential pulse voltammetry (DPV) are utilized for quantitative analysis, with DPV providing

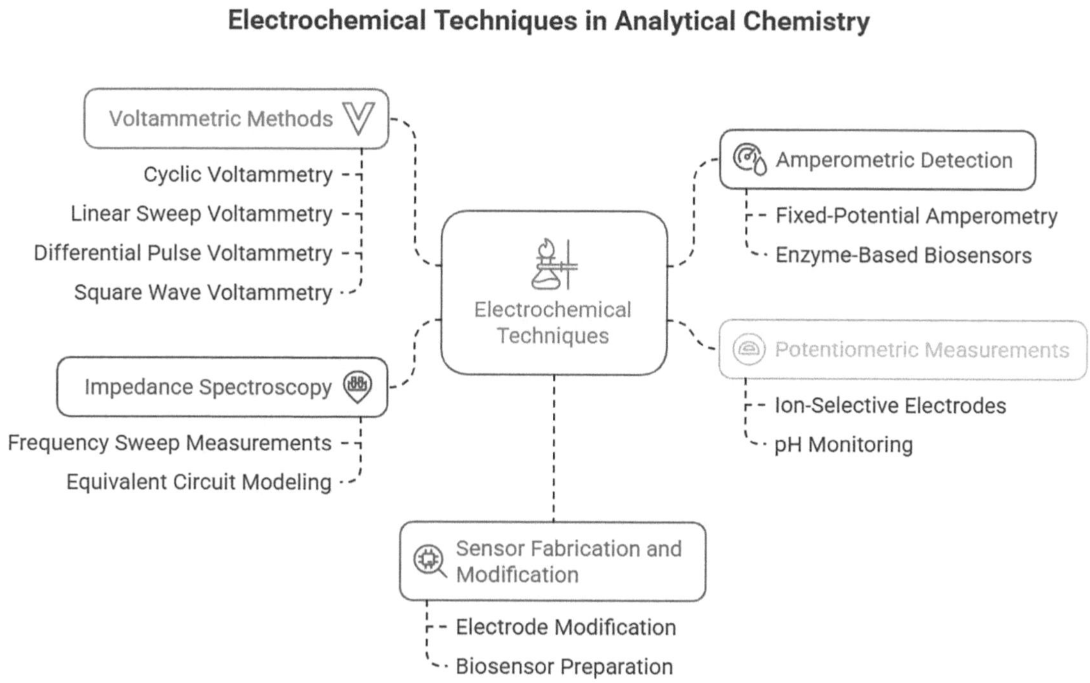

Fig. 1 Four distinct electrochemical methods used for detecting food contamination

enhanced sensitivity through background current suppression [2]. Square wave voltammetry (SWV) offers rapid analysis capabilities with improved resolution for overlapping peaks in complex food samples. *Amperometric Detection:* Fixed-potential amperometry is implemented for continuous monitoring applications, particularly suitable for enzyme-based biosensors detecting pesticide residues and mycotoxins. The methodology involves establishing optimal working potentials through preliminary voltammetric studies, followed by chronoamperometric measurements under stirred conditions to ensure mass transport-limited responses [3]. *Potentiometric Measurements:* Ion-selective electrode (ISE) methodologies are employed for detecting ionic contaminants and monitoring pH changes during enzymatic reactions. The protocol involves electrode calibration using standard solutions, followed by direct potentiometric measurements with appropriate correction for ionic strength effects and temperature variations [4]. *Impedance Spectroscopy:* Electrochemical impedance spectroscopy (EIS) is utilized for label-free detection of biological contaminants and monitoring biomolecular interactions. The methodology involves frequency sweep measurements across the range of 0.1 Hz–100 kHz, with data analysis using equivalent circuit modeling to extract relevant parameters such as charge transfer resistance and double-layer capacitance [5]. *Sensor Fabrication and Modification:* Electrode modification protocols include drop-casting, electrodeposition, and self-assembly techniques for incorporating nanomaterials and biological recognition elements. Surface characterization is performed using scanning electron microscopy (SEM) and atomic force microscopy (AFM) to confirm successful modification [6]. Biosensor preparation involves immobilization of enzymes, antibodies, or aptamers using crosslinking agents or entrapment within polymer matrices.

Sample Preparation and Analysis Food samples undergo appropriate pretreatment, including homogenization, extraction, and cleanup procedures to remove interfering substances. Standard addition methods and external calibration curves are employed for quantitative analysis, with recovery studies conducted to validate analytical performance [7]. Quality control measures include blank measurements, replicate analyses, and certified reference material validation to ensure method reliability and accuracy.

3 Methods

3.1 Voltammetry: Unveiling Reaction Dynamics Through Current-Potential Relationships

Voltammetry is an electrochemical technique that includes the measurement of current as a function of voltage (or potential) over time; in this case, a potential is applied to a working electrode (WE) and it is "scanned" over a defined range, measuring the current (I) in return. As the electrode is scanned, voltammograms—the graphs of I vs V—are produced, which contain information about the species and their pathways. The information varies depending on the electroactive species and the conditions under which reactions occur in the system [6]. The I value obtained directly relates to the speed of electron transfer taking place at the electrode surface, which depends on the concentration of the analyte and the potential set on the electrode. Voltammetry is defined by one main property: mass transport. Mass transport is responsible for the flow of active materials to the surface of the electrode, where they undergo electrochemical reactions. While usually the primary form of transport is diffusion (the movement of particles from high concentration areas to low), migration (transport of charged particles under influence of electric field) and convection (move as a result of stirring) can also occur and are often controlled to provide conditions of unrestricted diffusion [6].

Cyclic Voltammetry (CV) is arguably the most versatile of the known voltammetric techniques. In CV, the potential is scanned linearly along one direction (usually oxidative), and then reversed to a different direction (reductive). This leads to the generation of a cyclic voltammogram, which has distinct peaks for both oxidation and reduction. The position of these peaks states that the value of formal redox potentials (E^0) and the form and intensity of those peaks provide information regarding the irreversibility of the reaction, rate of electron transfer, and the existence of coupled chemical reactions. CV is a powerful method in studying the mechanisms at work and for qualitatively assessing the presence of different electroactive compounds within a system. To perform quantitative analysis and increase sensitivity in the signal, especially for trace amounts, Pulse Voltammetry techniques such as DPV and SWV are applied [7]. In these techniques, potential pulses are applied on top of linear ramps, and current is sampled at certain points, frequently as a difference, in order to reduce the effect of non-faradaic (charging) currents and raise the signal-to-noise ratio. An even more sensitive form is known as Stripping Voltammetry, which is known to be very sensitive for trace metal detection [7].

The process is two-step:

- Preconcentration (Deposition): The medium is concentrated (deposited) onto a working electrode, e.g., as a metal film or amalgam, by means of a fixed potential. This focuses in effect the analyte from a large dilute volume to a small electrode area.
- Stripping: The potential is then swept to reoxidize (strip) the adsorbed by-product(s) back into solution, resulting in a sharp current peak. The peak area or height is linear to the initial concentration of the analyte, which results in an extremely low limit of detection (LOD).

3.2 Amperometry: Real-time Monitoring at Constant Potential

Amperometry is a modification of the voltametric technique where the electrode potential is held constant as the current is measured. The selected potential is usually one at which the analyte participates in a diffusion-limited redox reaction, that is, current is rate-limited by the rate at which the analyte diffuses to the electrode surface [7, 8]. This is the basis for the obvious relationship between the steady-state current and the concentration of the electroactive species under these conditions. This direct current-concentration dependence renders amperometry a highly relevant method for real-time monitoring and sensing. There are many examples in different domains:

- Medical devices: For diabetes care, amperometric glucose sensors are commonly used. They use enzymes such as glucose oxidase that catalyze the glucose oxidation process, where electron transfer takes place and current is produced.
- Water quality assessment: The measurement of oxygen, chlorine, and heavy metals is also performed in a continuous way using amperometric principles. This data contributes toward the analysis of water quality or air pollution.
- Volumetric Methods: As with other detectors, amperometric electrodes are always used in conjunction with separation or

flow techniques such as high-performance liquid chromatography (HPLC) or Flow Injection analysis (FIA) to measure the activated species in a flowing stream.

Technical accuracy: The merits of amperometry cover these structures with utmost precision, expedience, and sensitive response, which is essential for constant, rapid measurements.

3.3 Potentiometry: Equilibrium Measurements for Concentration Determination

Unlike voltammetry and amperometry, potentiometry measures the potential difference between an indicator electrode and a reference electrode with almost zero current flow. This "zero current" condition is crucial for potentiometric measurements because it is an equilibrium measurement. The potential being measured is in relation to the activity (and thus concentration) of a specific ion in the solution as it is with the Nernst Equation stated below: One of the most important use cases of potentiometry is measuring pH levels by the glass electrode which serves as an ISE with high sensitivity for hydrogen ions. There are other ISEs such as las membrane ISE, solid state ISE, polymer membrane ISE etc. specifically designed to measure the concentration of different ions (such as hydrogen ions, fluoride/chloride ions, calcium/ potassium/ nitrate ions respectively) interference for multi complex matrices that go beyond pH. Potentiometric Titrations—as an illustrative example of another key application, in this method, the potential of an electrode is tracked as a titrant is added stepwise to the analyte solution. Equivalence point is determined by a steep increase (<step change> inflection point) in the potential vs. titrant volume graph. This technique is particularly advantageous in cases where visual indicators are not helpful or extreme precision is a necessity [7, 8]. Express the value of potentiometry in the aspect of easily disposable tools, low-cost equipment, no physical alteration to the sample being examined, suitable for real-time measurement, and ample sensitivity employing the right indicator electrode.

3.4 Electrochemical Impedance Spectroscopy (EIS): Probing Interfacial Phenomena

EIS is a nondestructive method which provides valuable information regarding the complex processes associated with the electrode and electrolyte interface. Whereas other techniques apply a constant direct current (DC) signal, EIS measures response functions using small voltage (potential) oscillations (or current) over a larger frequency range (spectrum) and observes the corresponding output alternating current (AC) signal of current (or voltage) response to the input stimulus. Z, which has both magnitude and phase and is measured, captures the relation between the applied AC signal and the obtained (AC) response.

EIS is unique in that it can tell apart different electrochemical processes based on how quickly they occur (referred to as their response time or frequency). Different frequencies result in the processes of charging and discharging, diffusion, double-layer charging, solution resistance, charge, surface film formation, and charge transfer. The impedance data collected is commonly illustrated through Nyquist plots, which depict imaginary impedance against real impedance, or Bode plots, which show the logarithm of impedance magnitude and phase angle against logarithmic frequency. These graphical representations are subsequently analyzed by fitting them to an equivalent electrical circuit model that encapsulates the physical processes at the electrode interface. Each element within the circuit, such as resistors, capacitors, and constant phase elements, is associated with a distinct electrochemical phenomenon [6–8].

The solution resistance (Rs) denotes the resistance encountered by the electrolyte, while the charge transfer resistance (Rct) provides insight into the kinetics of the electron transfer reaction, with a lower Rct signifying more rapid kinetics. Additionally, the double layer capacitance (Cdl) is associated with the charging of the electrical double layer at the interface between the electrode and electrolyte. Warburg impedance (ZW) reflects the diffusion resistance of species moving to and from the electrode surface. Lastly, film resistance or capacitance may be relevant in cases where passive films or coatings are present on the electrode surface [9].

3.5 Electrochemical Sensors for Food Contaminants

The world deals with major food safety problems that influence the health of its citizens, the economy, and world trade at the same time. The food industry often uses heavy metals (such as cadmium, lead, and mercury), illegal additives (melamine and Sudan I), leftover pesticides, many different veterinary drugs, biological poisons (mycotoxins), and food poisoning germs such as *E. coli* and Salmonella [10]. Gas chromatography (GC), HPLC, gas chromatography–mass spectrometry (GC–MS), liquid chromatography–mass spectrometry (LC–MS), and enzyme-linked immunosorbent assay (ELISA) have excellent sensitivity and specificity for spotting contaminants, but they face several severe problems [11]. Standard techniques for finding diseases are expensive, need extra time, involve complicated steps, and rely on advanced equipment and experienced personnel [12].

The difficulties mentioned have led to electrochemical sensors becoming a popular choice for discovering food contaminants. Such sensors take chemical actions and convert them into both sensitive and selective signals by using amperometry, square wave voltammetry, differential pulse voltammetry, and cyclic voltammetry [13]. Compared to other methods, DPV and SWV are preferred to find contaminants in traces because machines can easily deal with background currents and pick up the smallest concentrations [15]. When anodic stripping voltammetry (ASV) is used, in combination with nanostructured bismuth film electrodes (nsBiFE) and screen-printed carbon electrodes (SPCEs), these techniques can measure Cd^{2+} at a detection limit of 0.4 μg L^{-1} and Pb^{2+} as low as 0.1 μg L^{-1}, better than many traditional methods. Similarly, devices produced using nanotechnology, such as AuNPs/reduced graphene oxides (rGO)-decorated sensors, can effectively identify Sudan I in food items that contain various types of ingredients [14].

Electrochemical sensors are praised for working quickly, having very high sensitivity, being convenient for transportation, monitoring results instantly, clearly, and using small, single-use tools. Advances in biorecognition parts and nanotechnology have increased their possible applications [16]. By using this mix of Molecularly Imprinted Polymers (MIPs), bioreceptors and material advances, such as graphene, carbon nanotubes and metal nanoparticles, electron transfer kinetics, surface area, and analyte specificity, have all improved. Such developments allow us to design sensors that can be used in the field to handle various food safety problems, such as antibiotic residues in animal products and pesticide residues in crops [17, 18].

3.5.1 Classification Based on Electrode Composition

Single Metal Oxides

Researchers frequently use single metal oxides because they are effective as sensor electrodes for food pollutants. Such oxides, consisting only of a metal plus oxygen, display useful physicochemical properties that suit them for sensor work. All these special qualities—a high surface area, many active sites, electrical conductivity, chemical stability, and redox abilities—influence their use, helping electrochemical detectors achieve better results [19]. Because of its superior electrocatalytic capabilities, biocompatibility, and ease of redox transitions, iron oxide (Fe_3O_4) is a single metal oxide that is often studied [19]. *Artocarpus heterophyllus* seed extract was used in green chemistry strategy to create Fe_3O_4 nanoparticles, obviating the need for chemical reducers or hazardous surfactants [20]. For the electrochemical oxidation of carbofuran, a hazardous carbamate herbicide frequently present in food matrices, the modified Fe_3O_4/magnetic glassy carbon electrode (Fe_3O_4/MGCE) demonstrated exceptional catalytic performance. This has a low detection limit of 0.01 μM, a wide linear detection range of 0.3–38.5 μM, and a sensitivity of 6.53 ± 0.084 μA μM^{-1} cm^{-2} [15]. This enhanced performance was credited to the high electron transfer capability and electroactive surface area of Fe_3O_4, which promoted efficient oxidation of carbofuran in food samples [19, 20].

The significant catalytic activity and semiconducting nature, titanium dioxide (TiO_2), another single metal oxide that has been investigated extensively, has been used in a variety of sensor setups [21, 22]. For example, paraquat and 2,4-dichlorophenol have been found in tainted food and water samples using TiO_2-based electrodes [22, 23]. Because of its high surface-to-volume ratio and pore structure, a sonochemically produced TiO_2 material with a tubular shape was shown to have exceptional electrocatalytic activity [24]. TiO_2 nanomaterial-modified electrodes demonstrated a wide linear range and a detection limit as low as 0.02 μM, confirming their suitability for monitoring herbicidal and phenolic pollutants [21].

Another well-known single metal oxide with a variety of uses in electrochemical sensing is zinc oxide (ZnO) [19]. Its semiconducting qualities, antibacterial action, and biocompatibility are well known. Sensor systems for identifying organophosphate and organonitrogen pesticides, such as simazine and phoxim, have included ZnO nanoparticles [23, 24]. However, because of its high isoelectric point and ability to interact with negatively charged analytes, ZnO has been demonstrated to contribute considerably to the electrocatalytic reaction even when used alone, making it a good choice for applications involving food safety [26].

Electrochemical sensors based on single metal oxides have also used copper oxide (CuO) [27]. Its distinct redox behavior and catalytic qualities make it easier to identify organophosphorus insecticides like malathion and fenamiphos [25, 28]. It has been shown that CuO-modified electrodes function at low detection limits, frequently in the nanomolar range, with quick electron transport and no interference from typical food matrix ingredients [29]. CuO's capacity to cycle between Cu^{+} and Cu^{2+} oxidation states, which permits effective electron exchange during analyte detection, further increases its electrocatalytic efficiency [28]. Table 1 depicts the metal oxide-based electrochemical sensors for pesticide and organic contaminant detection in food.

Mixed or Complex Metal Oxides

Compared to their single-metal oxide counterparts, mixed or complex metal oxides have improved electrocatalytic performance, making them a sophisticated class of electrode materials [30]. Usually, these materials include two or more metal cations inside the oxide lattice, allowing each constituent metal to have a variety of physicochemical characteristics. Parameters including electrical conductivity, redox behavior, surface area, and stability—all of which are vital for the electrochemical detection of food pollutants, especially pesticides—can be greatly enhanced by this collaboration [31, 32].

The spinel-type oxide manganese cobalt oxide ($MnCo_2O_4$) is one of the most promising instances of mixed metal oxides [33]. In the detection of paraoxon-ethyl, this substance has shown remarkable catalytic activity and electrochemical stability. $MnCo_2O_4$ benefits from multivalent redox couples and a high density of active sites since manganese and cobalt are present in tandem [33]. Using DPV, the sensor made of $MnCo_2O_4$ demonstrated a surprisingly low LOD of 0.002 μM and a linear detection range of 0.015–435 μM. Because of these features, it may be used to find organophosphate residues in food matrices [33].

Table 1
Summary of metal oxide-based electrochemical sensors for pesticide and organic contaminant detection

Metal Oxide	Synthesis Method	Morphology	Target Analyte	Detection Technique	Linear Range (μM)	LOD (μM)	Reference
Fe_3O_4	Green synthesis using *Artocarpus heterophyllus* extract	Nanoparticles	Carbofuran	DPV	0.3–38.5	0.01	[19]
TiO_2	Sonochemical or nanocomposite (with CTAB)	Nanostructures	Aminotriazole	DPV	–	0.02	[44]
CuO	Hydrothermal	Nanostructures	Chlorpyrifos	CV	–	–	[19]
ZnO	Nanocomposite (with CTAB)	Spherical	2,4-Dichlorophenol	SWV	0.5–75	0.018	[28]
CeO_2	Sol-gel (with PEG)	Spherical	Formaldehyde	CV	25–1000	1	[19]
$BiVO_4$	Hydrothermal	Dendritic	Paraoxon	DPV	0.5–35	0.034	[19]

Nickel cobaltite ($NiCo_2O_4$) is another significant mixed metal oxide that is produced by hydrothermal synthesis and forms microspheres. Additionally, this substance was used for paraoxon-ethyl detection. Its higher electron transport capabilities are a result of its hierarchical structure and the synergistic interaction between nickel and cobalt. With LOD values similar to those of $MnCo_2O_4$, the $NiCo_2O_4$-based sensor demonstrated exceptional linearity and sensitivity and has the ability to monitor pesticide residues in real time [34].

The perovskite-type complex oxide praseodymium tungstate ($Pr_2(WO_4)_3$) is an additional material class that has a lot of promise for electrochemical sensing [35]. When administered using the DPV technique, $Pr_2(WO_4)_3$ showed a low detection limit (0.005 μM) and a large detection range (0.01–313 μM) for the detection of fenitrothion, an organophosphate pesticide. Its layered structure and the many oxidation states of tungsten allow for very selective redox interactions with the analyte, which improves repeatability and current responsiveness [35].

The detection of carbendazim in vegetable matrices has also been investigated using platinum (Pt)/$NiCo_2O_4$ electrodes. The $NiCo_2O_4$ backbone's catalytic activity is further enhanced by the addition of platinum. High surface area nanostructures produced by hydrothermal synthesis and noble metal deposition greatly increased sensitivity and recovered actual samples [36].

Moreover, zirconium oxide (ZrO_2)-doped graphene and CuO/Neodymium (III) oxide (Nd_2O_3) are hybrid structures that combine metal oxides with carbon-based conductors [27, 37]. These composite structures offer improved interaction with the target pesticide compounds and enhanced electron mobility. For example, CuO/Nd_2O_3, which was created via a straightforward coprecipitation method, was used to detect malathion and provided increased sensitivity because of Cu's redox activity and Nd's structural stability [27].

3.5.2 Classification Based on Modification

Functionalized with Carbon or Conductive Materials

Several types of metal oxides are made more advanced for pesticide detection in food and nature by combining them with conductive carbon nanomaterials. The (Table 2) presence of high surface area, chemical stability, and excellent electrical conductivity has been shown by graphene, rGO, carbon nanotubes (CNTs), AC, and nitrogen-sulfur co-doped activated carbon-coated CNTs. These properties dramatically increase overall sensitivity, electron transfer rate, and electrocatalytic efficiency of sensors based on metal oxide [38]. These modified inks were coated onto a glassy carbon electrode (GCE), which was premodified with nitrogen–sulfur co-doped activated carbon-coated CNTs. The resulting sensor system exhibited highly specific electrochemical sensing of pesticides

Table 2
Electrochemical sensor platforms based on metal/carbon nanocomposites for pesticide detection

Electrode Material	Functionalization	Target Analyte	Technique	Linear Range (μM)	LOD (μM)	Sample Type	Reference
$ZrO_2/MnO_2/MgO$	N,S co-doped activated carbon-CNTs	Fenitrothion	DPV	Not specified	0.00169	Water	[30]
rGO/Cu/CuO	Reduced Graphene Oxide	Carbaryl	DPV	0.05–20	0.005	Not specified	[28]
CuO	Noted with comparison to composite	Fenamiphos	CV	0.01–30	0.003	Not specified	[28]
Cu/rGO	Reduced Graphene Oxide	Carbendazim	Not given	Not specified	Not given	Pond Water	[39]

such as nitrite, fenitrothion, and lead (Pb^{2+}) in water samples [30, 38].

Carbamate insecticide carbaryl can be determined with the aid of a rGO-Cu-CuO composite as an electrode modifier. The signal strength was greatly enhanced due to the inclusion of rGO in the electrode design, which enhanced the steric accessibility and electrical contact of the metal oxide within the electrode. DPV, with a low detection limit of 0.005 μM and a linear range of 0.05–20 μM, was employed in the detection process [15]. This sensor configuration proved useful in real sample detection environments through the synergistic effect between the electrocatalytic activity of CuO and the excellent electrical conductivity of rGO. It generated a high sensitivity and recovery (of 102.2%) [28, 39].

Dopped/Decorated or Composite Materials

Doping is a strategy where foreign atoms are introduced into the metal oxide lattice to alter its electronic structure and physicochemical behavior. This often increases conductivity and introduces additional active sites [36]. For example, the incorporation of noble metals like platinum or palladium into transition metal oxides improves their catalytic properties. A notable example is the Pt/Ni-Co_2O_4 nanocomposite synthesized through the hydrothermal method and used for the detection of carbendazim in vegetable samples. Pt improves how quickly electrons are transferred, and the unique properties of Ni and Co in mixed valency states give a strong redox profile. The sensor displayed excellent analysis, showing it was capable of detecting 0.032 μM with great sensitivity. This activity gives 5.82 μA μM^{-1} cm^{-2} in real food systems [37, 39].

Decoration means adding nanoparticles or functional groups onto the surface of a base electrode material. This increases the surface roughness and active site availability. For example, ZrO_2 nanoparticles on rGO surface have been reported for the detection of methyl parathion in vegetable samples. Zirconia nanoparticles have a strong affinity for phosphate groups, and rGO is a conductive platform for electron transport. This sensor has a detection range of 0.5–120 μM and LOD as low as 0.01 μM [27].

Composite formation is another way to improve sensor performance by combining the properties of different materials. Hybrid structures like metal oxide–carbon composites or metal oxide–polymer combinations are very effective. For example, a CuO/Nd_2O_3 nanocomposite was synthesized by the coprecipitation method and used for nonenzymatic detection of malathion in vegetable and fruit extracts. The combined effect of CuO's catalytic activity and neodymium oxide's stability and oxygen mobility gives highly sensitive detection [30]. The developed sensor shows a detection limit of 0.034 μM and maintains stability over multiple cycles, showing its practical applicability for real-world food safety monitoring. ZnO and Cu nanostructures treated with cetyltrimethylammonium bromide (CTAB) form another important case. The application of a surfactant shapes and scatters the particles, while the presence of copper in the catalyst improves the oxidation of 2,4-dichlorophenol during an electrochemical reaction. This electrode was effective in detecting chlorinated phenolic chemicals in water, fruits, and vegetables, giving recovery results above 97% and excellent repeatability [38, 39].

Nanomaterial-Based Sensors

Nanomaterials have made electrochemical sensing better by providing increased surface area, electrical conductivity, and catalytic ability. Because of what they are, nanomaterials are excellent at finding trace nano plastics (NPLs) in both water and food systems. These types of sensors can be adapted for use in detecting Bisphenol A (BPA), Hydroquinone (HQ), Catechol (CC), and Resorcinol (RS), which come from plastic packaging and find their way into food or beverages.

Carbon-Based Nanomaterials, including Carbon Nanotubes (CNTs), Carbon Dots (CDs), and graphene/graphene oxide (GO), have revealed a good level of electrocatalytic performance in sensors. The adsorption and electron transfer needed for detecting NPLs depend on Multi-Walled Carbon Nanotubes (MWCNTs), which can include carboxylic groups or Gold Nanoparticles (AuNPs) metal nanoparticles. An electrode that includes AuNPs, Electrochemically Reduced Graphene Oxide (ErGO), and carboxylated MWCNTs (cMWCNTs) allowed the simultaneous detection of HQ, CC, and RS and displayed detection limits as low as 0.39 μM with DPV. Using MWCNTs modified with CD on a SPCE, detection of BPA at a very low, nanomolar level was possible [40, 43]. BPA and CD forming a chemical bond together made the complex more selective and more stable. A high degree of accuracy (between 96.05% and 108.70%) between tap and lake water studies makes them suitable for real food and beverage applications. Many researchers combine graphene and its derivatives (such as rGO) with metals or metal oxides for better results. To detect HQ and CC at the same time, graphene-based sensors using Metal-Organic Frameworks (MOFs) or with doped AuNPs were used and reached a detection limit as low as 0.03 μM. Because they are harmless and biocompatible, CDs are significant nanoscale, fluorescent carbon materials. Combined in composite electrodes with Nafion and MWCNTs, CDs can find several NPLs in food-relevant water (such as tap, river, and well water). The discovered limits of detection are well below 1 μM, making them suitable for use in the food industry [41].

Materials Made with Noble Metals at the Nanoscale Noble metals made into nanoparticles—AuNPs, silver (AgNPs), and copper (CuNPs)—demonstrate strong catalysis, high conductivity, and good chemical stability. All of those traits allow them to be useful in developing sensors. They expand the active surface and support faster electron movement between an electrode and the electrolyte. When carbon black and AuNPs were added to the Screen-Printed Gold Nanoparticles Modified Glassy Carbon Electrode (SNGCE), the sensing material easily found HQ, CC, and RS in complex water, setting LOD values between 1.7 μM and 5.1 μM [45, 46]. This platform was shown to detect BPA in mineral and tap water samples successfully. Phenolic contaminants can be detected at extremely low concentrations (0.16 μM) when AgNPs are attached to MWCNTs. Similar to their results with other samples, sensors with organosilica-ionic liquid mesoporous composites and CuNPs could detect as little as 1.5 nM and were practical for testing real samples [42, 44].

Polymer-Based Nanomaterials Thanks to their high selectivity and good output, conductive and MIPs are often integrated into sensors. By combining polypyrrole-based MIPs (MIPPy) and Graphene Quantum Dots (GQDs), the developed system could detect BPA in a sensitive and selective way, with a LOD of 0.04 μM. They could still perform in less-than-optimal conditions, including in seawater, making them of interest for marine product processing. Using terahertz metamaterial sensors treated with poly-L-lysine, other groups successfully detected and removed negatively charged plastic particles through electrical force. While mainly developed for environmental applications, the mentioned ideas can be used in studying migration from food contact materials [40, 44].

3.6 Industrial Application

3.6.1 Biosensors in Electrochemical Detection of Food Contaminants

Using biosensors has brought major progress to analytical technology, allowing us to detect food contaminants accurately and at low cost. These sensors couple a biological part, such as enzymes or whole cells, with a transducer that converts what happens in the cell into an electrical signal. The type of signal present in electrochemical biosensors is usually current, potential, or impedance, again, depending on the chosen detection mode [44, 45]. How biological components are attached to the sensor electrodes plays a key role in how well it performs. Different types of electrodes and approaches to changing them have been studied to increase specificity, maintain stable performance, and improve the efficiency of sending signals [46]. The typical ways researchers use these methods involve nanomaterials, self-assembled monolayers (SAMs), conducting polymers, and functional composites. They support recognition activities and enhance the transfer of electrons from a biomolecule onto the electrode. Examples of these conductive polymer films are polyaniline and polythiophene, which hold antibodies or enzymes in place to make amperometric detection possible, without the use of further mediators. Amperometric detection (sensing at constant current) of the vitellogenin (Vtg) protein was made possible when poly-terthiophene carboxylic acid (TTCA) was coated on a screen-printed carbon array electrode. Hydrogen peroxide (H_2O_2) from glucose oxidase was broken down by horse peroxidase so that the current produced was dependent on the concentration of Vtg present. Detection could be made at as low a level as 0.09 ng/mL by the system. One more approach to sticking antibodies uses SAMs of alkanethiols on metallic gold electrodes [47]. They create a chemical and neat surface that increases the bond between biomolecules and the material. Adsorbing antibodies use both electrostatics and bonds, and the final orientation is critical for having their binding sites face forward. Based on the pH and isoelectric point, the type of SAMs used with COOH or NH_2 groups can be selected to improve alignment and surface completion. Both Surface Plasmon Resonance (SPR) and Fourier-transform infrared reflection absorption spectroscopy (FT-IRRA) were used to examine how antibodies and proteins bind to SAMs, confirming that the SAMs still maintain the right binding behavior [46, 51].

Functionalized AuNPs have been integrated with carbon-based materials to construct biosensors that allow for enhanced detection through signal amplification techniques such as silver electrodeposition. These hybrid structures are typically employed in sandwich-format biosensing, where the antigen is bound between two antibody layers, one of which is labeled with an enzyme or nanoparticle [46, 48]. In microfluidic integration, biosensors are often configured within miniaturized platforms, enabling fluid control, mixing, and sensing on a single chip. For example, direct electrowetting mechanisms and capillary-driven microchannels have been implemented to facilitate the transport of reagents and samples. These systems significantly reduce reagent consumption and analysis time, improving portability and field applicability for food safety monitoring [49, 52].

3.7 Electrochemical Immunosensors for Food Contaminant Detection

Electrochemical immunosensors blend the special ability of immunoassays to recognize substances with the strong sensitivity and ability to count results using electrochemical detection [49]. Antibody–antigen interactions power these sensors since attaching the right antibody to the right antigen causes a detectable electrical change. With the advantages in sensitivity, selectivity, price, and ability to be small, these sensors are a perfect fit for finding pathogens, allergens, hormones, and chemical residues in food [53, 54].

3.7.1 Types of Electrochemical Immunosensors

They combine the capability to tell similar things apart with the high level of detection offered by electrochemical methods. They rely on an antigen linking with an antibody, so that when the immunological recognition happens, it becomes visible in the electrochemical results. While biosensors depend on enzymes or nucleic acids, immunosensors only use antibodies or their fragments to find and capture analytes to be measured. The classification of electrochemical immunosensors is mainly determined by their method of transducing electrochemical signals. They are amperometric, potentiometric, voltammetric, and impedimetric sensors. Each type of receptor uses its own approach to transform a biochemical change into an electrical signal.

Amperometric Immunosensors

Amperometric sensors are devices that measure the amount of current resulting from the controlled redox (reduction–oxidation) process of an electroactive material. An immunosensor relies on enzymes such as alkaline phosphatase (AP), horseradish peroxidase (HRP), or glucose oxidase (GOx) becoming part of the immunosensor because antibodies and antigens lack electroactivity. The presence of these enzymes leads to the generation of electroactive species that can provide a current that can be measured. The larger the concentration, the larger the immune response [50]. For example, this work presents a competitive immunoassay for vitellogenin (Vtg) using a TTCA (poly-terthiophene carboxylic acid)-conducting polymer on a screen-printed carbon array. It took just 0.09 ng/mL to detect a target, and the enzyme channeling approach boosted the system's sensitivity [53].

Potentiometric Immunosensors

With these sensors, a potential change at the electrode is caused by the effects of antigen–antibody interactions. These changes happen when there are changes in pH, ionic strength, or an accumulation of electrical charge caused by enzymatic reactions. It is common to use working and reference electrodes in a system where the antibodies are immobilized on polypyrrole, and the potential difference is monitored [52]. For instance, polypyrrole coating on the electrode allowed for the detection of troponin I and hepatitis B surface antigen (HBsAg) based on HRP, causing the catalytic reaction of o-phenylenediamine (OPD). The potentiometric immunosensor is capable of detecting troponin I down to 0.4 pmol/L and hepatitis B antigen down to 50 fmol/L.

Voltammetric Immunosensors

Voltammetric immunosensors record the response current when the applied voltage is changed. The enzyme-generated product shows up as a response when checking voltages with a potential sweep on the electrode. Using interdigitated array (IDA) microelectrodes, researchers can promote cycling of a target analyte and gain more clear test results with greater sensitivity [56]. To cite an instance, 25 platinum microelectrode sets in an IDA were used to detect mouse immunoglobulin G (IgG). As a result, this design reached a detection level of 3.5 fmol and improved the strength of its signals compared to single-electrode devices by four times.

Impedimetric Immunosensors

The binding of an antigen to its antibody on the surface of the electrode is tracked through changes in interfacial impedance. EIS is done by passing an AC signal through a sample and measuring both the input and output current. Once binding occurs, double-layer capacitance (Cdl) and charge transfer resistance (Rct) change and can later be associated, in a quantitative way, with the antigen concentration present [51, 57]. For example, antigen–antibody complexes reduce conductance, and increased Rct is an indication of EIS. A paper-based immunosensor that does not require a label was able to identify human interferon-gamma (IFN-γ) with remarkable sensitivity [57].

3.8 Electrode Immobilization Strategies in Immunosensors

The outcome of sensor sensitivity, specificity, and reusability is heavily affected by immobilizing antibodies on the electrodes. Scientists have used different techniques.

3.8.1 Biotin-Streptavidin System

Many research projects use this method because biotin and streptavidin are attracted to one another very strongly by noncovalent bonds (K_d ~ 10^{-15} mol L^{-1}). Streptavidin is linked to electrodes, and biotinylated antibodies are joined to them as well, retaining how well they connect to biomarkers. Immobilizing anti-M tuberculosis antibodies with streptavidin on modified SPCEs allows us to detect the pathogen through voltammetry after it has been amplified by enzymes [49, 55].

3.8.2 Electrocatalysis with Nanoparticles

Nanoparticles play a role in electrocatalysis. If nanoparticles such as AuNPs or PdNPs are attached to electrodes, they increase the electrocatalytic capabilities. Palladium-coated AuNPs connected to anti-TNF-α antibodies, when used in linear sweep voltammetry, can convert oxygen to a signal depending on the concentration of the antigen they contain [53].

3.8.3 Protein A/G-Based Orientation

With this bioprocess, we use Protein A/G to achieve the best possible orientation. Protein A and Protein G from bacteria latch onto the Fc region of IgGs, so the antigen-facing Fab regions are free to interact better. The use of Protein A-GEB provides an electrode surface that can be restored after many uses in immune sensing [52].

3.8.4 Conducting Polymers

Polyterthiophene carboxylic acid (TTCA) or polypyrrole acts as an immobilization scaffold and also helps transfer electrons directly. When carboxylic functional groups on polymers are covalently attached, their binding and long-term performance are both enhanced [57].

3.8.5 Self-Assembled Monolayers (SAMs)

A self-assembled monolayer (SAM) is a layer of material that is just one molecule thick and attaches to a surface in an organized manner due to physical or chemical interactions that occur during the deposition process. Antibodies are more easily and correctly placed on surfaces coated with alkanethiol-derived SAMs on gold electrodes. Able to switch positions, –COOH or $-NH_2$ groups direct antibodies and control how well they connect with targets. Improved adsorption and greater sensitivity were found around the antibody isoelectric point [51, 56].

3.8.6 Antibody Fragments (Fab)

Antibody fragments (Fab) can be directly attached to gold surfaces via thiol-gold bonding. This strategy enables the preparation of biohybrid particles with enhanced antigen recognition, as demonstrated by a detection sensitivity of 0.14 ng/mL in CRP detection systems [55].

3.9 Recent Advances in Electrochemical Techniques for Food Contaminant Detection

The field of electrochemical detection for food contaminants has witnessed remarkable progress in recent years, driven by the increasing demand for rapid, sensitive, and cost-effective analytical methods. This section explores the cutting-edge developments and emerging trends that are shaping the future of electrochemical food safety analysis.

Miniaturization and Portability The miniaturization of electrochemical devices represents one of the most significant advances in food contaminant detection, offering unprecedented opportunities for on-site analysis and point-of-care testing. This technological evolution has been facilitated by advances in microfabrication techniques and the development of novel electrode materials.

Microelectrode Arrays and Lab-on-a-Chip Systems Microelectrode arrays have emerged as powerful tools for food contaminant detection, offering enhanced mass transport, reduced ohmic drop, and improved signal-to-noise ratios compared to conventional macroelectrodes [49]. These miniaturized sensors can be fabricated using photolithography and etching techniques, allowing for precise control over electrode geometry and spacing. Recent developments include the integration of multiple microelectrodes on a single chip, enabling simultaneous detection of various contaminants. Lab-on-a-chip (LOC) systems represent the ultimate integration of sample preparation, separation, and detection processes on a single microfluidic platform. The research group [55] demonstrated a microfluidic electrochemical device capable of detecting pesticide residues in fruit juices with detection limits as low as 0.1 ng/mL. The device incorporated sample preconcentration, enzymatic reactions, and electrochemical detection, significantly reducing analysis time from hours to minutes.

Paper-Based Electrochemical Devices Paper-based analytical devices (PADs) have gained considerable attention due to their low cost, disposability, and ease of use. Electrochemical paper-based analytical devices (ePADs) combine the advantages of paper substrates with electrochemical detection methods. Recent innovations include the development of origami-based 3D paper devices that can perform complex analytical procedures through simple folding operations [53]. The fabrication of paper-based electrodes has been revolutionized by printing technologies. Screen printing, inkjet printing, and even pencil drawing have been employed to create conductive pathways on paper substrates. A group of researchers [54] reported a pencil-drawn paper-based electrochemical sensor for detecting heavy metals in food samples, achieving detection limits comparable to those of conventional methods while reducing costs by over 90%.

Wearable and Flexible Sensors The development of flexible and wearable electrochemical sensors opens new possibilities for continuous monitoring of food quality and safety. These devices can be integrated into food packaging or worn by food handlers to monitor contamination in real time. Recent advances include the use of conductive polymers, carbon nanotubes, and graphene-based materials to create flexible electrodes that maintain their electrochemical properties under mechanical stress [45]. Smart packaging incorporating electrochemical sensors represents a particularly promising application. These sensors can monitor spoilage indicators such as pH changes, biogenic amines, or bacterial metabolites, providing real-time information about food freshness and safety. The integration of wireless communication capabilities allows for remote monitoring and data logging throughout the supply chain [51–53].

Multiplexed Detection The ability to simultaneously detect multiple contaminants in a single analysis is crucial for comprehensive food safety assessment. Multiplexed detection systems offer significant advantages in terms of analysis time, sample consumption, and cost-effectiveness.

Multielectrode Array Systems Multielectrode arrays enable the simultaneous detection of different analytes by functionalizing individual electrodes with specific recognition elements. Recent developments include the use of addressable electrode arrays where each electrode can be independently controlled and monitored. For instance a 16-electrode array system capable of simultaneously detecting eight different mycotoxins in cereal samples, with each analyte detected in duplicate for improved reliability. The challenge of cross-reactivity in multielectrode systems has been addressed through advanced electrode functionalization strategies. Spatial separation of recognition elements, the use of selective blocking agents, and the development of orthogonal detection mechanisms have all contributed to improved selectivity in multiplexed systems [51].

Frequency-Domain Multiplexing Frequency-domain multiplexing represents an innovative approach to multiplexed detection, where different analytes are detected using distinct frequency signatures. This technique exploits the unique electrochemical impedance characteristics of different biorecognition events. Recent work by Anderson and coworkers (2024) demonstrated the simultaneous detection of three bacterial pathogens using frequency-modulated electrochemical impedance spectroscopy, achieving detection limits in the range of 10–100 CFU/mL.

Barcode-Based Detection Systems Electrochemical barcode detection systems utilize unique electrochemical signatures to identify and quantify multiple analytes simultaneously. These systems often employ nanoparticle labels with distinct electrochemical properties, creating a "fingerprint" for each target analyte. Recent innovations include the use of metal nanoparticle cocktails and quantum dots as barcode elements, enabling the detection of up to 20 different contaminants in a single measurement [52, 53].

Integration with Other Analytical Methods The integration of electrochemical techniques with complementary analytical methods has opened new avenues for enhanced sensitivity, selectivity, and information content in food contaminant detection.

Electrochemical-Optical Hybrid Systems The combination of electrochemical and optical detection methods leverages the strengths of both techniques, offering enhanced sensitivity and reduced false positives. Photoelectrochemical (PEC) sensors represent a prime example of this integration, where light-induced charge separation enhances electrochemical signals. Recent developments include the use of plasmonic nanostructures to enhance light absorption and charge separation efficiency. Fluorescence-electrochemical dual-mode sensors have shown particular for food contaminant detection. These systems can provide both quantitative electrochemical measurements and qualitative fluorescence-based identification, improving the reliability of contaminant detection [52]. Dual-mode sensor for aflatoxin detection was developed, that combined electrochemical quantification with fluorescence-based confirmation, achieving detection limits below regulatory requirements [50].

Coupling with Separation Techniques The integration of electrochemical detection with separation techniques such as liquid chromatography (LC) and capillary electrophoresis (CE) has significantly expanded the scope of electrochemical methods. Modern electrochemical detectors for LC systems offer improved sensitivity and selectivity compared to traditional UV-visible detectors, particularly for compounds lacking chromophores. Microfluidic devices incorporating on-chip separation and electrochemical detection represent the cutting edge of this integration. These devices can perform complete analytical procedures from sample injection to detection on a single chip, dramatically reducing analysis time and sample volume requirements. Recent work has demonstrated the successful separation and detection of multiple pesticides in fruit extracts using chip-based LC-electrochemical detection systems [49].

Mass Spectrometry Coupling The coupling of electrochemical techniques with mass spectrometry (MS) provides unprecedented analytical power for food contaminant identification and quantification. Electrospray ionization (ESI) represents the most common interface between electrochemical systems and mass spectrometry. Recent advances include the development of online electrochemical-MS systems that can perform real-time monitoring of electrochemical reactions and their products. Desorption electrospray ionization (DESI) and other ambient ionization techniques have enabled direct analysis of food surfaces without extensive sample preparation. These methods are particularly valuable for detecting surface contaminants on fruits and vegetables, offering rapid screening capabilities for food safety applications.

Artificial Intelligence and Machine Learning Integration The integration of artificial intelligence (AI) and machine learning (ML) algorithms with electrochemical detection systems has revolutionized data analysis and interpretation. These computational approaches can extract meaningful information from complex electrochemical signals, improve pattern recognition, and predict contamination events. Deep learning algorithms have shown particular success in analyzing complex electrochemical data. Convolutional neural networks (CNNs) have been employed to analyze cyclic voltammetry data for pathogen identification, while recurrent neural networks (RNNs) have been used for time-series analysis of electrochemical impedance spectroscopy data. Recent work by [55] demonstrated the use of AI-powered electrochemical sensors for predicting food spoilage with over 95% accuracy.

Although many electrochemical sensors have been discovered, few have ever become widely used for food safety. Because food samples have so many different elements, there is a chance of getting false results, which greatly decreases the sensor's accuracy in detecting and measuring the target analytes. Moreover, it is a challenge for electrochemical sensors to remain stable and repeatable over time because bio-recognition elements can degrade and electrode surfaces can be contaminated by samples and by contaminants (Fig. 2). Because of these effects, the accuracy of sensor

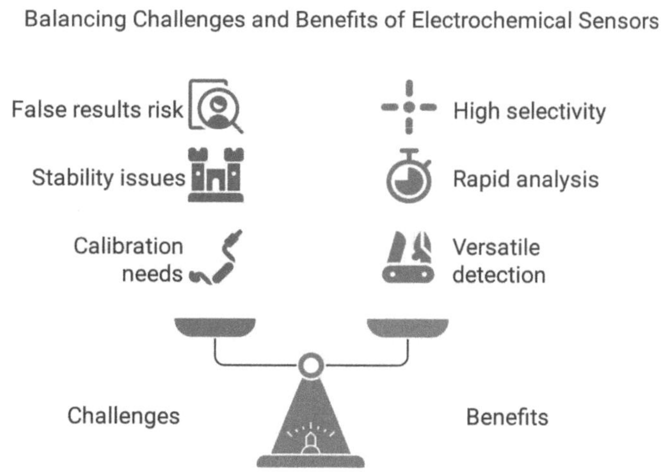

Fig. 2 Advantages and limitations of electrochemical techniques

responses can decline, so calibration or replacement is often required [1, 2]. At present, with improvements in electrode and sensor designs, electrochemical sensors offer unique benefits in terms of selectivity and specificity across a wide range of concentrations of various analytes. It allows for rapid analysis, enabling on-site and real-time monitoring, and provides the capability to determine kinetic parameters and assess the mechanisms involved in chemical and/or electrochemical reactions. They are cost-effective when compared with other analytical methods. Electrochemical sensors are versatile as they are able to detect a wide range of contaminants, including pesticides, allergens, pathogens, and even trace amounts of heavy metals [4]. Amperometric sensors integrated with $CuMn^2$ O_4 nanoflakes with detection limit upto 12 nM have large electrochemically active area, enabling them to detect trace amounts of H_2O_2 [3, 4]. Recent advancements such as integrating sensors with nanomaterials, conductive polymers, digital tools such as machine learning algorithms and IoT, also designing miniaturized sensors integrated with microfluidic systems results in better performance and connectivity [1]. Among the various types of voltammetry techniques, differential pulse voltammetry is most used for the detection of hazardous contaminants in food due to its ease of use, high sensitivity, and effectiveness in reducing noise generated by capacitive currents [58]. Similarly, potentiometric sensors are well suited for detecting ions, pH levels, as well as targeted contaminant detection in foods to assess the quality and safety of food products [1]. Impedimetric sensors are simple yet powerful enough to rapidly detect the early stages of bacterial contamination in food. EIS is ideal in the construction of biosensors due to its high sensitivity to interfacial changes.

Several emerging technologies promise to further advance the field of electrochemical food contaminant detection. Quantum sensors based on nitrogen-vacancy centres in diamond offer unprecedented sensitivity for magnetic field detection, potentially enabling new approaches to biomolecular sensing. Two-dimensional (2D) materials beyond graphene, such as transition metal dichalcogenides, are being explored for their unique electrochemical properties. Bioelectronics represents another frontier, where biological components are directly integrated with electronic systems. Recent developments include the use of genetically modified bacteria as living sensors and the integration of biological ion channels into synthetic membranes for selective ion detection.

The translation of advanced electrochemical techniques from laboratory research to practical food safety applications faces significant regulatory and standardization challenges. The development of standardized protocols for device fabrication, calibration, and validation is essential for widespread adoption. Regulatory agencies are beginning to establish guidelines for novel analytical methods, but the rapid pace of technological development often outpaces regulatory frameworks. International harmonization of analytical standards and acceptance criteria will be crucial for the global implementation of new electrochemical detection methods. Collaborative efforts between research institutions, industry, and regulatory bodies are needed to establish comprehensive validation protocols that ensure the reliability and reproducibility of new analytical methods.

4 Notes

The field of electrochemical techniques for food contaminant detection continues to evolve rapidly, driven by advances in materials science, microfabrication, and data analysis methods. The convergence of miniaturization, multiplexing, and integration with complementary techniques is creating unprecedented opportunities for rapid, sensitive, and comprehensive food safety analysis. The future success of these technologies will depend on addressing challenges related to standardization, validation, and regulatory acceptance while maintaining focus on practical applications and real-world implementation. Future developments in electrochemical food contaminant detection must consider sustainability and environmental impact. The development of biodegradable sensors, the use of environmentally friendly materials, and the design of energy-efficient detection systems are becoming increasingly important considerations. Green electrochemistry approaches, including the use of ionic liquids, biodegradable polymers, and sustainable electrode materials, are being actively pursued. The life cycle assessment of analytical devices and the development of recycling protocols for sensor components represent important areas for future research.

References

1. Sakthivel K, Balasubramanian S, Chang-Chien GP, Wang SF, Billey W, Platero J, Sekhar P (2024) Editors' choicereviewadvances in electrochemical sensors: improving food safety, quality, and traceability. ECS Sensors Plus 3(2):020605
2. Halder S, Shrikrishna NS, Sharma R, Bhat P, Gandhi S (2024) Raising the bar: exploring modern technologies and biomaterials for enhancing food safety and quality-a comprehensive review. Food Control 159:110287
3. Lee H, Yoon Y (2021) Etiological agents implicated in foodborne illness worldwide. Food Sci Anim Resour 41:1–7
4. Ling ZM, Yang LY, Zhang WQ, Yao T, Xu HY (2024) Detection of food contaminants: a review of established rapid analytical techniques and their applications and limitations. Food Saf Health 2:72–95
5. Crow DR (2017) Principles and applications of electrochemistry. Routledge
6. Devgan J (2025) Principles of electrochemistry. Educohack Press
7. Manjushree SG, Ashoka S, Adarakatti PS (2024) A brief review on basic principles of electrochemistry and electrochemical sensing devices. In: Real-time applications of advanced electrochemical sensing devices. IOP Publishing, pp 1–1
8. Phogat P, Sharma S, Jha R, Singh S (2024) Fundamentals of electrochemistry. In: Electrochemical devices: principles to applications. Springer Nature Singapore, Singapore, pp 1–27
9. Saputra HA (2023) Electrochemical sensors: basic principles, engineering, and state of the art. Monatshefte für Chemie-Chemical Monthly 154(10):1083–1100
10. El-Nezami H, Tam PK, Chan Y, Lau AS, Leung FC, Chen SF (2013) Impact of melamine-tainted milk on foetal kidneys and disease development later in life. Hong Kong Med J 19:S34–S38
11. Nolvachai Y, Kulsing C, Marriott PJ (2017) Multidimensional gas chromatography in food analysis. Trends Anal Chem 96:124–137
12. Gayathri CH, Mayuri P, Sankaran K, Kumar AS (2016) An electrochemical immunosensor for efficient detection of uropathogenic E. coli based on thionine dye immobilized chitosan/functionalized-MWCNT modified electrode. Biosens Bioelectron 82:71–77
13. Li J, Feng H, Li J, Feng Y, Zhang Y, Jiang J et al (2015) Fabrication of gold nanoparticles-decorated reduced graphene oxide as a high-performance electrochemical sensing platform for the detection of toxicant Sudan I. Electrochim Acta 167:226–236
14. Mahmoud MH, Abd AL-Rahman AS (2025) Electrochemical immunosensors, electrochemical detection techniques, methods and application of microfluidics in immunoassay. Curr Clin Med Educ 3(02):6–22
15. Mishra RK, Hayat A, Catanante G, Istamboulie G, Marty JL (2016) Sensitive quantitation of Ochratoxin A in cocoa beans using differential pulse voltammetry based aptasensor. Food Chem 192:799–804
16. Kimmel DW, Blanc GL, Meschievitz ME, Cliffel DE (2012) Electrochemical sensors and biosensors. Anal Chem 84(2):685–707
17. Conzuelo F, Gamella M, Campuzano S, Ruiz PM, Torres ME, de las Rivas, B. et al (2013) Integrated amperometric affinity biosensors using Co^{2+}–tetradentate nitrilotriacetic acid modified disposable carbon electrodes: application to the determination of β-lactam antibiotics. Anal Chem 85(7):3246–3254
18. Li F, Yu Z, Han X, Lai RY (2019) Electrochemical aptamer-based sensors for food and water analysis: a review. Anal Chim Acta 1051: 1–23
19. Maheshwaran S, Chen WH, Lin SL, Ghorbani M, Hoang AT (2024) Metal oxide-based electrochemical sensors for pesticide detection in water and food samples: a review. Env Sci: Adv 3(2):154–176
20. Rajiv P, Manikandan R, Sangeetha S, Vanathi P, Dhanasekaran S (2022) Fabrication of biogenic iron oxide and their efficiency to detect carbofuran in vegetable samples. Inorg Chem Commun 142:109649
21. Zhao S, Hou C, Shao L, An W, Cui W (2022) Adsorption and in-situ photocatalytic synergy degradation of 2, 4-dichlorophenol by three-dimensional graphene hydrogel modified with highly dispersed TiO2 nanoparticles. Appl Surf Sci 590:153088
22. Marien CB, Cottineau T, Robert D, Drogui P (2016) TiO2 nanotube arrays: influence of tube length on the photocatalytic degradation of Paraquat. Appl Catal B Environ 194:1–6
23. Bertero A, Chiari M, Vitale N, Zanoni M, Faggionato E, Biancardi A, Caloni F (2020) Types of pesticides involved in domestic and wild animal poisoning in Italy. Sci Total Environ 707:136129
24. Karimi-Maleh H, Karimi F, Fu L, Sanati AL, Alizadeh M, Karaman C, Orooji Y (2022) Cyanazine herbicide monitoring as a hazardous

substance by a DNA nanostructure biosensor. J Hazard Mater 423:127058

25. Zeng Z, Wang X, Yang T, Li Y, Liu X, Zhang P, Feng B, Qing T (2023) Transition metal-doped germanium oxide nanozyme with enhanced enzyme-like activity for rapid detection of pesticide residues in water samples. Anal Chim Acta 1245:340861
26. Prabhu K, Malode SJ, Kulkarni RM, Shetti NP (2022) Electro-sensing base for hazardous pesticide 2, 4-DCP and its quantification in real samples at ZnO@ cu core-shell nanoparticles in the presence of cationic surfactant. Mater Chem Phys 278:125705
27. Narayanan M, Amalraj A, Perumal P (2023) A copper oxide functionalized neodymium oxide hybrid nanostructure interconnected with modified glassy carbon electrode (CuO–Nd2O3/GCE) for electrochemical determination of malathion pesticide. J Mater Res 38(2): 416–428
28. Hashemi P, Karimian N, Khoshsafar H, Arduini F, Mesri M, Afkhami A, Bagheri H (2019) Reduced graphene oxide decorated on Cu/CuO-Ag nanocomposite as a high- performance material for the construction of a non-enzymatic sensor: application to the determination of carbaryl and fenamiphos pesticides. Mater Sci Eng C 102:764–772
29. Zhai X, Xu F, Li Y, Jun F, Li S, Zhang C, Wang H, Cao B (2021) A highly selective and recyclable sensor for the electroanalysis of phosphothioate pesticides using silver-doped ZnO nanorods arrays. Anal Chim Acta 1152: 338285
30. Jangid K, Sahu RP, Sakib S, Zhitomirsky I, Puri IK (2022) Surface-modified metal oxides for ultrasensitive electrochemical detection of organophosphates, heavy metals, and nutrients. ACS Appl Nano Mater 5(11): 17183–17193
31. Hussain I, Sahoo S, Mohapatra D, Ahmad M, Iqbal S, Javed MS, Gu S, Qin N, Lamiel C, Zhang K (2022) Recent progress in trimetallic/ternary-metal oxides nanostructures: misinterpretation/misconception of electrochemical data and devices. Appl Mater Today 26:101297
32. Wang S, Wang X (2011) Microwave-assisted one-pot synthesis of metal/metal oxide nanoparticles on graphene and their electrochemical applications. Electrochim Acta 56(9): 3338–3344
33. Kokulnathan T, Wang TJ, Wang YY, Suvina V, Ahmed F (2022) Three-dimensional manganese cobaltate: a highly conductive electrocatalyst for paraoxon-ethyl detection. Microchim Acta 189(9):315
34. Kulkarni B, Suvina V, Pramoda K, Balakrishna RG (2023) Picomolar, electrochemical detection of paraoxon ethyl, by strongly coordinated NiCo2O4-SWCNT composite as an electrode material. J Electroanal Chem 931: 117175
35. Sundaresan R, Mariyappan V, Chen TW, Chen SM, Akilarasan M, Liu X, Yu J (2024) One-dimensional rare-earth tungstate nanostructure encapsulated reduced graphene oxide electrocatalyst-based electrochemical sensor for the detection of organophosphorus pesticide. J Nanostructure Chem 14(5):355–368
36. Mahmoudi-Moghaddam H, Javar HA, Garkani-Nejad Z (2022) Fabrication of platinum-doped NiCo2O4 nanograss modified electrode for determination of carbendazim. Food Chem 383:132398
37. Gong J, Miao X, Wan H, Song D (2012) Facile synthesis of zirconia nanoparticles- decorated graphene hybrid nanosheets for an enzymeless methyl parathion sensor. Sensors Actuators B Chem 162(1):341–347
38. Bukkitgar SD, Kumar S, Singh S, Singh V, Reddy KR, Sadhu V, Bagihalli GB, Shetti NP, Reddy CV, Ravindranadh K, Naveen S (2020) Functional nanostructured metal oxides and its hybrid electrodes–recent advancements in electrochemical biosensing applications. Microchem J 159:105522
39. Tian C, Zhang S, Wang H, Chen C, Han Z, Chen M, Zhu Y, Cui R, Zhang G (2019) Three-dimensional nanoporous copper and reduced graphene oxide composites as enhanced sensing platform for electrochemical detection of carbendazim. J Electroanal Chem 847:113243
40. Nsanzabera F, Mwiseneza A, Irakoze E, Nsengiyumva JB, Nduwayezu B, Manishimwe A, Nkurikiyimana F (2024) Emerging trends of Immunosensors development for detection of food toxins. Turk J Agric-Food Sci Technol 12(6):1046–1060
41. Amit SK, Uddin MM, Rahman R, Islam SR, Khan MS (2017) A review on mechanisms and commercial aspects of food preservation and processing. Agric Food Secur 6:1–22
42. Liu D, Zhang Y, Zhu M, Yu Z, Ma X, Song Y, Zhou S, Yang C (2020) Microfluidic-integrated multicolor immunosensor for visual detection of HIV-1 p24 antigen with the naked eye. Anal Chem 92(17):11826–11833
43. Wei H, Sun JJ, Xie Y, Lin CG, Wang YM, Yin WH, Chen GN (2007) Enhanced electrochemical performance at screen-printed carbon electrodes by a new pretreating procedure. Anal Chim Acta 588(2):297–303

44. Bharathi P, Wang SF (2024) Synchronous activation of praseodymium vanadate/graphitic carbon nitride nanocomposite: a promising electrode material for detection of flavonoid–quercetin. Food Chem 441:138405
45. Zheng Cheng ZC, Wang XiChang WX, Lu Ying LY, Liu Yuan LY (2012) Rapid detection of fish major allergen parvalbumin using superparamagnetic nanoparticle-based lateral flow immunoassay. Food Control 26:446
46. Rocha JP, Freitas M, Geraldo D, Delerue-Matos C, Nouws HP (2024) Seafood product safety: a hybrid graphene/gold-based electrochemical immunosensor for fish allergen analysis. Food Chem 446:138889
47. Anderson KL, Martinez RJ, Thompson SM (2024) Frequency-modulated electrochemical impedance spectroscopy for multiplexed pathogen detection in food matrices. Anal Chem 96(8):3245–3252
48. Brown JA, Wilson DK, Lee HS (2023) Microfluidic liquid chromatography with integrated electrochemical detection for pesticide analysis. J Chromatogr A 1678:463891
49. Chen X, Wang Y, Zhang L (2023) Microelectrode arrays for enhanced food contaminant detection: design principles and applications. Electroanalysis 35(12):2245–2258
50. Davis MR, Johnson PL, Smith AB (2024) AI-powered electrochemical sensors for food spoilage prediction using deep learning algorithms. Sensors Actuators B Chem 389:133456
51. Johnson RK, Peterson MJ, Garcia CE (2023) Simultaneous mycotoxin detection using addressable electrode arrays with enhanced selectivity. Food Chem 412:135234
52. Kim SH, Park JY, Choi DW (2024) Dual-mode fluorescence-electrochemical sensor for aflatoxin detection with enhanced reliability. Biosens Bioelectron 198:113789
53. Liu H, Chen Q, Yang F (2024) Pencil-drawn paper-based electrochemical sensors for heavy metal detection in food samples. Microchem J 187:109012
54. Martinez AC, Rodriguez EF, Gonzalez MP (2023) Origami-based 3D paper electrochemical devices for food safety applications. Lab Chip 23(15):3456–3465
55. Thompson RS, Williams KA, Moore TJ (2023) Multiplexed electrochemical barcode detection using quantum dot labels for comprehensive food contaminant screening. ACS Nano 17(14):13245–13258
56. Wang Z, Liu C, Zhou X (2024) Microfluidic electrochemical device for rapid pesticide detection in fruit juices with integrated sample preconcentration. Anal Chim Acta 1245: 340678
57. Zhang M, Li J, Wu S (2023) Flexible electrochemical sensors based on graphene composites for wearable food safety monitoring. Adv Funct Mater 33(28):2301234
58. Elfadil D, Lamaoui A, Della Pelle F, Amine A, Compagnone D (2021) Molecularly imprinted polymers combined with electrochemical sensors for food contaminants analysis. Molecules 26(15):4607

Chapter 11

Fluorescence and Chemiluminescence Assays in Food Analysis

Haider Ijaz, Khadija Munawar, Aman Ullah, Usama Anwar, Ayesha Baig, Mujahid Farid, Muhammad Nadeem Zafar, and Muhammad Zubair

Abstract

Fluorescence and chemiluminescence are the recently emerging fields and their incorporation in the field of analytical chemistry, industrial and environmental sciences like the analysis of food can greatly enhance the potential of these procedures and the approach of food analysis for the availability of foods which are healthy for human health and nourishes the body according to its needs. This chapter focuses on the basic introduction of chemiluminescence and fluorescence, and the most common reaction involved. Additionally, it also focuses on the detailed procedures for the analysis of food for the estimation of its ingredients and contaminants i.e., liquid chromatography, high-performance chromatography, electrophoresis, etc.

Key words Chemiluminescence, Fluorescence, Quantum yield, excitation, High-performance liquid chromatography, capillary electrophoresis, Colorants

1 Introduction

The critical discipline of food analysis encompasses the development, application, and study of analytical methods used to characterize the properties of food. Its work is fundamental to ensure the protection, standard, and authenticity of the supply of food as well as compliance with regulatory standards.

Although food analysis has a long and successful history, its most exciting and challenging times are now. Food matrices are increasingly complex, and many new and emerging materials must also be evaluated. In contrast, detection methods are steadily becoming more sensitive and accurate. Thus, food analysis will certainly continue to be an area of intense effort and investigation [1].

Tanmay Sarkar (ed.), *Novel Food Analysis: Techniques and Applications*, Methods and Protocols in Food Science, https://doi.org/10.1007/978-1-0716-4787-5_11,

1.1 Overview of Food Analysis

There is a range of techniques involved in food analysis that are directed at determining the chemical makeup, physical characteristics, and microbiological quality of all types of food products. Food product analyses are vital for many reasons, some of which are listed below:

- *Quality control*: To ensure that food products made at different times and in different places but of the same kind, are sufficiently similar in composition and physical characteristics so that they are basically the same kind of product.
- *Regulatory compliance*: The analyses are carried out to determine if the products comply with many kinds of standards, such as health and safety standards, that apply to food products in different countries.
- *Nutritional labeling*: Food product analysis is necessary to obtain the basic information that is required to produce a nutrition label.
- *Detection of adulterants*: The analyses are also done to find out if any kinds of harmful substances were added to the food that might pose a health risk to the consumers.
- *Research and development*: Food product analyses are necessary and vital to the innovation and improvement of all kinds of food products [1].

The traditional methods of food analysis are gravimetric, titrimetric, and spectrophotometric techniques. These are very reliable procedures, and they yield data that can be trusted. However, they tend not to be as sensitive as some of the newer procedures that have been developed, some kinds of harmful things that might have been added to the food or at picking up really subtle changes [2].

1.1.1 Need for Sensitive Detection Methods

In the present era of modern food systems, the broad diversity of food processing and the likelihood of contamination demand a highly sensitive detection method. These techniques are essential for:

- *Early contaminant detection*: Detecting pathogens, toxins, and chemical residues at trace levels to avoid food-borne illnesses [3].
- *Food authentication*: Determination of food adulteration and mislabeling using compositional differences.
- *Nutrient content monitoring*: Precise analysis of vitamins, minerals, and other nutritional components.
- *Evaluating shelf life and spoilage*: Monitoring biochemistries as spoilage indicators.

1.1.2 Advanced Method of Detection

Enzyme-Linked Immunosorbent Assay (ELISA) Based on a biological detection method that, while not perfect, is good enough that it hasn't been replaced yet. Nevertheless, five to ten times lower is not significantly more sensitive than what was commonplace 50 years ago, and in a world with 7.7 billion people, the ELISA test simply performs too slowly if we want to ensure the protection and originality of our sample [4].

1.2 Fluorescence

Fluorescence is a type of photoluminescence in which molecules receive light of a certain wavelength and give it off at a longer wavelength with negligible delay (in the nanosecond range). This occurs in many substances and is put to work in all sorts of scientific applications to detect and quantify [5].

- *Strong sensitivity*: Fluorescence-based methods can recognize very small amounts of the target material [6].
- *Not destroying the sample*: The analysis does not harm the sample being analyzed, allowing the sample to be used again if necessary.
- *Real-time biological observation* is a major use of fluorescence, used in Clinical Observation.
- *Versatility*: Fluorescence is widely used in observing health-care-related samples and is a basis for some types of medical diagnostics, used in Environmental Observation, Fluorescence is widely applied to watch for biological hazards; it is also used in studying other environmental problems [7].

Fluorescence and chemiluminescence provide significantly better sensitivity and selectivity for trace-level analysis than many classical techniques such as titration, gravimetry, and spectrophotometry.

1.2.1 Basic Principles and Fundamentals of Fluorescence

Excitation When a photon of light is absorbed by a molecule called a fluorophore, it usually comes from the ultraviolet or visible range, and it jumps from the lower or ground state (S_0) to the upper or excited singlet state (usually S_1). This is fluorescence, and it happens instantaneously within femtoseconds, actually. Absorbed photon's energy must be equal to energy difference between the two states, and for a fluorophore to be a good one, it must not have too many close-lying states, as that would make the material non-radiative (or very lossy) and less likely to emit light during the transition from state S_1 back to state S_0.

Emission A molecule in the excited state loses some energy, but not as light, while it reconfigures into the vibrational ground state. This is known as vibrational relaxation. When that happens, the molecule has already lost some energy and is now in a much lower energy state than the one in which it started when it absorbed the excitation photon. But the next step in the process is to drop all the way down to the ground state. And it does that by emitting the photon, which is the light that we see. This process is known as fluorescence. And what's really interesting is that the whole thing happens very quickly [8].

Quantum Yield The quantum yield (Φ) is defined as the ratio of emitted photons to absorbed photons and shows how efficiently the fluorescence process occurs:

$$\Phi = \text{Photons emitted} \backslash \text{photons absorbed}$$

A high quantum yield indicates that a truly fluorescent molecule is present, and it is influenced by environmental factors such as pH, solvent polarity, and molecular interactions [9].

Illustration: A commonly used tool to depict these transitions is the Jablonski diagram, which shows the absorption, internal conversion, and emission processes in the energy states *(Fig. 1).*

1.3 Chemiluminescence Assay

The release of EMRs (UV or IR) produced in biochemical or other chemical changes or biotransformation during the deterioration of an unstable chemical body to a more stable specie of comparatively low energy is identified as chemiluminescence [10]. CL was familiarized with analytical discipline at the end of 1970, while refined CL compounds like luminol and its products ($C_2H_4O_2$, and $C_{13}H_{10}N^+$ esters) substituted radioactive and fluorescent tags used in immunoassays. Nevertheless, numerous glitches came to light from the faintness of the CL visual output and the quick degradation kinetics, and CL was not become a widely used detection practice till the revolutionary effort of Thorpe and Kricka, assisting to the introduction of the luminol improved CL reaction facilitated by enzyme named as horseradish peroxidase (HRP) used

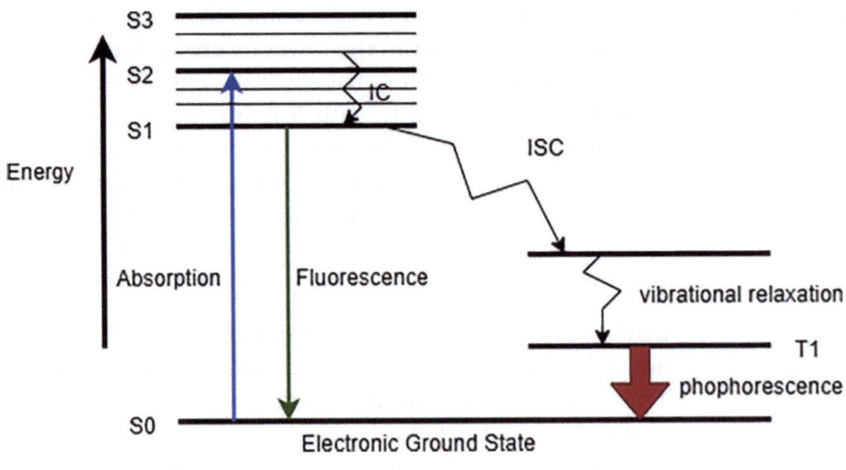

Fig. 1 Jablonski diagram

Table 1
Comparison with traditional analytical techniques

Feature	Fluorescence	Chemiluminescence	Traditional techniques
Excitation source	External light (e.g., UV)	Chemical reaction	Light, heat, and reagents
Sensitivity	High	Very high	Moderate
Background noise	Moderate	Low	Variable
Instrumentation	Fluorometer	Luminometer	Spectrophotometer, GC, HPLC
Sample integrity	Non-destructive	Usually non-destructive	Often destructive
Cost and complexity	Moderate	Low to moderate	Variable

for oxidation. The routine of enzyme assisted chemiluminescence tags as an alternative of CL particles considerably upgraded detection ability by the signal strengthening (by an element of approximately 1000–10,000) arising from the enzyme enhanced chemical changes, and the balanced discharge kinetics attained in the proximity of surplus chemiluminescence enzyme substratum prominently simplified analytical techniques and CL event handling [11] (Table 1).

1.3.1 Requirements for Chemiluminescence Emission

1. The change needs to be energy-releasing to harvest adequate energy to formulate the high-energy excited state by electronic means. To calculate whether the chemiluminescence transformation will take place or not, it is feasible to utilize Gibbs free energy (ΔG):

$$\Delta G = \Delta H - T\Delta S$$

Here,
T = temperature
ΔH = enthalpy
ΔS = entropy.

Enthalpy is the definite heat content for creating the activated state atom or molecule by electronic transmission to higher orbitals. To start the CL reaction, the energy of activation (ΔH_A) is delivered to the specie to luminescent. The obtainable energy to yield the activated atom or molecule (ΔE_{EX}) has to be the same or more than the difference between the energy during CL and the energy of excitation. If this happens, the chemiluminescence will be observed. Consider,

$$\text{Energy of the exciring specie or photon} = \Delta H_A - \Delta H_E \geq \Delta E_{EX}$$

In numerous CL processes, the disorderliness is minor, so Gibbs free energy and enthalpy have similarity in level, and the energy expenditure can be recognized in relation to (Kcal/mol). In this scenario, for chemiluminescence to proceed, the chemical change must be adequately exothermic, i.e.

$$\Delta G \geq \frac{h_c}{\lambda_{ex}} = \frac{2.86 \times 10^4}{\lambda_{ex}}$$

Here, λ_{ex} = wavelength boundary or limit (in nm) for activation of the luminous atom or molecule. Since the chemiluminescence mostly yields energy radiation in the range of 400 (violet)–750 (red) nanometers in UV and visible regions, the formation of the activated compound and its disintegration to give CL phenomenon in the visible region needs approximately 40–70 Kcal/mol. This exoergic circumstance is related to oxidation-reduction reactions via O_2 and H_2O_2 or oxidizing agents of similar capability.

2. The route for chemical conversion has to be satisfactory to supply the energy for the development of an activated energy state by the transfer of an electron. In the situation in which the energy is emitted as heat, for instance, through vibrational and rotating energy loss in orbitals, the CL will not be observed.
3. The release of radiation should be a promising de-excitation procedure of the unstable atom or molecule relative to supplementary competing radiationless course that can give the impression in little quantity in the situation of sensitized chemiluminescence, both the proficiency of transmission from the activated particle to the fluorophore and the fluroesence competence of the concluding must be significant [12].

1.3.2 Common Reactions of Chemiluminescence

The most common reactions in the phenomenon of chemiluminescence are as follows:

- Oxidation reactions.
- Direct chemiluminescent reactions.
- Energy transfer reactions.
- Electrochemiluminescence.

Oxidation Reactions The oxidation reaction is the process of removal of electrons, adding oxygen, increase in oxidation number under the effect of an oxidizer like hydrogen peroxide, oxygen, or potassium permanganate (Fig. 2). The typical and usual example of chemiluminescence based on oxidation is the luminol oxidation. Luminol ($C_8H_7N_3O_2$) (and its products) is until now the most often applied reagent in oxidative chemiluminescence. The chemiluminescent discharge of luminol is constructed on its reaction with oxidizing reagents, by H_2O_2, $[Fe(CN)_6]^{3-}$, MnO_4^-, IO_4^-, BrO^- generated electronically, and further oxidizing species in a basic environment or with its electrochemical oxidation using an appropriate electrode. Permitting the accepted CL procedure, the theoretical emitter is the unstable $C_8H_5NO_4^{2-}$(3-aminophthalate anion), which yields its stable form with radiation emission via chemiluminescence at 425 nanometers. The sample can behave as a reactant, promoter, reaction retarder, catalyst, or accelerator in chemiluminescence [13].

The generally accepted mechanism of CL of luminol is depicted in Fig. 2:

The excited state of 3-aminophthalate anion then de-excites to give luminescence as illustrated in Fig. 3:

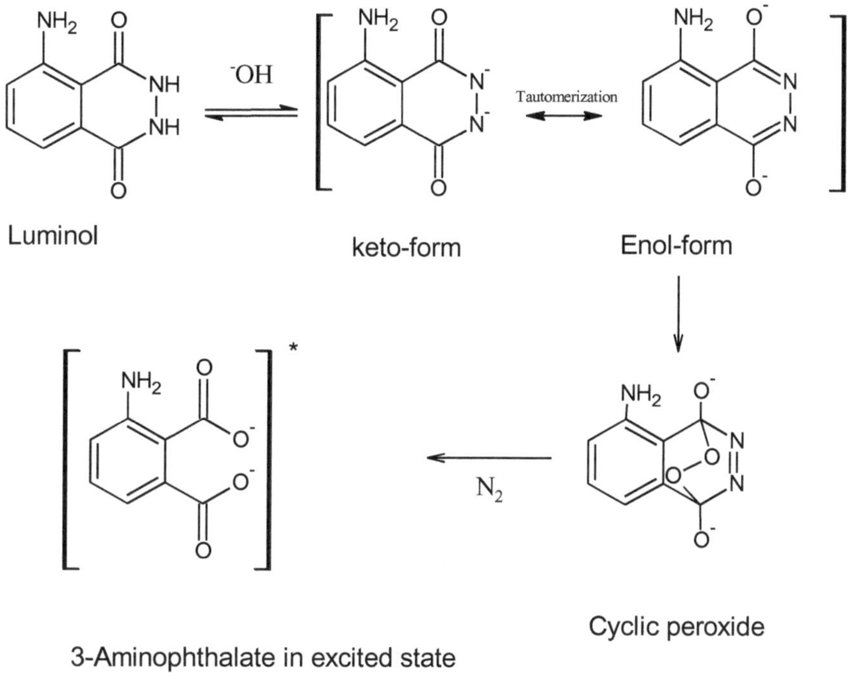

Fig. 2 Mechanism for oxidation of luminol by hydrogen peroxide

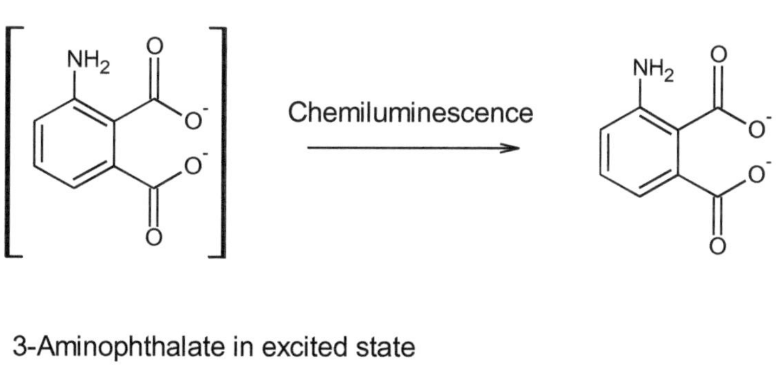

Fig. 3 Chemiluminescence from 3-aminopthalate ion

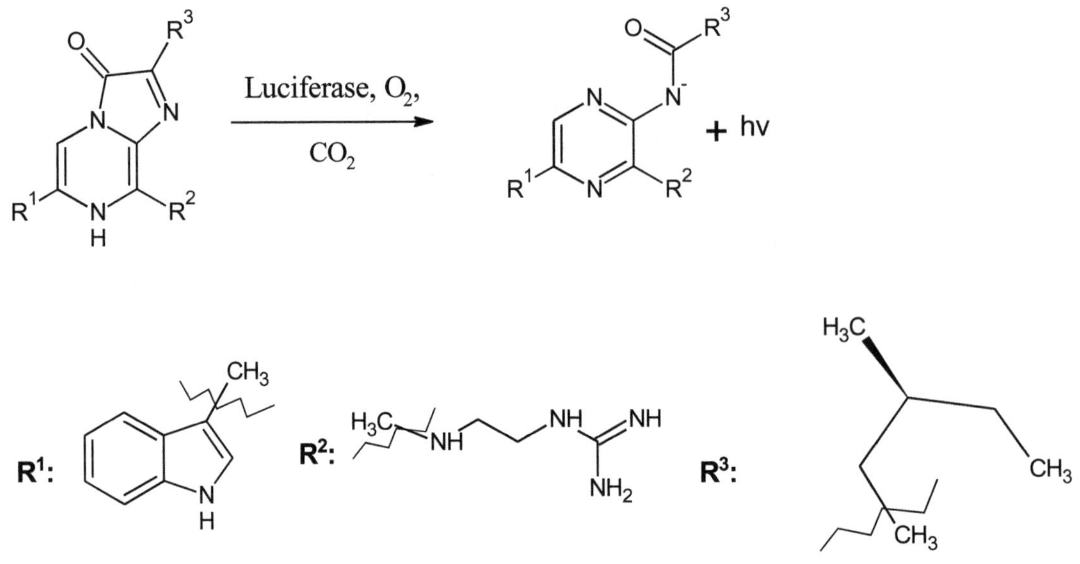

Fig. 4 Mechanism of oxidation of *Cypridina luciferin*

Direct Chemiluminescence Cypridina luciferin emits blue colour radiations when oxidized by molecular O_2, accelerated by an enzyme named *C. luciferase* as shown in Fig. 4. Furthermore, release was also documented next to the oxidation by molecular oxygen radical ($O_2^{\bullet -}$) of 2-methyl-6-phenyl-3, 7-dihydroimidazo. At the period of its finding, this discriminating oxidation delivered an exclusive chance for direct chemiluminescence exposure of molecular oxygen radical and its additional expansion of further comparable chemiluminescent investigations [14].

Energy Transfer Reactions In this development, the reactants experience a chemical change to produce an unstable high-energy product let's say U*, which afterwards emits energy by transmission to a fluorophore, say X, and developing a high-energy luminophore, say X^*. X^* to the end produces radioactivity, which is called as indirect chemiluminescence, and this process is defined as CL resonance energy transmission (CRET) [15].

$$S + T \rightarrow U^* + V$$

$$U^* + X \rightarrow U + X^*$$

$$X^* \rightarrow X + h\nu$$

Electrochemiluminescence Reactions Electro-CL (ECL) is a type of chemiluminescence which is observed in the electrically assisted electrochemical change of electrochemiluminescent and unstable particles. As an influential investigative apparatus, ECL has established substantial consideration owed to its hypersensitivity, ease in action, and governable radiation production. Widely experiential electrochemiluminescence excited state species hold transition metal complexes [i.e., Ru, Ir, and Pt (adam's catalyst)] and polycyclic aromatics (PCAs) ($C_8H_7N_3O_2$ and acrinidium esters). Nanomaterials (e.g., transputors, ultrasmall noble metals, and carbon-based matter) likewise show noteworthy electrochemiluminescence activities. Normally, the electrochemiluminescent atoms or molecules can produce apparent electrochemiluminescence results with the assistance of reactant collaborators, i.e., $S_2O_8^{2-}$, tri-n-propylamine, dibutylaminoethanol (DBAE), hydrogen peroxide, and oxalate ion.

2 Materials

2.1 Fluorescent Materials

Materials that fluoresce absorb light of one wavelength and give it off again at a longer wavelength. They are found all over the place in imaging, sensing, and display technologies. The following are the dyes used in the process:

- Fluorescein: A brilliant green dye utilized in the biological imaging process.
- Rhodamine is recognized for its red fluorescence and is utilized in numerous applications [16].
- Blue fluorescence is emitted by coumarins, which are used in lasers and sensors.
- Cyanine dyes (e.g., Cy3, Cy5): Fluorescent probes that are near-infrared, which makes them ideal for biological labeling.
- Quantum Dots (e.g., CdSe, ZnS): Nanoparticles that fluoresce and that can be made to do so in any color, with their fluorescence being size-tunable; used in bioimaging and optoelectronics [17].
- Materials doped with rare earth elements: These include Eu^{3+} (red), Tb^{3+} (green), and Er^{3+} (green), which are used extensively in lasers and phosphors.
- Phosphors. Materials such as ZnS: Mn^{2+} (orange-red) are used in display screens and paints.
- CNPs: Nano-crystalline particles that display green fluorescence when subjected to UV light and that find application in the field of bioimaging.
- Carbon Quantum Dots (CQDs): High-stability, tunable-emission, yet fluorescent nanomaterials that fit biomedicine and sensing applications well [18].

- Polymer Gels That Emit Light: Fluorescent, soft materials for use in bioimaging, chemical and environmental sensing, and for photonics [19].

2.2 Chemiluminescent Materials

Chemiluminescent materials emit light as the result of a chemical reaction, with no requirement for any external light sources. The following compounds are used for the process of chemiluminescence:

- Luminol: An established chemiluminescent compound that gives off blue light when it undergoes oxidation, and that finds frequent application in forensic science.
- Acridinium esters—Compounds that give off intense light signals in chemiluminescent reactions and are utilized in the field of biomedical detection and analysis [20].
- Carbon nanodots that are chemiluminescent: Nanomaterials based on carbon that show chemiluminescence and can be used to encrypt information and make sensors [21].

3 Methods

3.1 Fluorescence Instrumentation

A fluorescence measurement apparatus (fluorometer or spectrofluorometer) excites a sample with a light source and then detects the emission of fluorescence at the different (usually longer) wavelengths. The key parts include the following components as depicted in Fig. 5:

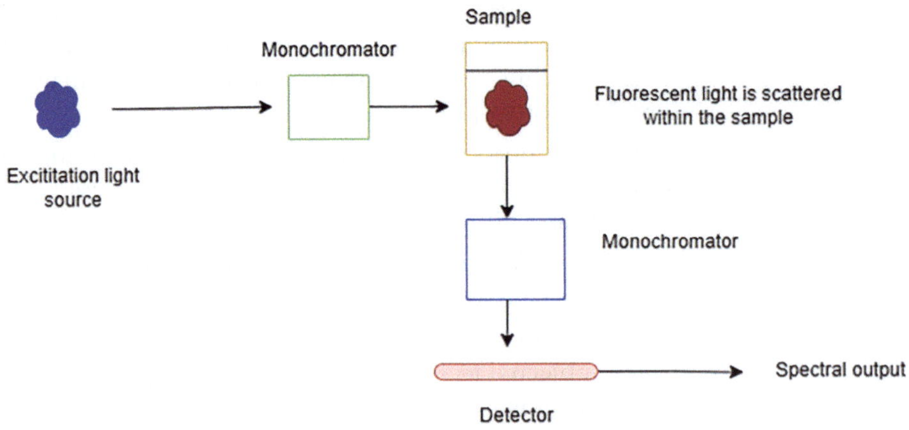

Fig. 5 Basic instrumentation of fluorescence

Light Source
To excite the fluorophores, a high-intensity light source is crucial. Used commonly as sources are:

- Xenon arc lamps (for broadband UV–visible excitation).
- Mercury vapor lamps.
- LEDs and lasers for specific wavelength excitation.

Excitation Monochromator or Filter
This part chooses a particular light wavelength from the source to stimulate the specimen. Filters or monochromators guarantee that solely the wished-for, stimulating wavelength hits the specimen.

Sample Holder
The quart cuvette is where the food sample is kept, and geometry matters for the outcome of any experiment run within its confines. We typically use a 90-degree geometry to reduce the detection of scattered excitation light.

Emission Monochromator or Filter
This pinpoints the fluorescence emission wavelength from the sample and eliminates scattered excitation light, thus enhancing the signal-to-noise ratio.

Detector
A fluorescence detector needs a photomultiplier tube (PMT) or photodiode for the detection of emitted fluorescence. The PMT is a light-sensitive detector that can count the number of photons hitting its surface.

Data Processor
Today, fluorometers are modern devices that have a computer interface and software to control the experiment and to collect data. They can even generate spectra [22].

3.1.1 Fluorescence Assays

Assays of fluorescence are analytical methods that harness the light emitted by fluorophores to detect and measure the presence of biological or chemical substances (Table 2). They have a broad set of applications in life sciences, diagnostics, environmental testing, and drug discovery. Assays of fluorescence are among the most versatile and dependable methods used for the detection and quantitation of a host of different classes of biological and chemical targets [23].

Fluorescence can be broadly classified into the following types:

- *Intrinsic fluorescence*: Results from the presence of naturally occurring fluorescent molecules (like tryptophan, NADH, or flavins) in biological samples.

Table 2
Time-resolved vs. steady-state fluorescence

Aspect	Steady-state fluorescence	Time-resolved fluorescence
Measurement	Total intensity over time	Fluorescence decay kinetics over nanoseconds
Sensitivity to environment	Moderate	High (lifetime is sensitive to changes in local environment)
Instrumentation	Simpler (continuous light source)	Requires pulsed lasers and fast detectors
Applications	Quantitative assays, imaging	FRET, conformational analysis, quenching studies

- *Extrinsic fluorescence*: Results from synthetic fluorescent dyes or probes that attach to certain targeted molecules.
- *Fluorescence resonance energy transfer (FRET)*: The way to study molecular interactions that involves the transfer of energy between two very close fluorophores.
- *Two-photon excitation fluorescence*: A method that allows the visualization of certain samples at greater depths without as much damage as other methods might cause [24].

Labeling is fundamental to the detection of nonfluorescent analytes. Common strategies are:

- *Covalent labeling*: Fluorophores are chemically bound to biomolecules such as proteins, nucleic acids, or antibodies. This is extensively used in immunoassays and gene expression studies.
- *Non-covalent labeling*: Fluorophores bind via hydrophobic or electrostatic interactions, as seen in intercalating dyes for DNA (e.g., ethidium bromide).
- *Enzyme-linked fluorescence*: Enzymes convert nonfluorescent substrates to fluorescent products (e.g., fluorescein diacetate hydrolysis in viability assays) [25].

Fluorescence Quenching and Enhancement

Quenching is a decrease in intensity of fluorescence due to various mechanisms:

- *Dynamic quenching*: Occurs through collisions between the quencher and a fluorophore.
- *Static quenching*: This includes the ground state complex formation between the fluorophore and quencher.
- *Inner filter effect*: Absorption of excited light or emitted light by other substances in the solution [26].

Quenching is exploited in sensor development and in binding affinity studies.

Fluorescence enhancement occurs when environmental changes increase emission:

- *Metal-enhanced fluorescence (MEF)*: Metal nanoparticles enhance the local electromagnetic field [27].
- *Binding-induced enhancement*: Fluorophores show higher emission upon binding (e.g., SYBR Green with DNA).

Overall emission intensity is measured by

- *Steady-state fluorescence*, an absorption luminescence technique that is ideal for routine assays.
- *Time-resolved fluorescence* leads to detailed spectroscopy, which provides an excellent understanding of molecular interactions and the local environment in which they exist [28].

3.2 Methods Used in Food Analysis via Chemiluminescence

The chemiluminescence in food analysis involves a number of different techniques that are focused on the detection of food ingredients and their estimation. These techniques may include;

- Chemiluminescence based on liquid chromatography.
- Chemiluminescence based on flow injection.
- HPLC and chemiluminescence.
- Electrophoresis and chemiluminescence.

Chemiluminescence Based on Liquid Chromatography

In current ages, liquid chromatography (LC) has been developed into an elite approach for the examination of an eclectic variety of medicinal and ecological models. The objective of chemiluminescent-centered chemical patterns is to improve the detection ability of a group of composites or of a certain selected analytes, presented by intensification in perceptiveness and/or discrimination. Benefits of detection of chemiluminescence based on liquid chromatography are;

- Low detection limits
- Selectivity
- Simple instrumentation [10]

Figure 6 gives the visual display for LC based on chemiluminescence [29, 30].

Chemiluminescence Based on Flow Injection

Primary information on the practice of microfabrication chemiluminescence started into 1990s, when the MF of electro chemicals by means of Si as a substrate was described for the production and recognition of the (ECL) of tris(bipyridine)ruthenium(II) ($[Ru(bipy)_3]^{2+}$) which necessitates a chamber capacity of merely 85 microliters. A number of other chemiluminescence schemes have been assimilated in-target; nevertheless, the luminol

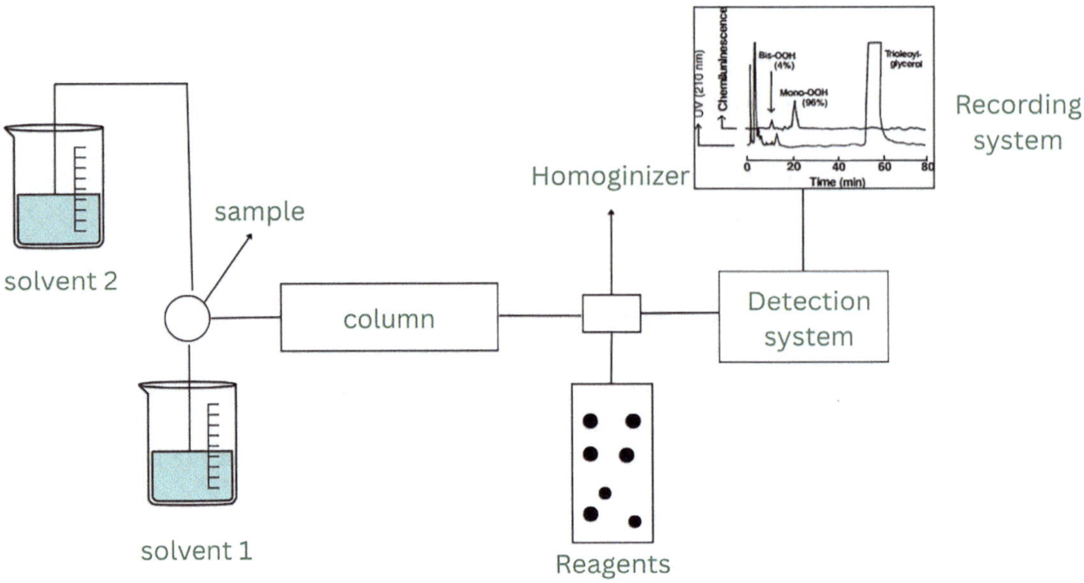

Fig. 6 Visual representation of instrumentation and procedure for LC based on CL

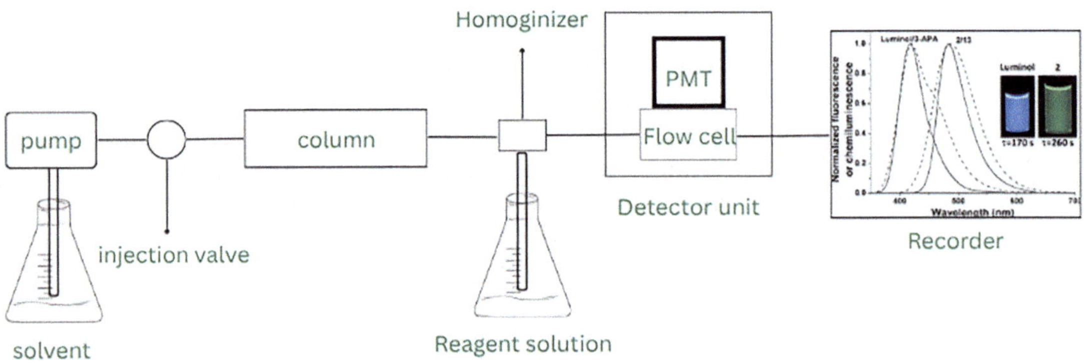

Fig. 7 Instrumentation for flow injection chemiluminescence

chemiluminescence method is typically taken into consideration, with approximately 50% of the described procedure constructed on this principle. The chemiluminescence discharge of luminol is established on its oxidation in the presence of MnO_4^-, or other oxidizers like hydrogen peroxide, vanadium pentaoxide, etc. It is acknowledged that the oxidation of luminol yields the 3-aminophthalate anion in its activated state, which has a determined release at a wavelength of 425 nm [31]. Figure 7 illustrates the general construction of chemiluminescence based on flow [32, 33]. This system involves the detection system of the flow cell and the photomultiplier tube, which amplifies the light signals from the column.

HPLC and Chemiluminescence

High-performance/pressure liquid chromatography based on chemiluminescence (CL) exposure has additionally been functional for measuring several categories of naturally significant complexes counting (–COOH). The chemiluminescence method has been lately modified to high-performance/pressure liquid chromatography for the synchronized calculation of Nickel Manganese Cobalt (propoxur, bendiocarb, carbaryl, and promecarb), which are chiefly found in fruits, attaining a lower level for analytical sensitivity fluctuating between 3.9 and 36.7 nanograms/litter [10]. It has likewise been brought into being that certain carbamates were capable of boosting the intensity of electrogenerated chemiluminescence of $[Ru(bipy)_3]^{2+}$ significantly at a MWCNT (Multi-Walled Carbon Nanotube) or a sulfonated tetrafluoroethylene-based fluoropolymer film reformed probe. Grounded on which, a technique for estimation of pirimicarb, methomyl, aldicarb, and carbofuran was industrialized, employed in the examination of landscape water and food samples [11]. The lowest detectable range lies between 2.0×10^{-9} and 1.7×10^{-7} moles [34]. Therefore, decreasing the concentration of the analytes when diluted with reagent(s) at times causes the results to become insensitive. The chemiluminescence reaction circumstances, such as concentrations, pH, liquid formulation and heat, impact of the chemiluminescence emission strength [35]. Figure 8 gives the visual display for an HPLC based on chemiluminescence [36].

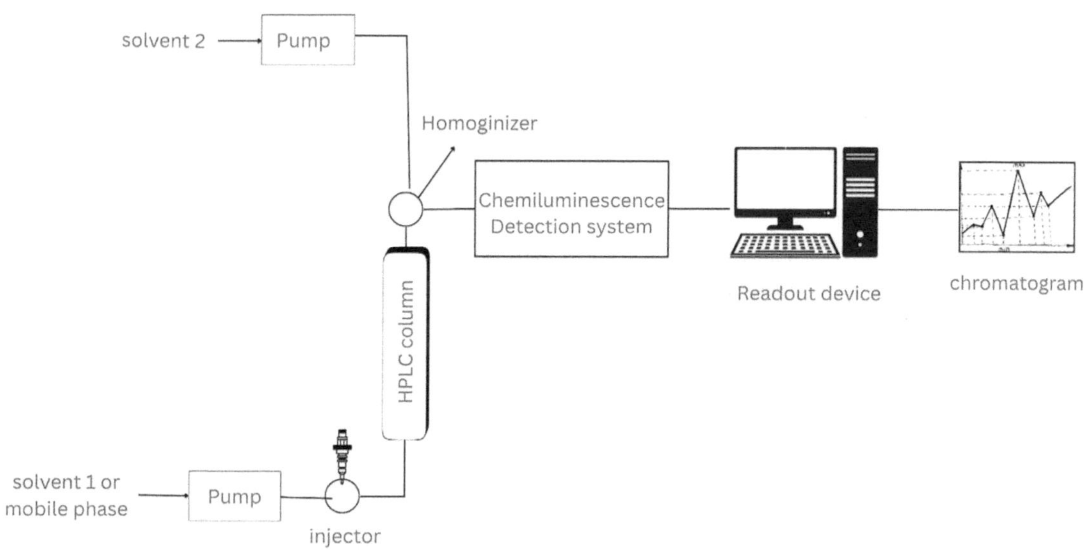

Fig. 8 Arrangement of instrumental parts of chemiluminescence based on HPLC

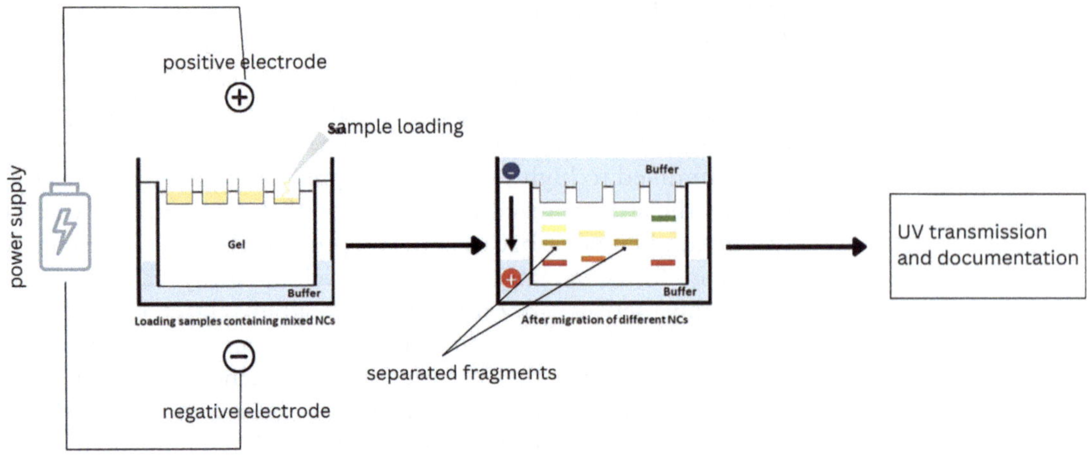

Fig. 9 Instrumentation of chemiluminescence-based CE

Electrophoresis and Chemiluminescence

The procedure for Electrophoresis provides multiple numbers of plates at once and is very suitable for the accurate and precise estimation of several samples of chemical ingredients in foods and their products. Due to the narrower capillary diameter and observation chamber dimensions, along with the trace concentrations of food contaminants and compositional units to be analyzed, sensitized estimation systems related to a number of different chemical and physiological principles are being formulated. One logical technique to enhance the lower detection limit in CE analysis of food is the insertion of CL-assisted detectors [37]. Figure 9 depicts the general construction of chemiluminescence-based electrophoresis [38].

4 Note

Ensurance of food safety, standard, and compliance with both national and international regulations relies heavily on the science of food analysis. Food products are assessed for all sorts of important characteristics: composition, contamination, labeling, and general (not product-specific) safety by structured schemes set up under regulatory frameworks and quality control systems [39].

4.1 Regulatory Aspects of Food Analysis

Legal benchmarks and guiding principles for food safety, as well as for labeling, additives, and contaminants, are established by regulatory frameworks. Often, these frameworks are the work of a national or international body. Some of the most important are:

- *Codex Alimentarius Commission (FAO/WHO)*: Establishes global food norms that safeguard the health of consumers and underpin the fairness of trade in food (FAO/WHO, 2020).

- In the USA, it is *Food and the Drug Administration (FDA)*, whereas the European Food Safety Authority (EFSA) in the EU deals with: Regulating food labeling, additives, and safety limits.

As an example, we define maximum residue limits (MRLs) for the pesticides in samples of food to protect consumers, and we monitor them through routine testing.

4.1.1 Quality Control in Food Analysis

Quality control (QC) specifies the operational procedures and the activities used to meet the quality standards for food testing. It includes:

- Ensuring that laboratories produce valid and reliable results requires Good Laboratory Practices (GLP) and accreditation to ISO/IEC 17025.
- Proficiency Testing (PT) and Internal Quality Control (IQC), are regularly conducted to guarantee precision and consistency.
- To reduce testing variability, use standard operating procedures and methods that have been shown to work [39].

Effective quality control makes certain that the results of all the different analyses conducted on a drug substance or drug product meet the established specifications.

In analytical science, fluorescence and chemiluminescence methods are moving in the right direction. They are becoming more sensitive, more selective, and more suitable for miniaturization, which is necessary if we are to achieve real-time, on-site analysis with these techniques and the types of samples we normally deal with. The key technologies that are driving these developments are nanotechnology, microfluidics, and automation of instrumentation [40].

Nanomaterial-based probes, especially quantum dots and upconversion nanoparticles, are proving to be game-changers in making fluorescence assays much more capable, in terms of being brighter, more photostable, and more easily used in assays where multiple different kinds of signals have to be disentangled. In a parallel development, the kinds of new molecules that are becoming available for use in chemiluminescent assays are moving this detection modality into the realm where it can be applied to ultra-trace analysis in areas like clinical diagnostics and food safety [41].

Regulatory agencies from around the world are adopting *risk-based approaches* and *performance-based validation* to ensure the protection of products for international trade. They are moving away from rigid validation templates that stifle innovation; this is especially true for fields that are developing rapidly, such as biosensors and lab-on-a-chip technologies, and pharmaceuticals [42].

The increasing call for sustainable analytical chemistry is driving fluorescence and chemiluminescence techniques to use environmentally friendly solvents, conserve reagents, and employ energy-efficient instruments.

To sum up, the combination of *advanced materials*, *data science*, and *regulatory evolution* is expected to turn fluorescence and chemiluminescence methods from traditional laboratory techniques into robust, smart analytical tools for a range of applications—point-of-care diagnostics, environmental surveillance, food authentication, and more. These methods represent some of the most exciting possibilities in the rapidly evolving toolkit for ensuring public safety.

4.2 Applications

The analysis of food contaminants is done via the analytical estimation of mycotoxins. Traditional techniques involve:

- HPLC based on either luminescence or DAD (diode array detectors).
- TLC based on chemiluminescence or fluorescence.
- Gas chromatography (GC) is assisted with a detector named mass spectrometer (MS) or electron capture detector (ECD), collectively called as GC–MS and GC–ECD, respectively.

These techniques are utilized for estimation of numerous kinds of mycotoxins, i.e., fumonisin and deoxynivalenol (DON) aflatoxins, ochratoxin A, in multiple foodstuffs, i.e., grains, legumes, cereals, and coffee. Nevertheless, such procedures require controlled and specific instrumentation and expertise, and the method for mycotoxin separation from the food samples should have to be used in accordance with the kind of mycotoxin and food sample that is going to be analyzed [43].

As for cosmetics, food is also adulterated with colorants, for instance, colorants in meat to make it look fresh and attractive, in cakes to make them more attractive, and in candies. It is observed, colorants in food samples are not only used for attractiveness in colors, but also utilized for the preservation of water. Although in accordance with food colorants policies, lab-synthesized colorants are unable to be added to meat and its preserved products. For the detection of colorants, it is observed that thin layered double hydroxides (LDHs) amplify the results of the (CL) of the Luminol via hydrogen peroxide procedure greater than 950-fold. The increased CL is because of the synergistic effect of the cationic layer and the carbonates, which are intercalated [44]. It can also be used in the following fields:

- Enzyme assay in chemiluminescence [45].
- Detection of pathogens and microbial contamination [46].
- Analysis of antioxidant [47].

Figure 10 gives the visual display for these applications.

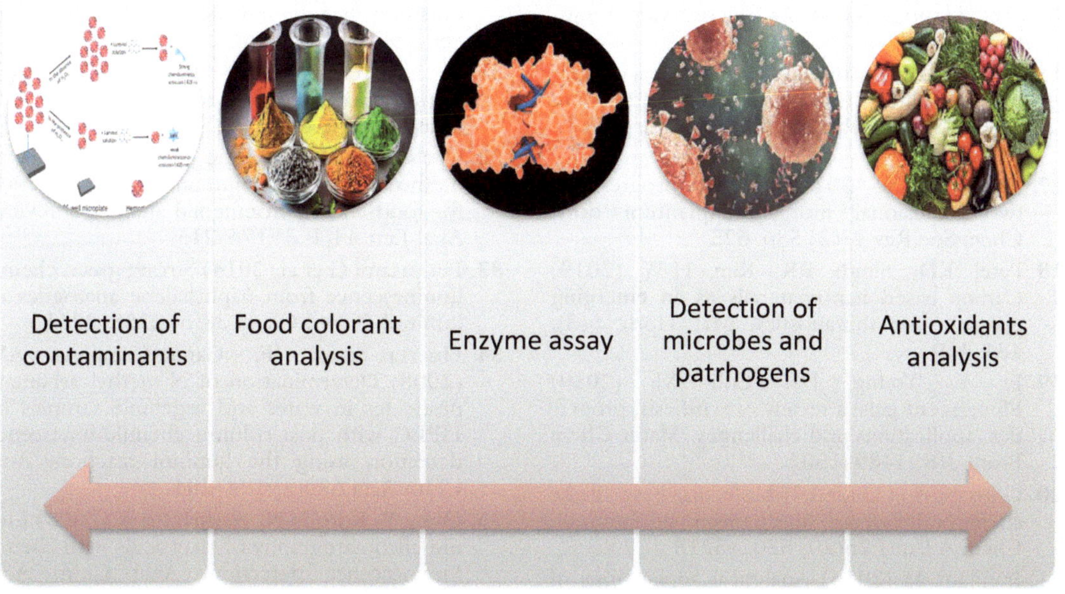

Fig. 10 Applications of chemiluminescence and fluorescence assay

References

1. Muñoz-Olivas R (2004) Screening analysis: an overview of methods applied to environmental, clinical and food analyses. TrAC Trends Anal Chem 23(3):203–216
2. Kharbach M et al (2023) Current application of advancing spectroscopy techniques in food analysis: data handling with chemometric approaches. Foods 12(14):2753
3. González-Domínguez R (2020) Food authentication: techniques, trends and emerging approaches. Foods 9(3):346
4. Hnasko R (2015) Elisa. Springer
5. Lakowicz JR, Lakowicz JR (1999) Introduction to fluorescence. In: Principles of fluorescence spectroscopy, pp 1–23
6. Ahmad MH, Sahar A, Hitzmann B (2017) Fluorescence spectroscopy for the monitoring of food processes. In: Measurement, modeling and automation in advanced food processing, pp 121–151
7. Karoui R, Blecker C (2011) Fluorescence spectroscopy measurement for quality assessment of food systems—a review. Food Bioprocess Technol 4:364–386
8. Ogundele OM et al (2013) Basic principles of fluorescence microscopy. d J Young Researchers 2 3(1)
9. Birks J (1976) Fluorescence quantum yield measurements. J Res Natl Bureau of Standards. Sect A Phys Chem 80(3):389
10. Bowie AR, Sanders MG, Worsfold PJ (1996) Analytical applications of liquid phase chemiluminescence reactions—a review. J Biolumin Chemilumin 11(2):61–90
11. Roda A, Guardigli M (2012) Analytical chemiluminescence and bioluminescence: latest achievements and new horizons. Anal Bioanal Chem 402:69–76
12. Fereja TH, Hymete A, Gunasekaran T (2013) A recent review on chemiluminescence reaction, principle and application on pharmaceutical analysis. Int Scholarly Res Not 2013(1): 230858
13. Tsaplev YB (2012) Chemiluminescence determination of hydrogen peroxide. J Anal Chem 67:506–514
14. Tzani MA et al (2021) Direct and indirect chemiluminescence: reactions, mechanisms and challenges. Molecules 26(24):7664

15. Song H et al (2019) Quantum dots-based chemiluminescence probes: an overview. Luminescence 34(6):530–543
16. Sabatini DA, Austin TA (1991) Characteristics of rhodamine WT and fluorescein as adsorbing ground-water tracers. Groundwater 29(3): 341–349
17. Xu Y et al (2018) Recent progress in two-dimensional inorganic quantum dots. Chem Soc Rev 47(2):586–625
18. Patel KD, Singh RK, Kim H-W (2019) Carbon-based nanomaterials as an emerging platform for theranostics. Mat Horiz 6(3): 434–469
19. Li Y, Young DJ, Loh XJ (2019) Fluorescent gels: a review of synthesis, properties, applications and challenges. Mater Chem Front 3(8):1489–1502
20. Giussani A et al (2019) Molecular basis of the chemiluminescence mechanism of luminol. Chem A Eur J 25(20):5202–5213
21. Iranifam M (2016) Analytical applications of chemiluminescence systems assisted by carbon nanostructures. TrAC Trends Anal Chem 80: 387–415
22. Lakowicz JR, Lakowicz JR (1999) Instrumentation for fluorescence spectroscopy. In: Principles of fluorescence spectroscopy, pp 25–61
23. Stuker F, Ripoll J, Rudin M (2011) Fluorescence molecular tomography: principles and potential for pharmaceutical research. Pharmaceutics 3(2):229–274
24. Wang X, Lai Y (2021) Three basic types of fluorescence microscopy and recent improvement. In: E3S web of conferences. EDP Sciences
25. Schneider AFL, Hackenberger CPR (2017) Fluorescent labelling in living cells. Curr Opin Biotechnol 48:61–68
26. Eftink MR (2002) Fluorescence quenching: theory and applications. In: Topics in fluorescence spectroscopy: principles. Springer, pp 53–126
27. Geddes CD, Lakowicz JR (2002) Metal-enhanced fluorescence. J Fluoresc 12:121–129
28. Pradhan A et al (1995) Steady state and time-resolved fluorescence properties of metastatic and non-metastatic malignant cells from different species. J Photochem Photobiol B Biol 31(3):101–112
29. Barbatsi M, Economou A (2023) Programmable low-pressure chromatographic sub-90 s assay of parabens in cosmetics with post-column chemiluminescence detection. Separations 10(6):350
30. Miyazawa T (2021) Lipid hydroperoxides in nutrition, health, and diseases. Proc Jpn Acad Ser B 97(4):161–196
31. Al Lawati HA (2013) Flow-based analysis using microfluidics–chemiluminescence systems. Luminescence 28(5):618–627
32. Christodouleas D et al (2011) Flow-based methods with chemiluminescence detection for food and environmental analysis: a review. Anal Lett 44(1–3):176–215
33. Periyasami G et al (2014) Strong green chemiluminescence from naphthalene analogues of luminol. New J Chem 38(6):2258–2261
34. Huertas-Pérez JF, García-Campaña AM (2008) Determination of N-methylcarbamate pesticides in water and vegetable samples by HPLC with post-column chemiluminescence detection using the luminol reaction. Anal Chim Acta 630(2):194–204
35. Ohba Y, Kuroda N, Nakashima K (2002) Liquid chromatography of fatty acids with chemiluminescence detection. Anal Chim Acta 465(1–2):101–109
36. Deng M, Chen F-N, Long X-Y (2013) A novel method of chemiluminescence detection coupled with high-performance liquid chromatography and its application in direct determination of tartaric, malic and citric acids in fruit juice. Anal Methods 5(4): 982–986
37. Liu Y-M, Cheng J-K (2002) Ultrasensitive chemiluminescence detection in capillary electrophoresis. J Chromatogr A 959(1–2):1–13
38. Lores-Padin A, Fernandez B, Pereiro R (2022) General purification methods of metal nanoclusters. In: Luminescent Metal Nanoclusters. Elsevier, pp 161–186
39. Prylipko TM et al (2021) Control and identification of food products under EC regulations and standards. Int J Agric Ext 9(5):83–91
40. Zhang K, Flannery B, Zhang L (2024) Challenges and future state for mycotoxin analysis: a review from a regulatory perspective. J Agric Food Chem 72(15):8380–8388
41. Qiao X et al (2022) Recent advances in nanomaterial-based sensing for food safety analysis. Processes 10(12):2576
42. De la Torre BG, Albericio F (2024) The pharmaceutical industry in 2023: an analysis of FDA drug approvals from the perspective of molecules. Molecules 29(3):585
43. Sarter S, Zakhia N (2004) Chemiluminescent and bioluminescent assays as innovative prospects for mycotoxin determination in food and feed. Lumin J Biol Chem Lumin 19(6): 345–351

44. Zhang J et al (2020) Improved food additive analysis by ever-increasing nanotechnology. J Food Drug Anal 28(4):622
45. Magliulo M et al (2005) Development and validation of an ultrasensitive chemiluminescent enzyme immunoassay for aflatoxin M1 in milk. J Agric Food Chem 53(9):3300–3305
46. Charlermroj R et al (2011) A chemiluminescent antibody array system for detection of foodborne pathogens in milk. Anal Lett 44(6):1085–1099
47. Pamunuwa GK, Atapattu SN (2023) Chemiluminescence methods for antioxidant analysis in food matrices. J Chromatogr Open 4:100096

Chapter 12

Electronic Nose and Tongue Technologies for Flavor and Quality Control

Sammra Maqsood, Muhammad Asif, Sadaf Shakoor, Muhammad Tayyab Arshad, Asif Zarnab, Kübra Sağlam, and Mondher Mzoughi

Abstract

The pioneering role of biomimetic sensor systems in modern food quality assessment is explored in Electronic Nose and Tongue Technologies for Flavor and Quality Control. Although they are valuable, conventional sensory assessment techniques suffer from subjectivity, reproducibility, and on-line analysis. To mimic the human senses of taste and smell, this chapter introduces the E-nose and E-tongue systems, which are new, objective tools. It gives an overview of sensor technologies such as conducting polymers, ion-selective electrodes, lipid/polymer membranes, and metal-oxide semiconductors, and how these may be incorporated into arrays of sensors. This chapter categorizes a number of applications for a great range of samples, such as beverages, meat, milk, spices, drugs, water and other environmental matrices. The processes of data collection and interpretation are explored through the prism of the software tools and hardware components, with pattern recognition algorithms and chemometric methods being emphasized. The operating, calibration, analysis, and validation methods of sensors are also included, as well as examples of flavor profiling and spoilage detection as case studies. The chapter finishes with discussion of the practical deployment aspects, including the sensor limitations, environmental sensitivity, integration with the Internet of Things, and industrial law compliance. The effect of E-nose and E-tongue technologies on sensory analysis and quality control within the food and related industries is examined in depth within this comprehensive treatment.

Key words Electronic nose (E-nose), Electronic tongue (E-tongue), Flavor analysis, Food quality control, Chemometrics, Pattern recognition, Sensor arrays, IoT integration, Artificial intelligence (AI), Biomimetic sensors

1 Introduction

Brand integrity and consumer satisfaction in the contemporary food industry rely heavily on stable and high-quality flavor profiles [1]. Consumer enjoyment is largely flavor-dependent, which is the integration of taste, aroma, and chemical sensations; therefore, it forms impressions of freshness, authenticity, and safety

Tanmay Sarkar (ed.), *Novel Food Analysis: Techniques and Applications*, Methods and Protocols in Food Science,
https://doi.org/10.1007/978-1-0716-4787-5_12,

[2]. Sustaining sensory consistency is an ongoing challenge with the world's food production rising to meet growing demand, particularly in processed and packaged foods. Relying frequently on human sensory panels and routine analysis equipment, traditional flavor and quality control methods are coming under greater scrutiny for their drawbacks in subjective nature, time-intensive, and inability to detect minor compositional differences [2, 3].

While central to traditional food quality measurement, human sensory panels are inherently subject to biological variation, fatigue, mood, and environmental influences [4]. While extremely well-trained panels can provide subtle evaluations, particularly over extended periods, their answers are not objective or reproducible. Most analytical techniques, such as gas chromatography–mass spectrometry (GC–MS) or gas chromatography–ion mobility spectrometry (GC–IMS), provide dense chemical information but are limited by their requirement of sample preparation, lengthy processing time, and dependence on laboratory facilities [1, 5]. These traditional processes might not even be capable of completely spectrum of sensory-active food chemicals that dictate consumer perception once the complexity in food matrices is increased and fermented, plant-based, as well as functional foods have increasing importance. The field of flavor science is witnessing a paradigm shift in the face of such challenges as biomimetic technologies, more particularly, the Electronic Nose (E-nose) and Electronic Tongue (E-tongue) systems [1] have emerged. Modeled after human olfactory and gustatory systems, such instruments offer fascinating alternatives to rapid, repeatable, objective assessment of food quality. Through the use of sensor arrays, e-nose technologies mimic the mammalian olfactory system by sensing volatile organic compounds (VOCs), thus converting intricate scent patterns into digital fingerprints [6, 7]. By means of arrays of chemical sensors that react to ions and molecules dissolved in a solution, E-tongues also mimic the ability of the tongue to perceive non-volatile taste molecules like bitterness, sweetness, and umami [8, 9].

Technological expansions in nanotechnology, biosensing, and machine learning permit the development of such artificial sense systems. To better mimic the selectivity and sensitivity of human receptors, bioinspired sensors joined with triboelectric and electrochemical exposure systems are being discovered [10, 11]. E-nose and E-tongue sensors allow signal processing and data investigation approaches such as principal component analysis (PCA), artificial neural networks (ANNs), and deep learning algorithms to interpret enormous multidimensional data sets into meaning, individual small differences among food samples with greater accuracy [12, 13]. From observing spoiling and fermentation methods to noticing adulteration and certifying steadiness in production batches, these technologies have previously established worth in several applications. For illustration, the E-nose has been efficiently applied to the analysis of off-flavors in cocoa, contributing an objective substitute to human assessment in sensing smoky or rancid flavors, influenced by poor fermentation or storing conditions [14]. Likewise, E-tongue systems have been rummage-sale in authentication and superiority determination of multifaceted matrices such as tea, wine, and fish roe, detecting subtle differences in flavor profiles associated with geographic position, maturity, or adulteration [15–17].

In addition, highlighted by the COVID-19 pandemic is the acute necessity of steady, non-human dependent sensory assessment methods [18]. Ranging as common COVID-19 symptoms, olfactory and gustatory dysfunctions emphasized the weakness and heterogeneity of human sensory assessment [18, 19]. To this end, the integration of E-nose and E-tongue machines into routine food testing protocols not only minimizes risk but also enhances uniformity and throughput [19]. Despite the latest developments in downsizing sensors, signal transduction, and pattern recognition having pushed these technologies nearer to their deployment in real-world applications, scientific curiosity surrounding artificial sensing has been around [20]. From proof-of-concept devices in laboratory setups to industrially compliant machinery, E-nose and E-tongue devices have advanced, depicting ongoing improvement in sensor design and multidisciplinary cooperation between chemists, data scientists, and food technologists [20, 21]. In addition, how they communicate with Internet of Things (IoT) platforms and cloud analytics is a harbinger of an era where taste quality monitoring will be continuous, proactive, and integral all along the food supply chain [22].

Electronic sensing systems are subject to some limitations, although their potential is immense. Still, hurdles to wider adoption are sensor drift, temperature or humidity sensitivity, and the requirement for extensive calibration [23]. Additionally, in contrast to the human sensory system, which integrates olfactory, gustatory, tactile, and visual information, E-nose and E-tongue systems typically focus on chemical signals only, thus often necessitating complementary methods or sensor fusion strategies for more comprehensive assessments [23, 24]. However, ongoing investigation is working to correct these shortcomings and discover the usage of multimodal devices such as biosensors and electronic eyes to imprisonment more wide-ranging dimensions of food perception [23]. Lastly, the quest for taste and quality control for foodstuff has propelled the expansion of sophisticated electrical sensory systems with massive future prospects for food science. Drawing on hominoid sensory physiology, e-nose and e-tongue know-hows proposals robust, high-throughput devices accomplished of transforming food industry quality assurance. Encouraged by improvements in sensory biology, data science, and nanotechnology, these evolving technologies hold the potential of revolutionizing the techniques we assess, monitor, and promise the sensory quality of food produced in an increasingly complex worldwide marketplace.

This chapter attempts to discover the emerging area of biomimetic sensor know-hows precisely, Electronic Nose (E-nose) and Electronic Tongue (E-tongue) systems and its revolutionary capability to alter food quality regulation and flavor analysis. It tries to deliver a comprehensive consideration of how these technologies deliver objective, reproducible, and speedy analysis of multifaceted food matrices, thus duplicating human smell and taste. The chapter stresses the significance of E-nose and E-tongue systems as groundbreaking substitutes to traditional sensory panels and chemical analyses by bearing in mind the fundamental ideas of sensory perception, noteworthy sensor types, processing methods for data, and practical applications in a wide range of industries in the food segment. Placing these know-hows at the forefront of future food quality monitoring, it also shields integration methods, technological benefits, restrictions, and upcoming advancements concerning nanomaterials, artificial intelligence, and IoT-enabled smart devices.

2 Materials

A novel biomimetic device in the form of an electronic nose (e-nose) or an electronic tongue (e-tongues) could duplicate the human sense of taste and smell, respectively. Recent advancements in sensor technology, electronics, materials science, and AI have enabled the development of these technologies to a level where they could probe matrices, food-related as well as otherwise, with unparalleled sensitivity and specificity [2, 4]. E-Nose and E-Tongue Technology Overview (Table 1): Materials, Sensor Types, Applications, and References.

2.1 Sensor Types Used in E-Nose and E-Tongue

Due to its high sensitivity to volatile organic compounds (VOCs) such as ammonia and hydrogen sulfide, which are critical in establishing the level of food spoilage, metal-oxide semiconductors (MOS) remain to hold the market share of e-nose sensors [1, 36]. Examples of materials that can sense trace amounts of volatile organic compounds (VOCs) by varying electrical resistance upon exposure to gas include tin oxide (SnO_2), zinc oxide (ZnO), and tungsten oxide (WO_3). Environmental pollutant detection, beverage authentication of drinks like wine and tea, and freshness monitoring in meat are some of the several typical applications of these sensors [33, 39, 46]. Their sensitivity to moisture and high operating temperatures (200–500 °C) limits their widespread application, however. With a view to enhancing selectivity and energy efficiency, recent research on nanostructured MOS has

Table 1
Overview of materials, sensor types, applications, and references for E-Nose and E-Tongue technologies

Section	Description	Sensor/component Type	Application area	References
Sensor types	Metal-oxide semiconductors	Gas sensors (MOS)	Food spoilage detection	[4]
Sensor types	Conducting polymers	Chemiresistors	Beverage quality control	[6]
Sensor types	Piezoelectric sensors	Quartz Crystal Microbalance (QCM)	Dairy product monitoring	[8]
Sensor types	Lipid/polymer membranes	Taste sensors	Meat freshness analysis	[9]
Sensor types	Ion-selective electrodes	Potentiometric E-tongue	Mineral water quality assessment	[10]
Sample types	Food samples	Beverages, dairy, meat, spices	Flavor profiling	[3]
Sample types	Non-food applications	Pharmaceuticals, water quality	Contaminant detection	[33]
Hardware components	Sensor arrays	Multi-sensor setup	Fruit grading	[34]
Hardware components	Sample delivery systems	Flow control modules	Real-time monitoring	[47]
Hardware components	Data acquisition systems	Signal processing units	Shelf-life prediction	[35]
Software tools	Pattern recognition software	PCA, LDA, ANN algorithms	Food classification	[7]
Software tools	Chemometric methods	Multivariate analysis	Meat adulteration detection	[28]
Software tools	Machine learning integration	Deep learning models	Food traceability	[42]
Sensor types	Triboelectric sensors	Integrated e-tongue system	Beverage authentication	[10]
Sample types	Fermented food analysis	HS-SPME-GC-MS linked sensors	Traditional food flavor mapping	[15]
Hardware components	Flexible multi-channel platforms	Versatile sensor operation	Quality assurance in food processing	[34]
Software tools	IoT and cloud platforms	Data storage and remote access	Smart food logistics	[12]

(continued)

Table 1 (continued)

Section	Description	Sensor/component Type	Application area	References
Software tools	Sensor fusion approaches	Combining E-nose and E-tongue data	Tomato juice freshness evaluation	[29]
Software tools	Cross-validation techniques	Model validation	Comparative sensory analysis	[11]
Software tools	AI-enhanced chemometric tools	Real-time predictive analytics	Shelf-life extension research	[42]

sought to address these issues [40]. Another significant class of sensors is conducting polymers (CPs), i.e., polypyrrole, polyaniline, and polythiophene [10]. A power-efficient alternative to MOS, these operate at room temperature [8, 20]. CP-based sensors have been successfully applied in non-invasive medical diagnosis, e.g., breath analysis, in spice and dairy adulteration detection, and in fruit and vegetable freshness monitoring [6, 13]. Their selectivity can be chemically fine-tuned, but the breakdown of polymer chains means long-term stability is in question [9]. Surface acoustic wave (SAW) sensors and quartz crystal microbalances (QCMs) are two kinds of piezoelectric sensors that quantify variations in mass resulting from gas adsorption onto a sensor surface [39]. Due to their ultra-high sensitivity, the sensors have been found to be of particular application in fragrance profiling of costly foodstuffs such as tea and coffee, detecting explosives, and quality testing of pharmaceuticals [5, 13, 39]. However, they require time-consuming calibration and are prone to ambient noise [30]. The main interface in e-tongue devices mimicking taste receptors is lipid/polymer membranes. Such membranes enable applications like pharmacological taste masking, water analysis, and taste profiling of drinks through the production of electrical signals upon response to interactions with sweet, sour, salty, bitter, and umami compounds [9, 10, 44]. Biohybrid membranes comprising taste receptor proteins are a new development that significantly enhances sensitivity and specificity [10]. One additional significant type of e-tongue sensor is the ion-selective electrode (ISE), and it is able to detect particular ions in a liquid, such as calcium, potassium, and sodium [43, 46]. Measurement of mineral content in drinks, analysis of cheese for salt content, and observation of nitrate contamination in sewage are just some of the many applications where they are used ubiquitously [6, 43, 46]. Response rate and detection sensitivity have been optimized in recent times with the utilization of nanomaterials for ISE [24, 25]. Figure 1 illustrates about materials most frequently used to detect traces of volatile organic compounds:

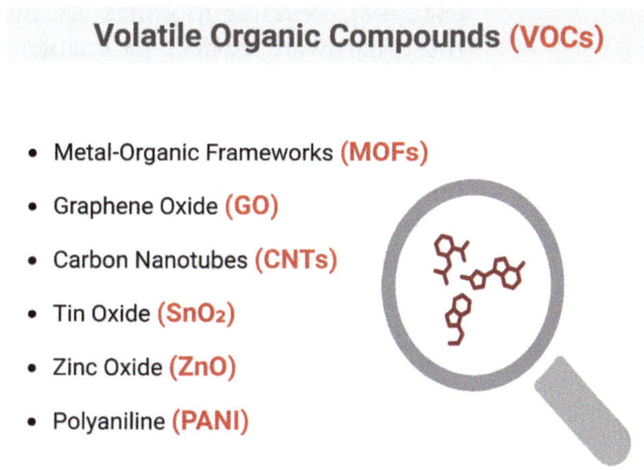

Fig. 1 Materials most frequently used to detect traces of volatile organic compounds

2.2 Sample Types

A variety of food and non-food samples is exposed to electronic sensing systems [30]. The food sector employs electronic noses and tongues for various purposes, including the examination of drinks (like the taste characterization and authenticity confirmation of wine, beer, and tea), fruit juices' freshness assessment, and microbiological contamination detection in milk [1, 6, 26]. For dairy products, they evaluate ripening stages and flavor profiles of cheese [46]. These sensors can detect adulteration, e.g., mutton and pork blended, and spoilage in seafood and meat by measuring volatile organic compound (VOC) content [27, 28]. When dealing with high-value products such as soy sauce, saffron, black pepper, and other fermented foods and spices, e-sensing is an excellent method to ascertain authenticity and consistency [6, 32]. Electronic nose and electronic tongue technologies are not only significant in the food sector, but also in environmental monitoring (e.g., to detect heavy metals and microorganisms in water), medical diagnostics (e.g., to examine breath for COVID-19-associated anosmia) and pharmaceutical applications (e.g., bitterness assessment of drugs) [13, 19, 44].

2.3 Hardware Components

Sensor arrays, sample delivery systems, and data collecting platforms constitute the hardware backbone of e-nose and e-tongue systems [2, 3]. To capture a broad VOC profile, E-nose arrays consist of 6–16 various MOS or CP sensors [4]. With the inclusion of voltammetric sensors, lipid membranes, and ion-selective electrodes, E-tongue arrays can replicate the whole range of human taste experience [9]. Static headspace sampling is utilized for volatile organic compound (VOC) samples in the gaseous phase, and flow injection techniques are utilized for liquid samples [30]. Platforms for data gathering could range from as basic as an Arduino or Raspberry Pi to as sophisticated as a high-speed FPGA platform

[31, 49]. Wireless modules are increasingly being coupled with these hardware devices to enable sensing applications using the Internet of Things [34, 45].

2.4 Software and Data Processing Tools

Electronic sensing mechanisms are based on sophisticated data processing [41, 48]. Decrease dimensionality and enhance classification using pattern recognition techniques such as Principal Component Analysis (PCA) and Linear Discriminant Analysis (LDA) [8, 31]. Machine learning models such as Convolutional Neural Networks (CNN), Support Vector Machines (SVM), and Artificial Neural Networks (ANN) offer improved prediction and decision capabilities in real-time systems [41, 42]. Furthermore, chemometric techniques like cluster analysis and Partial Least Squares Regression (PLS) can be employed to cluster similar data and relate sensor signals to sensory attributes [3, 39]. New cloud-based platforms for real-time food diagnostics and monitoring are being utilized, as well as platforms such as MATLAB and Python (with libraries such as scikit-learn and TensorFlow) [37, 42]. Exquisite powers of food quality monitoring, environmental protection, and healthcare diagnostics are provided by e-nose and e-tongue technologies supported by large numbers of sensor types, strong hardware systems, and extended software tools [6, 38]. These systems are already highly flexible and beneficial, and they will only improve with the advancement of nanotechnology, AI integration, and internet of things connectivity [1, 6, 12]. Figure 2 describes about advantages of machine learning models (cnn, svm, and ann) for data processing and prediction tasks:

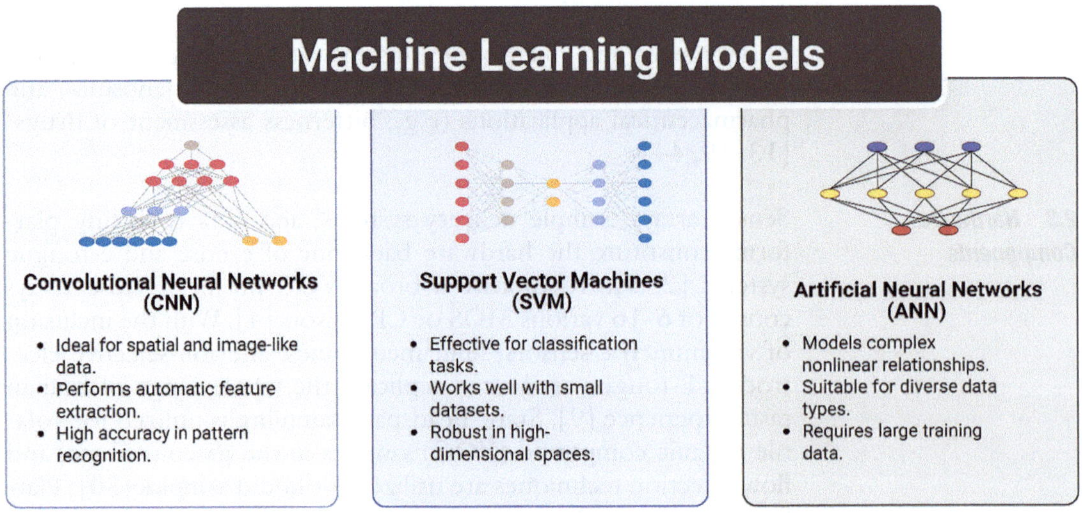

Fig. 2 Advantages of machine learning models (CNN, SVM, and ANN) for data processing and prediction tasks

3 Methods

There has been a recent achievement in the food industry employing technology such as the electronic nose (e-nose) and the electronic tongue (e-tongue) for flavor analysis and quality control. With these sensors, rapid objective, and reproducible evaluations become possible, which analyze volatile and non-volatile compounds in a manner analogous to human perception. Here you can read in-depth information on these technologies, from the operation process to data processing, design for experiments, validation methods, and application in real-world environments. The e-nose and e-tongue are bio-inspired analytical tools whose goal is to subjectively assess flavors by mimicking human gustation and olfaction perception [70]. The e-nose operates through a network of gas sensors such as quartz crystal microbalances (QCM), conducting polymers (CP), and metal oxide semiconductors (MOS) [3, 8, 36]. Every sample possesses a unique "smellprint" formed through the interactions of volatile organic compounds with sensor surfaces, inducing measurable changes in electrical resistance, capacitance, or frequency [20, 30]. Before applying multivariate analysis, the data undergo sophisticated signal processing methods like baseline correction and filtering out noise [12, 33]. Experimental procedures and the fundamental principles behind E-nose and E-tongue systems are explained in Table 2.

Table 2
Fundamental Principles and Experimental Procedures in E-Nose and E-Tongue Systems

Section	Topic	Key points	Techniques/tools	References
Operating principles	Signal generation	Chemical interaction with sensor arrays generates unique patterns	Metal oxide sensors, polymer sensors	[4, 6, 8, 35]
	Signal detection	Conversion to electrical signals for digital processing	Potentiometric, voltammetric, optical	[9, 10, 34]
	Signal interpretation	Pattern recognition for aroma/ taste profiles	PCA, ANN, machine learning	[8, 37, 42]
Experimental setup	Calibration	Standard solutions used for system accuracy	Reference gases, standard solutions	[6, 11, 33]
	Baseline correction	Drift removal and normalization	Pre-processing algorithms	[12, 34, 37]
	Sensor conditioning	Stabilization and pre-use initialization	Heating cycles, exposure to blank samples	[4, 8, 30]

Conversely, the e-tongue is capable of identifying sugars, organic acids, and bitterants non-volatile compounds in liquid or soft foods through electrochemical sensors [9, 25]. Potentiometric, voltammetric, and impedance-based models are frequently employed sensors, each of which is capable of sensing various tastes (e.g., sweet, sour, salty, bitter, and umami) [43, 48]. For instance, ion-selective electrodes track sodium or potassium ions for saltiness assessment [9, 30], and lipid membrane sensors can identify umami compounds such as monosodium glutamate (MSG). Pattern recognition algorithms are then used for the multidimensional taste profile produced by the sensor array to determine correlation with human sensory perception [25, 52]. To interpret complex sensor readings, both systems employ multivariate data analysis. Dimensionality reduction by principal component analysis (PCA) is widespread; it compresses high-dimensional data into two or three principal components, which are plotted to display clusters in samples [51, 60]. For enabling quality grading or adulteration detection, linear discriminant analysis (LDA) enhances class discrimination further [55, 57]. Classification for large sets of data can be improved with machine learning approaches such as ANN and SVM [55, 59]. An example is the discrimination between the geographical locations of tea and coffee. For real-time monitoring of industrial applications, recent advances have coupled deep learning with e-nose/e-tongue devices [69, 72]. Because many chemicals can produce similar responses in sensors, cross-reactivity is an important challenge. An answer to this issue is the employment of machine learning models that are designed to recognize subtle shifts in patterns and sensor arrays constructed with wide selectivity [36, 37]. Different from ANN algorithms, which can detect hexanal (an indicator of lipid oxidation) and ethanol (an indicator of fermentation), MOS sensors in e-noses react to ketones, alcohols, and aldehydes [20, 59]. Likewise, electronic tongues enhance resolution for multicomponent matrices such as fruit juices or wine through a combination of signals from different types of electrodes using sensor fusion algorithms [25, 48].

To make measurements repeatable, calibration and standardization are required. Reference gases (e.g., ethanol for alcoholic drinks) and standard solutions (e.g., HCl for acidity) are employed for calibrating e-noses and e-tongues, respectively [12, 43]. Baseline drift correction is performed before measuring to account for environmental changes or sensor age [33]. Some of these are, for instance, rinsing e-tongue electrodes in buffer solutions to remove carryover effects and zero-air exposure to reset e-nose sensors [4, 9]. Sample preparedness varies based on the diet matrix. Homogenization and headspace equilibration are required steps in e-nose analysis of solid foods (e.g., meat and cereals), while e-tongues are able to analyze liquids (e.g., milk and juice) immediately [5, 26]. Coupling e-noses with volatile extraction methods such as solid-phase microextraction (SPME) enhances sensitivity to delicate odors [7, 17]. Through the detection of phenolic chemicals, SPME-GC-MS was in a position to rapidly screen cocoa for smoky off-flavors [14]. This information was then correlated with data from e-noses. Smart farming and food processing sectors are witnessing the development of real-time monitoring use cases. Anticipating when perishables like dairy products and fruits would spoil, wireless e-nose networks track volatile organic compound (VOC) emissions during storage [45, 46]. Through their integration with Internet of Things platforms, electronic tongues make it possible to ensure ongoing quality control in beverage manufacturing, such as tracking the development of the fermentation process of vinegar or beer [48, 72]. These technologies reduce requirements for labor-intensive and costly laboratory techniques such as HPLC or GC–MS [3, 7].

3.1 Experimental Setup and Sample Preparation

Target analytes determine the selection of sensors. For e-noses, CP sensors offer faster response times, while MOS sensors are suitable for broad-range VOC detection [3, 36]. Calibration is necessary for QCM sensors when handling wet samples like fruits because they are extremely sensitive to moisture [8, 33]. Biosensors to detect intricate flavors (e.g., bitterness in beer) and ion-selective electrodes (ISE) to detect simple tastes are widely employed in electronic tongues [9, 25]. Calibration of e-noses and e-tongues uses standard solutions (e.g., 0.1 M NaCl for salt) and reference gases of known concentration (e.g., 100 ppm ethanol), respectively [12, 43]. R^2 >0.98 is acceptable when employing multi-point calibration curves to measure linearity [4, 33]. To ensure that all of the solid foods are reduced to a uniform particle size less than 2 mm and the volatile organic compounds (VOCs) are released evenly, homogenization of samples is a crucial step in food sample preparation for analysis. To effectively release the volatiles, enzymatic digestion can be done on meat and milk products before analysis [5, 46]. Static headspace sampling is often utilized for the generation of headspace, an essential aspect of electronic nose (e-nose) systems. This entails letting the sample acclimate to a temperature environment of 40–60 °C for approximately 30 min to enable volatile organic compounds to accumulate in the headspace [7, 17]. Dynamic headspace purging can also be applied to increase sensitivity, especially with the analysis of VOCs at low concentration [14].

To desorb particles, liquid samples have to be filtered by a membrane before being ready for electronic tongue (e-tongue) analysis. To further standardize data and ensure precise sensor readings, pH correction may be required [9, 48]. Before application, e-nose sensors are stabilized through exposure to nitrogen or artificial air [4], whereas e-tongue electrodes are polished and hydrated in buffer solutions [43]. Sensor conditioning is essential to achieve stable responses. To record accurate signal dynamics, data acquisition often involves measurement cycles running over 2–5 min and sensor responses sampled at intervals of 1 s [8, 33]. Three repetitions of each measurement are the norm to ensure repeatability and reliability of data [12]. During the process of analysis, control over environmental conditions is exercised with care to reduce signal drift and enhance the stability of sensors [33, 36]. Finally, the outcomes of e-nose and e-tongue tests are validated by applying benchmark analytical methods such as gas chromatography–mass spectrometry (GC–MS) or high-performance liquid chromatography (HPLC). Sensor-based information is supposed to reflect the true nature of the sample; therefore, these methods agree with the ground-truth chemical composition [7, 17].

3.2 Data Analysis Techniques

For the extraction of meaningful information about food flavor profiles and food quality from the data generated by e-nose and e-tongue instruments, sophisticated analytical methods have to be employed [32]. To enhance signal quality, the raw sensor data have to first be preprocessed, including normalizing, baseline correction, and noise reduction [51, 60]. To remove electrical noise and maintain important signal characteristics, most utilize wavelet transformations and moving average smoothing [33, 52]. In an effort to quantify volatile organic compound concentrations using e-nose measurements, response curves are commonly transformed to maximum response values or area-under-curve metrics [8, 36]. The electrochemical sensors employed in e-tongue data are often required to correct drift using algorithms; techniques like orthogonal signal correction or component correction have been found effective [25, 43]. In working with the high-dimensional data produced by sensor arrays, dimensionality reduction is important [3]. PCA has enough information to explain 75–95% of the variance for any given dataset with just two or three principal components [51, 60]. Through this, we are able to identify how samples group together in a reduced space, which is important since varying types of products or quality classes are likely to group together [55, 57]. Building on PCA, linear discriminant analysis (LDA) works to maximize variation between classes and minimize variation within them, creating even more differentiated separation for classification purposes [52, 58]. Nonlinear interaction in progressively more complex data can be accommodated with kernel-based methods like KPCA (Kernel PCA) [57]. To show patterns of high dimensions in a two-dimensional visualizable space, there have been developments such as t-SNE and UMAP in recent times [56].

The analysis of e-nose/e-tongue data has been revolutionized by machine learning algorithms. Supervised learning algorithms such as support vector machines (SVM) have been reported to classify tasks such as meat freshness grading and discrimination of tea varieties with over 90% accuracy [55, 58]. Although they require larger training sets, artificial neural networks (ANN) and deep learning architectures, especially, can model complex nonlinear relationships in sensor data [59, 61]. Due to their strong performance and inherent feature importance metrics, Random Forest classifiers are effective in identifying which sensors are most critical to categorization [53]. Identifying unknown samples or detecting outliers is where the unsupervised methods, like k-means clustering, truly excel [53, 56]. Sensor data can be linked with measurable quality attributes using multivariate regression techniques. Chemical levels (e.g., alcohol, pH) can be precisely calibrated from sensor signals with partial least squares regression (PLSR) [17, 55]. Forecasting sensory scores from e-tongue measurements has been most successfully done for artificial neural networks with R^2 >0.85 for attributes such as bitterness or sweetness in drinks [59, 61]. In the case of food matrices that are heterogeneous, ensemble methods that employ numerous algorithms work better than individual models [53, 55].

To receive an overall picture, sensor fusion algorithms integrate the readings of multiple sensors. Mid-level fusion is done through fusing derived features [29, 32], while low-level fusion is done by fusing raw e-nose and e-tongue data before modeling. For example, e-nose and e-tongue fusion identified adulteration of minced mutton with pig at a 98% accuracy rate [28, 32], showing how decision-level fusion combines predictions from individual models, generally produces better results. Sensor arrays and spectroscopic information (NIR, Raman) are fused in advanced fusion techniques to enhance specificity [23, 39]. Real-time processing structures are becoming increasingly important for industrial applications. Cloud-based systems do the complex modeling, whereas simple PCA and classification can be performed at the measurement site by edge computing devices [69, 72]. Due to streaming algorithms and adaptive models that learn from new data, production lines are now able to track quality in real-time [45, 72]. Products may be sorted on high-speed conveyors utilizing FPGA instantiations of machine learning models, which classify in milliseconds [34]. Model reliability is ensured through validation procedures [37]. The aim of maximizing the usage of data and delivering sound performance estimates is achieved by k-fold cross-validation, where k is typically 5 or 10 [37, 51]. To assess practicality, external validation based on independent test sets must be performed [12, 37]. Single predictions can have uncertainty quantified using confidence measures such as distance-to-model measures or prediction probabilities [55, 58]. Handling sensor drift and shifting conditions is the domain of model maintenance. Models can adapt to minor sensor changes without retraining completely using incremental learning techniques [37, 45]. Transfer learning techniques can be used so that models developed on one device can be readily moved to another device of the same make [34, 72]. Accuracy over time is guaranteed by periodically checking performance with control samples [12, 33]. Two of the unresolved issues are the interpretability of complex models and the handling of naturally occurring products with a large amount of variability [72]. To enhance the explainability of machine learning decisions, explainable AI methods like SHAP values and LIME are being employed [53, 72]. Especially effective for expensive or scarce food products are few-shot learning models that try to reduce the demand for large training datasets [40, 72].

3.3 Validation Methods

For quantifying how well a model performs, cross-validation techniques are necessary. Accurate estimates of predictive accuracy are readily generated using the k-fold method [37, 51] with all available information. Even though computationally expensive, leave-one-out cross-validation (LOOCV) is particularly useful for small data sets [51, 55]. Ensure that results are not dependent on a single data partitioning by performing repeated random subsampling validation [37]. For cases where feature selection and model tuning are both required, nested cross-validation provides unbiased performance estimations [55, 58]. Utilization of E-Nose and E-Tongue for Food Analysis: Data Processing, Validation, and Applications (Table 3).

The most trustworthy method to assess performance in the real world is by holdout validation on independent test sets [12, 37]. Temporal resilience can be best assessed with test samples that were not produced in the same production batch or harvest year as the training data [17, 37]. Generalizability of a growth condition can be verified by spatial validation, where samples from various geographic locations are used [32, 55]. Validation samples for shelf life should cover all probable storage durations likely to be met in practical situations [46, 47]. E-sensing results can be confirmed by comparing them with reference methods. Quantification ability of volatile organic compounds (VOCs) is corroborated through correlation with GC–MS or HPLC results [7, 14]. Sensory panel test results indicate that classifications done by sensors are consistent with what human senses perceive [17, 27]. Accuracy of the quality traits that have been predicted is validated by physical-chemical analyses, e.g., pH and °Brix [25, 48]. A method of determining how sensor readings compare with reference methods is Bland-Altman analysis [12, 37]. Reproducibility testing is used to measure the accuracy of measurements. For the majority of applications, an RSD of below 5% is considered excellent when assessing intra-day repeatability, which assesses variance under identical conditions [12, 33]. To check whether there is a sensor drift issue, you can see it by observing inter-day reproducibility [33, 43]. In the context of industrial implementation, inter-instrument repeatability is critical. Usually, devices with identical devices can have a relative standard deviation (RSD) of below 10%.

Table 3
Data processing, validation, and application of E-nose and E-tongue in food analysis

Section	Topic	Key points	Techniques/tools	References
Data analysis techniques	PCA	Dimensionality reduction and data clustering	Principal component analysis	[51, 54], [60]
	LDA	Supervised classification approach	Linear discriminant analysis	[54, 55, 57]
	ANN	Adaptive classification and nonlinear pattern recognition	Artificial neural networks	[59, 61, 62]
	SVM	Efficient for small-sample high-dimensional data	Support vector machine	[52, 55, 58]
	Clustering methods	Unsupervised data grouping	K-means, hierarchical clustering	[53, 63, 64]
Validation methods	Cross-validation	Ensures robustness and avoids overfitting	k-fold, leave-one-out	[55, 58, 59]
	Comparison with human panels	Confirms relevance of sensor outputs	Sensory panel evaluation	[8, 25, 27]
	Reproducibility	Assessment of result consistency	Repeatability tests	[12, 47, 50]
3.5 case studies	Flavor profiling of beverages	Discrimination and grading of liquors, teas, and sauces	E-nose and E-tongue with chemometrics	[27, 32, 48]
	Spoilage detection in meat and dairy	Monitoring freshness and microbial activity	GC–MS, E-nose, sensor fusion	[43, 46, 50]
	Quality grading in fruits and vegetables	Ripeness, sweetness, spoilage, and aroma classification	Combined E-nose/tongue with machine learning	[26, 28, 29]

Robustness testing examines the performance under actual conditions of use [49]. Temperature and humidity stress tests may ensure reliability in different environments [33, 36]. Inconsistencies in sample preparation, such as differences in grinding times, can be employed to test the robustness of a procedure [5, 26]. To establish if common food matrix constituents affect measurements, interference tests are conducted [9, 25]. Detailed performance evaluation is supplied through multivariate validation measures. In classification, accuracy measures such as sensitivity, specificity, and F1-score offer more information than accuracy in isolation [55, 58]. Regression predictive strength can be assessed through a variety of means, such as R^2, RMSE, and RPD [17, 55]. In binary classification tasks, such as detecting spoilage, the area under the ROC curve (AUC) is particularly insightful [46, 58]. The highest level of validation is obtained through blind testing by use of unknown samples. Detection capability is validated through challenge tests with purposefully contaminated or defective items [28, 32]. Testing laboratories can match their findings to those of other colleagues by means of proficiency testing programs [12, 44]. The most valid gold standard is actual deployment in field use [45, 72]. Stable reliability is assured through long-term monitoring of performance. Accuracy to reference samples is monitored over time with control charts [12, 33]. Periodic recalibration of sensors keeps them functioning correctly [4, 43]. If retraining is required, model degradation monitoring can identify it [37, 45]. Business adoption relies on standardization activity. Measurement alignment from equipment is simplified with reference materials [12, 23]. Reproducibility limits are established by collaborative studies [44]. Organizations like AOAC International provide method validation criteria that can be utilized as a guideline to demonstrate reliability [12]. Digital twins, which simulate how an instrument performs in various environments, represent one new form of validation [72]. To make sure that the validation outcomes remain intact and unchangeable, researchers are considering using blockchain technology [69, 72]. Technicians are now capable of performing validation operations with greater accuracy due to augmented reality interfaces [34].

3.4 Case Studies

3.4.1 Flavor Profiling of Beverages

With developments in electronic tongue and nose technology, which scan for flavors comprehensively, the quality control of beverages has been revolutionized. To identify Keemun black tea by its quality, scientists employed a combination of HS-SPME-GC-MS and e-nose. They identified that linalool and geraniol were the most critical indicators of fragrance [17]. A decrease in analysis time from hours to minutes, the e-nose achieved a classification accuracy of 92%, compared to expert panels [17, 39]. With PLSR models successfully predicting sensory ratings ($R^2 = 0.89$) [3, 8], MOS-based e-noses successfully discriminated Arabica and Robusta coffee varietals based on pyrazine and furan profiles. The application of e-tongues for the vinegar aging classification in alcoholic drinks was incredibly precise. Potentiometric sensors were used by scientists to measure acetic acid and sugar profiles in balsamic vinegars of various maturation times with a 95% level of accuracy [27]. Also, in the vine industry, sensor fusion of e-tongue and e-nose data was utilized for geographical origin verification with greater reliability compared to single-modality methods at 97% [48]. Minor differences in phenolic chemicals responsible for defining terroir and organic acids were found by the technique [25, 48]. Figure 3 depicts the flavor profiling: technological tools and their functions.

3.4.2 Spoilage Detection in Perishable Foods

For rapid monitoring of spoiling, the meat and dairy sectors have adopted e-sensing. Through the monitoring of increasing amounts of alcohols and sulfur compounds, pioneering research on chicken demonstrated that MOS e-noses can detect microbial spoilage

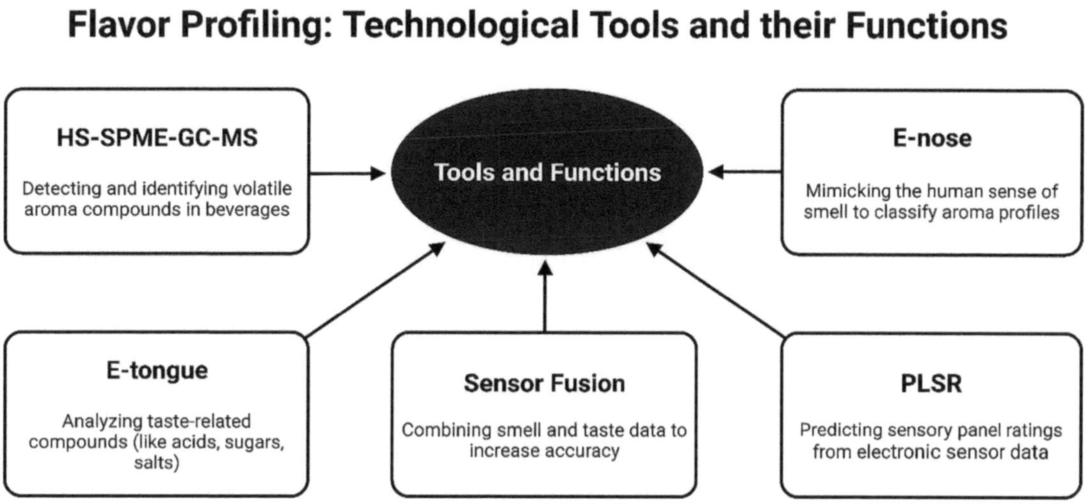

Fig. 3 Flavor profiling: technological tools and their functions

24–48 h prior to the detection of visible signs [46]. Storage room quality can be tracked in real-time, as the system achieved a sensitivity of 94% using SVM classifiers [46, 47]. Voltammetric e-tongues have also proven to be useful in the food industry, and a study identified one predicting yogurt storage life by monitoring the accumulation of lactic acid; the technique was strongly correlated ($r = 0.91$) with conventional pH measurements [15, 25]. When used in seafood, a hybrid GC–IMS/e-nose apparatus blending VOC profiling with sensor array data reduced false positives in freshness testing [7]. Compared to more traditional sensory methods, this method enhanced the discrimination between fresh (between 0 and 2 days) and spoiled (within 5 days) fish by an impressive degree, with 98% accuracy [7, 46]. By monitoring hexanal levels, QCM sensor-equipped e-noses have been capable of detecting lipid oxidation in potato chips 7 days prior to rancidification, a first for packaged food [14, 36].

3.4.3 Fruit and Vegetable Quality Control

Non-destructive quality analysis can now be performed throughout the supply chain due to e-sensing technologies. For instance, PLSR models accurately forecasted soluble solids content and hardness in tomatoes, while flavor changes during ripening and storage were tracked by a combination of e-nose and e-tongue systems [29]. High-throughput sorting was achieved by the device with measurements taken as short as 30 s [29, 32]. Electronic noses detected with 90% accuracy in apples before visible symptoms appeared [26, 36], distinguishing between storage issues like superficial scald and bitter pit based on ester and alcohol profiles.

3.5 Process Monitoring and Authentication

Industrial food production can ensure quality in real-time with the assistance of e-sensing [65]. Through monitoring key fragrance molecules such as 2-methylpropanal and 3-methylbutanal, a chocolate producer was able to utilize an e-nose system to monitor conching effectiveness [14]. Process variation was reduced while consistent flavor profiles were sustained using the approach [14, 23]. A nano-sensor enriched e-nose may identify squalene and tocopherol markers in olive oil, enabling it to identify extra virgin versus contaminated ones with certainty [40, 47]. One of the new applications is tailored nutrition, where portable e-tongues are utilized to measure individuals' taste response and return customized meal recommendations [9, 66, 72]. Through monitoring volatile organic compound (VOC) concentrations, smart packaging can incorporate integrated e-nose sensors that give real-time freshness markers [45, 46]. Internet of things (IoT) connectivity is becoming more the norm in innovative systems, enabling cloud-based quality monitoring along global supply chains [68, 69, 72]. Through providing rapid, objective, and affordable taste analysis solutions that complement or replace traditional sensory and chemical methods, these case studies demonstrate how e-nose and e-tongue technologies are transforming food quality management—from farm to fork [3, 8, 23]. Almost all sectors of the food industry are experiencing the advantages of this technology as it evolves to accommodate new AI and miniaturization standards [39, 72].

4 Notes

4.1 Practical Considerations in Sensor Selection

The target analytes and working environment have to be appropriately regarded while selecting sensors for e-nose and e-tongue systems [23, 67, 71]. Metal oxide semiconductor (MOS) sensors are sensitive to a wide range of volatile organic compounds (VOCs) and are thus widely used in e-nose applications [72]. Nevertheless, they possess low selectivity and consume a lot of power [8, 36]. While they are less susceptible to humidity influence, CP sensors operate at lower temperatures and respond faster [3, 23]. QCM sensors are excellent for water samples due to their high sensitivity to moisture and to heavy molecules, but they must be frequently calibrated [33, 40]. In the case of e-tongue systems, potentiometric sensors are ideal for straightforward and unchanging basic taste analysis, while voltametric sensors are ideal for complex matrices such as wine or coffee due to their higher sensitivity [9, 25]. Greater specificity for target compounds such as bitterants or umami molecules has also been obtained with more recent advances in biosensors, including lipid membrane-based electrodes or enzyme-based electrodes [10, 11]. When selecting sensors, the food matrix plays a fundamental role. Surface acoustic wave (SAW) sensors, which are fouling-resistant, are required for high-fat products like meat and cheese [46, 47]. Acidic foods like citrus fruits need electrodes to be made from pH-resistant materials like gold or platinum [26, 29]. When selecting sensors, environmental conditions must be taken into account. Due to the possibility of significant signal drift with changes in humidity, integrated humidity compensation algorithms must be used for operation with MOS sensors [36, 45]. Due to the effect of temperature changes on the electrochemical sensor responses of e-tongues, temperature stabilization or correction models must be incorporated into the system [9, 43]. The need for dust- and mechanical shock-resistant ruggedized sensors in field usage was emphasized in recent research in portable e-noses for orchard fruit quality monitoring [45, 50, 73].

In complex food matrices, cross-sensitivity remains a significant issue. With a blend of optical and electrochemical transducer concepts, for instance, attention can be enhanced, as shown through a study capable of differentiating between 15 various cheese types using a 94% accuracy rate [12, 47]. To reduce interference and maintain useful sensor signals, machine learning techniques like attention mechanisms are employed [53, 72]. Performance metrics of sensors are not yet standardized. It is typical for research to report significant metrics with different values, such as the limit of detection (LOD), response time (t_{90}), and recovery time (t_{10}) [12, 23]. New guidelines for e-sensor validation for food authentication have been issued by the European Food Safety Authority (EFSA). These regulations provide minimum requirements for sensitivity (LOD <1% of target concentration) and repeatability (RSD <5% for 30 consecutive measurements) [12, 47].

4.2 Limitations and Challenges

The reliability and widespread applicability of e-nose and e-tongue technologies are hampered by several significant limitations, despite their benefits. It has been found that the signal drift in MOS-based e-noses reaches 10–20% after merely 30 days of use, indicating that sensor drift is one of the most enduring issues [36, 45]. Aging sensors, chemical poisoning through reactivity, and morphological modification at active sites are some of the reasons behind the drift [23, 36]. The electro-tongue sensitivity can be cut by half after performing as few as 20 measurements in complex matrices such as milk or meat extracts because of fouling of the electrodes by lipids or proteins [9, 43]. This problem can be reduced, but not removed, by using sophisticated drift compensation algorithms utilizing reference gases for e-noses or reference solutions for e-tongues [33, 37]. Machine learning models with time-dependent correction factors could retain their classification accuracy greater than 85% for 6 months without recalibration, as a 2023 study reported [37, 45]. For MOS sensors specifically, where a relative humidity change can lead to a 30–50% signal variability [36, 45], humidity interference is ruinous to e-nose performance. At a sample humidity of 90% RH, as is with some wet foods such as fresh meat or fruit, this becomes an extremely critical problem [26, 46]. There are now two solutions for this challenge: either the costly and complex humidity-controlled sample chambers or the dual-sensor arrays, which may monitor volatile organic compounds and humidity simultaneously [36, 45]. An innovative covering made of hydrophobic polymer proved promising results in recent experiments with fresh produce, reducing humidity interference [40, 45]. Corresponding problems exist for electronic tongues when the ionic strength in liquid samples varies, and they need to be calibrated routinely or corrected with advanced chemometric treatment [9, 25].

There are operational challenges due to maintenance requirements [45]. For industrial applications, e-noses typically need to be calibrated once a week and have sensors replaced once a month. It is advisable to recalibrate the device every 6–12 months [23, 33]. It is advisable to change the E-tongue membrane every 50–100 readings and the electrodes are polished every 10–20 usage [9, 43]. Due to these requirements, there is a considerable amount of downtime; one 2022 case study found that an industrial meat processing plant lost 15% of their production due to e-nose maintenance [46]. There is a considerable cost premium for the new automated self-cleaning systems employing ultraviolet light or electrochemical cycling [34, 43]. A vineyard field test indicated that the e-nose sensitivity decreased by 40% in one growing season owing to the deposition of dust, which emphasized the additional challenge of sensor contamination for field-deployable systems [45, 50]. Applications are restricted because of limited sensitivity to individual chemicals. Electronic noses excel at detecting large volatile organic compound (VOC) concentrations (>1 ppm), but they fail to detect subtle aroma components that have a significant impact on flavor, such as pyrazines in coffee or sulfur compounds in wine [8, 14]. For these applications, GC-MS remains 100–1000 times more sensitive [7, 17]. One study in 2024 showed that electronic tongues would only be able to recognize 60% of the bitter compounds that could be recognized by human panels [10, 25], showing their limitation in recognizing sub-threshold taste molecules. Although pre-concentration methods (e.g., SPME) and nano-based sensors are closing this gap, these raise the system complexity as well [40, 72]. Complex matrix effects complicate practical applications. Based on the food matrix, the identical volatile organic compound (VOC) could garner different sensor responses; e.g., the ethanol signals from beer and wine differed by 30% due to differences in sugar content [48]. The performance of the e-tongue was 40% worse in high-fat milk samples than low-fat ones [12, 25]. The present strategy is to develop matrix-specific calibration models, which is time-consuming and costly [37, 46]. To overcome these limitations, multimodal strategies that combine e-sensors with spectroscopic techniques (e.g., NIR, Raman) hold great promise [23, 39].

5 Conclusion and Future Perspectives

Flavor analysis and food quality control strategies used by the food industry have been revolutionized with the coming together of e-tongue and e-nose technology. Through the availability of quick, objective, and reproducible results, these biomimetic tools make a credible alternative to the traditional sensory test. These technologies provide high-resolution chemical fingerprints that capture the complexity of food matrices by using a broad variety of sensor types, including lipid membranes and metal-oxide semiconductors, along with sophisticated hardware platforms and smart data processing tools. Their significance across the food value chain is established by their proven utility in the detection of spoilage, freshness determination, and authentication of goods. Apart from food application, e-sensing systems have been used in pharmaceuticals, environmental monitoring, and medical diagnostics, reflecting their versatility. Their applicability and versatility are evidenced by the fact that the systems cut across disciplines such as materials science, electronics, data science, and biology. Some of the most significant challenges facing the industry, such as human panel subjectivity, test delays, and low sensitivity of traditional methods, are solved with electronic sensing technologies, as this chapter demonstrates. A significant evolution within modern analytical methodologies is e-nose and e-tongue devices, which can mimic human sensory perception while being able to transcend its limitations.

Miniaturization, integration into IoT platforms, and the development of strong artificial intelligence (AI) algorithms will shape electronic sensing technologies of the future. Compact, energy-saving designs that could track production lines, store shelves, and even domestic kitchens in real-time are expected in future devices. New possibilities for digital traceability, precision agriculture, intelligent food supply chains, and mass-scale, remote data gathering and decision-making will emerge from the convergence of e-nose/e-tongue systems with cloud computing and wireless connectivity. In addition, bioinspired and nanostructured materials will also find application in the next-generation sensor. They will be better suited to reproduce the human receptor and be sensitive, robust, and selective. Adaptive calibration algorithms and self-calibrating AI models are set to combat challenges such as sensor drift, interference from the environment, and long-term stability. Food safety regulations, quality grading techniques, and consumer confidence will increasingly be influenced by these technologies as regulatory systems begin to recognize and incorporate data from electronic sensory systems. Possibly transforming the future of sensory science in industrial and therapeutic applications is the coupling of electronic sensing with artificial intelligence and robots.

References

1. Lv B, Mo W, Jian C, Li S, Guo Y (2024) Recent progress in food quality control through flavor analysis using gas chromatography–ion mobility spectrometry. J Food Meas Charact 18(9): 8002–8015
2. Sung J, Frost S, Suh JH (2024) Progress in flavor research in food: flavor chemistry in food quality, safety, and sensory properties. Food Chem: X 25:102071
3. Calín-Sánchez Á, Carbonell-Barrachina ÁA (2021) Flavor and aroma analysis as a tool for quality control of foods. Foods 10(2):224
4. Wesoły M, Przewodowski W, Ciosek-Skibińska P (2023) Electronic noses and electronic tongues for the agricultural purposes. TrAC Trends Anal Chem 164:117082
5. Song H, Liu J (2018) GC-O-MS technique and its applications in food flavor analysis. Food Res Int 114:187–198
6. Mahanti NK, Shivashankar S, Chhetri KB, Kumar A, Rao BB, Aravind J, Swami DV (2024) Enhancing food authentication through E-nose and E-tongue technologies: current trends and future directions. Trends Food Sci Technol 150:104574
7. Wang S, Chen H, Sun B (2020) Recent progress in food flavor analysis using gas chromatography–ion mobility spectrometry (GC–IMS). Food Chem 315:126158
8. Tan J, Xu J (2020) Applications of electronic nose (e-nose) and electronic tongue (e-tongue) in food quality-related properties determination: a review. Artif Intell Agric 4: 104–115
9. Podrażka M, Bączyńska E, Kundys M, Jeleń PS, Witkowska Nery E (2017) Electronic tongue—a tool for all tastes? Biosensors 8(1):3
10. Liu J, Qian J, Adil M, Bi Y, Wu H, Hu X et al (2024) Bioinspired integrated triboelectric electronic tongue. Microsyst Nanoeng 10(1): 57
11. Chen Y, Du L, Tian Y, Zhu P, Liu S, Liang D et al (2022) Progress in the development of detection strategies based on olfactory and gustatory biomimetic biosensors. Biosensors 12(10):858
12. Galvan D, Aquino A, Effting L, Mantovani ACG, Bona E, Conte-Junior CA (2022) E-sensing and nanoscale-sensing devices associated with data processing algorithms applied to food quality control: a systematic review. Crit Rev Food Sci Nutr 62(24):6605–6645
13. Fitzgerald J, Fenniri H (2017) Cutting edge methods for non-invasive disease diagnosis using e-tongue and e-nose devices. Biosensors 7(4):59
14. Perotti P, Cordero C, Bortolini C, Rubiolo P, Bicchi C, Liberto E (2020) Cocoa smoky off-flavor: chemical characterization and objective evaluation for quality control. Food Chem 309:125561
15. Cai W, Tang F, Wang Y, Zhang Z, Xue Y, Zhao X et al (2021) Bacterial diversity and flavor profile of Zha-Chili, a traditional fermented food in China. Food Res Int 141:110112
16. Farag MA, Abib B, Tawfik S, Shafik N, Khattab AR (2021) Caviar and fish roe substitutes: current status of their nutritive value, bio-chemical diversity, authenticity and quality control methods with future perspectives. Trends Food Sci Technol 110:405–417
17. Su D, He JJ, Zhou YZ, Li YL, Zhou HJ (2022) Aroma effects of key volatile compounds in Keemun black tea at different grades: HS-SPME-GC-MS, sensory evaluation, and chemometrics. Food Chem 373:131587
18. Mehraeen E, Behnezhad F, Salehi MA, Noori T, Harandi H, Seyed Alinaghi S (2021) Olfactory and gustatory dysfunctions due to the coronavirus disease (COVID-19): a review of current evidence. Eur Arch Otorrinolaringol 278:307–312
19. Lechien JR, Chiesa-Estomba CM, De Siati DR, Horoi M, Le Bon SD, Rodriguez A et al (2020) Olfactory and gustatory dysfunctions as a clinical presentation of mild-to-moderate forms of the coronavirus disease (COVID-19): a multicenter European study. Eur Arch Otorrinolaringol 277(8):2251–2261
20. Kaur G, Bhari R, Kumar K (2022) Electronic noses and tongue-based sensor systems in food science. In: Nanosensing and bioanalytical technologies in food quality control. Springer Singapore, Singapore, pp 357–384
21. Bhattacharyya N (2024) Evolution of electronic perception systems: from conventional electronic systems to. Uddalak 1(1):1
22. Ordoñez-Araque R, Rodríguez-Villacres J, Urresto-Villegas J (2020) Electronic nose, tongue and eye: their usefulness for the food industry. Vitae 27(3). https://doi.org/10.17533/udea.vitae.v27n3a01
23. Aouadi B, Zaukuu JLZ, Vitális F, Bodor Z, Fehér O, Gillay Z et al (2020) Historical evolution and food control achievements of near infrared spectroscopy, electronic nose, and electronic tongue—critical overview. Sensors 20(19):5479

24. Pal S, Kumar D, Ulucan-Karnak F, Narang J, Shukla SK (2024) Bio-inspired electronic sensors for healthcare applications. Chem Eng J:155894. https://doi.org/10.1016/j.cej.2024.155894
25. Cho S, Moazzem MS (2022) Recent applications of potentiometric electronic tongue and electronic nose in sensory evaluation. Prev Nutr Food Sci 27(4):354
26. Qiu S, Wang J (2015) Application of sensory evaluation, HS-SPME GC-MS, E-Nose, and E-Tongue for quality detection in citrus fruits. J Food Sci 80(10):S2296–S2304
27. Jo Y, Chung N, Park SW, Noh BS, Jeong YJ, Kwon JH (2016) Application of E-tongue, E-nose, and MS-E-nose for discriminating aged vinegars based on taste and aroma profiles. Food Sci Biotechnol 25:1313–1318
28. Tian X, Wang J, Ma Z, Li M, Wei Z (2019) Combination of an E-nose and an E-tongue for adulteration detection of minced mutton mixed with pork. J Food Qual 2019(1): 4342509
29. Hong X, Wang J (2015) Use of electronic nose and tongue to track freshness of cherry tomatoes squeezed for juice consumption: comparison of different sensor fusion approaches. Food Bioprocess Technol 8:158–170
30. Zou Y, Wan H, Zhang X, Ha D, Wang P (2015) Electronic nose and electronic tongue. In: Bioinspired smell and taste sensors, pp 19–44
31. Wang J, Wei Z (2015) The classification and prediction of green teas by electrochemical response data extraction and fusion approaches based on the combination of e-nose and e-tongue. RSC Adv 5(129):106959–106970
32. Yu S, Huang X, Wang L, Ren Y, Zhang X, Wang Y (2022) Characterization of selected Chinese soybean paste based on flavor profiles using HS-SPME-GC/MS, E-nose and E-tongue combined with chemometrics. Food Chem 375:131840
33. Moufid M, Tiebe C, El Bari N, Fakra DAH, Bartholmai M, Bouchikhi B (2022) Pollution parameters evaluation of wastewater collected at different treatment stages from wastewater treatment plant based on E-nose and E-tongue systems combined with chemometric techniques. Chemom Intell Lab Syst 227:104593
34. Stighezza M, De Munari I, Caselli M, Boni A, Bianchi V (2025) A flexible Multi-Channel platform for versatile E-nose and E-tongue operations. IEEE Trans Instrum Meas 74:1
35. Tibaduiza D, Anaya M, Gómez J, Sarmiento J, Perez M, Lara C et al (2024) Electronic tongues and noses: a general overview. Biosensors 14(4):190
36. Chen H, Huo D, Zhang J (2022) Gas recognition in E-nose system: a review. IEEE Trans Biomed Circuits Syst 16(2):169–184
37. Abi-Rizk H, Bouveresse DJR, Chamberland J, Cordella CB (2023) Recent developments of e-sensing devices coupled to data processing techniques in food quality evaluation: a critical review. Anal Methods 15(41):5410–5440
38. Jing W, Zhao X, Li M, Hu X, Cheng X, Ma S, Wei F (2022) Application of multiple-source data fusion for the discrimination of two botanical origins of magnolia officinalis cortex based on E-nose measurements, E-tongue measurements, and chemical analysis. Molecules 27(12):3892
39. Gharibzahedi SMT, Barba FJ, Zhou J, Wang M, Altintas Z (2022) Electronic sensor technologies in monitoring quality of tea: a review. Biosensors 12(5):356
40. Bhangare BK, Sinju KR, Patil SJ, Ramgir NS, Debnath AK (2023) Introduction to nano-e-nose. In: Nanotechnology-based E-noses. Woodhead Publishing, pp 29–55
41. Baruah S, Mazumder DH (2025) A review on application of machine learning techniques coupled with E-nose in healthcare, agriculture and allied domains. IEEE Trans Instrum Meas 74:1
42. Yadav U, Khurana S, Rawal K, Arora B, Yadav A (2025) Integrating AI with electronic tongue (E-tongue) for real-time detection in the food industry. In: Artificial intelligence in the food industry. CRC Press, pp 277–296
43. Moufid M, Hofmann M, El Bari N, Tiebe C, Bartholmai M, Bouchikhi B (2021) Wastewater monitoring by means of e-nose, VE-tongue, TD-GC-MS, and SPME-GC-MS. Talanta 221:121450
44. Gómez JKC, Acevedo CMD, García R (2021) Methodology for the bacteria detection in drinking water through an e-nose and e-tongue. INGE CUC 17(1):5
45. Moshayedi AJ, Sohail Khan A, Hu J, Nawaz A, Zhu J (2023) E-nose-driven advancements in ammonia gas detection: a comprehensive review from traditional to cutting-edge systems in indoor to outdoor agriculture. Sustainability 15(15):11601
46. Munekata PE, Finardi S, de Souza CK, Meinert C, Pateiro M, Hoffmann TG et al (2023) Applications of electronic nose, electronic eye and electronic tongue in quality, safety and shelf life of meat and meat products: a review. Sensors 23(2):672

47. Gil M, Rudy M, Duma-Kocan P, Stanisławczyk R (2025) Electronic sensing Technologies in Food Quality Assessment: a comprehensive literature review. Appl Sci 15(3):1530
48. Leon-Medina JX, Anaya M, Tibaduiza DA (2023) New electronic tongue sensor array system for accurate liquor beverage classification. Sensors 23(13):6178
49. Carrillo Gómez JK, Durán Acevedo CM, Ovidio García-Rico R (2021) Methodology for the bacteria detection in drinking water through an e-nose and e-tongue. INGE-CUC 17(1):165
50. Vanaraj R, Bincy IP, Mayakrishnan G, Kim IS, Kim SC (2025) A systematic review of the applications of electronic nose and electronic tongue in food quality assessment and safety. Chemosensors 13(5):161
51. Choubey DK, Kumar M, Shukla V, Tripathi S, Dhandhania VK (2020) Comparative analysis of classification methods with PCA and LDA for diabetes. Curr Diabetes Rev 16(8): 833–850
52. Basak H, Roy A, Lahiri JB, Bose S, Patra S (2021) SVM and ANN based classification of EMG signals by using PCA and LDA. arXiv preprint arXiv:2110.15279
53. Neeraj KN, Maurya V (2020) A review on machine learning (feature selection, classification and clustering) approaches of big data mining in different area of research. J Crit Rev 7(19):2610–2626
54. Sachin D (2015) Dimensionality reduction and classification through PCA and LDA. Int J Comput Appl 122(17):4
55. Mohammadi N, Esteki M, Simal-Gandara J (2024) Machine learning for authentication of black tea from narrow-geographic origins: combination of PCA and PLS with LDA and SVM classifiers. LWT 203:116401
56. Reddy GT, Reddy MPK, Lakshmanna K, Kaluri R, Rajput DS, Srivastava G, Baker T (2020) Analysis of dimensionality reduction techniques on big data. IEEE Access 8: 54776–54788
57. Jiménez-Narváez AD, Vaca VDC, Loor-Duque JJ, Martín IRA, Reyes-Chacón IG, Vizcaíno P, Morocho-Cayamcela ME (2025) Predictive modeling for fetal health: a comparative study of PCA, LDA and KPCA for dimensionality reduction. IEEE Access 13: 59687
58. Egwom OJ, Hassan M, Tanimu JJ, Hamada M, Ogar OM (2022) An LDA–SVM machine learning model for breast cancer classification. BioMedInformatics 2(3):345–358
59. Sachdeva J, Kumar V, Gupta I, Khandelwal N, Ahuja CK (2016) A package-SFERCB-"Segmentation, feature extraction, reduction and classification analysis by both SVM and ANN for brain tumors". Appl Soft Comput 47:151–167
60. Hasan BMS, Abdulazeez AM (2021) A review of principal component analysis algorithm for dimensionality reduction. J Soft Comput Data Min 2(1):20–30
61. Saeed B, Zia-ur-Rehman M, Gilani SO, Amin F, Waris A, Jamil M, Shafique M (2021) Leveraging ANN and LDA classifiers for characterizing different hand movements using emg signals. Arab J Sci Eng 46:1761–1769
62. Pujari D, Yakkundimath R, Byadgi AS (2016) SVM and ANN based classification of plant diseases using feature reduction technique. IJIMAI 3(7):6–14
63. Maione C, Barbosa RM (2019) Recent applications of multivariate data analysis methods in the authentication of rice and the most analyzed parameters: a review. Crit Rev Food Sci Nutr 59(12):1868–1879
64. Hausmann M, Breimann R, Fett M, Kraus B, Schmitt F, Welzbacher P, Kirchner E (2023) Systematic approaches for sensor selection and integration–a systematic literature review. Procedia CIRP 119:687–692
65. Harari GM, Lane ND, Wang R, Crosier BS, Campbell AT, Gosling SD (2016) Using smartphones to collect behavioral data in psychological science: opportunities, practical considerations, and challenges. Perspect Psychol Sci 11(6):838–854
66. Phinyomark A, Campbell E, Scheme E (2019) Surface electromyography (EMG) signal processing, classification, and practical considerations. In: Biomedical signal processing: advances in theory, algorithms and applications. Springer Singapore, Singapore, pp 3–29
67. Swarna S, Kolluru VR (2024) Active channel selection by sensors using artificial neural networks. Int J Electr Electron Res 12(4): 1466–1473
68. Sun X, Kang CH, Kong M, Alkhazragi O, Guo Y, Ouhssain M et al (2020) A review on practical considerations and solutions in underwater wireless optical communication. J Lightwave Technol 38(2):421–431
69. Alahi MEE, Sukkuea A, Tina FW, Nag A, Kurdthongmee W, Suwannarat K, Mukhopadhyay SC (2023) Integration of IoT-enabled technologies and artificial intelligence (AI) for smart city scenario: recent advancements and future trends. Sensors 23(11):5206
70. Mohanty AK, Ahamed S, Kamra R, Junnarkar AA (2023) Challenges and future prospects of

IoT and AI integration in education. In: Progress in language, literature and education research, p 94

71. Gouiza NISSRINE, Jebari HAKIM, Reklaoui KAMAL (2024) Integration of Iot-enabled technologies and artificial intelligence in diverse domains: recent advancements and future trends. J Theor Appl Inf Technol 102(5):1975–2029
72. Knickerbocker JU, Budd R, Dang B, Chen Q, Colgan E, Hung LW et al (2018, May) Heterogeneous integration technology demonstrations for future healthcare, IoT, and AI computing solutions. In: In 2018 IEEE 68th electronic components and technology conference (ECTC). IEEE, pp 1519–1528
73. Bibri SE, Alexandre A, Sharifi A, Krogstie J (2023) Environmentally sustainable smart cities and their converging AI, IoT, and big data technologies and solutions: an integrated approach to an extensive literature review. Energy Inform 6(1):9

Chapter 13

Hyperspectral Imaging for Nondestructive Food Quality Evaluation: A Protocol-Based Study

Rukshana Irani, Suchandra Datta, Amrita Chakraborty, Ayan Chatterjee, Aminu Abdullahi, and Estela F. Diaz

Abstract

Hyperspectral imaging (HSI) is an emerging technology that uses imaging and spectroscopy to obtain spatial and spectral information for each pixel in a food sample. This technique enables comprehensive and non-destructive analysis of food quality. The HSI technique is quite rapid, robust, reliable, and has a systematic stepwise framework. The methodology consisted of sample preparation, system setup, calibration, illumination setup, image acquisition, data storage, image preprocessing, spectral feature extraction, model development, visualisation, and interpretation. All the steps are quite intricate and should be followed cautiously. Hyperspectral images are captured under controlled conditions, and spatial as well as spectral features associated with different attributes of food quality are generated. Multivariate statistical methods or machine learning algorithms are used to relate spectral features to the quality parameters of foods. It applies to a variety of food products, including fruits, vegetables, dairy products, meats, and grains. These techniques can support quality control in the food industry and provide safe food to consumers.

Key words Hyperspectral Imaging, Hypercube, Illumination, Spectral, Spatial, Food quality

1 Introduction

Nowadays, the quality and safety of food have become key factors for the modern food industry, because consumers always prefer products with superior quality at an affordable price. The quality and safety of food are generally explained by different attributes such as physical (texture, colour, marbling, tenderness), chemical (fat content, moisture, protein content, pH, drip loss), and biological (total bacterial count) [1].

Food quality also ensures these products meet the standards associated with nutrition, freshness, and taste, while consistency appeals to consumers. Traditionally, assessment of quality and safety involved human visual inspection and chemical or biological

Tanmay Sarkar (ed.), *Novel Food Analysis: Techniques and Applications*, Methods and Protocols in Food Science,
https://doi.org/10.1007/978-1-0716-4787-5_13,

determination investigation, which were tedious, time-consuming, destructive, and sometimes not environmentally friendly [2].

At present, analysis and assessment of food quality and safety are one of the most crucial aspects of food products in all the processes of the food industry. With recent advancements in computer technology and instrumentation engineering, there have been significant advancements in techniques for the assessment of food quality and safety. As a result, some intrinsic and extrinsic parameter checking methodologies have been developed. These are accurate, fast, real-time, and non-destructive, designed to optimise quality and assure the safety of food [3].

There are chiefly two types of techniques that have been used for the detection of food quality and safety, i.e. destructive and non-destructive. Destructive techniques are defined as those in which the sample is only used for the analysis and cannot be further used for other purposes. The non-destructive processes are those in which the sample can be further used after being analysed. Except for a few spectroscopic techniques like Near-infrared spectroscopy (NIRS) and Fourier transform infrared (FTIR), the majority of traditional and experimental techniques, such as High-performance liquid chromatography (HPLC), Gas chromatography (GC), and Enzyme-linked Immunosorbent Assay (ELISA), are known to be destructive techniques. However, almost all of the destructive techniques are time-consuming, laborious, and have a higher chance of human or machine error. Furthermore, the chemical reactions in these techniques result in the generation of several waste products [4].

It also has a considerable possibility of creating biological hazards. As a result, studies on non-destructive techniques have gained tremendous popularity over the past few years among scientists and researchers. It has been observed that the non-destructive techniques are promising analytical methods for evaluating the safety and quality of food products. Hyperspectral imaging (HSI) is one of the most favourable approaches amongst them. It is a combination of imaging and spectroscopic techniques, and offers advantages over several other methods, like Multispectral Imaging technology (MSI), Near Infrared Spectroscopy (NIRS), and Red-Green-Blue (RGB) imaging [4, 5].

Hyperspectral imaging is a non-destructive technique that appears to have many applications in food safety for validation as well as microbial, chemical, and physical contaminant identification in the food industry. It is a technique of imaging technology that acquires spatial and spectral information of objects ranging from 400 to 2500 nm in the electromagnetic spectrum, generating a three-dimensional data set called a hypercube. It includes various image channels of the objects, each corresponding to a specific wavelength or a narrow range of wavelengths. This wavelength range includes both types of wavelengths, visible and nonvisible [6].

This HSI technique is widely applicable for research and practical applications in different fields due to its ability to detect both visible and non-visible features. It blends image technology with infrared wave analysis in spectroscopy. As a result, the assessment of both internal and external characteristics of materials becomes easier. The acquired hypercubes are processed using different methods to extract efficient wavelengths, calculate image features, and then analyse the extracted features to achieve beneficial results, which helps in decision-making processes [7].

Food quality can be assessed precisely using a hyperspectral imaging system. The spectral reflectance of food items across a wide range of wavelengths can be captured by using this technique. Identification of chemical and physical properties related to moisture, freshness, and the presence of contamination can be carried out using this technique. This technique can also record comprehensive spectral information at each pixel, which produces spatially resolved maps. As a result, it becomes easier to differentiate between distinct food components or other quality parameters. This permits real-time non-destructive analysis without delay. The results are obtained instantly without damaging the product [8].

The major challenge of hyperspectral imaging is the extraction of relevant information from high-dimensional data. It often contains redundant information. Other significant drawbacks include the sensor's noise, variations in illumination, environment, heterogeneity of the sample, and anisotropy. Efficient preprocessing and chemometric techniques can lower the dimensionality of hyperspectral data. Thus, aiding real-time food monitoring using HSI. These techniques reduce computational time, improve the performance of the model, and increase overall effectiveness by eliminating irrelevant or non-informative variables. This paper mainly focuses on establishing a standardised protocol of the HSI technique in food quality assessment. It demonstrates the potential of HSI to detect spoilage, chemical composition, and defects in food products. It also provides a structured framework for image acquisition, data processing, and model development [9].

2 Materials [4, 10]

2.1 Hyperspectral Cameras (Pushbroom, Whiskbroom, or Snapshot Type)/Detectors

The primary detectors generally used in HSI are given below.

1. Charged-coupled device (CCD) Cameras contain silicon sensors.
2. Complementary metal-oxide semiconductor (CMOS) cameras.
3. Indium gallium arsenide (InGaAs) sensors.

2.2 Light Source/ Illumination Unit

1. Tungsten-halogen lamps: 340–2500 nm for UV, visible, and NIR regions.
2. Light emitting diodes (LEDs): 870–980 nm infrared range.
3. Lasers.

   The selection of the illumination source is determined by its application.

2.3 A Stage for Movement/Translation Stage.

2.3.1 Wavelength Dispersion Device

1. Prisms and gratings: Optical elements that disperse incident light into different wavelengths.
2. Imaging spectrograph: It disperses broad-spectrum light into distinct spectral wavelengths.
3. Filters: Bandpass filters that isolate specific wavelength ranges from others.
4. Filter Wheels: Disks that have multiple bandpass filters that rotate to select the required spectral band.
5. Acousto-optic tunable filters (AOTF) based on tellurium dioxide (TeO2) crystals for electronically controlled wavelength tuning.
6. Liquid crystal tunable filters (LCTF) for selective wavelength filtering.

2.3.2 Computer with HSI Software

Examples include ENVI (Environment for Visualising Images, MATLAB, Unscrambler, and Image acquisition software.

1. *Sample holder:* It is made up of aluminium, stainless steel.
2. *Calibration reference panels (white and dark).*
3. *Tripod Stand for Mounting the Camera.*

3 Method

3.1 Step 1: Sample Preparation

The sample surface is cleaned and prepared, and has to be placed on a stable platform. Care should be taken so that no external reflective surfaces interfere with imaging.

There is not much sample preparation in HSI. HSI is not suitable for liquid foods; however, the foods that need to be tested should be converted into a powdered form. The sample preparation is minimal, with no complex preparation involved. The representative sample of the food to be tested is used. It should be free from any external contaminant. The sample is kept in stable, non-reflective holders, i.e., aluminium plates having square wells. Most solid samples can be used directly for analysis without any physical or chemical treatment [4, 11].

3.2 Step 2: System Setup and Calibration

The hyperspectral camera is placed on a tripod stand. The height and angle of the camera should be adjusted properly to capture the region of interest. The illumination source is also adjusted to provide an even and continuous source of light on the sample. Black (zero reflectance) is performed and used as a reference standard. It is achieved either by covering the lens or using a black surface in the imaging area, and the resulting image is captured.

White (near 100% reflectance) reference standards are also estimated to calibrate the HSI system. A white reference panel is used in the imaging area, and then the image is captured.

The detectors in the HSI system can only detect spectral intensity, not the original reflectance, after image acquisition and the formation of hyperspectral images. As a result, calibration is extremely important.

The following equation can be used for the calibration of reflectance.

$$I_c = (I_o - B) \times (100)/(W - B)$$

Here, Ic is the corrected hyperspectral image, Io is the original hyperspectral image, B is the black image with nearly 0% reflectance, and W is a white reference image with approximately 99.9% reflectance. By using the above equation, we will get calibrated hyperspectral images [4, 7, 12].

3.3 Step 3: Illumination Setup

The sample is properly illuminated with an appropriate light source, like arrowband LEDs, broadband tungsten halogen lamps, or tunable lasers. This optimisation is carried out for even and sample-safe lighting. The source of illumination is generally placed at a 45-degree angle to the sample [4].

3.4 Step 4: Image Acquisition

Detailed spectral information is captured using hyperspectral imaging across a broad range of wavelengths for every pixel of the sample.

This step is based on the integration of wavelength dispersion devices and the detectors that lead to the formation of a three-dimensional hypercube. The hypercube consists of two spatial dimensions and one spectral dimension. In this step, wavelength dispersion devices separate the incoming light emitted or reflected from the sample into its constituent wavelengths.

The system uses different modes, which are mentioned below.

1. *Point Scanning or Whisker broom imaging:* It produces single-dimensional spectral data for each measurement. It can successively scan each spatial point, thereby providing high spectral resolution. Here, an area is scanned by a hyperspectral sensor by moving from one point to another point or from one pixel to another pixel. Here, the sensor moves along one axis to obtain spectral information from a single point. It then moves to the next point and repeats the same process [13].

2. *Line Scanning or Pushbroom Imaging*

   The spectral data is obtained along a line. An entire line of an image is recorded by line-scanning rather than a single pixel at a time. Here, a two-dimensional dispersing element and a two-dimensional detector array are used [13, 14].

   The movement of the camera or sensor takes place across the scene, and thus, a line of pixels at each moment is acquired. A shift in line is done, and then the process repeats to form a 2D image [13, 15].

3. *Area Scanning or Snapshot Imaging*: Here, a complete two-dimensional image of an area is captured in a single shot or exposure. The whole frame is captured simultaneously. In other words, we can say that a 2-dimensional image is captured at multiple wavelengths in a single shot simultaneously. There is no need to filter wavelength sequentially or scan line by line [16].

An appropriate spectral range and resolution have to be selected depending on the sample to be analysed. Scanning parameters like exposure time, scan speed, and frame rate have to be set depending on the mode or method used for imaging [10] [17]. The dispersed light is recorded through sensitive cameras. The cameras are equipped with CCD or CMOS sensors. The high quantum efficiency of the detectors ensures an accurate measurement of intensity for each wavelength per pixel [18].

3.5 Step 4: Data Storage and Image Preprocessing

The acquired spectral cube is stored properly. It should be comprised of spatial (x, y) and spectral (λ) dimensions. Preprocessing techniques such as noise reduction, calibration, and correction of illumination are applied to it to enhance the quality of the data. Outliers and dead pixels are removed [12, 17].

Spatial and spectral preprocessing involves the removal of unwanted image artefacts. Preprocessing also consists of unfolding, dead pixel removal, background removal, spike removal, and area of interest (ROI) selection [4].

3.6 Step 5: Spectral Feature Extraction

Spectral signatures from different regions or pixels are extracted. Feature extraction is a very crucial phase of HSI. It transforms high-dimensional spectral spatial data into low low-dimensional representation which contains the most discriminative and informative features. It reduces computational complexity due to the removal of highly redundant data. Many techniques are widely used to convert raw spectral data for analysis. The techniques that are commonly used include Principal Component Analysis (PCA), Independent Component Analysis (ICA), Self-Organising Maps (SOM), Hierarchical Cluster Analysis (HCA), Factor Analysis, Linear Discriminant Analysis (LDA), Partial Least Squares Discriminant Analysis (PLS-DA), Support Vector Machines (SVM),

Artificial Neural Networks (ANN), Random Forests (RF) etc. Relevant spectral features are extracted from the hyperspectral data cube that correspond to chemical and physical characteristics of the food samples. It can be done through dimensionality reduction techniques or spectral signature analysis [4, 10, 19].

3.7 Step 6: Model Development

For model calibration and validation, component concentrations will be examined using conventional techniques such as chemical assays, chromatography, or spectrophotometry. These reference values are used to train models that are capable of relating spectral features to sample properties. Methods for classifying algorithms and pattern recognition are employed to identify and estimate quality parameters of the extracted spectral features. It also detects defects in it. Examples of regression algorithms include Partial Least Squares Regression (PLSR), Principal Component Regression (PCR), and Support Vector Regression (SVR), while classification algorithms include Support Vector Machine (SVM), Random Forest, and Artificial Neural Network (ANN). Statistical or machine learning methods are used for classification or Regression [4, 10].

3.8 Step 7: Visualisation and Interpretation

Classification results obtained from HSI are translated into comprehensive insights regarding food quality.

Classification or chemical maps are generated, displaying the spatial distribution of quality parameters like moisture, sugar content, presence of microbial contaminants, etc.

Spectral data of known objects or foodstuffs are compared with unknown to find out the difference. These comparisons can be used to understand the composition of the sample, assessment of freshness, and detect any contaminant, etc. [10] [20].

4 Notes

HSI combines imaging and spectroscopy techniques, and has been demonstrated to have tremendous potential in the assessment of internal and external properties of materials. External characteristics are assessed using visible reflectance, whereas internal characteristics are determined using near-infrared reflectance from the surface of the objects. These abilities have made the hyperspectral imaging technique successful in applications across different fields, including industry, medicine, agriculture, natural resources, and food [21].

The uses of HIS in the evaluation of food quality and its safety control have been discussed below.

4.1 Quality and Safety Assessment of Fruits and Vegetables

Nowadays, consumers are willing to pay more for prime fruit and vegetables that have guaranteed quality and safety. Hyperspectral imaging has proven its capability in the assessment of the quality and safety of fruit and vegetables. Several factors, such as contamination, bruises, surface defects, starch index, firmness, sugar content, bitter pit, and chilling injury/freeze damage, etc. can be evaluated using HSI [7].

The presence of any defect is one of the key factors which influence the quality and price of fresh fruit and vegetables. Hyperspectral imaging has been used for the detection of surface defects of apple, cherry, and citrus fruits, as well as internal defects of cucumber. This investigation is based on two strategies. The first one is the identification of spectral feature wavelengths of the defect (s) and using these spectral signatures to develop discrimination models. Most of these studies were done in reflectance mode, i.e., within the visible to near-infrared (VIS-NIR) range of (400–1100 nm) [22].

Bruise damage is one of the common external defects that causes quality loss and degradation of fruits and vegetables. These damages usually happen during the harvesting or transportation. Many studies used apples for the detection of bruises using hyperspectral imaging. HIS has been used for apples to assess many quality attributes, such as firmness, soluble solids content, and mealiness. These studies demonstrated that the hyperspectral scattering technique is potentially useful for nondestructive detection of quality attributes of apples. A study also investigated the potential use of hyperspectral imaging to identify citrus canker lesions. Citrus canker is a serious disease that affects most commercial citrus species, caused by the bacterium *Xanthomonas axonopodis.* The hyperspectral images have been processed and classified to differentiate citrus canker lesions from normal healthy peels and other diseased conditions of peels, like greasy spot, insect damage, melanose, scab, and wind scar. The analytical method was based on Spectral Information Divergence (SID), and the overall accuracy was found to be 96% [23].

Hyperspectral imaging has also been used for the detection of bruise damage on white mushrooms and to identify freeze damage of mushrooms using Principal Component Analysis (PCA). The deterioration of the quality of sliced mushrooms during storage is also measured using HSI by analysing moisture content, colour, and texture. These studies showed the potential of HSI in the detection of damage and quality assessment of mushrooms. Taghizadeh et al. (2010) investigated the shelf life using parameters such as weight loss, colour, and maturity index under different packaging of polymer films. This research demonstrated that hyperspectral imaging has potential as an analytical tool for the evaluation of the shelf life of fresh mushrooms. It also indicated that HSI can be used to evaluate the effect of different packaging materials.

Hyperspectral imaging was also used to predict the activity of *polyphenol oxidase* (PPO) in mushrooms within the VIS/NIR range. The findings of this study revealed the possibility of developing a sensor that could rapidly identify mushrooms with a higher risk of enzymatic browning [24].

Decay is another major defect that affects fruit during post-harvest storage and transport. Hyperspectral imaging has been used in the early automatic detection of fungal infections in post-harvest citrus fruits, instead of labourious manual detection carried out by trained workers. Recently, increased attention has been given to selecting optimal spectral features from hyperspectral images. Some researchers have proposed a methodology of the Receiver Operating Characteristic (ROC) curve to identify the most relevant spectral features for identifying decay in citrus fruits caused mainly by two common fungi, i.e. *Penicillium digitatum* and *Penicillium italicum*. The ROC procedure was found to be better than other common feature selection procedures in the identification of spectral features for the detection of decay [25].

Detection of faecal contamination on fruit and vegetables is very important as faeces are the source of many pathogens. Several studies indicate that hyperspectral fluorescence imaging is more capable in the determination of determining trace faecal spots than hyperspectral reflectance imaging. In a recent study, violet 410-nm light-emitting-diode (LED) line lights were selected to provide near-uniform illumination to the linear field of view at a low cost. Besides LEDs, the hyperspectral line-scan imaging system has also been developed. It assembled an electron multiplying charge-coupled device (EMCCD) camera, which is capable of acquiring high-resolution images at high speed under low illumination environments. The algorithms that utilised the fluorescence intensities at four wavebands (680, 684, 720, and 780 nm) have been found to detect more than 99% of the faecal spots [26].

Detection of insect damage in fruit and vegetables is another challenge, as they can cause serious economic loss. Hyperspectral imaging has also been used to identify not only external insect infections, such as in jujube fruit, mango, cucumber, and mung bean, but also internal infections, such as in vegetable soybeans. Most of the studies selected a spectral range of approximately 400–1000 nm for the acquisition of hyperspectral images, except for the study on jujube fruit, which primarily acquired images within the visible range (400–720 nm), and the study on mung bean (1000–1600 nm). Reflectance mode was employed in the analyses of jujube fruit and mung bean, while transmittance mode was utilised in the research on vegetable soybeans. In the study on cucumbers, both reflectance (450–740 nm) and transmittance (740–1000 nm) modes were merged into a single system for image acquisition. The HSI technique offered better detection ability than manual inspection. In the analysis of cucumber overall accuracy of HSI was found to be 82–93%, than the manual inspection, which detected with 75% accuracy. It also gave good detection results for jujube (97.0%), mung bean (82–85%), and soybeans (95.6%) [27].

Mealiness is a texture attribute in fruits characterised by abnormal softness of tissues along with a lack of juiciness. Various studies were conducted to detect mealiness in apples using hyperspectral imaging. The desired result of the hyperspectral images was obtained using the scatter mode in the spectral range of 600–1000 nm. Surface colour and moisture content are also important features to be studied for evaluating the quality of fruit and vegetables. The colour measurements were mostly achieved by obtaining the hyperspectral images with reflectance mode in the range between 400 and 1000 nm, which includes the visible spectral range. The prediction of colour was carried out in two ways. The first one is by using tristimulus values that are calculated from the hyperspectral images, and the other one is using multivariate algorithms such as Partial Least Squares Regression (PLSR) to determine colour attributes. Hyperspectral imaging was found to have a similar ability for colour determination when compared with the traditional colourimeter. HSI is believed to provide more spatial and spectral information about the surface than Red, Green, Blue (RGB) imaging. In addition, hyperspectral imaging in reflectance mode in the range of 400–1000 nm has been used in the determination of the moisture content of bananas [28].

Firmness and soluble solid content (SSC) are two important quality attributes commonly used to determine the maturity and harvest time of fruit and vegetables. Many efforts have been made to use hyperspectral imaging for the determination of the firmness of fruits and vegetables, including apples, bananas, strawberries, blueberries, pears, and peaches. Usually, the hyperspectral or multispectral imaging system is used with the scatter mode for firmness determination [9, 29, 30].

4.2 Quality and Safety Assessment of Meat and Fish

The production of premium-quality meat and meat products is very important for the modern meat industry. The demand for good-quality meat products among consumers is always considered a key factor that determines success in today's highly competitive market. Most of the research associated with hyperspectral imaging has been performed on pork, beef, lamb, and ham. As per the perception of consumers, the most important factor, such as palatability or quality, tenderness of meat is widely desired [31].

Hyperspectral imaging has been used to classify pork based on quality and to also predict pork quality traits such as marbling scores. In a study, NIR hyperspectral imaging (900–1700 nm) was used for the grading and classification of pork quality. Automatic and objective measurement methods were studied for classifying pork in terms of quality, and quality trait prediction was also done by employing advanced image processing techniques such as Gabor filters and the wide line detector. The positive results indicated that the HSI has great potential in meat quality analysis, especially when it is combined with advanced image processing techniques. In an experiment, the determination of tenderness of pork was also carried out by using Warner-Bratzler Shear Force (WBSF) of hyperspectral scattering imaging technique with the spectral range of 400–1100 nm [32].

Water content is a critical parameter of the quality of meat and meat products, and its determination is frequently carried out in the meat industry. A hyperspectral imaging system with the range of 900–1700 nm has been used to determine the water content of lamb, pork, beef, pre-sliced turkey hams, and Spanish cooked ham. Apart from water content, water holding capacity (WHC), commonly measured as drip loss, is another water-related attribute that is not only required for the food-processing industries but also for consumers during the purchase of meat. Hyperspectral imaging in the range (900–1700 nm) has been used to determine WHC/drip loss in beef, pork, and lamb meats. Fat is another important quality criterion for production control from slaughter to final product in the meat industry. In the past few years, hyperspectral imaging has also been applied for the prediction of fat content in meats. In the short-wave near infrared spectral (SW-NIR) region (760 to 1040 nm), an imaging technique has been developed for online measurement of fat content in homogeneous beef and pork trimmings. Apart from assessing fat content, hyperspectral imaging has also been used for evaluating the proportions of total saturated fatty acid (SFA), total unsaturated fatty acid (UFA), and fatty acids like myristic, palmitic, stearic, myristoleic, palmitoleic, oleic and linoleic acids in beef cuts [33].

Meat products with contaminants can cause illness in humans when no automatic inspection protocols are followed, and only human vision detection is done during slaughter and further processing. In the past few years, the USDA's (United States Department of Agriculture) Agricultural Research Service has used hyperspectral and multispectral imaging techniques to determine different contaminants on poultry carcasses. Besides contamination detection, in recent years, the research on food safety inspection using hyperspectral imaging has been extended to the area of microbial spoilage detection. In the past few years, the feasibility of hyperspectral imaging for determining total viable count (TVC) of pork meat has been investigated. Hyperspectral imaging has also been used to determine the microbial spoilage in chicken filets, such as TVC of *Pseudomonas*, *Enterobacteriaceae* and psychrotrophic plate count (PPC) [34].

In addition to the above-mentioned quality and safety attributes, in recent times, hyperspectral imaging has also been applied to evaluate other meat qualities, such as colour, pH, and protein content. Colour is the primary quality attribute of food that significantly affects consumers' acceptance and market acceptance. Besides colour, pH is another important physical characteristic of meat. Hyperspectral imaging with the wavelength range of 900–1700 nm has been studied for colour and pH determination.

Fish and fish products have been highly beneficial to human health. The first application of hyperspectral imaging for the quantitative measurement of fish quality was carried out in 2006 for moisture determination, followed by determining fat distribution. Since 2010, more attributes of fish have been assessed by using hyperspectral imaging technology, which includes colour, moisture, texture, pH, WHC, ice fraction, astaxanthin content, and microbial spoilage. Among these attributes, most of them were well predicted by analysing hyperspectral images. It was also observed that, except for determining colour, other attributes were also measured using hyperspectral imaging systems within the wavelength ranges of visible and short-wave near infrared region (400 nm to 1000 nm) [35].

4.3 Quality and Safety Assessment of Dairy and Dairy Products

The evaluation of compositional quality parameters such as fat, carbohydrates, moisture, and protein contents of dairy products (e.g., cheese, yoghurt, cream, and butter) has been accomplished in a rapid, accurate, and non-destructive way by using near-infrared spectroscopy (NIRS). The spatial distribution of different quality parameters needs to be evaluated. Hyperspectral imaging plays a strategic role in this field of application. Hyperspectral imaging systems can create distribution maps of quality parameters during processing, and they can be applied for monitoring the ingredient uniformity, inconsistencies in processing, and identifying areas of poor mixing, overheating, and microbial spoilage [36].

Furthermore, distribution maps of moisture content generated by HSI can be used as indicators to evaluate the effectiveness of the baking or drying process during the production of certain food products. To demonstrate the potential of hyperspectral imaging in the classification of cheese products, researchers tested a set of 13 commercially produced cheeses containing varying amounts of protein, fat, and carbohydrate using a hyperspectral imaging system (960–1662 nm). The spectral features were used to classify each pixel of the hyperspectral image into one or more groups using the spectral angle mapping algorithm. Cheese samples with the same fat content appeared in the same colour zone in the visualisation map. It was also demonstrated that use of a similar system, but in the range of 950–1650 nm, can detect contaminants (e.g., plastic residues) in cheese samples based on their spectral pattern, which are otherwise invisible to the human eye [13].

4.4 Quality and Safety Assessment of Grains

The major challenge in the grain industry is to classify and identify grains of specific classes or cultivars. Certain classes of grain are used as primary ingredients for some specific products. Therefore, an accurate and rapid classification technique for grains is needed. The first experiment for grain classification was attempted by Archibald et al. (1998) using a shortwave near infrared (SWNIR) hyperspectral imaging system in the wavelength range of 632–1098 nm for distinguishing different cultivars. Later, Mahesh et al. (2008) classified eight types of wheat grown in western Canada with 94–100%, 86–100%, and above 90% accuracy by LDA (Linear Discriminant Analysis), QDA (Quadratic Discriminant Analysis), and ANN (Artificial Neural Networks) models respectively, with the use of an NIR hyperspectral imaging (960–1700 nm) system. Later, the same team developed the classification model using wavelet texture analysis and obtained the highest classification accuracy of 99.1% using by LDA model [37].

Recently, Singh et al. (2009) achieved high classification accuracy (85.0–100.0%) in differentiating healthy from insect-damaged wheat kernels using a NIR hyperspectral imaging (1000–1700 nm). Furthermore, it assessed the potential of short-wave NIR hyperspectral imaging (700–1000 nm) for detecting midge-damaged wheat kernels and obtained a high accuracy of 95.3–99.3% in distinguishing healthy and midge-damaged wheat kernels. Macroscopic near-infrared (NIR) reflectance hyperspectral imaging (950–1700 nm) was developed to analyse the chemical composition of grains. Image and principal component analysis (PCA) were also used to differentiate the kernel germ from endosperm regions within individual kernels. Partial least squares (PLS) analysis has also been used to predict oil or oleic acid concentrations from either germ or from the entire kernel spectra. Compared with traditional spectroscopy, a hyperspectral imaging system has the advantage of enabling analysis of individual kernels because traditional methods are difficult to apply in these conditions due to their heterogeneous structure [38].

For the non-destructive classification of grains based on their hardness, Williams et al. (2009) described an approach for maize kernel classification by using two hyperspectral imaging systems (960–1662 nm and 1000–2498 nm) and multivariate data analysis. In another application related to food safety, a hyperspectral imaging system was tested for detecting aflatoxin mycotoxin produced by *Aspergillus flavus* and *Aspergillus parasiticus* fungi that affect corn, peanuts, and a wide variety of other grain products. The conventional approach to determine aflatoxin contamination is usually performed by the destructive methods that require grain kernels to be ground and then chemically analysed. The preliminary results revealed that hyperspectral imaging has the potential to quantify the levels of aflatoxin contamination in a non-destructive and non-invasive way [39].

In recent years, the application of hyperspectral imaging has shown a promising perspective in food quality and safety assessment. The main advantages of this system have been associated with the ability to obtain one or more images at specific wavelengths in the NIR, MIR, and Vis range. This allows for the detection of individual characteristics or attributes that are directly related to the quality of a wide range of products and raw materials used in the production of foods. Nonetheless, the most attractive benefit of these methods is obtaining spatial, spectral, and multi-constituent information about the sample that helps in the analysis of food. Various vibrational spectroscopy technologies, in addition to infrared (IR), have been documented as fingerprinting methods and have been applied for the analysis of various foods and agricultural goods. Over the past few years, such techniques have been widely reintroduced by several researchers for a variety of applications. The primary advantages of this technique over conventional chemical and chromatographic methods are their timeliness and ease of use in everyday operations [3, 40, 41].

However, many key issues need to be addressed in further studies. These are mainly related to the development of more efficient, reliable, fast, and low-cost modalities that can be applied both in research and practical industrial applications. The large volume of spectral–spatial data is often characterised by several irrelevant or noisy signals that may seriously challenge the processing of the data and the extraction of meaningful information. In general, models are built with limited data sets, acquired under controlled conditions in a laboratory. So, these models may not function properly for new samples measured in the natural or real-world setting. Therefore, it is essential to first rigorously calibrate and validate such models against a wide range of samples and conditions. Extensive industry-wide research is required to develop reliable and efficient models that will function well in real-world scenarios in a robust manner [6].

Furthermore, the high cost of commercial HSI systems limits its application within the food supply chain for quality assessments. Hyperspectral imagery techniques are promising for assessing the composition, physicochemical properties, and sensory characteristics of food. However, the lack of commercial and robust instrumentation, along with a shortfall of academic training, stands as a potential barrier for the worldwide application of these technologies at the industrial level. Apart from it, the potential of hyperspectral imaging for food quality and safety analysis within food processing and packaging is heavily influenced by the sensitivity and resolution of the cameras used and the data processing methods employed [30].

Recent studies indicate that hyperspectral imaging has a high prospect in the quality inspection of a variety of food products. The successful application of hyperspectral imaging can improve quality and safety measurement of finished products to meet consumer expectations. It ensures correct labelling, mentioning quality attributes of the product on retail packaging. This will help consumers to get quality products, and the industries will also be able to produce foods with enhanced quality and safety.

In addition, the hyperspectral image classification methods can be further explored towards classifying complicated images with high accuracy. The collaborations between the academic and industry communities are needed to create customised systems that meet the specific applications at affordable prices. This will enable the food industry to implement the highly capable HSI technology on a large scale and comprehend its full potential in terms of improving food safety and quality. This will lead to reduced food losses throughout the supply chain. Additionally, to simplify data processing and feature identification in the food industry, standardised methods and procedures for raw data processing and feature categorisation must be developed [41].

References

1. Faucitano L, Huff P, Teuscher F, Gariepy C, Wegner J (2005) Application of computer image analysis to measure pork marbling characteristics. Meat Sci 69(3):537–543. https://doi.org/10.1016/j.meatsci.2004.09.010
2. Huang H, Liu L, Ngadi MO (2014) Recent developments in hyperspectral imaging for assessment of food quality and safety. Sensors 14(4):7248–7276. https://doi.org/10.3390/s140407248
3. Nikzadfar M, Rashvand M, Zhang H, Shenfield A, Genovese F, Altieri G, Matera A, Tornese I, Laveglia S, Paterna G, Lovallo C, Mammadov O, Aykanat B, Di Renzo GC (2024) Hyperspectral imaging aiding artificial intelligence: a reliable approach for food qualification and safety. Appl Sci 14(21):9821. https://www.mdpi.com/2076-3417/14/21/9821
4. Patel D, Bhise S, Kapdi SS et al (2024) Non-destructive hyperspectral imaging technology to assess the quality and safety of food: a review. Food Prod Process Nutr 6:69. https://doi.org/10.1186/s43014-024-00246-4
5. Amigo JM (2010) Practical issues of hyperspectral imaging analysis of solid dosage forms. Anal Bioanal Chem 398(1):93–109. https://doi.org/10.1007/s00216-010-3828-z
6. Gurrala KR (2021) Hyperspectral imaging for food quality assessment. Int Res J Eng Sci Technol Innov 7(1):1–14
7. Lorente D, Aleixos N, Gómez J, Cubero S, García OL, Blasco J (2012) Recent advances and applications of hyperspectral imaging for fruit and vegetable quality assessment. Food Bioprocess Technol 5(4):1121–1142. https://doi.org/10.1007/s11947-011-0725-1
8. Bhargava A, Sachdeva A, Sharma K, Alsharif MH, Uthansakul P, Uthansakul M (2024) Hyperspectral imaging and its applications: a review. Heliyon 10(12). https://doi.org/10.1016/j.heliyon.2024.e33208
9. Kheiralipour K, Jayas DS (2024) Current and future applications of hyperspectral imaging in agriculture, nature, and food. Trends Tech Sci Res 7(2):1–9. https://doi.org/10.19080/TTSR.2024.07.555708
10. Gowen AA, O'Donnell CP, Cullen PJ, Downey G, Frias JM (2007) Hyperspectral imaging—an emerging process analytical tool for food quality and safety control. Trends Food Sci Technol 18(12):590–598. https://doi.org/10.1016/j.tifs.2007.06.001
11. Qin J, Burks TF, Kim MS (2013) Chapter 3—Hyperspectral and multispectral imaging for evaluating food safety and quality. In: Sun D-W (ed) Computer vision technology for food quality evaluation. Academic Press, pp 59–89. https://doi.org/10.1016/B978-0-12-388436-1.00003-2

12. Miko (2024) Food quality and composition analysis with hyperspectral imaging. Specim. https://www.specim.com/food-quality-and-composition-analysis-with-hyperspectral-imaging/
13. Elmasry G, Kamruzzaman M, Sun D-W, Allen P (2012) Principles and applications of hyperspectral imaging in quality evaluation of agro-food products: a review. Crit Rev Food Sci Nutr 52(11):999–1023. https://doi.org/10.1080/10408398.2010.518691
14. Liu Z, Yu H, MacGregor JF (2007) Standardization of line-scan NIR imaging systems. J Chemom 21(2):88–95. https://doi.org/10.1002/cem.1065
15. West PC (2025, April 17) The fundamentals of line scan imaging: What it is and when to use it. Vision Systems Design. https://www.vision-systems.com/factory/article/55266728/the-fundamentals-of-line-scan-imaging-part-1-what-it-is-and-when-to-use-it
16. Photonics Media (n.d.) Area scan. Photonics Dictionary. Retrieved May 31, 2025, from https://www.photonics.com/EDU/area_scan/d8300
17. Burger J, Gowen A (2011) Hyperspectral imaging for food quality analysis and control. Elsevier
18. Holman EA, Fang Y, Chen L, DeWeese M, Holman H, Sternberg PW (2020) Autonomous adaptive data acquisition for scanning hyperspectral imaging. Commun Biol 3(1): 1–7. https://doi.org/10.1038/s42003-020-01385-3
19. Shenming Q, Xiang L, Zhihua G (2022) A new hyperspectral image classification method based on spatial-spectral features. Sci Rep 12(1):1–16. https://doi.org/10.1038/s41598-022-05422-5
20. Feng YZ, Sun DW (2012) Application of hyperspectral imaging in food safety inspection and control: a review. Crit Rev Food Sci Nutr 52(11):1039–1058. https://doi.org/10.1080/10408398.2011.651542
21. Crichton SOJ, Kirchner SM, Porley V, Retz S, von Gersdorff G, Hensel O, Weygandt M, Sturm B (2017) Classification of organic beef freshness using VNIR hyperspectral imaging. Meat Sci 129:20–27. https://doi.org/10.1016/j.meatsci.2017.02.005
22. Ariana DP, Lu RF (2010) Evaluation of internal defect and surface color of whole pickles using hyperspectral imaging. J Food Eng 96(4):583–590
23. Baranowski P, Mazurek W, Wozniak J, Majewska U (2012) Detection of early bruises in apples using hyperspectral data and thermal imaging. J Food Eng 110:345–355
24. Taghizadeh M, Gowen AA, O'Donnell CP (2011) Comparison of hyperspectral imaging with conventional RGB imaging for quality evaluation of Agaricus bisporus mushrooms. Biosyst Eng 108(2):191–194
25. Lorente D, Aleixos N, Gómez-Sanchis J, Cubero S, Blasco J (2013) Selection of optimal wavelength features for decay detection in citrus fruit using the ROC curve and neural networks. Food Bioprocess Technol 6:530–541
26. Yang C-C, Kim MS, Kang S, Cho B-K, Chao K, Lefcourt AM et al (2012) Red to far-red multispectral fluorescence image fusion for detection of fecal contamination on apples. J Food Eng 108(2):312–319
27. Kamruzzaman M, Barbin D, ElMasry G, Sun D-W, Allen P (2012a) Potential of hyperspectral imaging and pattern recognition for categorization and authentication of red meat. Innovative Food Sci Emerg Technol 16:316–325
28. Rajkumar P, Wang N, Eimasry G, Raghavan GSV, Gariepy Y (2012) Studies on banana fruit quality and maturity stages using hyperspectral imaging. J Food Eng 108(1):194–200
29. Leiva-Valenzuela GA, Lu R, Aguilera JM (2012) Prediction of firmness and soluble solids content of blueberries using hyperspectral reflectance imaging. J Food Eng 115(1): 91–98
30. Wu D, Sun D (2013) Advanced applications of hyperspectral imaging technology for food quality and safety analysis and assessment: a review — part II: applications. Innovative Food Sci Emerg Technol 19:15–28. https://doi.org/10.1016/j.ifset.2013.04.016
31. Wu JH, Peng YK, Chen JJ, Wang W, Gao XD, Huang H (2010) Study of spatially resolved hyperspectral scattering images for assessing beef quality characteristics. Spectrosc Spectr Anal 30(7):1815–1819
32. Talens P, Mora L, Morsy N, Barbin DF, ElMasry G, Sun D-W (2013) Prediction of water and protein contents and quality classification of Spanish cooked ham using NIR hyperspectral imaging. J Food Eng 117(3): 272–280
33. Kobayashi K, Matsui Y, Maebuchi Y, Toyota T, Nakauchi S (2010) Near infrared spectroscopy and hyperspectral imaging for prediction and visualisation of fat and fatty acid content in intact raw beef cuts. J Near Infrared Spectrosc 18(5):301–315
34. Barbin DF, Elmasry G, Sun D-W, Allen P (2012) Near-infrared hyperspectral imaging

for grading and classification of pork. Meat Sci 90(1):259–268

35. Sivertsen AH, Heia K, Hindberg K, Godtliebsen F (2012) Automatic nematode detection in cod fillets (Gadus morhua L.) by hyperspectral imaging. J Food Eng 111(4):675–681
36. Barreto A, Cruz-Tirado JP, Siche R, Quevedo R (2018) Determination of starch content in adulterated fresh cheese using hyperspectral imaging. Food Biosci 21:14–19
37. Burger J, Geladi P (2006a) Hyperspectral NIR imaging for calibration and prediction: a comparison between image and spectrometer data for studying organic and biological samples. Analyst 131:1152–1160
38. Singh CB, Jayas DS, Paliwal J, White NDG (2010) Detection of midge-damaged wheat kernels using short-wave near-infrared hyperspectral and digital color imaging. Biosyst Eng 105:380–387
39. Williams P, Geladi P, Fox G, Manley M (2009) Maize kernel hardness classification by near infrared (NIR) hyperspectral imaging and multivariate data analysis. Anal Chim Acta 653: 121–130
40. Yao K, Sun J, Chen C, Xu M, Zhou X, Cao Y, Tian Y (2022) Non-destructive detection of egg qualities based on hyperspectral imaging. J Food Eng 325:111024
41. Lv W, Wang X (2020) Overview of hyperspectral image classification. J Sens 2020(1): 4817234. https://doi.org/10.1155/2020/4817234

Chapter 14

Microbial Fermentation Profiling: Tools and Techniques for Food Analysis

Selva Kumar T, S. Saranyadevi, Aminu Abdullahi, and Balakrishnan Navina

Abstract

Food manufacturing relies heavily on microbial fermentation, which affects nutritional qualities, shelf life, flavor, and texture. A crucial method for understanding and optimizing fermentation processes in various food systems is microbial fermentation profiling. The main instruments and methods for characterizing microbial communities and their metabolic processes in fermented foods are examined in this chapter. An introduction to microbial fermentation and its importance in food processing opens the chapter. It draws attention to important microbial groups that support various fermentation processes, such as lactic acid bacteria, yeasts, and acetic acid bacteria. The role of advanced analytical tools in identifying and characterizing microbial communities is reviewed, encompassing both conventional culture-based approaches and state-of-the-art spectroscopic, chromatographic, and molecular biology techniques. The chapter also explores omics-based methodologies that offer a thorough grasp of microbial interactions and metabolic outputs, including transcriptomics, proteomics, metabolomics, and genomics. The potential of real-time monitoring technologies, like biosensors and AI-driven automation, to improve fermentation control and guarantee product uniformity is investigated. Concerns like safety, regulations, and quality control methods related to fermented foods are also covered in this chapter. Case studies on vegetable fermentations, dairy, cereal-based products, and fermented drinks demonstrate applications of microbial profiling in the food business. Lastly, new developments in sustainable fermentation techniques, machine learning, and synthetic biology are explored, opening the door for more advancements. For researchers, food technologists, and industry professionals looking to maximize microbial fermentation for better food quality and safety, this chapter offers an extensive resource.

Key words Microbial fermentation, Fermentation profiling, Omics technologies, Food microbiology, Metabolic pathways, Quality control, Analytical techniques

1 Introduction

1.1 Microbial Fermentation Profiling

Microbial fermentation profiling encompasses the in-depth analysis of microbial populations, patterning of metabolites, and the subsequent change of the fermenting system's biochemistry. This technique plays a crucial role in the unveiling of the complex

Tanmay Sarkar (ed.), *Novel Food Analysis: Techniques and Applications*, Methods and Protocols in Food Science,
https://doi.org/10.1007/978-1-0716-4787-5_14,

interaction of microbes and the resulting end fermentation product [1, 2]. What makes microbial fermentation profiling crucial is the feasibility of describing the functional and dynamic parameters of the microbial community in fermenting systems. Using advanced technology in the form of high-throughput sequencing, metabarcoding, and metabolomics, it is possible to define the dominant microbe species, metabolic pathways, and resulting products of biochemistry that confer the fermented products their organoleptic characteristics and nutritional worth [3, 4]. One of the more fascinating aspects of microbial fermentation profiling is identifying the spatial and temporal heterogeneity of fermenting systems. For instance, it was revealed by studies that microbial community and metabolite profile may vary considerably along different levels of fermented products and different levels of the fermenting period [5]. Such heterogeneity is controlled by comprehensive factors in the manner of environmental conditions, microbial interaction, and artificial manipulation, whereby the ferment ecosystem's complexity occurs [4, 5]. In addition, microbial fermentation profiling also showed the necessity of identifying the strain-level differences of microbial populations. These minute differences in genes have a tremendous effect on the fermenting system and quality of the product [6]. Through the identification and characterization of specific strains, it is possible to design more precise approaches in optimizing fermenting processes and product uniformity [7].

1.1.1 Microbial Fermentation in Food Production

Having several advantages and uses, microbial fermentation is a vital stage in the manufacture of food. It has always been used since ancient times to make staple foods like bread and cheese; in the last decades, it has attracted great attention in the manufacture of a range of food products [8]. Apart from the rise in the quality, taste, and shelf life of the food, it also helps to create healthier foods in the face of growing knowledge of nutrition and health [9, 10]. In traditional fermented foods, lactic acid bacteria (LAB) have a unique influence on the quality, safety, and flavour evolution [11]. Production of fermented drinks, including vinegar, kombucha, and lambic beer, also depends on other microorganisms, including acetic acid bacteria (AAB) [12]. In the manufacture of fermented drinks and foods, yeasts also play a preliminary part and are responsible for the release of volatile organic compounds (VOCs), which shape the sensory profile [13]. Curiously, recent research has concentrated on new bacteria for technical fermentation, including *Clostridium butyricum* and *Akkermansia muciniphila*, which could be quite important for future product manufacture [9]. Furthermore, high-throughput sequencing techniques such as metagenomics have enabled further elucidation of microbial communities associated with fermented foods, which benefits knowledge in production practices and potential health benefits.

1.1.2 Applications in Food Quality and Safety

Microbial fermentation processes have many beneficial applications for food safety and ensuring quality. For example, fermentation may enhance the bioavailability of nutrients, reduce the number of undesirable substances, and modify the chemical makeup of food, increasing the nutritional profile of a plant source of protein [14]. Aside from food safety by inhibiting pathogenic bacteria and shelf-life extension by producing antibacterial compounds [3], the fermentation process is also important for quality and flavor advancement in traditional fermented foods [5, 15]. LAB and AAB are the microbes responsible, and through the conversion of metabolites in conjunction with other microorganisms, LABs and AABs contribute to flavor development in, and tastes of fermented foods. Using synthetic microbial communities (SynComs) has also been a good way to simplify the complexity of microbial systems and customize the microbial community to the specific needs of the food industry for both quality and safety improvements [16].

1.2 Key Microbial Players in Fermentation

Microorganisms that represent fermentation-based products, as well as indicating what the fermented foods and beverages will ultimately exist in many classes. LABs are indispensable in many fermentation processes as they contribute significantly to the latter in terms of flavor production, preservation of the product, and possibly contribute to health benefits. The most common LAB species are usually *Lactobacillus*, *Leuconostoc*, *Lactococcus*, and especially *Streptococcus* [17]. LAB communities are commonly monospecific in indigenous drinks, such as Peruvian chicha de jora, with *Lactobacillus* typically being the most identified genus [17]. LAB contributes to the nutritional quality of fermented food products, which greatly improves shelf life [18].

1.2.1 Yeasts and Fungi

Yeasts, mainly Saccharomyces cerevisiae, are the main contributors of alcoholic fermentation for many products, such as wine and beer [19]. Non-Saccharomyces yeasts were once seen as undesirable, but now communities value them for their potential to improve product quality and their biocontrol potential [19]. In the case of kombucha and similar fermentations, yeasts like *Brettanomyces bruxellensis* and *Zygosaccharomyces parabailii* are important to produce volatile compounds, which add complexity to the overall flavor [20].

1.2.2 Acetic Acid Bacteria

AAB are essential to producing vinegar and some acidic fermented foods, providing safety, nutritional, and sensory elements through acetic acid fermentation [15]. The AAB genera of importance are primarily *Acetobacter* and *Gluconobacter*, and these two genera are seen in vinegar, kombucha, and some beers [15, 21].

1.2.3 Beneficial Microorganisms

Other potential microbial contributors may include enterobacteria, Bacillus species, and molds. Depending on circumstances, these microorganisms may also be contributors to fermentation or product characteristics of certain fermentations [18, 21]. For example, molds may be involved with product characteristics when maize undergoes fermentation [18]. These interactions can be complex given the variety of microbial groups (LAB, yeasts, AAB) and regions and states, as well as by strain, nutritional condition, duration of fermentation, etc. [15]. Considerations of these interactions would certainly aid the ability to control or optimize fermentation processes that lead to the product characteristics sought. Examination of the contribution of microbial groups, LAB, yeasts, and AAB, for example, was essential in developing a stable fermented product that is helpful [20].

2 Materials

2.1 Laboratory Equipments

Autoclave—for sterilizing media and equipment

Incubators—for controlled microbial growth conditions

Laminar flow hood—for aseptic work

Fermenters/Bioreactors—for controlled fermentation experiments

pH meters and DO meters—for monitoring fermentation conditions

Centrifuge—for sample preparation and cell separation

Spectrophotometer—for microbial growth and metabolite analysis

Microscope—for microbial morphology observation

2.2 Microbial Cultures and Media

Pure cultures of yeasts, lactic acid bacteria, molds (e.g., *Saccharomyces cerevisiae*, *Lactobacillus* spp., *Aspergillus* spp.)

Selective and differential media for microbial isolation and identification.

Fermentation substrates (milk, vegetables, grains, sugar solutions).

3 Methods

3.1 Metabolic Pathways in Fermentation

3.1.1 Glycolysis and Lactic Acid Fermentation

Cells can obtain energy from glucose through two interrelated metabolic mechanisms called glycolysis and lactic acid fermentation, particularly in anaerobic (oxygen-limited) environments. The breakdown of glucose begins with glycolysis, which occurs in almost all cells' cytoplasm. Through a sequence of 10 enzyme-catalyzed events, this pathway transforms one glucose molecule, a six-carbon sugar, into two pyruvate molecules, a three-carbon substance. Glycolysis results in the reduction of two molecules of

NAD^+ (nicotinamide adenine dinucleotide) to NADH and the net production of two ATP (adenosine triphosphate) molecules via substrate-level phosphorylation [22]. Pyruvate normally reaches the mitochondria in the presence of oxygen to undergo oxidative phosphorylation and further oxidation in the citric acid cycle. But in anaerobic environments, like working muscle tissue or some microbes, oxygen is either non-existent or very limited, which hinders the mitochondria's ability to function effectively. Cells use fermentation pathways to replenish the NAD^+ required for glycolysis to proceed. Lactic acid fermentation is a typical anaerobic route where the enzyme lactate dehydrogenase oxidizes NADH back to NAD^+ and converts pyruvate to lactate [23]. Despite the absence of oxygen, this process guarantees a steady flow of NAD^+, enabling glycolysis to continue and provide the cell with ATP. Although less effective than aerobic respiration, this fermentation route is necessary for short bursts of energy and survival in anaerobic settings.

3.1.2 Alcoholic Fermentation

In the absence of oxygen, some microorganisms, especially yeast (*Saccharomyces cerevisiae*), turn glucose into ethanol and carbon dioxide through a biological process known as alcoholic fermentation. The universal metabolic pathway known as glycolysis starts this process by converting one glucose molecule, a six-carbon sugar, into two pyruvate molecules, a three-carbon substance, producing two ATP molecules and two NADH molecules as a net gain. Cells must produce NAD^+ from NADH to maintain glycolysis in anaerobic circumstances, where oxidative phosphorylation is not possible. Pyruvate is converted in two steps during alcoholic fermentation [24]. Acetaldehyde, a two-carbon molecule, is first created when the enzyme pyruvate decarboxylase extracts a carbon dioxide molecule from each pyruvate. Then, by taking electrons from NADH, the enzyme alcohol dehydrogenase converts acetaldehyde to ethanol while also replenishing NAD^+. Glycolysis needs this NAD^+ recycling to continue generating ATP in anaerobic environments. Yeast utilizes alcoholic fermentation to produce alcoholic beverages (where ethanol is the desired product) and bread (where the dough rises due to the carbon dioxide). Fermentation is an essential process for energy production in oxygen-deficient conditions, even though its energy yield is much lower than that of aerobic respiration [25].

3.1.3 Acetic Acid and Mixed Acid Fermentation

Two significant anaerobic microbial processes that transform carbohydrates into different organic acids and other products are acetic acid fermentation and mixed acid fermentation. Acetobacter and Gluconobacter species are aerobic or microaerophilic bacteria that are mainly responsible for acetic acid production. In this process, ethanol, which is typically made by yeasts from glucose during the first fermentation step, oxidizes to acetic acid when oxygen is present. Alcohol dehydrogenase and aldehyde dehydrogenase enzymes, respectively, catalyze the two-step process that converts ethanol to acetaldehyde and then to acetic acid [26]. This kind of fermentation is frequently employed in the production of vinegar and stands out from other types of fermentation due to its oxygen requirement.

Conversely, facultative anaerobic bacteria like *Escherichia coli* and other intestinal bacteria perform the strictly anaerobic process of mixed acid fermentation. Lactic acid, acetic acid, succinic acid, ethanol, carbon dioxide, and hydrogen gas are among the complex combination of end products produced by the fermentation of glucose or other carbohydrates in this process. Utilized as a diagnostic technique in microbiology to identify bacteria based on their fermentation patterns (e.g. the methyl red test), the mixed acid fermentation process is important in microbial ecology [27]. Although the mixed acid fermentation pathway produces less ATP than aerobic respiration, its metabolic variety enables bacteria to maximize energy extraction from substrates in oxygen-limited settings. All things considered, acetic acid and mixed acid fermentation are essential to many industrial processes and are vital to both natural and artificial microbial communities.

3.1.4 Secondary Metabolites Role in Flavor

Although they are not directly involved in the growth, development, or reproduction of microorganisms, secondary metabolites—organic compounds produced by the microbes during fermentation are essential in determining the flavor, aroma, and mouthfeel of fermented foods and beverages. Secondary metabolites, as opposed to main metabolites like ethanol, lactic acid, or acetic acid, are frequently generated later in the fermentation process and smaller amounts. Esters, ketones, aldehydes, higher alcohols, phenolic compounds, sulfur compounds, and different organic acids are some examples of these substances [28]. Different flavor notes are contributed by each kind of secondary metabolite; for instance, phenolic chemicals like 4-ethylphenol can lend smokey or spicy overtones, while esters like ethyl acetate impart fruity and floral scents. The microbial species involved (such as yeasts, lactic acid bacteria, and molds), substrate composition, fermentation temperature, oxygen availability, and fermentation length are some of the variables that affect the formation of these metabolites. Yeasts like Saccharomyces cerevisiae are especially well-known for generating a diverse range of volatile chemicals during alcoholic fermentations, which contribute to the complexity and personality of wines, beers, and spirits [29]. Similar to this, lactic acid bacteria produce diacetyl, acetoin, and other by-products during dairy and vegetable fermentations, which help to provide buttery, tangy, or earthy flavors. Secondary metabolites are highly prized in both conventional and commercial fermentation procedures because they are crucial to the uniqueness and caliber of fermented goods.

3.2 Analytical Tools for Microbial Fermentation Profiling

3.2.1 Culture-Based Methods

To evaluate the metabolic activities and contributions of microorganisms to fermentation, culture-based techniques for microbial fermentation profiling rely on the growth and isolation of these organisms on selective or differential media. Because of their affordability, ease of use, and capacity to offer direct insights into living, active microbial communities, these conventional methods continue to be fundamental in microbiology. Analytical tools used in culture-based methods include biochemical assays, such as sugar fermentation tests, catalase and oxidase tests, and IMViC tests (Indole, Methyl Red, Voges-Proskauer, and Citrate), which help characterize the metabolic capabilities of isolated strains; and colony morphology analysis, which provides preliminary identification through visual inspection of the size, shape, color, and texture of colonies [2]. While gas collection methods like Durham tubes track the evolution of gases, indicating fermentation, pH indicators in media can identify the generation of acid. Furthermore, species are frequently distinguished using chromogenic or fluorogenic media according to certain enzyme activity. Culture-based techniques can also be used in conjunction with microscopy, Gram staining, and quantitative plate counts to assess microbial load and community composition to profile microbial communities in greater depth. Culture-based profiling is still very useful for functional research, strain separation, and fermentation metabolic trait validation, even if it may understate diversity because some bacteria cannot be grown in vitro [30]. These approaches improve our comprehension and control of microbial ecosystems for safer and higher-quality fermentation when combined with molecular techniques.

3.2.2 Microscopy and Imaging Techniques

To analyze and track microbial activity and structural changes during the fermentation process, microscopy and imaging techniques are essential analytical tools. These methods make it possible to see microbial cells, biofilms, spores, and morphological changes brought on by growth, metabolism, and stress reactions up close. Light microscopy is frequently used for routine microbial morphology observation, cell count estimation, and basic staining methods like Gram staining, which aid in bacterial type differentiation. Phase-contrast and differential interference contrast (DIC) microscopy offer improved viewing of live, unstained cells for more in-depth structural investigation. This allows for real-time monitoring of cell division, motility, and aggregation during fermentation. Frequently used in conjunction with fluorescent dyes or genetically modified reporter proteins (like GFP), fluorescence microscopy enables the precise labeling of cellular constituents, monitoring of metabolic conditions, and evaluation of viability using live/dead staining methods [31]. More sophisticated instruments, such as confocal laser scanning microscopy (CLSM), provide three-dimensional, high-resolution imaging of microbial communities, particularly biofilms and co-cultures in intricate fermentation settings. Transmission electron microscopy (TEM) and scanning electron microscopy (SEM) offer nanometer-scale resolution pictures for ultrastructural investigations, exposing internal components, cell surface structures, and microbial-substrate interactions. Monitoring microbial behavior, contamination problems, morphological consistency, and fermentation optimization is all made possible by these imaging techniques. They are crucial instruments in both research and commercial fermentation operations because, when paired with image analysis software, they provide a quantitative assessment of cell size, shape, clustering, and spatial distribution within the fermentation matrix [32].

3.2.3 Spectroscopic and Chromatographic Methods

Analytical tools such as spectroscopy and chromatography are crucial for tracking and describing the intricate biochemical alterations that take place throughout the fermentation process. By accurately identifying and measuring substrates, metabolites, by-products, and other chemical components, these techniques offer vital information about the effectiveness of fermentation, the caliber of the final product, and microbial metabolism. Biomass content, enzyme activity, and the presence of substances based on their absorbance or emission properties are all frequently measured using spectroscopic techniques, such as UV–visible spectroscopy [33]. Key fermentation products like ethanol, lactic acid, and volatile organic compounds can be detected thanks to infrared (IR) spectroscopy, especially Fourier-transform infrared (FTIR) spectroscopy, which makes it possible to identify functional groups in organic molecules. Without damaging the sample, nuclear magnetic resonance (NMR) spectroscopy can be used to track metabolic fluxes in real time and provides comprehensive structural information on fermentation metabolites [34]. Table 1 exemplifies the spectroscopic methods used and their applications in the fermentation process.

Chromatographic methods, on the other hand, offer high-resolution separation and quantification of distinct elements within a complicated fermentation mixture. Particularly in alcoholic fermentations, gas chromatography (GC) is frequently employed for the analysis of volatile substances, including ethanol, esters, and organic acids. It makes it possible to identify flavor and aroma molecules with sensitivity and specificity when combined with mass spectrometry (GC–MS). Sugars, amino acids, organic acids, and phenolic compounds are examples of non-volatile materials that are commonly analyzed using high-performance liquid chromatography (HPLC). Lactic and acetic acids are examples of ionic species that can be detected and measured using ion chromatography (IC). Because they provide repeatable, precise, and comprehensive chemical profiles of fermentation systems, these methods are essential in quality control, process optimization, and research contexts. A complete toolkit for comprehending the chemical

Table 1
Different spectroscopic

Method type	Technique	Principle	Application in fermentation	Advantages	Limitations
Spectroscopic	UV–vis spectroscopy	Absorption of UV/visible light by molecules	Biomass, enzyme, metabolite quantification	Fast, non-destructive, simple setup	Limited specificity, interference from other compounds
	IR (FTIR) spectroscopy	Infrared light absorption by molecular vibrations	Monitoring metabolic profiles, substrate changes	Non-invasive, real-time monitoring	Requires calibration, water interference
	Raman spectroscopy	Inelastic scattering of monochromatic light (Raman effect)	Real-time monitoring of metabolites, substrates	Real-time, minimal sample prep, works in aqueous environments	Weak signal, fluorescence interference
	NMR spectroscopy	Magnetic resonance of nuclei in a magnetic field	Identifying and quantifying intracellular metabolites	Non-destructive, detailed structural info	Expensive, high expertise required
	Fluorescence spectroscopy	Emission of light by a substance that has absorbed light	Detecting NADH, proteins, and specific metabolites	High sensitivity, selective	Requires fluorophores or intrinsic fluorescence

landscape of fermentation and guaranteeing the uniformity and safety of the finished product is offered by the combination of spectroscopic and chromatographic techniques [35].

3.2.4 Molecular Biology Approaches

Molecular biology approaches are a broad category of theoretical and experimental methods used to investigate, control, and comprehend the molecular foundations of biological processes. These methods mostly concentrate on how proteins, RNA, and DNA interact and how these interactions affect how cells function. The polymerase chain reaction (PCR), one of the fundamental methods, enables the amplification of DNA sequences, facilitating in-depth study even from small samples. Gel electrophoresis is another popular technique that helps separate proteins or nucleic acids according to size and charge, giving information on their abundance, size, and purity. Gene cloning, the development of genetically modified creatures, and the manufacturing of medicinal proteins like insulin have all been made possible by recombinant DNA technology, which joins DNA from various organisms [36]. Genomics has been transformed by DNA sequencing techniques, especially next-generation sequencing (NGS), which enables researchers to quickly and affordably decode whole genomes. Furthermore, CRISPR-Cas9 genome editing has developed into a robust instrument for targeted gene change, with enormous promise for both scientific and medical uses. Functional investigations of gene expression and regulation are made possible using Northern, Southern, and Western blotting techniques to identify certain proteins or nucleic acids. Additionally, microarrays and real-time PCR (qPCR) are used to measure the levels of gene expression under different circumstances. When taken as a whole, these molecular biology techniques offer crucial resources for comprehending the mechanisms underlying disease, creating novel diagnostics and therapies, and promoting genetic engineering and biotechnology [37].

3.3 Omics-Based Techniques for Fermentation Analysis

3.3.1 Genomics for Microbial Identification

Because it makes it possible to precisely characterize and track the microbial communities engaged in the fermentation process, genomics for microbial identification is essential to fermentation studies. While genomic approaches offer thorough and high-resolution insights, traditional microbial identification techniques, such as culturing and biochemical testing, frequently fail to uncover non-culturable or low-abundance species. Whole genome sequencing (WGS) is a crucial method that enables the thorough examination of a microorganism's genetic material, allowing for species-level identification and the discovery of functional genes in charge of metabolic pathways crucial to fermentation. Furthermore, the whole range of microbial diversity in a fermenting system can be revealed through the direct examination of genetic material from environmental samples, known as metagenomic sequencing, which eliminates the requirement for culturing and allows the study of complicated microbial communities [38]. Methods such as 16S rRNA gene sequencing are frequently employed for phylogenetic analysis and bacterial identification, providing a trustworthy way to monitor microbial population changes over time. These genomic methods can detect contamination, monitor microbial succession, and ensure the existence of beneficial bacteria, which are crucial for constant product quality and safety in industrial fermentation. Furthermore, the selection or creation of microbial strains with desired characteristics, such as increased metabolite synthesis or stress resistance, can be guided by genetic data [39]. Researchers and producers can enhance yields, preserve the integrity of microbial cultures utilized in the food, beverage, pharmaceutical, and biofuel industries, and optimize fermentation settings by incorporating genomics into fermentation studies.

3.3.2 Transcriptomics and Gene Expression

It is a vital field of molecular biology that focuses on the thorough examination of RNA transcripts generated by the genome under certain circumstances, providing information about gene activity and regulation. Transcriptomics depicts the dynamic aspect of gene expression, indicating which genes are activated or inactive in response to different biological stimuli or environmental changes, in contrast to genomics, which displays the static genetic code. RNA sequencing (RNA-Seq), a potent next-generation sequencing (NGS) technique that allows for the very accurate and sensitive quantification and comparison of transcript levels across several samples or periods, is the primary tool in transcriptomics [40]. Using this method, scientists might find non-coding RNAs that might have important regulatory functions, new transcripts, alternative splicing events, and genes that are differentially expressed. Scientists can gain a better understanding of developmental phases, disease progression, cellular responses to stress, and the effects of genetic changes by examining the transcriptome. Transcriptomics is utilized in practical domains including biotechnology and medicine to optimize metabolic pathways in created organisms, find biomarkers, and create tailored medicines. Moreover, a systems-level knowledge of biological processes is made possible by combining transcriptome data with other omics technologies like proteomics and metabolomics. All things considered, transcriptomics is a vital tool for clarifying the intricacies of functional genomics and gene control [41].

3.3.3 Proteomics: Profiling Microbial Proteins

It is a potent method that concentrates on the extensive investigation of proteins in microbes, offering vital information on their composition, roles, and interactions. Proteomics studies the dynamic expression and changes of proteins, which are the functional by-products of genes and are essential to nearly all biological functions, in contrast to genomics, which studies genetic material.

Proteomics aids in identifying the proteins in microbial systems that are directly involved in pathogenicity, metabolism, stress responses, and other important microbial processes. Mass spectrometry (MS), especially liquid chromatography–tandem mass spectrometry (LC–MS/MS), is the most widely used technique for proteome profiling because it enables the high-throughput identification and quantification of thousands of proteins in a single experiment [42]. With this method, proteins—even those with extremely low abundances—can be found in environmental samples and complicated microbial communities. Furthermore, even though MS-based methods have essentially replaced 2-dimensional gel electrophoresis (2-DE), 2-DE is still often employed in proteomics to separate proteins according to their molecular weight and isoelectric point. The study of post-translational modifications (PTMs), such as phosphorylation, acetylation, or glycosylation, is another aspect of proteomics. These PTMs have a significant impact on protein activity, location, and interactions, which in turn control microbial behaviour. Proteomics can be used in fermentation to track the production of metabolites, identify stress signals, monitor the physiological status of microbial cultures, and enhance the performance of industrial microorganisms [43]. Researchers can learn a great deal about metabolic pathways, microbial adaptability, and possible targets for genetic or biochemical engineering by studying the proteome. Proteomics, in the end, offers a thorough, real-time picture of microbial activity, facilitating improved fermentation process optimization, increased product yields, and the creation of new microbial strains for a range of uses in biotechnology, food production, and medicine.

3.3.4 Metabolomics: Assessing Fermentation By-products

A vital analytical method for examining the entire collection of metabolites found in a biological system, Metabolomics: Assessing fermentation by-products offers a real-time picture of cellular metabolism and the biochemical processes taking place during fermentation. Numerous environmental conditions, such as temperature, pH, oxygen levels, and nutrient availability, affect fermentation, a metabolic process by which bacteria transform substrates into energy-rich by-products like ethanol, lactic acid, or gases. By providing information on the metabolic pathways that microorganisms use during fermentation, metabolomics aids in the monitoring and evaluation of these by-products [44]. Nuclear magnetic resonance (NMR) spectroscopy and mass spectrometry (MS) are the most widely utilized methods for metabolomic analysis because they provide very sensitive and accurate identification and quantification of a broad variety of metabolites. Liquid chromatography–mass spectrometry (LC–MS) is better suited for polar metabolites like sugars, amino acids, and nucleotides, whereas gas chromatography-mass spectrometry (GC–MS) is commonly used for volatile metabolites like ethanol and organic acids. Metabolomics can detect metabolic changes, pinpoint biosynthetic pathway bottlenecks, and evaluate the effectiveness of microbial strains employed in industrial fermentation processes by examining the profiles of fermentation byproducts. Optimizing fermentation conditions, increasing yield, and reducing undesirable byproducts that could impede the intended production can all be made possible with the help of this information (Fig. 1). Metabolomics also makes it possible to find new biomarkers or pathways that can be modified to produce more lucrative metabolites, including food additives, antibiotics, and biofuels [45]. Metabolomics is a vital tool for improving fermentation processes, boosting productivity, and spurring innovation in industrial microbiology and biotechnology by offering a thorough and dynamic understanding of microbial metabolism.

3.4 Real-Time Monitoring of Fermentation

3.4.1 Biosensors for Fermentation Tracking

During fermentation processes, this cutting-edge analytical equipment allows for the real-time monitoring of critical biochemical parameters, guaranteeing increased control, effectiveness, and product consistency. A transducer that transforms the biological reaction into a quantifiable electrical, optical, or thermal signal is usually combined with a biological recognition element, such as enzymes, antibodies, or microbial cells, to form a biosensor. Important factors such as pH, glucose, ethanol, organic acids, dissolved oxygen, and other metabolic byproducts are detected and measured during fermentation using biosensors. A glucose biosensor that uses the enzyme glucose oxidase, for instance, can track sugar intake by microorganisms constantly and provide real-time feedback on substrate use [46]. Similarly, real-time tracking of alcohol synthesis is possible using ethanol biosensors, which is especially useful in sectors like pharmaceuticals, biofuel production, and brewing. While electrochemical biosensors are valued for their portability, quick reaction, and affordability, optical biosensors, such as fluorescence and surface plasmon resonance (SPR)-based systems, offer non-invasive and extremely sensitive detection. Continuous data collection made possible by the incorporation of biosensors into automated fermentation systems enables dynamic modifications to fermentation parameters like temperature, aeration, or nutrient supply. This real-time feedback loop streamlines the production process and lowers costs by drastically decreasing the need for laboratory testing and manual sampling [47]. Furthermore, more reliable and selective biosensors designed for industrial applications have been developed because of developments in synthetic biology and nanotechnology. All things considered, by providing accurate, quick, and real-time insights into microbial activity and process efficiency, biosensors are revolutionizing fermentation monitoring and improving productivity and product quality across a variety of biotechnological applications.

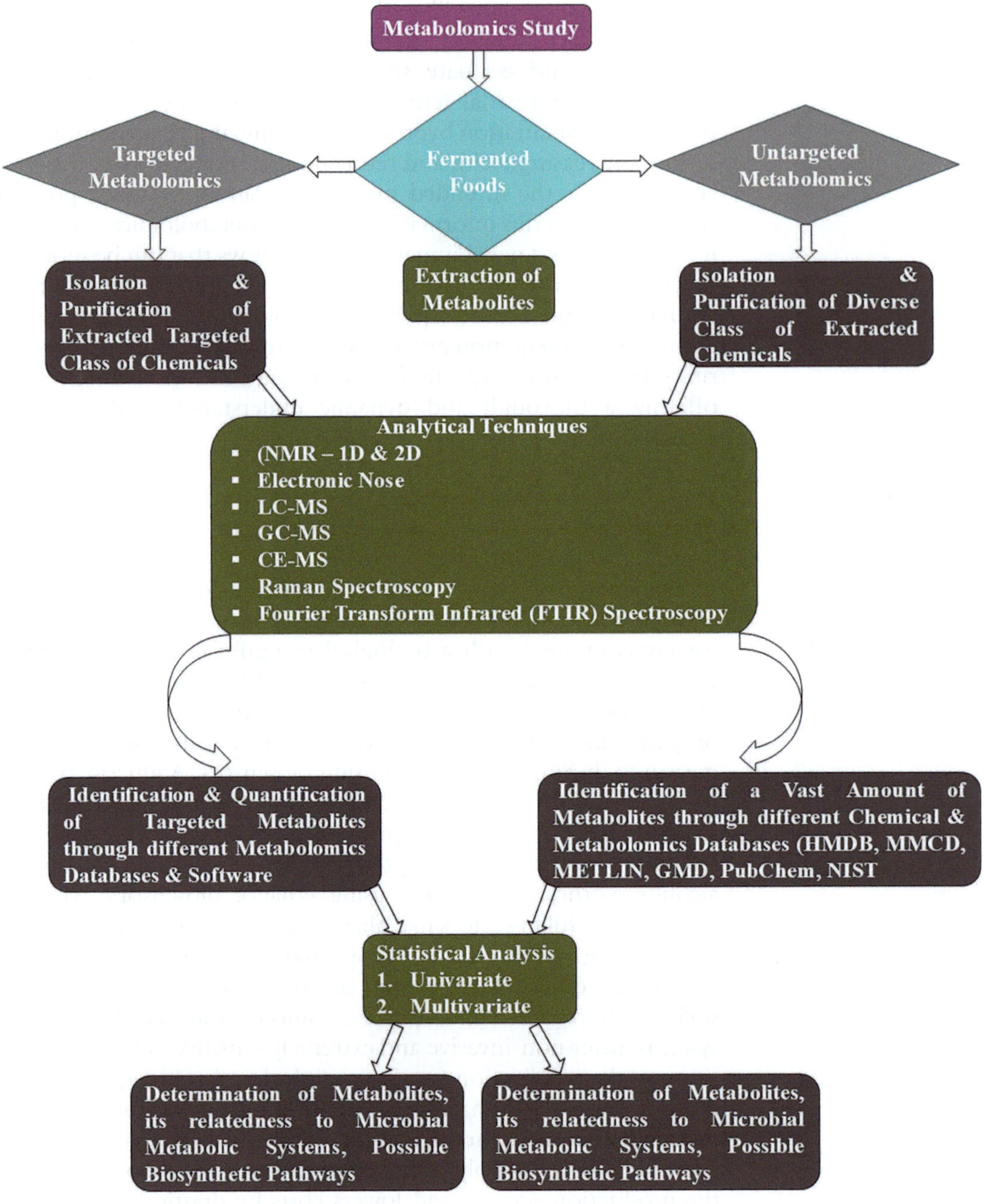

Fig. 1 A summary of the steps involved in the metabolomics method

3.4.2 Electronic Nose and Tongue Technologies

They are sophisticated sensor-based systems that are being used more and more for fermentation process monitoring and quality control. They are made to resemble human gustatory and olfactory senses, respectively. To identify and distinguish between complex mixes of volatile organic compounds (VOCs), which are frequently created during microbial fermentation, the electronic nose, or e-nose, uses a variety of chemical sensors in conjunction with pattern recognition algorithms. Analyzing taste-related characteristics is made possible by the electronic tongue, or e-tongue, which employs taste sensors that react to non-volatile compounds such as organic acids, salts, and sugars in liquids. These technologies are ideal for real-time fermentation monitoring in the food, beverage, and pharmaceutical industries because they offer quick, non-invasive, and repeatable studies [48]. For example, e-tongues can track the emergence of flavour traits or the presence of off-flavors, while e-noses can identify spoiling, contamination, or changes in fragrance profiles. By providing objective, quantitative data, both technologies lessen dependency on conventional sensory panels, which can be erratic and subjective. These techniques can also be combined with machine learning algorithms to enhance pattern recognition, allowing for the early identification of fermentation profile aberrations and the prompt implementation of corrective measures. Therefore, electronic nose and tongue technologies offer substantial value in precision fermentation and sensory analysis while improving quality assurance, streamlining product creation, and supporting process optimization [49].

3.4.3 Automation and AI in Fermentation Analysis

With unmatched efficiency, accuracy, and consistency in industrial biotechnology, these are transforming the monitoring, control, and optimization of fermentation operations. Automation is the act of carrying out repetitive operations like data gathering, parameter adjustments, and sampling without the need for human intervention by utilizing sophisticated sensors, robotic systems, and process control technology. By doing this, human error is reduced, throughput is increased, and vital parameters like pH, temperature, dissolved oxygen, substrate concentration, and byproduct production may be continuously and in real-time monitored [50]. Furthermore, the analysis of large and intricate datasets produced during fermentation is made possible by artificial intelligence (AI), namely machine learning (ML) and deep learning algorithms. AI can recognize trends, forecast results, and suggest or even carry out the best process modifications on its own. To improve decision-making and save downtime, AI models, for example, can predict fermentation yields based on early process indicators or identify possible errors before they happen. Researchers may now test and improve processes in silico before implementing them in real-world settings thanks to the emergence of digital twins, or virtual representations of fermentation systems. Additionally, the combination of AI and automation enables adaptive control systems that react dynamically to changes in the environment and microbial behavior, producing fermentation results that are more reliable and effective [51]. By lowering operating expenses and resource consumption, these technologies not only improve product quality and consistency but also increase the sustainability and scalability of fermentation operations. Automation and artificial intelligence (AI) are increasingly essential tools in contemporary fermentation analysis and industrial biotechnology as the need for accurate and high-performing bio-production increases.

3.5 Quality Control and Food Safety in Fermented Products

3.5.1 Monitoring Contaminants and Spoilage Microorganisms

A crucial component of fermentation analysis, it is necessary to guarantee the uniformity, safety, and quality of fermented products in a variety of sectors, such as food, beverages, pharmaceuticals, and biofuels. Unwanted bacteria, yeasts, or molds are examples of contaminants and spoilage organisms that can outcompete helpful microbes, change the dynamics of fermentation, lower the quality of the final product, and even generate toxic substances. Low-abundance or non-culturable organisms may not be detected by conventional detection techniques, which include microscopic inspection and culture on selective media. These techniques are frequently laborious [52]. Metagenomic analysis, next-generation sequencing (NGS), and quantitative PCR (qPCR) are examples of contemporary molecular approaches that provide quicker, more thorough, and precise identification of microbial contamination at the species or even strain level. With the use of these instruments, spoilage organisms can be identified early on, before they result in noticeable or perceptible changes, enabling prompt action. Furthermore, machine learning algorithms may evaluate both historical and current data to forecast contamination risks based on process conditions, while biosensors and fast immunoassays can offer near-real-time monitoring of microbial hazards. To guarantee that ambient and product samples are routinely tested for microbiological integrity, routine monitoring programs frequently involve the evaluation of crucial control points in the production line [53]. In addition to preventing financial losses from spoiling and recalls, efficient monitoring also guarantees regulatory compliance and fosters consumer confidence. Ultimately, preserving the integrity of fermentation processes and producing end products that are dependable, safe, and of high quality depend on effective methods for identifying and controlling impurities.

3.5.2 Standardization and Regulatory Considerations

For products to be safe, effective, and reproducible, particularly in industries like food manufacturing, pharmaceuticals, and biotechnology, standardization and regulatory considerations in fermentation analysis are crucial. From acquiring raw materials to analyzing the finished product, standardization entails applying uniform procedures, verified analytical techniques, and quality control measures throughout the fermentation process. Frameworks and recommendations for the use of microbial strains, production conditions, contamination control, labeling, and acceptable amounts of byproducts or residuals are provided by regulatory bodies such as the

Food and Drug Administration (FDA), European Medicines Agency (EMA), and Codex Alimentarius. Particularly when fermentation products are meant for human consumption or therapeutic use, adherence to Hazard Analysis and Critical Control Points (HACCP) and Good Manufacturing Practices (GMP) is frequently required [54]. Furthermore, fermentation procedures that use genetically modified organisms (GMOs) must follow biosafety guidelines and go through thorough evaluations of the effects they may have on the environment and human health. While regulatory control safeguards consumers and promotes global trade, standardization guarantees that fermentation results are repeatable across batches and facilities. A growing focus on digital documentation, traceability, and data integrity supported by automation and AI-based systems accompanies the complexity of contemporary fermentation technology. By taking these steps, audits are made easier, openness is increased, and fermented products gain credibility. All things considered, the safe, effective, and internationally recognized use of fermentation technology is based on strict standardization and compliance with legal criteria [55].

3.5.3 Ensuring Consistency in Fermented Foods

It is an essential component of food manufacturing that has a direct impact on brand reputation, consumer trust, and product quality. Temperature, pH, oxygen levels, substrate composition, microbial strains, and fermentation duration are some of the variables that affect the intricate biological process of fermentation. Significant changes in flavor, texture, aroma, and nutritional value can result from even small adjustments to these characteristics. Manufacturers utilize controlled fermentation facilities with real-time monitoring systems, carefully choose and preserve starter cultures, and apply standard operating procedures (SOPs) to ensure consistency between manufacturing batches. To guarantee consistent microbiological performance and product results, technologies like biosensors, automated bioreactors, and AI-driven data analysis assist in monitoring and controlling important factors [56]. Furthermore, samples are regularly examined for microbiological content, pH, titratable acidity, and sensory characteristics by quality assurance (QA) and quality control (QC) programs. While metabolomics and proteomics offer more profound insights into the biochemical changes that occur during fermentation, the use of genetically defined and stable microbial strains further minimizes variability. Consistency may also entail upholding certain cultural and environmental customs for artisanal and traditional fermented foods. Maintaining consistency improves scalability and worldwide marketability in addition to meeting regulatory requirements and client expectations [57]. It is more crucial than ever to have strong measures in place to guarantee batch-to-batch consistency as consumer demand for premium, healthy, and probiotic-rich fermented foods increases.

3.6 Applications of Fermentation Profiling in Food Industry

Fermentation profiling has many uses in the food industry, where it is applicable to enhance food safety, quality, and functionality. Essentially, fermentation profiling examines and evaluates the metabolic activity of microorganisms during fermentation, enabling enhanced product development and quality control [58]. A primary use of fermentation profiling and analysis relates to developing flavours and enhancing flavour quality. A food manufacturer can evaluate the metabolic pathways associated with the fermentation microbes to manipulate fermentative conditions, generating desired flavour and aroma profiles for fermented foods [59, 60]. Further, where fermented beverages are concerned and where high-quality fermented beverages are produced, such as wine, pulsed electric field (PEF) techniques may be used to speed up aroma substances formation and enhance colour [59]. A further application of fermentation profiling is assessing nutritional and functional properties of foods. Fermentation profiling may be useful to assess the degradation of anti-nutritional factors, anti-nutritional components, or allergenic components, as well as the conversion of phytochemicals into more bioactive/bioavailable forms [61]. Functional foods with probiotic activity will seek to use fermentation profiling relationships, such as food and polyphenol interactions and probiotic bacteria, to enhance functional capacity in fermented products [2].

3.6.1 Dairy and Cheese Fermentation

Fermentation profiling is integral to entrapping information in dairy and cheese making, the effectiveness of microbial activity, product quality, and process optimization. Several studies have explored fermentation profiling-related topics with important applications and findings. In dairy fermentation, lactic acid bacteria (LAB) are the main functional microorganisms used, particularly in cheese and yogurt, as they convert carbohydrates to organic acids (e.g. lactic acid), which help to preserve nutrients, provide unique flavor, and enable texture abilities [62]. More specifically, fermentation profiling has identified that LABs, mostly inclining to Lactobacillus species, release bioactive peptides via the proteolysis of the casein fractions of milk proteins, suggesting that fermented dairy products contain health-promoting properties [63]. Examples of other topics of research around the use of ultrasound functionality in fermentation, fermentation profiling, amongst other terms, studies have mentioned that ultrasound can modify the metabolism in LABs. Consequently, ultrasound during fermentation can decrease the time to ferment while having favorable effects on the characteristics of the fermented milk products. Effectively, studies suggest that ultrasound on microorganisms has both activatory/inhibitory effects, being dependent on power, frequency, and time of ultrasonic treatment [64]. In the cheese production process, the section on fermentation profiling has been able to predict times to end of life and highlight throughput opportunities to improve cheesemaking processes. A study combining mechanistic modelling and artificial neural networks (LSTM) demonstrated excellent accuracy for predicting termination times for fermentation, and possibly better scheduling downstream, thus improving efficiency in the production of cheese on an industrial scale [65]. Further, research conducted on traditional Kazak cheese indicated how variance in LAB use was contributing to the flavor, texture, and nutritional profile changes experienced at fermentation [65]. Fermentation profiling is certainly an important aspect of cheese production and dairy production in general for understanding microbial dynamics and integrating this knowledge into our protocols to improve production processes and to understand our contributing factors when developing new products. On the other end of the scale, these studies also demonstrate the potential for fermentation profiling to increase probiotic potential in products, develop flavours, and improve efficiency in production processes.

3.6.2 Bread and Cereal-Based Fermentation

Microbial fermentation profiling has importance in the development of fermented food products, from breads to cereals. There are several applications for microbial fermentation, and its quality influences product quality, nutritional composition, and product safety. For instance, in bread production sourdough fermentation is a traditional process that has roots in the oldest fermentation process, sourdough fermentation is dynamic, which can change by a variety of variables including wheat flour gluten specifically the gluten content, sourdough type, the type of flour during fermentation, lactic acid bacteria and yeast structural collapsed microbiome disposition from the prior fermentation, and environmental impacts [66]. Microbial fermentation in sourdough fermentation can reorganize gluten, retarded digestibility of starch, bioaccessible vitamin and mineral content. The combination of different bacteria, yeasts, and their metabolites generates new metabolites and chemical reactions that contribute to changes in the physical and sensory aspects of the final bread product [66]. Microbial fermentation can produce many different food ingredients such as enzymes, antimicrobial agents, vitamins and minerals, organic acids, and thickeners that may improve cereal-based products, creating new products [8]. In cereal-based fermented products, microbial communities can vary considerably based on product and regional context. As an example, traditional African fermented beverages will be made with various species of lactic acid bacteria (LAB), yeasts, and molds to give safety, shelf life, and nutritional properties to products [67]. The bacterial communities of Munkoyo, a fermented cereal-based beverage from Zambia, may have regional variation in

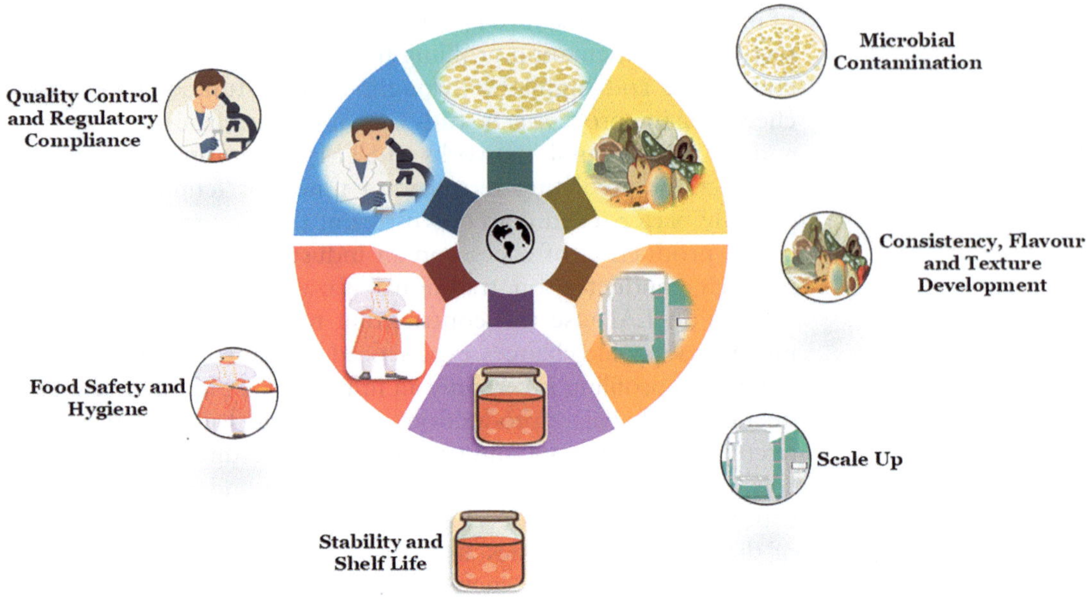

Fig. 2 Challenges faced during the processing of the food

composition, but were fermented successfully [68]. In general, to gain an understanding of how microbial fermentation profiling in cereal-based fermentation, such as bread, can affect quality, safety, and nutritional qualities of products, also helps in changing the starch characteristics of products and may open more applications for other products [69]. Also, understanding potential microbial communities in fermentation processing is crucial when considering starter cultures, and there is a standardized means of production for commercial-scale fermented cereal-based foods and beverages development [1, 68]. Figure 2 exemplifies the challenges involved in processing the fermented food.

3.6.3 Fermented Beverages (Beer, Wine, Kombucha)

Fermentation profiling is useful for understanding and improving the production of fermented products, with potential examples of products including wine, kombucha, beer, kefir, etc. Through fermentation profiling, it is possible to gain a more holistic view of the microbial community, microbial interactions, metabolite production, and overall fermentation attributes. With this knowledge, a fermentation producer can begin to target the quality and consistency of the end product. Thus, fermentation profiling has made visible the interactions of bacteria and yeast in the SCOBY (Symbiotic Culture of Bacteria and Yeast) during kombucha production, and prompted the development of new starter cultures, such as, *Pichia kluyveri*, that may enhance aroma profile through acetate esters, which primarily include isoamyl acetate and 2-phenethyl acetate [70]. Likewise, in kefir production, fermentation profiling has facilitated the retention of metabolic activity signatures of mixed cultures of bacteria and yeast that uniquely support claims of health benefits and probiotic characteristics of kefir [71]. Also, fermentation profiling has been employed to enhance the consistency and quality of kombucha. The study of species diversity, substrate consumption, and metabolite production has provided important information that informs product adaptation to consumer demands for quality consistency in kombucha. Research on non-Saccharomyces yeasts has shown their relevance to wine and the ability to contribute to the wine's composition by making health-promoting compounds such as melatonin and serotonin [72]. This approach not only enhances the quality and consistency of traditional products but has the potential to create new products with not only new characteristics but also new flavors, aromas, and health-promoting compounds.

3.6.4 Pickles, Sauerkraut, and Other Vegetable Fermentations

Profiling microbial fermentation has various applications in pickles, sauerkraut, and vegetable fermentations: Microbial communities and succession can be monitored through time in the fermentation process using high throughput sequencing techniques. For example, 16S rRNA amplicon sequencing showed that the sauerkraut microbiome was rapidly established and persisted through fermentation and packaging [73]. Similar techniques have been employed to profile the bacteria of fermented vegetables from China, revealing dominant phyla and genera [74, 75]. Fermentation profiling can identify microorganisms responsible for fermentation, and Lactobacillus or closely related genera are important and typically dominant contributors in vegetable fermentations [76–78]. Defining microorganisms in the fermentation process will depend on the types of vegetables used and the fermentation process [74]. It is also possible that the microbial community of starting ingredients or production conditions may be quite different to the final fermented product. One study showed that the lactic acid bacteria that dominated fermented sauerkraut were present relatively sparingly in both the starting ingredients and production facility [73]. In conclusion, microbial fermentation profiling can provide a complete picture of a fermentation process, assist in identifying potential probiotic strains, be utilized to improve fermentation and product quality, and allow for studying the relationship between microbial communities and metabolite profiles, including understanding flavor development and safety of fermented vegetables [74, 79].

3.7 Future Innovations in Fermentation Profiling

The field of fermentation profiling has many advances being made, with several new avenues being explored to increase efficiency and sustainability. Incorporating artificial intelligence (AI), synthetic biology, and novel fermentation processes are three hot topics that are leading the charge in fermentation profiling advances. AI will be centre stage in changing how fermentation processes are conducted, using machine learning algorithms to analyze large datasets, optimize fermentation conditions, and predict outcomes [80]. This AI approach will be a game changer to increase fermentation processes' efficiency, to identify and create usable optimal parameters without traditional trial-and-error. In addition, AI is being used to find health-promoting molecules from functional food ingredients that lead to products produced via fermentation [81]. Synthetic biology is another hot topic in fermentation profiling. Synthetic biology allows researchers to enter a new frontier for biotechnology when researchers "build to use" and create new fermentation processes [78]. Synthetic biology methods are being employed to modify microorganisms to produce certain compounds or improve fermentation. For example, metabolic engineering and synthetic biology methods are in development to tackle challenges associated with continuous fermentation processes, such as contamination, strain degradation, and low product titer [82]. Sustainable and alternative fermentation solutions are gaining attention as the interest in sustainable food production surges. The growing interest in precision fermentation—a technology that utilizes genetically modified microorganisms to produce proteins or other compounds—offers a potential solution for delivering sustainable food ingredients [81]. This approach has the potential to offer a replacement to traditional recovery methods, or chemical synthesis, with a greener option. To broaden the economic success of biological processes related to fermentation, continuous fermentation methods are being developed. These new technologies will have many advantages, versus batch or fed-batch fermentation, like: removal of substrate and product inhibition, time and length without the prolongation of a microbial organism, and lower operational costs [82]. As with other "bioreactors," challenges including contamination and strain degeneration must be resolved to allow for widespread applicability at an industrial level. In conclusion, the future of fermentation profiling is sure to converge with AI, synthetic biology, and sustainable methods. In short, these opportunities seem to facilitate a stride toward improved efficiency, sustainability, and versatility of fermentation processes.

4 Notes

Microbial fermentation profiling stands as a foundation in advanced food analysis, providing insights into the microbial ecosystems, which define the flavor, nutrition, safety of fermented foods. By combining traditional microbiological methods with innovative technologies such as omics, biosensors, and AI-driven platforms, food science researchers will benefit more about fermentation processes. The capability to monitor microbial interactions in present gives the way for precise control and optimization of fermentation, guaranteeing consistency and quality in several food products. Additionally, the compilation of synthetic biology and sustainable bioprocessing techniques indicates a encouraging shift toward efficient fermentation models. As food systems advance to meet consumer demand for functional, clean-label, and sustainable products, microbial profiling will play an increasingly vital role. This chapter highlights the significance of implementing multidimensional analytical technologies to know the full potential of microbial fermentation eventually leading innovations in the global food industry.

References

1. Elhalis H, Chin XH, Chow Y (2024) Soybean fermentation: microbial ecology and starter culture technology. Crit Rev Food Sci Nutr 64:7648–7670
2. Sharma R, Garg P, Kumar P, Bhatia SK, Kulshrestha S (2020) Microbial fermentation and its role in quality improvement of fermented foods. Fermentation 6:106
3. Ohwofasa A, Dhami M, Tian B, Winefield C, On SL (2023) Environmental influences on microbial community development during organic pinot noir wine production in outdoor and indoor fermentation conditions. Heliyon 9:e15658
4. Wang S, Xiong W, Wang Y, Nie Y, Wu Q, Xu Y, Geisen S (2020) Temperature-induced annual variation in microbial community changes and resulting metabolome shifts in a controlled fermentation system. Msystems 5:10–128
5. Zhao S, Niu C, Wang Y, Zan Y, Zheng F, Liu C, Wang J, Li Q (2023) Revealing the succession of spatial heterogeneity of the microbial community during broad bean paste fermentation. Appl Environ Microbiol 89: e00621–e00623
6. Yan Y, Nguyen LH, Franzosa EA, Huttenhower C (2020) Strain-level epidemiology of microbial communities and the human microbiome. Genome Med 12:1–6
7. Zhang G, Zhao L, Li W, Yao H, Lu C, Zhao G, Wu Y, Li Y, Kong G (2023) Changes in physicochemical properties and microbial community succession during leaf stacking fermentation. AMB Express 13:132
8. Iram A, Ozcan A, Turhan I, Demirci A (2023) Production of value-added products as food ingredients via microbial fermentation. Processes 11:1715
9. Ağagündüz D, Yılmaz B, Koçak T, Altıntaş Başar HB, Rocha JM, Özoğul F (2022) Novel candidate microorganisms for fermentation technology: from potential benefits to safety issues. Foods 11:3074
10. Yuan Y, Yang Y, Xiao L, Qu L, Zhang X, Wei Y (2023) Advancing insights into probiotics during vegetable fermentation. Foods 12:3789
11. Wang Y, Zhang C, Liu F, Jin Z, Xia X (2023) Ecological succession and functional characteristics of lactic acid bacteria in traditional fermented foods. Crit Rev Food Sci Nutr 63: 5841–5855
12. Lynch KM, Zannini E, Wilkinson S, Daenen L, Arendt EK (2019) Physiology of acetic acid bacteria and their role in vinegar and fermented beverages. Compr Rev Food Sci Food Saf 18: 587–625
13. Liszkowska W, Berlowska J (2021) Yeast fermentation at low temperatures: adaptation to changing environmental conditions and formation of volatile compounds. Molecules 26: 1035
14. Karlund A, Gomez-Gallego C, Korhonen J, Palo-oja OM, El-Nezami H, Kolehmainen M (2020) Harnessing microbes for sustainable development: food fermentation as a tool for improving the nutritional quality of alternative protein sources. Nutrients (Web) 12:1020
15. Han D, Yang Y, Guo Z, Dai S, Jiang M, Zhu Y, Wang Y, Yu Z, Wang K, Rong C, Yu Y (2024) A review on the interaction of acetic acid bacteria and microbes in food fermentation: a microbial ecology perspective. Foods 13:2534
16. Jin R, Song J, Liu C, Lin R, Liang D, Aweya JJ, Weng W, Zhu L, Shang J, Yang S (2024) Synthetic microbial communities: novel strategies to enhance the quality of traditional fermented foods. Compr Rev Food Sci Food Saf 23: e13388
17. Bassi D, Orrù L, Cabanillas Vasquez J, Cocconcelli PS, Fontana C (2020) Peruvian chicha:

a focus on the microbial populations of this ancient maize-based fermented beverage. Microorganisms 8:93
18. Chaves-López C, Rossi C, Maggio F, Paparella A, Serio A (2020) Changes occurring in spontaneous maize fermentation: an overview. Fermentation 6:36
19. Martín MC, Prendes LP, Morata VI, Merín MG (2024) Biocontrol and enzymatic activity of non-saccharomyces wine yeasts: improvements in winemaking. Fermentation 10:218
20. Ferremi Leali N, Binati RL, Martelli F, Gatto V, Luzzini G, Salini A, Slaghenaufi D, Fusco S, Ugliano M, Torriani S, Salvetti E (2022) Reconstruction of simplified microbial consortia to modulate sensory quality of kombucha tea. Foods 11:3045
21. De Roos J, De Vuyst L (2018) Acetic acid bacteria in fermented foods and beverages. Curr Opin Biotechnol 49:115–119
22. Chandel NS (2021) Metabolism of proliferating cells. Cold Spring Harb Perspect Biol 13: a040618
23. Naifeh J, Dimri M, Varacallo MA (2023) Biochemistry, aerobic glycolysis. In: StatPearls [Internet]. StatPearls Publishing
24. Reiter T, Montpetit R, Byer S, Frias I, Leon E, Viano R, Mcloughlin M, Halligan T, Hernandez D, Runnebaum R, Montpetit B (2021) Saccharomyces cerevisiae gene expression during fermentation of Pinot Noir wines at an industrially relevant scale. Appl Environ Microbiol 87:e00036–e00021
25. Eram MS, Ma K (2013) Decarboxylation of pyruvate to acetaldehyde for ethanol production by hyperthermophiles. Biomolecules 3: 578–596
26. Moens F, Lefeber T, De Vuyst L (2014) Oxidation of metabolites highlights the microbial interactions and role of Acetobacter pasteurianus during cocoa bean fermentation. Appl Environ Microbiol 80:1848–1857
27. Escanciano IA, Wojtusik M, Esteban J, Ladero M, Santos VE (2022) Modeling the succinic acid bioprocess: a review. Fermentation 8:368
28. Méndez-Hernández JE, Rodríguez-Durán LV, Páez-Lerma JB, Soto-Cruz NO (2023) Strategies for supplying precursors to enhance the production of secondary metabolites in solid-state fermentation. Fermentation 9:804
29. Ledauphin J, Guichard H, Saint-Clair JF, Picoche B, Barillier D (2003) Chemical and sensorial aroma characterization of freshly distilled calvados. 2. Identification of volatile compounds and key odorants. J Agric Food Chem 51:433–442
30. Chu CW, Zhu ZC, Bian HT, Jiang JC (2021) Design of self-heating test platform for sulfide corrosion and oxidation based on Fuzzy PID temperature control system. Meas Control 54: 1082–1096
31. Zhao A, Sun J, Liu Y (2023) Understanding bacterial biofilms: from definition to treatment strategies. Front Cell Infect Microbiol 13: 1137947
32. Neu TR, Lawrence JR (2014) Investigation of microbial biofilm structure by laser scanning microscopy. Productive. Biofilms 146:1–51
33. Lozada-Ramírez JD, Ortega-Regules AE, Hernández LR, Anaya de Parrodi C (2021) Spectroscopic and spectrometric applications for the identification of bioactive compounds from vegetal extracts. Appl Sci 11:3039
34. Kupina SA, Shrikhande AJ (2003) Evaluation of a Fourier transform infrared instrument for rapid quality-control wine analyses. Am J Enol Vitic 54:131–134
35. Louw S (2021) Recent trends in the chromatographic analysis of volatile flavor and fragrance compounds: annual review 2020. Anal Sci Adv 2:157–170
36. Clark DP, Pazdernik NJ (2012) Molecular biology. Elsevier
37. Guo N, Liu JB, Li W, Ma YS, Fu D (2022) The power and the promise of CRISPR/Cas9 genome editing for clinical application with gene therapy. J Adv Res 40:135–152
38. Walsh AM, Leech J, Huttenhower C, Delhomme-Nguyen H, Crispie F, Chervaux C, Cotter PD (2023) Integrated molecular approaches for fermented food microbiome research. FEMS Microbiol Rev 47:fuad001
39. La Rosa G, De Carolis E, Sali M, Papacchini M, Riccardi C, Mansi A, Paba E, Alquati C, Bestetti G, Muscillo M (2006) Genetic diversity of bacterial strains isolated from soils, contaminated with polycyclic aromatic hydrocarbons, by 16S rRNA gene sequencing and amplified fragment length polymorphism fingerprinting. Microbiol Res 161:150–157
40. Cunningham AS, Gorospe M (2024) Striving for clarity in language about gene expression. Nucleic Acids Res 52:10747–10753
41. Statello L, Guo CJ, Chen LL, Huarte M (2021) Gene regulation by long non-coding RNAs and its biological functions. Nat Rev Mol Cell Biol 22:96–118
42. Guo H, Wang L, Deng Y, Ye J (2021) Novel perspectives of environmental proteomics. Sci Total Environ 788:147588
43. Drews O (2012) High-resolution separation of cardiovascular proteomes: two-dimensional

electrophoresis and liquid chromatography. Circ Cardiovasc Genet 5:592

44. Nadeem MS, Razeeth M, Choudhry HM, Anwar F, Zamzami MA, Murtaza BN, Al-Abbasi FA, Khan MI, Shakoori AR (2020) LC-MS/MS-based metabolic profiling of Escherichia coli under heterologous gene expression stress. J Cell Biochem 121:125–134
45. Emwas AH, Roy R, McKay RT, Tenori L, Saccenti E, Gowda GN, Raftery D, Alahmari F, Jaremko L, Jaremko M, Wishart DS (2019) NMR spectroscopy for metabolomics research. Metabolites 9:123
46. Bhatia D, Paul S, Acharjee T, Ramachairy SS (2024) Biosensors and their widespread impact on human health. Sens Int 5:100257
47. Ghazali MF, Mustafa M (2025) Bioethanol as an alternative fuels: a review on production strategies and technique for analysis. Energy Convers Manag 26:100933
48. Vanaraj R, Bincy IP, Mayakrishnan G, Kim IS, Kim SC (2025) A systematic review of the applications of electronic nose and electronic tongue in food quality assessment and safety. Chemosensors 13:161
49. Baldwin EA, Bai J, Plotto A, Dea S (2011) Electronic noses and tongues: applications for the food and pharmaceutical industries. Sensors 11:4744–4766
50. Niyigaba T, Küçükgöz K, Kołożyn-Krajewska D, Królikowski T, Trząskowska M (2025) Advances in fermentation technology: a focus on health and safety. Appl Sci 15:3001
51. Bonanni D, Litrico M, Ahmed W, Morerio P, Cazzorla T, Spaccapaniccia E, Cattani F, Allegretti M, Beccari AR, Del Bue A, Martin F (2023) A deep learning approach to optimize recombinant protein production in Escherichia coli fermentations. Fermentation 9:503
52. Sawant SS, Park HY, Sim EY, Kim HS, Choi HS (2025) Microbial fermentation in food: impact on functional properties and nutritional enhancement—a review of recent developments. Fermentation 11:15
53. Gu W, Miller S, Chiu CY (2019) Clinical metagenomic next-generation sequencing for pathogen detection. Annu Rev Pathol 14(1): 319–338
54. Krzyżostan M, Wawrzyńczak A, Nowak I (2024) Use of waste from the food industry and applications of the fermentation process to create sustainable cosmetic products: a review. Sustainability 16:2757
55. Lensch A, Lindfors HA, Duwenig E, Fleischmann T, Hjort C, Kärenlampi SO, McMurtry L, Melton ED, Andersen MR, Skinner R, Wyss M (2024) Safety aspects of microorganisms deliberately released into the environment. EFB Bioeconomy J 4:100061
56. Praveen M, Brogi S (2025) Microbial fermentation in food and beverage industries: innovations, challenges, and opportunities. Foods 14: 114
57. Agakidou E, Agakidis C, Gika H, Sarafidis K (2020) Emerging biomarkers for prediction and early diagnosis of necrotizing enterocolitis in the era of metabolomics and proteomics. Front Pediatr 8:602255
58. Gao Y, Hou L, Gao J, Li D, Tian Z, Fan B, Wang F, Li S (2021) Metabolomics approaches for the comprehensive evaluation of fermented foods: a review. Foods 10:2294
59. Feng Y, Yang T, Zhang Y, Zhang A, Gai L, Niu D (2022) Potential applications of pulsed electric field in the fermented wine industry. Front Nutr 9:1048632
60. Wang Y, Wu J, Lv M, Shao Z, Hungwe M, Wang J, Bai X, Xie J, Wang Y, Geng W (2021) Metabolism characteristics of lactic acid bacteria and the expanding applications in food industry. Front Bioeng Biotechnol 9: 612285
61. Shiferaw Terefe N, Augustin MA (2020) Fermentation for tailoring the technological and health related functionality of food products. Crit Rev Food Sci Nutr 60:2887–2913
62. Widyastuti Y, Febrisiantosa A, Tidona F (2021) Health-promoting properties of lactobacilli in fermented dairy products. Front Microbiol 12: 673890
63. Kondrotiene K, Zavistanaviciute P, Aksomaitiene J, Novoslavskij A, Malakauskas M (2023) Lactococcus lactis in dairy fermentation—Health-promoting and probiotic properties. Fermentation 10:16
64. Akdeniz V, Akalın AS (2022) Recent advances in dual effect of power ultrasound to microorganisms in dairy industry: activation or inactivation. Crit Rev Food Sci Nutr 62:889–904
65. Li J, Huang Q, Zheng X, Ge Z, Lin K, Zhang D, Chen Y, Wang B, Shi X (2020) Investigation of the lactic acid bacteria in kazak cheese and their contributions to cheese fermentation. Front Microbiol 11:228
66. Akamine IT, Mansoldo FR, Vermelho AB (2023) Probiotics in the sourdough bread fermentation: current status. Fermentation 9:90
67. Madilo FK, Parry-Hanson Kunadu A, Tano-Debrah K, Saalia FK, Kolanisi U (2024) Diversity of production techniques and microbiology of African cereal-based traditional fermented beverages. J Food Qual 2024: 1241614

68. Phiri S, Schoustra SE, van den Heuvel J, Smid EJ, Shindano J, Linnemann A (2019) Fermented cereal-based Munkoyo beverage: processing practices, microbial diversity and aroma compounds. PLoS One 14:e0223501
69. Yan X, McClements DJ, Luo S, Ye J, Liu C (2024) A review of the effects of fermentation on the structure, properties, and application of cereal starch in foods. Crit Rev Food Sci Nutr 65:1–20
70. van Wyk N, Binder J, Ludszuweit M, Köhler S, Brezina S, Semmler H, Pretorius IS, Rauhut D, Senz M, von Wallbrunn C (2023) The influence of Pichia kluyveri addition on the aroma profile of a kombucha tea fermentation. Foods 12:1938
71. Dahiya D, Nigam PS (2023) Therapeutic and dietary support for gastrointestinal tract using kefir as a nutraceutical beverage: dairy-milk-based or plant-sourced kefir probiotic products for vegan and lactose-intolerant populations. Fermentation 9:388
72. Vilela A (2019) The importance of yeasts on fermentation quality and human health-promoting compounds. Fermentation 5:46
73. Zabat MA, Sano WH, Wurster JI, Cabral DJ, Belenky P (2018) Microbial community analysis of sauerkraut fermentation reveals a stable and rapidly established community. Foods 7:77
74. Liang T, Xie X, Wu L, Li L, Li H, Xi Y, Feng Y, Xue L, Chen M, Chen X, Zhang J (2021) Microbial communities and physiochemical properties of four distinctive traditionally fermented vegetables from North China and their influence on quality and safety. Foods 11:21
75. Zhang J, Song HS, Zhang C, Kim YB, Roh SW, Liu D (2021) Culture-independent analysis of the bacterial community in Chinese fermented vegetables and genomic analysis of lactic acid bacteria. Arch Microbiol 203:4693–4703
76. Liu L, She X, Chen X, Qian Y, Tao Y, Li Y, Guo S, Xiang W, Liu G, Rao Y (2020) Microbiota succession and chemical composition involved in the radish fermentation process in different containers. Front Microbiol 11:445
77. Rao Y, Qian Y, Tao Y, She X, Li Y, Che Z, Li H, Liu L (2019) Influence of oxygen exposure on fermentation process and sensory qualities of Sichuan pickle (paocai). RSC Adv 9:38520–38530
78. Touret T, Oliveira M, Semedo-Lemsaddek T (2018) Putative probiotic lactic acid bacteria isolated from sauerkraut fermentations. PLoS One 13:e0203501
79. Yang X, Hu W, Xiu Z, Jiang A, Yang X, Saren G, Ji Y, Guan Y, Feng K (2020) Microbial community dynamics and metabolome changes during spontaneous fermentation of northeast sauerkraut from different households. Front Microbiol 11:1878
80. Shah P, Kendall F, Khozin S, Goosen R, Hu J, Laramie J, Ringel M, Schork N (2019) Artificial intelligence and machine learning in clinical development: a translational perspective. NPJ Digit Med 2:69
81. Alasi SO, Sanusi MS, Sunmonu MO, Odewole MM, Adepoju AL (2024) Exploring recent developments in novel technologies and AI integration for plant-based protein functionality: a review. J Agric Food Res 15:101036
82. Dong Y, Zhang Y, Liu D, Chen Z (2023) Strain and process engineering toward continuous industrial fermentation. Front Chem Sci Eng 17:1336–1353

Chapter 15

SPR (Surface Plasmon Resonance) for Real-Time Food Interaction Studies

Flaviana Coelho Pacheco, Lorena Evangelista Fernandes, Ana Flávia Coelho Pacheco, and Paulo Henrique Costa Paiva

Abstract

A thorough manual for using surface plasmon resonance (SPR) in real-time investigations of biomolecular interactions pertinent to food systems is provided in this protocol chapter. By tracking variations in the refractive index close to a sensor surface, SPR is a label-free optical method that can identify molecular binding events with high sensitivity. The method has been used more and more in food science research because it is especially well-suited for measuring kinetic and affinity parameters (k_a, k_d, and KD) in close to physiological settings. Sample preparation, ligand immobilization, interaction analysis, surface regeneration, and quality control of data acquisition are all crucial steps in SPR experimentation that are covered by the methods presented here. Practical factors such as buffer optimization, ligand purity evaluation, chip selection tactics, analyte solubilization procedures, and regeneration scouting are covered in detail in each section. Additionally, advice is given on how to reduce nonspecific interactions, adjust for refractive index discrepancies, and guarantee sensor surface stability throughout several assay cycles. In addition to detailed technical instructions, this chapter contains application-specific subsections that show how SPR can be used to characterize emulsifier-polysaccharide binding, assess the release behavior of encapsulated nutraceuticals, detect allergens and pathogens, and study enzyme-substrate kinetics in dairy systems. Along with outlining the integration of SPR with complementary tools like mass spectrometry and HPLC, the protocol also highlights experimental designs for fingerprinting and traceability in food authentication. All things considered, this chapter is a fundamental tool for scientists and business experts who want to plan, carry out, and analyze SPR tests with accuracy and consistency. The techniques described emphasise reproducibility, analytical accuracy, and real-world application in the framework of contemporary food science, supporting both fundamental and applied research, including complex food matrices, biopolymers, and bioactive compounds.

Key words Surface plasmon resonance, Food analysis, Biomolecular interactions, Kinetic assays, Emulsifiers, Allergens, Encapsulation, Protein–polyphenol interactions

1 Introduction

Surface plasmon resonance (SPR) is a widely used optical technique for investigating biomolecular interactions in real-time without the need for fluorescent or radioactive labels. It is an optical technique

Tanmay Sarkar (ed.), *Novel Food Analysis: Techniques and Applications*, Methods and Protocols in Food Science,
https://doi.org/10.1007/978-1-0716-4787-5_15,

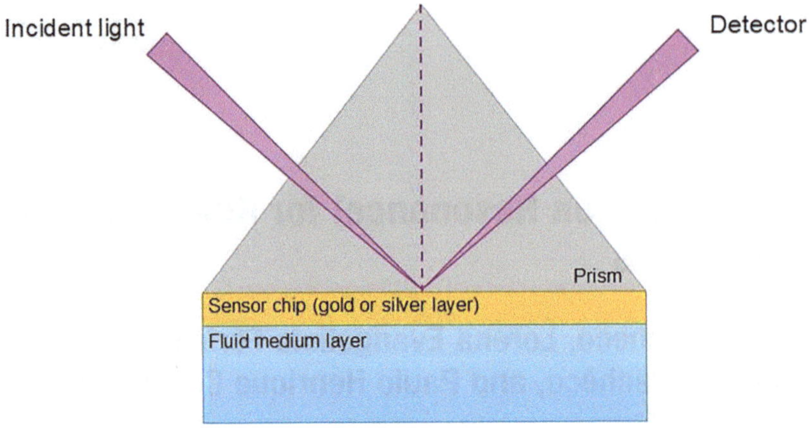

Fig. 1 Schematic representation of the surface plasmon resonance (SPR) principle. A polarized light beam reflects at the interface between a glass prism and a thin gold film, exciting surface gold electrons. Changes in the refractive index near the sensor surface, caused by molecular interactions, shift the resonance angle and are detected in real time

that detects changes in the refractive index near a metal surface, typically gold. When polarized light hits the surface at a specific angle, it excites surface plasmons, and any biomolecular interaction at the surface alters the resonance conditions (Fig. 1). These changes are measured in real time, allowing label-free detection of molecular binding events being highly sensitive for monitoring the association and dissociation processes between two or more molecules [1–3].

In recent years, SPR has become an essential tool in various scientific fields, including biotechnology, pharmacology, immunology, and clinical diagnostics [4, 5]. It's particularly valuable for determining kinetic and thermodynamic parameters of interactions, such as association constants (k_a), dissociation constants (k_d), and binding constants (K_b), in addition to being useful for analyte quantification and ligand screening [2, 6].

The principle of SPR-based systems relies on immobilizing one molecule (the ligand) onto the surface of a sensor chip, while another molecule (the analyte) is injected in a continuous flow, allowing for detailed analysis of their interaction [3, 7]. Under near-physiological conditions, this method allows the qualitative and quantitative study of molecular binding mechanisms [2, 8].

SPR is a non-invasive, highly sensitive, radioactive-label-free technique that is adaptable to a wide range of experimental designs, from single-analyte injections to sophisticated multi-step assays, including competitive binding, epitope mapping, and cross-reactivity tests [1, 6, 9]. For research on interactions involving proteins, peptides, nucleic acids, carbohydrates, lipids, and small molecules, its adaptability and flexibility set it as a gold standard [5, 10].

Within the field of food science, SPR has become increasingly important for evaluating interactions between functional compounds and food matrices, supporting formulation, quality control, and the creation of ingredients with health-promoting properties [11, 12]. Moreover, the adaptability of SPR with high-throughput and automated systems has expanded its utility in screening vast molecule libraries, which is necessary for industrial and regulatory environments [10, 13].

Particularly under conditions that resemble real-life processing and digestion, the demand for methods that can clarify the binding dynamics between bioactive ingredients and food matrix components drives the increasing interest in using SPR in food systems. This capacity helps design functional foods and evaluate ingredient bioavailability and efficacy [11, 12].

This chapter aims to present the main technical aspects of SPR, covering the necessary materials, experimental design, and methodological procedures. It concludes with a critical analysis of the technique's advantages, limitations, and applications.

2 Materials

The execution of experiments using SPR requires strict attention to the preparation of materials and reagents, as the quality of the inputs directly influences the reproducibility and reliability of the generated data [14]. Below are the main components required for conducting the experiments.

Sensor chips are responsible for accommodating the ligand on the surface and transmitting the optical signal corresponding to molecular interactions. The most widely used chip is the Sensor Chip CM5, which features a carboxymethyl dextran layer over a gold film, providing a versatile surface for the covalent immobilization of various types of molecules [2]. Other specialized chips include the Sensor Chip SA, which contains immobilized streptavidin for capturing biotinylated ligands; the Sensor Chip NTA, which uses nitrilotriacetic acid groups to capture histidine-tagged (His-tag) recombinant proteins through coordination with nickel ions, and the Sensor Chip HPA, with a hydrophobic surface designed for experiments involving lipids, glycolipids, and membrane-associated proteins [7].

Recent advances in chip chemistry have also brought low non-specific binding surfaces, including zwitterionic coatings and PEGylated layers. These changes minimize background noise, so improving signal specificity—a particularly important factor in the analysis of serum, plasma, or food-derived samples [3].

The ligand, that is, the molecule to be immobilized on the chip surface, must be of high purity—preferably greater than 90%—in order to ensure interaction specificity and avoid interference in kinetic analysis [15]. Various types of molecules can be used as ligands, including proteins, peptides, oligonucleotides, carbohydrates, and low molecular weight compounds. However, when commercial products are used as ligands, attention must be paid to the presence of additives such as albumin (BSA), buffers containing primary amines (such as Tris), or preservatives (such as sodium azide), all of which must be removed prior to immobilization, especially when using amine coupling strategies [16].

Another important consideration is ligand orientation. Site-directed immobilization, such as that achieved by His-tag or biotin-streptavidin systems, guarantees that the active site of proteins with well-defined binding domains stays accessible [15, 17]. Particularly in multimeric complexes, random immobilization through amine coupling may result in decreased activity because of steric hindrance or conformational loss [18].

The analyte, in turn, is the molecule that is injected in continuous flow to interact with the immobilized ligand. Its preparation must take into account compatibility with the running buffer used in the experiment to minimize the so-called "bulk effect," which refers to the difference in refractive index between the sample solvent and the reference buffer [14]. In quantitative experiments, it is essential that the analyte be diluted in the same buffer used in the system, and that, if necessary, buffer exchange be performed using techniques such as dialysis or gel filtration [17]. The ideal analyte concentration will depend on the expected affinity with the ligand and is commonly set near the estimated value of the dissociation constant (k_d) [3].

For analytes prone to aggregation or precipitation, particularly hydrophobic drugs or peptides, the use of solubilizing agents (e.g., DMSO) may be required [11]. However, DMSO can alter the refractive index, and its concentration must be carefully matched between samples and blanks [19]. The use of DMSO calibration standards is strongly recommended in these cases to ensure accurate signal interpretation [18].

Among the essential solutions for the experiments, the standard running buffer, known as HBS-EP, is noteworthy. It is composed of 10 mM HEPES (pH 7.4), 150 mM NaCl, 3 mM EDTA, and 0.005% Surfactant P20 [13, 20]. This buffer should always be filtered (using a 0.22 μm filter) and degassed before use to avoid bubble formation in the microfluidic system. Solutions used for immobilization include activating reagents such as EDC and NHS, which promote activation of carboxyl groups on the chip surface, and compounds like PDEA, used in thiol coupling to introduce reactive disulfide groups. Regeneration solutions, applied at the end of each experimental cycle to remove the analyte from the surface, may vary widely depending on the stability of the ligand and the nature of the studied interaction [12–14]. These include acidic or basic solutions (e.g., glycine-HCl or NaOH), high ionic strength buffers (e.g., 1–5 M NaCl), mild detergents (such as SDS or CHAPS), and organic solvents (e.g., ethylene glycol) [20].

To completely remove tightly bound analytes or avoid carry-over, some complex systems might need co-injection of chelating agents or sequential regeneration techniques [1–4]. Regeneration compatibility across several ligand surfaces must be guaranteed in multiplex assays to prevent harm to delicate immobilized structures. Robust regeneration protocols require the optimization of ionic strength, pH, and contact time [6, 7].

Finally, the experimental system also requires complementary equipment, such as high-speed centrifuges (for removing particles prior to injection), filters for solution sterilization, and, when necessary, instruments for liposome preparation, such as extruders or ultrasonic probes, especially in experiments involving HPA chips [9, 13]. The experimental system itself includes automated injection and flow control modules, as well as specialized software for kinetic data acquisition and analysis [18, 19].

Additional helpful resources include regeneration scouting kits, which offer pre-formulated solutions to find appropriate regeneration conditions, and calibration kits, which are supplied by the instrument manufacturer for confirming the optical response of the instrument [8]. Advanced users can create unique workflows and integrate SPR with robotic sample preparation by coupling it with external systems like temperature-controlled stages, fluidic switching valves, or autosamplers [10].

Summary of Materials:

1. *Sensor chips* (CM5, SA, NTA, HPA).
2. *Ligands* (proteins, peptides, oligonucleotides, carbohydrates, low molecular weight compounds).
3. *Analytes* (in compatible buffer, purified or in mixture).
4. *Running buffer* (e.g., HBS-EP: HEPES, NaCl, EDTA, Surfactant P20).
5. *Immobilization solutions* (EDC/NHS, PDEA, coupling buffers).
6. *Regeneration solutions* (glycine-HCl, NaOH, SDS, NaCl, ethylene glycol).
7. *SPR Biosensor System* (e.g. Biacore (Biacore A100, Flexchip, 3000, T100, X100, C); Metrohm Autolab (Autolab ESPRIT, Springle); ICX Technologies (SensiQ, Pioneer, Discovery); Reichert (SR 7000); DKK-TOA Corporation (SPR-20); Biosensing instrument (BI-SPR1000, 2000, 3000); Bio-Rad (ProteOn XPR36); Thermo Electron Corporation (SPR 100); Analytical μ-Systems (BIOSUPLAR 6) and IBIS Technologies (IBIS-iSPR)) [1].

8. *0.22 μm filters or centrifuge* (≥10,000× *g*).
9. *Sample and buffer reservoirs* (glass or polypropylene vials).
10. *Software* for kinetic data acquisition and analysis.
11. *Optional equipment* for liposome preparation (extruder, ultrasonic bath).

3 Method

3.1 Sample Preparation

For SPR analysis to yield accurate results, the ligand and analyte must be prepared properly (Fig. 2). To immobilize the ligand on the functionalized surface of the sensor chip, it should be diluted in a buffer that is compatible with the immobilization technique of choice and free of additives like glycine or tris that could disrupt the covalent coupling reaction and alter the chemistry of the carbodiimide (EDC/NHS) [1, 15]. It is advised to set the buffer's pH to values just below the ligand's isoelectric point in order to promote electrostatic interactions with the negatively charged dextran matrix and increase immobilization efficiency. To reduce the "volume effect" and guarantee signal stability, the analyte must be in the same buffer as the running buffer. The integrity of the flow in the microfluidic system can be jeopardized by bubble formation and microbial contamination, which can be avoided by using freshly made, filtered (0.22 μm or 0.45 μm), and degassed buffers [1, 12, 15].

Eliminating suspended particles is also crucial. Filtration or centrifugation (e.g., 10,000× *g* for 5–10 min) should be applied to both the ligand and the analyte in order to prevent clogging in the microchannels and lower the possibility of background noise. The expected dissociation constant (KD) should be used to optimize sample concentrations; higher analyte concentrations are needed for low-affinity interactions, and lower concentrations are needed for high-affinity interactions [15]. To precisely record the kinetic parameters of the interaction, dilution series in running buffer that span at least five points and are arranged in two- or three-fold serial dilution are typically employed [1, 6]. Extra caution is needed in food matrices. For instance, to reduce interferences, plant extracts need to be clarified and have their lipids removed, and fermented product supernatants might need special

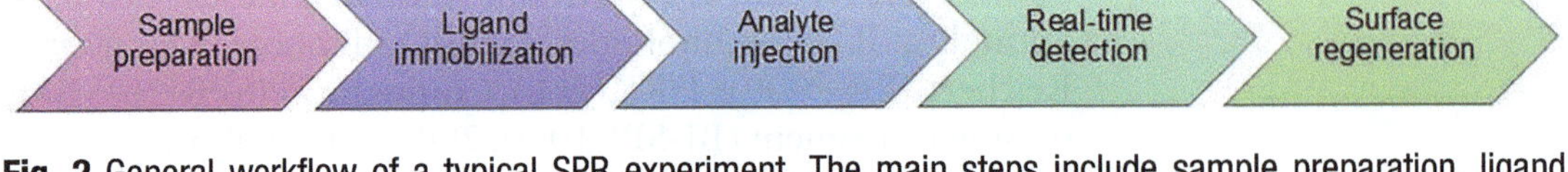

Fig. 2 General workflow of a typical SPR experiment. The main steps include sample preparation, ligand immobilization on the sensor chip, analyte injection, real-time interaction monitoring, and surface regeneration

pretreatments to get rid of bacteria or leftover particles. Analytes are frequently adapted to the running buffer using methods like dialysis or buffer exchange with desalting columns, which prevent refractive index deviations and enhance the quality of the data collected [1, 15, 18].

To preserve the structural integrity of the analytes, it is also advised to avoid freeze-thaw cycles for thermolabile proteins and, when required, to employ stabilizers. Aggregates can contaminate the sensor surface irreversibly and cause nonspecific responses or false positives. Blocking agents, like BSA at 0.1 mg/mL, can be added to the running buffer to greatly reduce nonspecific interactions and increase the selectivity of the molecular interactions that are of interest [1, 15].

3.2 Ligand Immobilization

Ligand immobilization is typically performed through covalent coupling via amine groups, using activating reagents such as EDC and NHS. After activation of the carboxyl groups on the chip surface, the ligand is injected, promoting the formation of covalent bonds. Excess reactive groups are then blocked with ethanolamine [1, 11]. Alternatively, thiol coupling can be used when the ligand contains sulfhydryl groups, or indirect capture methods may be applied, such as the biotin-streptavidin interaction or the binding of His-tagged proteins to Ni^{2+}-functionalized surfaces [15].

Attaching the ligand to the sensor chip's surface is known as immobilization, and it has a direct impact on the accuracy and repeatability of kinetic and affinity measurements. The most popular method usually uses activating reagents like NHS and EDC to accomplish covalent coupling via amine groups [1]. Ethanolamine is then used to block excess reactive groups. By injecting a combination of NHS (N-hydroxysuccinimide) and EDC (N-ethyl-N′--(3-dimethylaminopropyl) carbodiimide), which create reactive esters on the carboxyl groups of the dextran matrix, the sensor chip (such as the CM5 with a carboxymethylated dextran matrix) is activated [1, 15, 18]. After the ligand is injected onto the activated surface in acetate buffer at the ideal pH, the primary amines on the protein surface are able to create strong amide bonds with the dextran matrix. 1 M ethanolamine-HCl (pH 8.5) is used to block unreacted esters; this procedure usually takes seven minutes. The rise in response units (RU), where 1000 RU is equivalent to roughly 1 ng/mm^2 of bound protein, can be used to track the effectiveness of immobilization in real time [15].

Alternative immobilization methods include:

1. *Thiol coupling*, used when ligands contain free sulfhydryl groups, involves preactivation of the chip with PDEA (2-(2-pyridinyldithio)ethanamine hydrochloride).

2. *Affinity capture via tags*, such as His-tag/NTA-Ni^{2+}, where a Ni^{2+}-loaded NTA surface binds reversibly to His-tagged proteins, or biotin/streptavidin systems, known for irreversible high-affinity capture.
3. *Fc/anti-Fc interactions*, which provide specific immobilization for antibody orientation.
4. *Hydrophobic or lipophilic capture,* particularly for proteins or membrane vesicles, e.g. L1 chip surfaces capture lipid vesicles via hydrophobic interactions.

The interaction under study determines the ideal ligand immobilization density. Densities between 100 and 1000 RU are common for kinetic measurements, whereas lower densities lessen the constraints of mass transport and rebinding for small molecule analyses. Overly high densities can change the ligand conformation or make it more difficult to access the analyte, which can result in interactions that are not physiological.

A multi-step immobilization procedure typically includes:

1. Surface activation (EDC/NHS).
2. Ligand injection.
3. Deactivation with ethanolamine.
4. Stabilization phase with running buffer to equilibrate the surface.

Each step should be validated by monitoring RU changes and ensuring that immobilization levels are reproducible between experimental replicates.

3.3 Interaction Analysis

Once the ligand is immobilized, the interaction analysis with the analyte begins. The molecule in solution is injected over the functionalized surface, and the system records in real time the refractive index changes associated with the binding event [1, 15]. The SPR response is recorded in response units (RU), which involve changes in the mass close to the sensor surface (approximately 1 RU = 1 pg/mm^2). The resulting sensorgram, Fig. 3, allows monitoring of the association and dissociation phases of the interaction, enabling the calculation of the kinetic parameters (ka, kd), equilibrium layer (KD) or total binding capacity (Rmax). The flow rate, contact time, and concentration of the analyte must be adjusted according to the experimental objective—whether for single-binding detection, kinetic analysis, or quantification [15].

Depending on the experimental objectives, the interaction analysis can be:

1. *Kinetic analysis*, to determine the association (ka) and dissociation (kd) rate constants and the equilibrium dissociation constant KD.

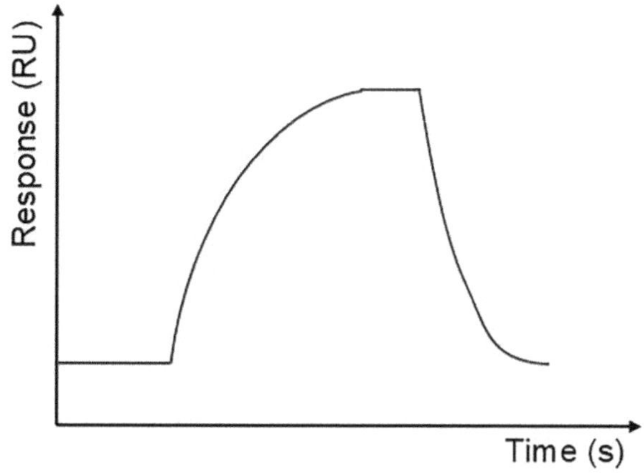

Fig. 3 Typical SPR sensorgram showing the response in resonance units (RU) over time. The curve illustrates the association phase (analyte binding to immobilized ligand), the dissociation phase (analyte release), and the regeneration step to restore the sensor surface

2. *Layer analysis*, injecting multiple concentrations of analyte and fitting the equilibrium responses to a Langmuir isotherm.
3. *Steady-state analysis*, useful for slow dissociation rates, where equilibrium binding levels are reached during injection.
4. *Competition assays*, in which pre-incubated mixtures of analyte and competitors are injected to assess binding protection or specificity.
5. *Sandwich assays*, to identify ternary complexes involving ligand-analyte-secondary analyte systems.

To ensure accurate interpretation:

1. *Double referencing* is recommended: subtracting responses from a reference surface (e.g., blank immobilization) and from injections only, with exceptions.
2. *Mass transport effects* should be minimized by adjusting the flow rate and ligand density. High flow rates (30–100 μL/min) exceed the boundary layer thickness and improve analyte delivery.
3. *Solvent correction* is essential when organic solvents such as DMSO are used, as they alter the refractive index.

Other important considerations include using long or sufficient injection times to capture both fast and slow binding events, assessing potential rebinding effects, and ensuring that equilibrium is reached during dissociation for accurate kinetic tuning.

3.4 Surface Regeneration

Surface regeneration is performed at the end of each analysis cycle to completely remove the bound analyte without damaging the immobilized ligand, allowing multiple binding cycles on the same chip surface. The choice of regeneration solution depends on the strength and nature of the molecular interaction and may include acidic or basic buffers, high-salt solutions, mild detergents, or organic solvents [15, 16]. Regeneration efficiency is assessed by the consistency of signals obtained from repeated injections of the same analyte and ensures that each binding cycle starts from a consistent baseline, critical for quantitative interpretation.

Regeneration conditions depend on the nature of the interaction and may involve:

1. *Low or high pH buffers* (e.g., glycine-HCl pH 2.0–3.0 or Tris pH 9.0).
2. *High salt concentrations* (e.g., 2–5 M NaCl), disrupt ionic interactions.
3. *Chaotropic agents* (e.g., 1–2 M guanidine-HCl or 6 M urea), effective in breaking hydrogen bonds.
4. *Organic solvents* (e.g., 10–50% acetonitrile, methanol, or DMSO), used for hydrophobic or small-molecule analytes.
5. *Detergents* such as SDS (0.05–0.5%) for membrane proteins or lipophilic ligands.

Every regeneration condition ought to undergo empirical testing. The ligand may become denatured or desorbable due to inadequate regeneration, which could jeopardize further analyses. In a standard regeneration step, the selected solution is injected for 30–60 s at a high flow rate (e.g., 50 μL/min), and then the baseline is re-equilibrated using running buffer until it stabilizes.

Restoring the RU signal to baseline while avoiding ligand loss—which is indicated by a progressive drop in the maximal response over multiple cycles—is the aim. Certain surfaces, like streptavidin chips, must be immobilized again for every assay because they are not regenerable.

Before kinetic experiments, regeneration analysis should be carried out, which entails the sequential injection of candidate solutions at various pH and ionic strength levels. The ideal condition is one that maintains ligand activity throughout cycles while quickly eliminating all bound analyte.

3.5 Control, Data Acquisition, and Analysis

All experimental steps are controlled via dedicated software, which allows programming of injection sequences, adjustment of operational parameters, and real-time data recording. Sensorgram analysis includes reference surface subtraction (blanks) and curve fitting to appropriate kinetic models [1]. The inclusion of proper controls, such as inactive surfaces or irrelevant proteins, is essential to validate the specificity of the observed interaction.

All experimental procedures are managed by specialized software that enables real-time data recording, operating parameter adjustment, and injection sequence programming. Reference surface subtraction (blanks) and curve fitting to suitable kinetic models are two aspects of sensorgram analysis. To confirm the specificity of the observed interaction and guarantee experimental reproducibility, appropriate controls—such as inactive surfaces or irrelevant proteins—must be included. Data quality is greatly enhanced by the addition of suitable blanks, reference surfaces, and solvent controls [15].

Recommended controls include:

1. *A reference channel* without ligand or with an irrelevant ligand, used to correct for non-specific interactions, changes in the bulk refractive index or system drift.
2. *Blank injections* (buffer only), typically performed before and after the analyte series to detect system artifacts or instrument drift.
3. *Solvent correction curves*, constructed by injecting a series of known concentrations of DMSO to correct for shifts in the refractive index when DMSO is used.
4. *Replicates* to assess the variability and repeatability of kinetic and affinity measurements.

SPR data acquisition produces sensorgrams that are analyzed using fitting algorithms provided by the instrument software (e.g., BIAevaluation, Scrubber, TraceDrawer). Key kinetic models include:

1. *1:1 Langmuir binding* (single reversible interaction).
2. *Two-state binding* (conformational change after initial binding).
3. *Heterogeneous ligand model*, for surfaces with multiple binding sites.
4. *Bivalent analyte models*, where one analyte molecule binds to two ligand sites.
5. *Steady-state affinity analysis*, for interactions too slow or complex for kinetic resolution.

Statistical parameters such as Chi^2 (<10% of Rmax), residuals (random, low amplitude), and U-values (<15%) indicate the quality of the fit. The fitting process should produce consistent parameters across all concentration series, with good overlap between the experimental and fitted curves.

Proper interpretation also requires consideration of potential artifacts such as:

1. *Mass transport limitation*, which causes apparently slower association rates.

2. *Ligand heterogeneity*, introducing complex binding behavior.
3. *Instrument drift*, impacting baseline stability.
4. *Non-specific binding*, which inflates RH values and distorts kinetics.

To ensure transparency and reproducibility, data should be exported, annotated, and archived. Reporting standards are raised by adding sensorgram overlays, calibration curves, and quality control reports.

Ultimately, for reliable SPR-based interaction studies in a range of research contexts, such as drug discovery, vaccine development, food safety, and biomolecular engineering, meticulous experimental design, exacting controls, surface stability validation, and well-informed model selection are essential.

3.6 Food Applications

The application of SPR in food science has gained significant momentum due to the technique's ability to perform real-time, label-free, and highly sensitive detection of molecular interactions. These features are particularly valuable in complex food matrices, where conventional analytical methods often struggle with selectivity, matrix interference, and time constraints. As the food industry moves toward more efficient, traceable, and safe processing and formulation practices, SPR offers a powerful analytical tool to support quality control, innovation, and regulatory compliance.

3.6.1 Characterization of Functional Ingredient Interactions

The molecular recognition mechanisms underlying protein–bioactive interactions that are pertinent to food systems can be better understood through kinetic and thermodynamic analyses employing SPR under various chemical and environmental conditions, including temperature, ionic strength, and pH [21, 22]. Numerous non-covalent forces, such as hydrogen bonds, van der Waals forces, and electrostatic and hydrophobic interactions, control these intermolecular events. By describing these forces, SPR makes it possible to clarify the mechanisms that propel the formation of supramolecular complexes and the dynamic dissociation and association patterns of interacting molecules. To evaluate the functional potential of these complexes in food matrices, a thorough comprehension of the interaction profile—including binding kinetics and thermodynamics—is necessary [21, 23].

Characterizing the interactions between bioactive substances like vitamins, polyphenols, essential oils, and peptides and macromolecules like proteins, polysaccharides, and lipids is a particularly exciting use of SPR in food science [1, 24]. These interactions affect the bioactives' transport, release profile, and bioavailability in addition to their physicochemical stability. For example, depending on the type and strength of the interaction, polyphenol–protein interactions can either increase or decrease antioxidant activity.

Whey proteins and casein micelles in dairy systems can serve as organic carriers of hydrophobic substances, creating complexes that enhance solubility and stop deterioration during storage or heat processing [12, 14, 22].

Furthermore, by selecting or altering macromolecules to maximize their binding capacity, customized delivery systems can be developed using the interaction patterns that SPR reveals [14, 25]. Research has demonstrated that the binding affinity between anthocyanins and β-lactoglobulin can be adjusted by varying the ionic strength and pH. This suggests that process conditions can be altered to promote the stability of these complexes in dairy-based supplements or functional beverages. Similar to this, flavonoid binding to plant proteins—like soy or pea protein isolates—has attracted attention in the context of plant-based food systems, where the protein's surface charge and structural conformation are important factors in determining the strength of the interaction [26].

SPR has the distinct benefit of real-time, label-free binding event monitoring, which is essential for detecting weak or transitory interactions that are otherwise challenging to identify. In addition to figuring out equilibrium constants, SPR enables researchers to differentiate between slow and fast binding kinetics, indicating whether the interaction is controlled by conformational changes or is diffusion limited [21–23]. Predicting how bioactive–macromolecule complexes will behave under physiological or processing circumstances, such as heating, acidification, or enzyme exposure, is made easier with this level of specificity [25].

Proteins can bind a wide range of bioactives with different kinetics and affinities because of their structural adaptability and functional diversity. These supramolecular assemblies are especially useful for protecting, stabilizing, and delivering sensitive compounds in a controlled manner [27]. By protecting labile sites or changing the compound's solvation environment, these interactions can also stop unwanted reactions like oxidation, hydrolysis, or precipitation. Notably, when linked to carrier macromolecules, bioactive molecules frequently demonstrate improved functional performance, including increased antioxidant or antimicrobial activity, improved solubility, chemical and oxidative stability, and longer shelf life [15, 27]. This improvement is particularly pertinent to the creation of functional foods and nutraceuticals, where preserving compound integrity during processing, storage, and digestion continues to be a major obstacle [22].

Additionally, knowing these molecular-level interactions aids in the logical development of food products with specific release profiles. For example, in encapsulation strategies, the release of the active compound in the stomach, small intestine, or colon is determined by the strength and reversibility of protein–bioactive binding [21, 25]. Therefore, by predicting the gastrointestinal fate of these complexes, SPR studies can help develop delivery systems that maximize bioefficacy by allowing for site-specific release [26].

The creation of food-grade complexes with bioactive substances involves not only proteins but also polysaccharides and lipid-based structures. However, because of their amphiphilia and conformational flexibility, proteins are particularly well-suited for binding investigations [22, 23]. When studied via SPR, the interaction profiles of different protein classes—such as globulins, albumins, or caseins—with various categories of bioactives provide a comparative framework for selecting optimal carriers under defined formulation constraints [24].

All things considered, SPR is a strong analytical tool for developing the field of molecular interactions in functional food systems. SPR provides a deeper understanding of the interactions between food components at the nanoscale by fusing precise kinetic and thermodynamic data with structural insights [26]. This knowledge is essential for creating next-generation food products with improved health-promoting qualities.

3.6.2 Monitoring of Enzyme Activities and Proteolysis

In dairy and fermented food products, the development of texture, flavor, functionality, and nutritional quality is greatly influenced by the enzymatic processing of food proteins. In industrial applications, enzymes like lipases, transglutaminases, and proteases are widely used to modify food matrices in a controlled way [1]. A potent method for tracking enzyme-substrate interactions and enzymatic cleavage events with high sensitivity and temporal resolution is SPR, a real-time, label-free analytical platform [1, 15].

Proteolytic activity is essential for processes like cheese ripening in dairy systems, where the enzymatic hydrolysis of caseins releases peptides and amino acids that determine the final product's flavor and texture. SDS-PAGE and spectrophotometric assays are two examples of traditional proteolysis monitoring techniques that frequently call for extensive sample processing, labeling, or discontinuous sampling [15]. SPR, on the other hand, makes it possible to continuously observe substrate conversion in situ. Proteolytic enzymes are injected into the sensor chip after casein molecules or peptide fragments have been immobilized in order to monitor dissociation events and affinity changes in real time.

Kinetic parameters that provide information about the specificity, efficiency, and stability of the enzymatic reaction under different pH, ionic strength, or temperature conditions include the rate of binding (k_a), dissociation (k_d), and equilibrium dissociation constant (KD) [1, 15]. These parameters can be obtained from the sensorgram data. This degree of resolution is especially useful for comparing the proteolytic activity of microbial versus animal-derived proteases and for screening new or recombinant enzymes for use in food applications.

Additionally, SPR makes it possible to study how endogenous inhibitors, like lactoferrin-derived peptides, or food additives and antimicrobial agents that are added to food in order to increase shelf life or alter bioactivity, inhibit enzymes. One can estimate IC_{50} values and determine mechanisms of inhibition (competitive, non-competitive, or uncompetitive) using kinetic modeling by pre-incubating the enzyme with a potential inhibitor and assessing its decreased binding to the immobilized substrate.

SPR has been successfully used to investigate transglutaminase (TGase)-mediated protein cross-linking in addition to proteolysis [15]. This process is commonly used to improve the elasticity of gluten-free bakery goods, fortify yogurt gels, and restructure meat analogs. Covalent cross-links that improve the mechanical and water-holding qualities of food proteins are created when TGase catalyzes an acyl-transfer reaction between the γ-carboxamide group of glutamine and the ε-amino group of lysine residues [1, 15].

By immobilizing the acyl-donor or acyl-acceptor protein on the sensor surface and flowing the corresponding reaction partner with or without TGase, SPR allows for the visualization of the real-time formation of cross-linked complexes. Covalent bond accumulation is correlated with an increase in response units (RUs), which makes it possible to measure enzymatic efficiency and assess the effects of variables such as temperature, substrate conformation, and calcium ions on cross-linking performance [1, 8].

Since plant-based protein systems frequently lack the viscoelastic qualities of animal proteins, this application is especially pertinent in these settings. SPR clarifies how enzymatic treatment alters interprotein associations, which aids in the logical design of textured plant-based products [15]. To improve formulations for plant-based cheeses or meat alternatives, for instance, the cross-linking of soy proteins or pea globulins can be tracked using SPR [2, 8].

More sophisticated methods combine SPR with models of simulated gastrointestinal digestion, allowing for the dynamic study of proteolytic degradation in physiologically relevant settings [8]. These models aid in predicting the release of bioactive peptides, evaluating the protective effects of encapsulating matrices or inhibitors on enzyme susceptibility, and assessing the reduction of allergenicity after digestion [1, 2].

The parallel testing of several enzyme-substrate systems has also been made possible by recent advancements in multi-cycle SPR protocols and high-throughput chip formats, which have facilitated effective screening workflows for formulation optimization or enzyme engineering [8, 15].

In conclusion, SPR offers a thorough platform for tracking proteolysis and enzymatic activity in intricate food systems. It is essential for both academic research and industrial quality control because it can provide kinetic, affinity, and mechanistic data in real time without labeling or destroying samples [15]. The technique's relevance in contemporary food biotechnology is further supported by its applications, which range from traditional dairy fermentation to innovative plant-based product development.

3.6.3 Investigation of Allergen–Antibody Binding

A major public health concern is food allergies, especially for those with IgE-mediated hypersensitivities. Ensuring food safety and regulatory compliance requires the sensitive and accurate detection of allergenic proteins, such as those in peanuts (Ara h 1–3), milk (β-lactoglobulin, caseins), eggs (ovalbumin), wheat (gliadins), and soy (glycinin, β-conglycinin). Because it allows for the real-time, label-free measurement of the molecular interactions between allergens and immunoglobulins (IgE and IgG subclasses), SPR has emerged as a crucial tool in allergen research [28].

SPR offers comprehensive kinetic and affinity data by immobilizing purified allergenic proteins or food extracts onto the sensor chip surface and allowing antibody-containing sera to flow across it. The dissociation rate constant (k_d) shows the stability of the antigen–antibody complex, whereas the association rate constant (*k*a) shows how rapidly antibodies identify and bind to allergenic epitopes [1, 29]. The equilibrium dissociation constant (KD), which can provide information on allergenic potency and cross-reactivity, is obtained by dividing these parameters by their ratio [15].

When researching the effects of food processing techniques like thermal treatment, fermentation, high-pressure processing, or enzymatic hydrolysis, SPR's ability to assess comparative allergenicity is especially pertinent. These procedures may produce neoallergens, break epitopes, or change the conformation of proteins [28]. SPR makes it easier to compare native and processed proteins side by side, which aids in determining how food processing affects the affinity of IgE for binding. This directly affects the creation of hypoallergenic formulations and the thresholds for allergen labeling [29].

Additionally, competitive inhibition formats can be used, in which serum antibodies are exposed to an immobilized allergen while candidate blocking agents or processed food extracts are present. This makes it possible to screen for possible hypoallergenic alternatives or hiding substances that could lessen the immune response.

3.6.4 Evaluation of Emulsifier and Stabilizer Functionality

Emulsifiers and stabilizers play a critical role in the consistency, mouthfeel, and shelf life of multiphase food systems, including emulsions, foams, and gels. These functions are frequently performed by proteins and polysaccharides, which frequently form complex coacervates or interfacial films that affect the rheology of products. Such interfacial interactions can be monitored in real time under physiologically relevant conditions thanks to SPR.

These interactions can be immobilized directly onto the sensor chip or examined using supported lipid bilayers (SLBs) and monolayers that resemble emulsion interfaces. For example, SPR offers information on the creation of soluble or insoluble complexes, competitive adsorption dynamics, and the function of electrostatic forces when whey protein isolate (WPI) is immobilized and pectin is injected under carefully regulated pH and ionic strength conditions.

These studies guide the creation of stabilized foams, plant-based dairy substitutes, and acidified drinks where phase separation or aggregation must be prevented. Additionally, the method can distinguish between cooperative and competitive binding events, which is helpful when evaluating emulsifier or additive blends.

Adsorption kinetics on oil–water interfaces can be studied in real time using SPR chips with hydrophobic or lipid-modified surfaces in sophisticated applications. This is especially useful for examining the synergistic or antagonistic effects of natural or synthetic emulsifiers (such as lecithins, saponins, and quillaja extracts) with proteins. Such studies are essential for clean-label product development, where synthetic stabilizers are being replaced by plant-based or biodegradable alternatives.

3.6.5 Encapsulation and Controlled Release Mechanisms

Sensitive bioactives (such as vitamins, polyphenols, omega-3 fatty acids, and probiotics) can be protected and their release can be regulated in specific areas of the gastrointestinal tract by using encapsulation technologies, which range from liposomes and microcapsules to nanoparticles and nanogels [7, 15]. When examining the dynamics of interactions between encapsulated systems and receptors, digestive enzymes, and other biomolecules, SPR provides a distinct advantage [1, 8].

Encapsulated substances like curcumin or catechins are exposed to pepsin or pancreatin at appropriate pH levels in order to monitor cargo release under simulated gastric or intestinal conditions. Real-time monitoring of bioactive release and interaction kinetics is made possible by immobilizing target proteins or digestive receptors on the chip [1, 15].

By tracking how intact capsules bind to particular ligands (such as bile salts or mucins) and then analyzing cargo leakage in response to environmental stimuli, SPR can also evaluate the integrity of capsules. This encourages the creation of delivery systems that are sensitive to pH or enzymes, like protein-lipid nanocarriers or alginate-chitosan beads [11].

Furthermore, by measuring the binding of surface-modified delivery systems (such as folic acid, mannose, or antibody-tagged capsules) to lectins, microbial adhesins, or immobilized epithelial cell receptors, their targeting effectiveness can be assessed. The logical development of smart foods and nutraceuticals with improved bioavailability and site-specific action is aided by these data [15].

3.6.6 Integration with Complementary Analytical Techniques

The interpretation of binding events and the structural validation of molecular interactions are significantly improved when SPR is combined with complementary analytical techniques, despite its potency when used alone. In food science, where complex matrices and undefinable mixtures make single-method analyses difficult, this multimodal approach is especially helpful [27, 30].

Direct identification of analytes that have been captured from the sensor surface is made possible by the SPR-MS (Mass Spectrometry) coupling [1]. This allows for the direct identification of allergens, peptides, or contaminants from food extracts by desorbing bound molecules after analysis and examining their mass-to-charge ratios [15, 31].

By fractionating complex mixtures before injecting them into the SPR system, SPR-HPLC (High-Performance Liquid Chromatography) integration enables the kinetic and affinity profiling of individual components [32]. This is useful for characterizing fermentation products, flavor precursors, and plant extracts.

By comparing heat changes with binding rates, the SPR-ITC (Isothermal Titration Calorimetry) pairing can reconcile thermodynamic and kinetic data and provide a more complete picture of the enthalpic versus entropic contributions to interaction strength [1, 31]. This is particularly helpful when assessing how well dietary fibers or polyphenols bind to mucus or digestive enzymes [15, 32].

Lastly, SPR in conjunction with spectroscopic methods (such as FTIR or circular dichroism) allows for the investigation of structural alterations in proteins upon interaction, assisting in the evaluation of allergenicity or the structure–function analysis of food-grade biopolymers.

4 Note

SPR is a robust and versatile analytical technique for studying biomolecular interactions. Its primary characteristic is its high sensitivity and real-time detection of refractive index changes at the interface between the sensor chip and the flowing solution. This makes it perfect for kinetic and affinity analyses since it allows for continuous monitoring of the ligand-analyte association and dissociation processes without the need for chemical labeling.

The technique's foundations and the experimental steps for sample preparation, ligand immobilization, interaction analysis, and surface regeneration were covered in this chapter. To guarantee the accuracy of the results, methodological rigor is required at every stage. Important elements that have a direct impact on the caliber of the response are the appropriate choice of injection parameters, the careful application of control surfaces, and the use of compatible buffers.

Real-time analysis, the potential for complete assay automation, high reproducibility, accurate molecular interaction quantification, and the ability to reuse sensor surfaces while preserving ligand stability are some benefits of SPR. The method is also applicable to a variety of chemical and biological systems since it enables the detection of interactions even in complex samples.

Some practical considerations are necessary to guarantee that SPR's full potential is realized. The biomolecule and intended immobilization method must be compatible with the chip selection. When working with complex biological matrices, such as serum and food-based systems, modified surfaces, like PEGylated or zwitterionic chips, provide better performance. If commercial ligands are used, they should go through purification processes like chromatography or dialysis to get rid of additives and stabilizers that might affect the accuracy of the data and the effectiveness of immobilization. Calibration curves must be created in the presence of solubilizing agents such as DMSO in order to rectify shifts in the refractive index and avoid misinterpreting interaction signals.

Another important consideration is chip surface regeneration. To guarantee analyte removal without jeopardizing ligand stability, regeneration buffers must be empirically chosen. This entails previous scouting with different pH and ionic strength. Good laboratory practices should include calibration procedures and the archiving of instrument performance data in order to protect traceability and repeatability.

Applications of SPR are expanding outside of biomedicine, particularly in the food sector. In this context, it is an effective tool for describing functional ingredients, confirming interactions between bioactive substances and proteins, and keeping an eye out for allergens or contaminants. It is very well suited for routine and high-throughput analyses in industrial settings due to its sensitivity and label-free nature.

Furthermore, a systematic approach is necessary for the experimental workflow in SPR. Verifying the thermal and conformational stability of ligands and analytes is crucial during sample preparation, especially when dealing with sensitive proteins. Prior to injection, methods like fluorescence spectroscopy and circular dichroism can guarantee molecular integrity. Sometimes, ligand immobilization through covalent coupling can decrease biological activity. Enzymatic or binding activity tests are examples of post-immobilization validation assays that should be used to verify that functional sites are still accessible and functional.

Selecting the right kinetic model is essential. Sensorgrams must be interpreted considering the underlying biology of the interaction in addition to the best statistical fitting. Inaccurate conclusions regarding affinity or mechanism may result from models that oversimplify intricate binding events. Standardizing surface regeneration is necessary to prevent ligand performance from declining over time. Consistent outcomes can be achieved by limiting the number of regeneration cycles and tracking the ligand response over time.

Monitoring the ligand's functional activity both before and after repeated use is another crucial aspect of the experimental design. Re-immobilization might be necessary when binding sites gradually become inaccessible, either as a result of denaturation or steric hindrance.

Reducing non-specific binding is crucial in complex matrices like food samples or biological fluids. Background signals can be decreased by adding blocking agents, like casein derivatives or BSA, to the running buffer. Furthermore, it has been demonstrated that applying zwitterionic chip coatings or PEG-modified surfaces greatly enhances signal-to-noise ratios, especially in the analysis of serum or fermented products.

Thermodynamic analyses employing SPR also provide more in-depth insights into interaction mechanisms by determining the enthalpic and entropic contributions to binding through experiments conducted at different temperatures. Despite being more sophisticated, these methods are being used increasingly in food science to investigate the effectiveness and stability of functional additives.

Finally, rigorous use of positive and negative controls—including competition assays and testing mutant ligands—fortifies data interpretation and helps discriminate specific from nonspecific binding effects. When combined, these methodological tips and best practices help SPR analyses produce reliable, repeatable, and biologically significant data.

In the investigation of molecular interactions pertinent to food systems, SPR has become a revolutionary tool. It is specially suited to address issues in food formulation, safety, and quality assurance because it provides label-free, real-time analysis with high sensitivity. As explained in this chapter, SPR provides insights that are essential for food industry innovation by making it easier to characterize intricate relationships between food ingredients, enzymes, contaminants, and functional ingredients.

This chapter seeks to enable researchers to fully utilize SPR by offering exacting experimental procedures, going over important technical issues, and presenting a variety of food applications. SPR is a strong and adaptable platform that can be used to assess enzyme kinetics, identify allergens and pathogens, or maximize ingredient functionality. Its status as a pillar of contemporary food interaction research is further supported by its incorporation with complementary methodologies and flexibility in responding to new analytical requirements.

Acknowledgments

The authors would like to thank the Cândido Tostes Dairy Institute, Minas Gerais Agricultural Company (EPAMIG), and the Minas Gerais Research Foundation (FAPEMIG, Brazil) for funding projects APQ-01782-23, APQ-05926-24, PPE-00045-24, PPE-00090-23, PPE-00079-24, APQ-01760-21.

References

1. de Mol NJ, Fischer MJE (2010) Surface Plasmon resonance: a general introduction. In: de Mol NJ, Fischer MJE (eds) Surface plasmon resonance: methods and protocols. Humana Press, New York, pp 1–14
2. Myszka DG (1999) Improving biosensor analysis. J Mol Recognit 12(5):279–284
3. Rich RL, Myszka DG (2006) Survey of the year 2005 commercial optical biosensor literature. J Mol Recognit 19(6):478–534
4. Mitchell JS, Wu Y (2010) Surface plasmon resonance signal enhancement for immunoassay of small molecules. In: de Mol NJ, Fischer MJE (eds) Surface plasmon resonance: methods and protocols. Humana Press, New York, pp 113–130
5. Karlsson R (2004) SPR for molecular interaction analysis: a review of emerging techniques and applications. J Mol Recognit 17(3):151–161
6. Fischer MJE (2010) Amine coupling through EDC/NHS: a practical approach. In: de Mol NJ, Fischer MJE (eds) Surface Plasmon resonance: methods and protocols. Humana Press, New York, pp 55–74
7. Katsamba PS, Park S, Laird-Offringa IA (2002) Kinetic studies of RNA–protein interactions using surface plasmon resonance. Methods 26(2):95–104
8. Schuck P, Zhao H (2010) The role of mass transport limitation and surface heterogeneity in the biophysical characterization of macromolecular binding processes by SPR. In: de Mol NJ, Fischer MJE (eds) Surface plasmon resonance: methods and protocols. Humana Press, New York, pp 15–54
9. Duverger E, Lamerant-Fayel N, Frison N, Monsigny M (2010) Carbohydrate–Lectin interactions assayed by SPR. In: de Mol NJ, Fischer MJE (eds) Surface plasmon resonance: methods and protocols. Humana Press, New York, pp 157–178
10. Homola J (2008) Surface plasmon resonance sensors for detection of chemical and biological species. Chem Rev 108(2):462–493
11. Gobi KV, Iwasaka H, Tanaka H, Miura N (2007) Highly sensitive and selective immunosensor for ochratoxin a using surface plasmon resonance-enzyme flow system. Sens Actuators B Chem 123(1):177–184
12. Aslan K, Lakowicz JR, Geddes CD (2004) Nanogold-plasmon-resonance-based glucose sensing. Anal Biochem 330(1):145–155
13. Nedelkov D (2010) Integration of SPR biosensors with mass spectrometry. In: de Mol NJ, Fischer MJE (eds) Surface Plasmon resonance: methods and protocols. Humana Press, New York, pp 261–268
14. Schasfoort RBM (2017) Handbook of surface plasmon resonance, 2nd edn. Royal Society of Chemistry, Cambridge
15. Nagata K, Handa H (2000) Real-time analysis of biomolecular interactions: applications of BIACORE. Springer Japan, Tokyo
16. Shank-Retzlaff M, Larkin M (2002) Estimating the purity of immunoglobulin G using UV absorbance and SDS-PAGE. Biotechnol Prog 18(6):1302–1306
17. Ohlson S (2016) Designing transient binding drugs using weak affinity chromatography and surface plasmon resonance. Methods Mol Biol 1439:257–272
18. Malmqvist M (1993) Surface plasmon resonance for detection and measurement of antibody-antigen affinity and kinetics. Curr Opin Immunol 5(2):282–286
19. Kim HJ, Lee SJ (2010) Optimization of buffer exchange and sample preparation for SPR assays. Biosens Bioelectron 26(2): 845–851
20. Karlsson R (1994) Affinity analysis of non-steady-state data obtained under mass transport limited conditions using BIAcore technology. J Mol Recognit 7(3):199–204
21. Li X, Gong P, Zhao Q, Zhou X, Zhang Y, Zhao Y (2022) Plug-in optical fiber SPR biosensor for lung cancer gene detection with temperature and pH compensation, vol 359. Sensors and Actuators B: Chemical, p 131596

22. Zhang L, Wang H, Zhang H, Zhang N, Zheng X, Li W et al (2022 Oct) Development of a portable multiplexed instrument for multi-proteins detection in human urine using surface plasmon resonance. Sensors Actuators B Chem 15(369):132272
23. Pattnaik P (2005) Surface plasmon resonance applications in understanding receptor ligand interaction. Appl Biochem Biotechnol 126:079
24. Myszka DG (1997) Kinetic analysis of macromolecular interactions using surface plasmon resonance biosensors. Curr Opin Biotechnol 8(1):50–57
25. Santa Rosa LN, de Paula Rezende J, Coelho YL, Mendes TAO, da Silva LHM, dos Santos Pires AC (2021) β-Lactoglobulin conformation influences its interaction with caffeine. Food Biosci 44:101418
26. Pacheco AFC, Nunes NM, de Paula HMC, Coelho YL, da Silva LHM, Pinto MS, dos Santos Pires AC (2020) β-Casein monomers as potential flavonoids nanocarriers: thermodynamics and kinetics of β-casein-naringin binding by fluorescence spectroscopy and surface plasmon resonance. Int Dairy J 108:104710
27. Gheonea I, Aprodu I, Râpeanu G, Stănciuc N (2018) Binding mechanisms between lycopene extracted from tomato peels and bovine β-lactoglobulin. J Lumin 203:582–589
28. Zhou J, Qi Q, Wang C, Qian Y, Liu G, Wang Y et al (2019) Surface plasmon resonance (SPR) biosensors for food allergen detection in food matrices. Biosens Bioelectron 142:111449
29. Zhou J, Wang Y, Qian Y, Zhang T, Zheng L, Fu L (2020) Quantification of shellfish major allergen tropomyosin by SPR biosensor with gold patterned biochips. Food Control 107:106547
30. Rezaei A, Fathi M, Mahdi S (2019) Nanoencapsulation of hydrophobic and low-soluble food bioactive compounds within di ff erent nanocarriers. Food Hydrocoll 88:146–162
31. Pertig D, Schäfer C, Ulrich J (2012 Jun) Stabilization of carotenoids. Chem Eng Technol 35(6):1045–1050
32. Morales P, Barros L, Ferreira ICFR (2016) Food colorants: challenges, opportunities and current desires of agro- industries to ensure consumer expectations and regulatory practices. Trends Food Sci Technol 52:1–15

Chapter 16

Advanced Rheological Techniques for Texture and Structural Analysis of Foods

Khwala Ben Jeddou and Amir Bouallegue

Abstract

Rheological instrumentation and rheological measurements have become essential tools in the analytical laboratories of food companies for characterizing ingredients and final products, as well as for predicting product performance and consumer acceptance. However, this characterization can be difficult to determine using a traditional rheometer due to the complex composition, structure of the food, as well as the difficulties in accounting for the changes to the food during oral processing. This study presents an advanced rheological technique that integrates large amplitude oscillatory shear, extensional rheology, interfacial rheology, microrheology, tribology, and texture profile analysis to provide a comprehensive assessment of the mechanical and sensory-relevant properties of food systems. The technique enables the deconvolution of linear and nonlinear viscoelastic responses, allowing for detailed insights into structural integrity, yield behavior, and breakdown mechanisms under processing and consumption-like conditions. By applying this methodology to a range of complex food matrices, including gels, emulsions, and semisolid foods, the study demonstrates its capability to correlate rheological parameters with sensory attributes such as creaminess, firmness, and spreadability. This approach offers a robust, non-destructive analytical framework for both food formulation and quality control, advancing the field of food texture and structure analysis.

Key words Rheology, Traditional rheometers, Advanced rheometers, Viscosity, Elasticity, Yield stress, Viscoelasticity

1 Introduction

Food texture, encompassing the sensory and functional properties of food as perceived through touch in the mouth, is a key quality factor determining consumer acceptance and overall product quality [1]. Indeed, quality attributes such as spreadability and creaminess are extremely important to the acceptance of semisolid food products by consumers. This complex attribute influences not only the eating experience but also the processing, handling, and storage of food products. Due to the diverse mechanical behavior that foods typically exhibit, ranging from low-viscosity fluids to

Tanmay Sarkar (ed.), *Novel Food Analysis: Techniques and Applications*, Methods and Protocols in Food Science, https://doi.org/10.1007/978-1-0716-4787-5_16,

semisolids and gels to hard, solid-like food products, rheology has become increasingly significant for foods [2].

The field of rheology studies flow behavior and, more specifically, the deformation of a certain material under the effects of an applied stress or strain [3, 4]. In fact, it studies the "yield stress fluid" that refers to any material or model that exhibits a dramatic change in viscosity (orders of magnitude) over a small range of applied stress [5]. A fundamental methodology is used by rheologists to characterize fluid flow characteristics crucial to food processing as well as comprehend interactions between food components and dynamics of structure development vital to food quality [2]. By examining how food materials respond to applied forces, rheology allows scientists and engineers to quantify characteristics such as firmness, viscosity, elasticity, and spreadability. This understanding is crucial for predicting food behavior during various stages, from manufacturing and packaging to the final consumption and digestion within the human body [6].

Rheological instrumentation and rheological measurements have become essential tools in the analytical laboratories of food companies for characterizing ingredients and final products, as well as for predicting product performance and consumer acceptance. In fact, it aids in the development of new products, assists in determining the physical characteristics of liquid and semi-solid foods, evaluates the quality of the raw materials to be used in the creation of a particular product while also assessing the quality of the food product at various stages of its development, determines the shelf-life of food products, evaluates the sensory properties of food products, and aids in the indirect study of food microstructure.

Historically, the assessment of food rheology often relied on traditional methods, primarily focusing on measuring the viscosity of liquid and semi-liquid foods using basic viscometers [7]. One significant limitation of basic viscometers is their inability to fully capture the viscoelastic nature of foods [8]. Many foods exhibit both viscous and elastic responses to deformation, and simple viscosity measurements primarily address the viscous component, failing to provide information about the material's elasticity or its ability to recover from deformation. Furthermore, traditional methods often lack the sensitivity to analyze the time-dependent properties of foods, such as thixotropy, where viscosity changes over time under constant shear, or viscoelastic relaxation, where stresses within a deformed material dissipate over time [9]. The behavior of foods can also vary significantly under different conditions of stress or strain, which is crucial for understanding their performance during various stages of processing, storage, and consumption. To overcome these limitations and gain a more profound understanding of food texture and structure, advanced rheological techniques have been developed. These advanced methods, including large amplitude oscillatory shear, extensional rheology, interfacial rheology, microrheology, tribology, and texture profile analysis, provide insights into material behavior across a wider range of deformation conditions and timescales, ultimately enabling the design and production of foods with desired textural qualities. As a result, rheological approaches are now being used more frequently [10].

This report aims to explore these advanced rheological techniques in detail, highlighting their principles, applications across various food categories, their role in understanding processing effects, and their correlation with sensory perception.

2 Materials

2.1 Food Samples

Food is a very complex, structured material, which is made up of water, proteins, carbohydrates, fats, and a significant amount of fiber. Food's structural behavior and flow are determined by its constituent parts, and structured fluids and stability are influenced by a number of variables [11]. All these constituents influence the flow and structural behavior of foods significantly, and therefore, the rheological properties of food are very dissimilar from those of conventional polymeric materials. Various factors affect the stability of structured fluids. The viscosity of the liquid phase in dispersions mostly plays a major role in the flow properties of the material. In fact, certain foods, such as food gels and emulsions, exhibit a unique and somewhat predictable rheological behavior that can be used to interpret their response to applied stress or strain [12]. One of the most important physical properties of an emulsion is its flow behavior, as rheology influences both its technical performance and its sensory attributes. Therefore, it is equally important to measure and to interpret the flow behavior of an emulsion [4, 13, 14]. Indeed, the viscosity of the liquid phase in dispersion affects a food material's ability to flow. The emulsion or suspension qualities, or dispersion characteristics, of a food product can be used to predict its behavior. Food dispersion, for instance, is crucial in some frequently consumed meals, including mayonnaise, tomato paste, sauces, and infant food [15].

The rheological behavior of dispersions is influenced by a number of factors, including particle size distribution, particle concentration, and antiparticle interactions. According to research, suspensions are made up of solid particles dispersed in a liquid medium. Emulsions, on the other hand, are made up of liquid droplets in a liquid medium (deformable particles). The temperature and length of the measurement have a major impact on the rheological characteristics of food as well [2]. Food behavior is greatly influenced by gel formation. Most foods containing proteins, carbohydrates, and polysaccharides become gels when exposed to heat or pressure. A gel's microstructure can be changed to create a variety of products with different physical properties, such as soft hydrogels and hard rubbery polymers [16].

Food perceives crispiness or crunchiness through an angular force displacement curve that appears after compression or cutting. Therefore, research that combines investigation of food structure, mechanical qualities, and oral processing with sensory texture attributes is essential to examine and comprehend the complex nature of food texture [17]. The ability to anticipate the quality of the final product from tests on raw materials or partially processed goods is of utmost importance to the food industry. The frozen dough, where a recent study was able to link the compression behavior of the dough to the quality of finished bread samples. Due to this, manufacturers can intervene and make the necessary changes to the bread-making process. Athawale et al. [18] demonstrates that rheological measurements performed on suspensions of cheese powder can be used to predict how well the cheese powders operate while making cheese crackers. It is frequently difficult to relate functionality to interfacial rheological properties, such as foam or emulsion stability. Interfacial rheology, however, can be utilized to explore the mechanics underlying phenomena seen in aggregate, even in highly complex systems [19].

2.2 Instrumentation

A rheometer is a tool for rheology research. It is an instrument widely applied to measure the rheological properties of polymer melt, polymer solution, suspension, emulsion, coating, ink, food materials, etc. [20]. It measures the stress-strain relationship to understand the flow/deformation properties of a material. A commercial rheometer can be divided into an analytical rheometer and an experiential rheometer. Most rheometers use the same measurement geometry (cone, plate, parallel plate, concentric cylinder, etc.) and have a comparable measurement range. However, the design difference between different rheometers varies greatly [20]. The measuring geometry is one of the most important components of the rheometer, since it allows us to adjust the deformation of the tested food material, according to the experimental design [21]. Depending on the physical properties of the sample being tested, several different geometries can be adjusted to the rheometric instruments. For example, cone-plate and parallel plate geometries are commonly used for liquid and semi-solid foods like as viscous pastes, gels, and concentrated suspensions, while concentric cylinder geometries are suitable for low-viscosity fluids and suspensions [22]. Vane rotors are often used for materials containing large particles to prevent slippage.

Advanced rheological analysis of foods relies on a variety of advanced instruments, each designed to apply specific types of deformation and measure the resulting material responses with high accuracy. Understanding the capabilities of these instruments is crucial for selecting the appropriate technique for a given food system and research question. Commercially, there are different types of rheometer available in the market; each rheometer follows a different principle, such as rotation, creeping, relaxation, swinging, and other methods [20].

1. Rotational rheometers are the most widely used rheological method that involve the utilization of a rotational device, which can be stress-controlled or rate-controlled, to study materials under a multitude of conditions [22]. The rotational rheometers are versatile instruments that apply a controlled rotational motion to a food sample, either by imposing a specific stress (controlled stress rheometers) or a specific strain (controlled strain rheometers), and measure the resulting deformation or stress, respectively [4]. These rheometers can perform a wide range of tests, including steady shear viscosity measurements, yield stress determination, creep and recovery analysis, and oscillatory shear tests (both SAOS and LAOS).
2. Oscillatory rheometers are a specialized type of rotational rheometer optimized for applying and measuring oscillatory shear deformations. They are essential for characterizing the viscoelastic properties of foods using SAOS and LAOS techniques, allowing for the separate determination of the storage and loss moduli as a function of frequency and amplitude [23].
3. Extensional rheometers are specifically designed to apply and measure the rheological properties of materials under extensional (stretching) flow. There are different types of extensional rheometers, including those that use uniaxial extension, biaxial extension (e.g., for dough inflation), and filament stretching techniques, catering to various food consistencies and applications [7].
4. Tribometers are specialized instruments used to measure the friction and lubrication properties of food samples under conditions that mimic the interactions between food and oral surfaces during consumption. They typically involve sliding a food sample between two model surfaces (representing the tongue and palate) and measuring the frictional force as a function of sliding speed and applied load [24].
5. Microrheometers encompass a range of instruments and techniques used to measure rheological properties at the microscopic level. These include optical tweezers (which use focused laser beams to manipulate micron-sized particles), atomic force microscopes (AFM), and particle tracking systems coupled with high-resolution microscopy [25]. These instruments allow for the investigation of local viscoelastic properties and spatial heterogeneity within food materials.

6. Texture analyzers are universal testing machines equipped with a variety of probes and fixtures designed to perform a wide range of texture measurements, including Texture Profile Analysis (TPA), compression tests, puncture tests, and shear tests [1]. These instruments simulate different aspects of food handling and consumption and provide quantitative data on textural attributes like hardness, springiness, and chewiness.

Advanced rheometers are highly capable of performing a variety of tests beyond simple viscosity measurements. They can conduct oscillatory shear tests (SAOS and LAOS) to determine viscoelastic moduli and non-linear behavior, perform creep and recovery experiments to assess time-dependent deformation and elasticity, and carry out frequency sweeps to investigate the material's response over a range of timescales.

The diverse array of advanced rheological instrumentation available to food scientists provides a powerful toolkit for characterizing the texture and structure of foods across a wide range of length scales and deformation conditions. Understanding the principles and capabilities of each instrument type is essential for selecting the most appropriate technique to address specific research questions or product development challenges in food science.

3 Methods

3.1 Fundamental Rheological Tests

3.1.1 Steady-State Flow Measurements

Steady shear (rotational) testing is typically performed on semisolid and fluid samples, as a complete rotation of the rheological attachment is required. Shear rate sweeps are the most common steady-shear test performed on foods, although yield stress tests and testing at a constant shear rate to determine thixotropic behavior are also used. Steady shear tests are used to determine viscosity profiles and shear-dependent behaviors. Data from these tests are often related to sensory viscosity, thickness, smoothness, and slipperiness [26].

In these tests, the goal is to get the general equation that relates stress and shear rate (or rate of deformation), which determines the rheological behavior of a fluid. To get this equation, a constant stress/strain rate is applied for a time, and the resulting strain rate/stress is measured. Although a range of stresses/strains is applied to complete a measurement, each measurement is obtained at a constant value, which explains why these tests are called steady tests [27].

Transient testing can be performed on fluid, semisolid, and solid foods, and includes stress relaxation, creep, and oscillatory testing. These tests, particularly strain and frequency sweeps, provide information about both the food's viscous and elastic components and, consequently, the viscoelastic behavior, which can be good indicators of sensory textures such as firmness, chewing resistance, smoothness, and adhesiveness [28]. They can also indicate long-time flow behaviors, which may be important for packaging selection [29] and shelf life determination. While empirical analyses can and have been used to determine food rheological behaviors, fundamental research is needed to build strong rheology sensory links that hold for more than one food product.

3.1.2 Dynamic Oscillatory Testing

Oscillatory tests are used to measure the viscoelastic and deformation behavior of materials. Such methods, also known as dynamic mechanical analysis (DMA), are applicable to liquid, semisolid, and solid materials [30]. They can be divided into two regimes: (1) Small Amplitude Oscillatory Shear (SAOS) and (2) Large Amplitude Oscillatory Shear (LAOS). Breakdown of food structure is the major limitation of the steady-state measurement, especially at higher shear rates. To overcome those limitations, SAOS, a nondestructive rheological technique, has been used extensively for the characterization of food materials. It is a powerful rheological technique used to probe the viscoelastic properties of food materials without damaging their structure [30].

Amplitude sweeps are often used for characterizing the structural nature of food materials. They may be conducted in strain or stress-controlled mode and are typically conducted at moderate frequencies [30]. Strain (amplitude) sweeps impart an oscillatory strain ramp on the sample and measure the resulting stress response wave; stress (amplitude) sweeps impart an oscillatory stress ramp and measure the resulting strain response wave. Both tests are performed at constant temperature and frequency. They can be used to measure viscoelastic moduli and phase angle and thus the rigidity of internal structure. Amplitude sweeps are generally conducted to determine the linear viscoelastic region (LVR) [30]. Indeed, a small sinusoidal stress or strain is applied to the sample, and the resulting strain or stress response is measured within the LVR [31]. The LVR is the range of stress or strain where the material's rheological properties remain independent of the applied deformation, ensuring that the test is non-destructive and reflects the intrinsic behavior of the food microstructure [32]. SAOS measurements yield several key rheological parameters that provide insights into the viscoelastic nature of food. Viscoelastic moduli, including the storage modulus (G', Pa), loss modulus (G'', Pa), and complex modulus ($G*$, Pa), can be calculated from stress-strain data. Storage and loss moduli are calculated using the phase angle (δ, rad)

1. Storage modulus (G', Pa) quantifies the elastic component, indicating the amount of energy stored in the material during deformation and recovered upon the removal of stress [33].

2. Loss modulus (G″, Pa) quantifies the viscous component, indicating the energy dissipated in each cycle as heat during deformation due to internal friction [33]. Storage modulus represents elastic (viscoelastic solids) behavior, capable of storing more energy, and loss modulus represents viscous (viscoelastic fluids) behavior, where more energy is lost during flow.
3. Complex modulus (G*, Pa) represents the overall resistance of the material to deformation, and it is calculated by dividing the amplitude of the stress wave by the amplitude of the strain wave.
4. Phase angle (δ, rad) provides a measure of the balance between the elastic and viscous behaviors of the material. It is defined as the arc tangent of the ratio of G″ to G′ ($\tan \delta = G''/G'$)0.7 A phase angle close to 0° indicates that the material behaves predominantly elastically (viscoelastic solids), while a phase angle close to 90° indicates predominantly viscous behavior (viscoelastic fluids) [33].

SAOS finds diverse applications in food science for characterizing the texture and structure of various food systems:

1. Frequency sweeps can be used to make qualitative comparisons of gel structural type, such as food gels formed by polysaccharides or proteins [34] and direct comparisons of gel viscoelastic behaviors as a function of temperature or other treatment variables [35]. In fact, frequency sweeps impart an oscillating frequency ramp to a sample at constant strain and temperature. Soft gels show viscoelastic moduli with a strong frequency dependence: they may have fluid-like behavior at low frequencies, then shift to solid-like behavior at higher frequencies. For instance, a strong gel typically exhibits a G′ that is much higher than G″ and relatively independent of frequency.
2. SAOS is also crucial for characterizing the stability and rheological properties of emulsions and other dispersed systems [31]. The viscoelastic moduli can provide information about droplet interactions and the overall structure of the emulsion.
3. Determining the LVR of food materials is essential for ensuring that rheological measurements are performed under non-destructive conditions. Indeed, stress sweeps are frequently used to determine the LVR. In the LVR, the stress increases linearly with deformation; consequently, the elastic G′ and viscous G″ moduli, which relate both magnitudes, are constants. Outside the LVR, both moduli decrease [36].
4. SAOS is valuable for analyzing the gelation process of biopolymer solutions by monitoring the changes in G′ and G″ over time or as a function of temperature. The point at which G′

becomes greater than G″ (gel point) indicates the formation of a continuous network structure [36].

5. The technique is also used to investigate the effect of temperature on food structure, such as the melting of fats in cheese or the gelatinization of starch [36]. Temperature sweep tests, where the temperature is gradually increased or decreased while measuring G′ and G″, can reveal important thermal transitions in food materials.
6. SAOS can be employed to evaluate the stability of food formulations over time or under different storage conditions by tracking changes in their viscoelastic properties. Significant changes in G′ or G″ can indicate instability, such as phase separation or structural breakdown [36].

SAOS provides a non-invasive and sensitive method for characterizing the fundamental viscoelastic properties and microstructure of foods. The frequency dependence of the storage and loss moduli can reveal the nature and strength of interactions within the food matrix, while temperature sweeps can elucidate critical phase transitions. The ability to work within the LVR ensures that the measurements reflect the inherent material properties without inducing structural damage.

3.1.3 Creep-Recovery

Creep tests are used for viscoelastic food materials (e.g., dough, cheese) which exhibit both viscous flow and elastic deformation to ascertain the texture stability. Creep and recovery analysis is a rheological technique used to investigate the time-dependent deformation and elastic recovery of food materials under a constant applied stress [37]. A creep test helps understand two situations: the degree of elastic deformation and rate of viscous flow on the application of a small (i.e., below any yield point) shear stress to a material, and the total recoverable elastic deformation on the removal of that stress, as shown in Fig. 1. This test takes into consideration responses to the application and/or removal of a

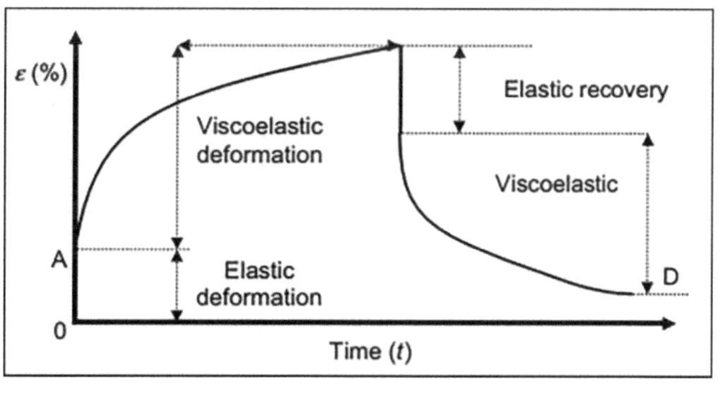

Fig. 1 Strain vs. time curve showing the creep and creep-recovery phenomena

constant defined shear stress by measuring shear strain over time. Typically, the small but defined shear stress is applied until an equilibrium shear strain rate is achieved. Once this constant rate is reached, the applied shear stress is removed, and the material is allowed to reach an equilibrium state. Such a test measures the rigidity/flexibility of a food product, and it plays a significant role in food product development.

Several key rheological parameters are derived from creep and recovery analysis, providing insights into the material's mechanical properties:

1. Creep compliance (J(t)) is a measure of the material's strain response over time to a constant applied stress. It is defined as the ratio of observed strain to stress applied (J(t) = strain/stress) and indicates the material's susceptibility to long-term deformation under load. A higher creep compliance signifies a greater tendency to deform.
2. Recovery percentage quantifies the extent to which the material recovers its original dimensions after the removal of the stress. It is calculated as the ratio of the recovered strain to the maximum strain achieved during the creep phase. A higher recovery percentage indicates a greater degree of elasticity [38].
3. Retardation time is a characteristic time constant that describes the rate at which the material's deformation is delayed or retarded during the creep phase. It is related to the viscoelastic elements within the material's structure [38].
4. Zero shear viscosity is the most commonly used piece of information generated by a creep test; it refers to the viscosity of the material observed under very low shear conditions [33].

Creep and recovery analysis have diverse applications in food science for understanding the textural and structural properties of various food systems:

1. It is used to evaluate the long-term stability and shelf life of food products by assessing their resistance to deformation under sustained stress and their ability to recover over time [32]. For example, it can predict whether a sauce will sag over time or if a gel will maintain its shape.
2. The technique is employed to determine the viscoelastic properties of semi-solid and solid foods, such as gels, cheese, and dough. The creep and recovery curves and the derived parameters provide information about the relative contributions of elastic and viscous components to the material's behavior.
3. Creep analysis can assess the ability of a food material to retain its shape under constant forces, such as gravity or its own weight, which is crucial for products like icings, dips, and thick sauces to prevent undesirable flow or sagging.

4. Researchers use creep and recovery tests to study the effects of different ingredients, such as polysaccharides in emulsion gels [39], and processing conditions, like temperature, on the viscoelastic behavior of food formulations [38].
5. The parameters obtained from creep and recovery analysis have been shown to correlate with sensory attributes related to texture, such as firmness, springiness, and elasticity, providing a link between instrumental measurements and human perception [37].

Creep and recovery analysis is a powerful tool for characterizing the time-dependent mechanical behavior of foods. By observing how a material deforms under constant stress and how it recovers upon stress removal, researchers can gain valuable insights into its elasticity, resistance to flow, and long-term stability. These insights are crucial for predicting the textural evolution of foods during storage and their behavior during consumption.

3.1.4 Stress Relaxation

Relaxation is the counterpart of creep. It is the decline in stress with time, in response to a constant applied strain, at constant temperature. This technique is particularly useful for understanding how internal stresses in a food dissipate over time and provides insights into the material's molecular mobility and structural rearrangements. Figure 2 shows schematically illustrates the stress relaxation behavior of three different types of materials: elastic solid, viscous fluid, and viscoelastic material. Elastic solids exhibit a perfect memory of their original state and have a constant stress and relaxation modulus over time, while the stress of ideal viscous liquids declines to zero immediately when strain is applied. Viscoelastic materials exhibit an intermediate behavior and relax gradually with the end point depending on the molecular structure of the tested material [30].

Several key rheological parameters are derived from stress relaxation studies, offering valuable information about the viscoelastic nature of foods:

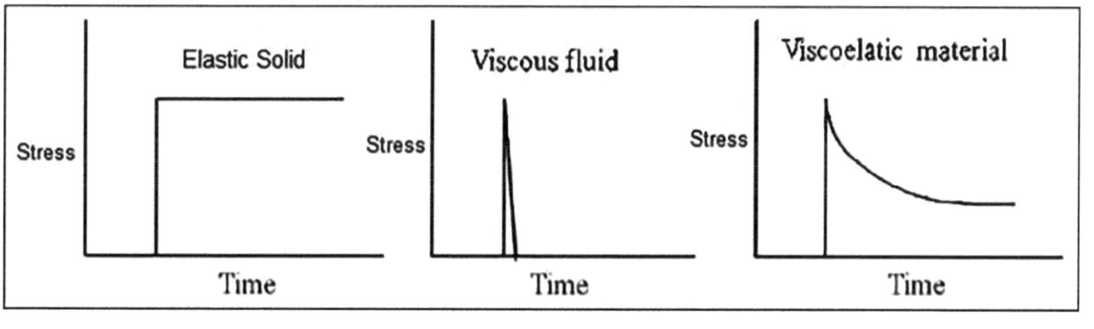

Fig. 2 Stress relaxation behavior of three different types of materials subjected to same amount of strain

1. The relaxation modulus (G(t)) is a time-dependent measure of the material's stiffness. It represents the stress required to maintain the constant strain at a given time and typically decreases as the material relaxes over time.
2. The relaxation time is a characteristic time constant that describes the rate at which the stress decays within the material.
3. The equilibrium stress is the residual stress that remains in the material after a sufficiently long period of time when the stress has reached a plateau.

Stress relaxation studies have a wide range of applications in food science for characterizing the textural and structural properties of various food systems:

1. They are used to characterize the viscoelastic properties of different food materials, including gels [40], dough [41], and solid and semi-solid cheeses [42]. The stress relaxation behavior can reveal information about the strength and stability of the material's network structure.
2. These studies help in understanding how internal stresses within a food material dissipate when it is subjected to a constant deformation, which is relevant to processes that occur during eating, such as the compression of food between the teeth or the tongue and palate during chewing [43].
3. Researchers utilize stress relaxation tests to evaluate the influence of different ingredients, such as salt concentration in gelatin gels [44], and processing conditions, like temperature in raw meat [41], on the viscoelastic behavior and textural attributes of foods.

The parameters obtained from stress relaxation studies, such as the relaxation time and the extent of stress decay, have been correlated with sensory attributes related to texture, including firmness and chewiness, providing a quantitative link between instrumental measurements and human perception [43]. Stress relaxation analysis provides valuable information about the time-dependent response of foods to a constant deformation, offering insights into their internal structure and molecular mobility. The rate and extent of stress decay reveal how effectively a food can relax under a sustained strain, which is crucial for understanding its textural behavior during handling and consumption.

Traditional rheological methods, such as steady-state viscometry and small amplitude oscillatory shear (SAOS), have long been used to measure basic properties like viscosity and viscoelastic moduli. Viscometry quantifies a material's resistance to flow, often applied to liquids like sauces or beverages. SAOS tests are considered the canonical method for probing the linear viscoelastic properties of foods because of its solid theoretical background and the ease of implementing suitable test protocols [45]. However, the deformations can be large and rapid for most of the food processing operations, providing an incomplete picture of the material's response. Consequently, the information obtained from traditional techniques was often insufficient to accurately predict the sensory attributes of food beyond simple descriptors like thickness or thinness. These shortcomings have driven the development of advanced rheological techniques that offer greater sensitivity and applicability to real-world scenarios.

3.2 Advanced Techniques

A transformation of measurement from rotational viscometry to either a controlled stress/strain rheometer or more advanced optimal Fourier transformation rheometry brought the accuracy, sophistication, and reliability to rheological data. Advanced rheometers are sophisticated instruments capable of applying a wider spectrum of controlled stresses and strains, including oscillatory motions that allow for the separate measurement of viscous and elastic components [4]. These instruments enable the investigation of dynamic rheological properties, such as viscoelastic moduli, yield stress, and time-dependent behavior, allowing for a more thorough characterization of flow and deformation behavior [4]. The evolution toward these advanced techniques reflects the increasing need for a deeper understanding of the complex material science underlying food texture.

3.2.1 Large Amplitude Oscillatory Shear (LAOS)

Large Amplitude Oscillatory Shear (LAOS) is an advanced rheological technique that extends beyond the limitations of SAOS by applying oscillatory deformations with amplitudes that are large enough to probe the non-linear viscoelastic response of food materials, which are closer to real processing and application conditions [46]. This is crucial because many food processing operations, such as mixing, kneading, and extrusion, as well as the oral processing of food during consumption (chewing, swallowing), involve large deformations that can significantly alter the food's microstructure and rheological properties [47]. LAOS measurements are related to the sensory and textural properties of food, which is a topic of great interest [48]. LAOS employs advanced analysis methods to characterize the non-linear viscoelastic behavior, including:

1. Higher harmonics analysis (e.g., Fourier Transform): When a material exhibits non-linear behavior under large amplitude oscillations, its stress response is no longer a simple sinusoidal wave but contains higher harmonic frequencies. Fourier Transform analysis can decompose this complex stress waveform into its fundamental and higher harmonic components, providing quantitative information about the nature and degree of non-linearity [47]. The amplitudes and phases of these higher

harmonics can be related to specific microstructural events occurring within the food material under large deformation.

2. Lissajous-Bowditch curves: These are graphical representations of the stress response as a function of the applied strain (or strain rate) within a single oscillatory cycle [23]. The shape of these curves deviates from the elliptical form observed for linear viscoelastic materials, providing a visual indication of the material's non-linear behavior, including phenomena like strain stiffening or softening.
3. Chebyshev coefficients: These mathematical coefficients can be used to decompose the non-linear stress response into a series of orthogonal polynomials, allowing for a more detailed and quantitative analysis of the elastic and viscous contributions to the non-linear behavior [23].
4. Strain stiffening (S) and shear thickening (T) indices: These dimensionless parameters provide a quantitative measure of the intra-cycle non-linear behavior. The strain stiffening index (S) reflects the increase in elastic stress with increasing strain amplitude within a cycle, while the shear thickening index (T) reflects the increase in viscous stress with increasing strain rate amplitude within a cycle [23].

LAOS has a wide range of significant applications in food science:

1. It is used to characterize the rheological behavior of foods under conditions that closely mimic industrial processing operations, such as the mixing of dough, extrusion of food materials, and other processes involving large deformations [47].
2. LAOS can also simulate the oral processing of food during consumption, including chewing and swallowing, providing insights into how food structure breaks down and reforms in the mouth [47].
3. This technique has become a standard tool for the description of the mechanical characteristics of soft matter and, as nondestructive tests, is used to determine the basic material constants.

Advances in the construction of rotary rheometers have allowed for a natural extension of the measurement range to the nonlinear range. This has resulted in further advancement of knowledge concerning the nonlinear rheological behaviors occurring in foods. The advanced analysis methods associated with LAOS provide a better prediction of food texture during processing and consumption [5].

3.2.2 Extensional Rheology

Extensional rheology is an advanced rheological technique that focuses on the behavior of food materials when subjected to stretching or elongation flow [49]. The knowledge of the rheological behavior in shear does not suffice to describe the behavior during extension. The rheological response of a complex fluid undergoing shear such as a polymer solution or melt injected into a die or mold, drawn through an extrusion die, blow molded, calendared, etc., will not correctly predict the processing behavior during such operations: inferring the extensional flow behavior from rheological tests performed in shear is usually impossible. There are three basic types of extensional (or elongational) flows: uniaxial, biaxial, and planar. In the uniaxial extensional flow, a material is stretched in one direction with a simultaneous reduction of the dimensions of that material in the two remaining directions. In the biaxial extensional flow, a material is stretched in two directions, with a change in sample thickness in the third direction. Finally, in the planar extensional flow, a material is contracted or elongated in one direction, which results in a change in its thickness (the width remains unaffected) [50]. Key rheological parameters measured in extensional rheology provide unique insights into a material's response to stretching:

1. Extensional viscosity is a measure of a material's resistance to extensional flow. It is defined as the ratio of the extensional stress to the extensional strain rate and is analogous to shear viscosity but under a different deformation mode [51].
2. The Trouton ratio is a dimensionless number defined as the ratio of the extensional viscosity to the shear viscosity. For Newtonian fluids, the Trouton ratio is equal to 3. Deviations from this value in non-Newtonian fluids indicate the presence of viscoelastic effects and can provide information about the material's molecular architecture [51].
3. Strain hardening is a phenomenon observed in some viscoelastic materials, particularly polymer melts and concentrated solutions like bread dough, where the extensional viscosity increases rapidly with increasing extensional strain [52]. This behavior is often indicative of the formation of a network structure or the entanglement of long molecules and is crucial for gas retention and volume development in baking [53].

Extensional rheology has diverse applications in food science:

1. It is extensively used to study the rheological properties of dough during mixing, proofing, and baking. The biaxial extension that occurs as gas cells expand is a critical process determining the final texture and volume of bread [52]. Measurements of extensional viscosity and strain

hardening provide valuable information about dough strength and gas retention capabilities [52].

2. Extensional rheology is essential for analyzing the "stringiness" or "spinnbarkeit" of certain food products, such as concentrated milk [54].
3. The technique is also used to understand the mechanical properties of edible films and coatings that may experience stretching during handling and consumption [49].
4. Extensional rheology plays a role in investigating the flow properties of saliva and saliva substitutes, as the elongation of these fluids is important for bolus formation and safe swallowing [55].

Researchers have begun to correlate extensional rheological properties with sensory attributes related to texture, such as texture length, ropiness, and overall mouthfeel [54]. This technique provides unique and critical information about how foods behave when subjected to stretching forces, which is often distinct from their response to shear. Measurements of extensional viscosity of liquid foods are important in structural characterization, process and quality control of products, and process modeling and design.

3.2.3 Interfacial Rheology

Interfacial rheology is the study of the deformation and flow of thin films of material at a liquid/gas or liquid/liquid interface, where interface indicates the boundary between two phases. This technique is a specialized approach to the study of viscoelasticity and interfacial tension at liquid interfaces and provides valuable insight into intermolecular interactions and surface forces. This knowledge is particularly important for the development of advanced formulations and systems with tailored properties and functionalities [56]. The Interfacial rheology allows for the measurement of several key parameters that characterize the mechanical behavior of these interfaces. Thus, measurements of interfacial viscosity and other interfacial rheological parameters are useful tools for sensitive detection of structural changes in the interphase that can affect the stability of the observed dispersion systems.

1. Interfacial tension is the force per unit length acting at the interface between two immiscible phases. It represents the energy required to increase the surface area of the interface. Lower interfacial tension generally leads to more stable emulsions and foams [57].
2. Interfacial viscosity measures the resistance to flow within the two-dimensional interfacial layer. It can contribute significantly to the stability of emulsions and foams by impeding the coalescence of droplets or the rupture of bubbles [57].

3. Interfacial moduli describe the viscoelastic behavior of the interfacial layer when it is subjected to deformation. The interfacial elastic modulus (E') represents the elastic component, indicating the ability of the interface to store energy and resist deformation. The interfacial viscous modulus (E'') represents the viscous component, indicating the energy dissipated as heat during interfacial deformation [57]. A high elastic modulus suggests a strong, gel-like interface that provides greater stability to the emulsion or foam.

Interfacial rheology has numerous applications in food science:

1. Interfacial rheology is critical for evaluating the stability of emulsions and foams. Measurements of interfacial tension, viscosity, and moduli (elastic and viscous) provide information about the strength and viscoelasticity of the interfacial films formed by emulsifiers and proteins, which prevent droplet coalescence or bubble rupture [56].
2. Researchers use interfacial rheology to characterize the adsorption behavior and interactions of molecules, such as proteins and lipids, at liquid–liquid and liquid–gas interfaces, which is fundamental to understanding how these molecules stabilize food colloids [57].
3. The technique helps in understanding the influence of interfacial properties on the texture and mouthfeel of emulsified and foamed food products. For instance, the creaminess of an emulsion is often related to the viscoelastic properties of the oil-water interface [56].
4. Interfacial rheology can also be used to study the formation and disruption of interfacial layers under various processing conditions, providing valuable insights for controlling the properties of food colloids during manufacturing.

Interfacial rheology plays a pivotal role in the development and optimization of industrial formulations by unraveling the complex relationships between the composition and stability of food emulsions and foams [56]. By probing the mechanical properties of the interfaces, researchers can gain a deeper understanding of how these systems are stabilized and how their interfacial characteristics contribute to the overall texture and sensory perception of the food product.

3.2.4 Microrheology

There is a growing interest in understanding the food microstructure and its correlation with the food textural and rheological attributes in food product development in academia and the food industry. This knowledge on food rheology and microstructure helps in minimizing textural defects in the processed foods and improving consumer satisfaction [30].

Microrheology is an advanced techniques that allow for the investigation of the viscoelastic properties and microstructure of food materials at the microscopic level by observing the motion of tracer particles embedded within them [58]. These tracer particles, typically in the micrometer or nanometer size range, can be either naturally occurring components of the food or intentionally introduced.

Microrheology techniques are broadly classified into two main categories:

1. Passive microrheology relies on tracking the spontaneous, thermally driven motion (Brownian motion) of the tracer particles [25]. By analyzing the statistical properties of this motion, such as the mean squared displacement (MSD), the local viscoelastic properties of the surrounding medium can be inferred using the Generalized Stokes-Einstein Equation [25]. Common passive microrheology techniques include Video Particle Tracking (VPT), dynamic light scattering (DLS), and diffusing wave spectroscopy (DWS) [58].
2. Active microrheology involves applying a controlled external force to a tracer particle and measuring its response, such as its displacement or the force required to move it [25]. Techniques like Optical Tweezers (Laser Tweezers) are used to manipulate individual particles with high precision, allowing for the probing of local mechanical properties and even the non-linear viscoelastic regime.

Microrheology offers several key advantages over traditional bulk rheological methods for characterizing food structure:

1. It requires significantly smaller sample volumes, often in the microliter range, making it ideal for studying precious or limited food samples [25].
2. Microrheological techniques can often probe a much wider range of frequencies than conventional rheometers, extending to very high frequencies that can reveal information about rapid molecular motions within the food matrix [25].
3. Certain microrheology methods, particularly particle tracking, allow for the examination of spatial heterogeneity within food materials by analyzing the local viscoelastic properties at different microscopic locations within the sample [25].
4. This is especially valuable for complex foods with non-uniform structures. Microrheology can be used to study fragile or weakly structured materials without causing significant disruption to their inherent structure, as the forces involved are often very small [59].

Microrheology has a wide array of applications in food science:

1. It is used to investigate the microstructure and local dynamics of various complex food fluids, including gels [31], emulsions [58], and colloidal suspensions [25]. For example, it can reveal the mesh size of a gel network or the interactions between droplets in an emulsion.
2. Microrheology can be employed to study interfacial properties at a microscopic level, such as measuring the local viscosity at the oil–water interface in emulsions stabilized by proteins [58].
3. It is used to analyze the formation of heterogeneous networks in mixed polysaccharide gels and to observe their aging processes at a microstructural level [58].

The theoretical and experimental study of microrheology has made significant progress during the last several years. In many cases, results have altered the picture that had developed from prior macro-scale studies [56]. Its ability to work with small sample volumes, access high frequencies, and reveal spatial heterogeneity makes it invaluable for gaining a fundamental understanding of the relationship between a food's microstructure and its macroscopic texture and sensory attributes. For instance, the mobility of tracer particles in a gel can be correlated with its perceived firmness.

3.2.5 Tribology for Lubrication Analysis

Tribology takes all theoretical and practical results from friction, wear, and lubrication behaviors of interacting surfaces that are in relative motion. When we eat, food interacts with various surfaces in our mouth, including the tongue, palate, and teeth. The frictional properties of the food against these surfaces play a crucial role in our perception of its texture, specifically the attributes related to mouthfeel such as creaminess, smoothness, slipperiness, astringency, and grittiness [60]. Tribological data are typically plotted in Stribeck curves and the friction regime (boundary, mixed, or hydrodynamic) assigned to different portions of the curve [61]. Soft surfaces (e.g., silicone polymer gels) are used to mimic the properties of oral surfaces for better ability to relate instrumental friction data to friction-related sensory attributes such as grittiness and astringency [62]. Human saliva or a solution of alpha-amylase and mucins can be added during testing to better replicate oral conditions [61]. Tribological tests can help improve understanding of chewdown and temporal sensory attributes [63]. Tribological measurements can provide objective data to quantify these interactions.

The primary focus of tribological studies of semisolid foods is their friction and lubrication behaviors. When two hard surfaces slide against each other in relative motion, the coefficient of friction (CoF) is the force that opposes sliding motion. Tribological tests often involve using model surfaces that mimic the tongue and palate, and the CoF is a function of the sliding surface material and physicochemical properties. The resulting data is often presented as a Stribeck curve, which is a plot of friction coefficient versus sliding speed. Friction between the two sliding surfaces is classified in one of three regimes: boundary, mixed, and hydrodynamic, which occur during oral processing [64].

Tribology has a wide range of applications in food science:

1. It is increasingly used for predicting and correlating the sensory perception of texture and mouthfeel, particularly those attributes related to the surface interactions of food in the mouth. For example, the creaminess of dairy products and emulsions has been successfully linked to their tribological properties.
2. Tribology is proving valuable in the development of low-fat food formulations that aim to replicate the desirable creamy mouthfeel of their full-fat counterparts. By understanding the lubrication mechanisms, food scientists can optimize formulations to achieve similar frictional properties in the mouth, even with reduced fat content [24].
3. It plays a role in designing food products with tailored textural properties for specific consumer groups, such as the elderly or individuals with swallowing difficulties. By controlling the frictional properties of food, it can be made easier and safer to swallow [55].
4. Tribological studies are also used to investigate the role of saliva and food components in oral lubrication and the formation of a food bolus that can be safely swallowed [55].
5. Emerging research suggests that tribology can even be used in identifying potential adulteration in foods by detecting deviations in their frictional behavior [60].

Food oral tribology is now emerging as a new experimental technique alongside food rheology in understanding the relationship among food structure, texture, and mouthfeel. By directly measuring the friction between food and model oral surfaces, researchers can use this knowledge to design and optimize food products for enhanced consumer satisfaction and specific dietary needs.

3.2.6 Texture Profile Analysis (TPA)

Texture is a critical shelf-life indicator and an important organoleptic characteristic for baked goods' overall acceptance. A food product is not only soft or firm but has other attributes like springiness, resilience, adherence, tooth-packing, and so on [65]. As a complex organoleptic parameter, texture is commonly determined by sensorial evaluation with trained panelists. However, these types of analysis are time-consuming and expensive. The elevated costs in both production and execution of sensory tests led to the development

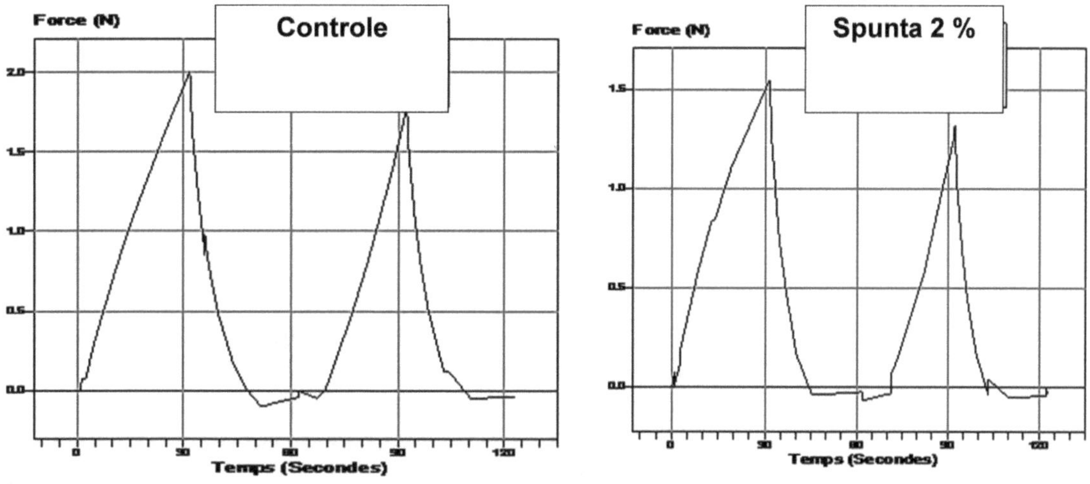

Fig. 3 Textural properties (hardness, cohesion, springiness, adhesion, chewiness, and breaking force) of cakes prepared with wheat flour fortified with potato peel powders

of instrumental techniques that reduce both time and long-term cost of food texture analysis.

Texture profile analysis (TPA) is a widely used empirical technique in food science for objectively measuring the textural properties of solid and semi-solid foods by simulating the chewing action that occurs during mastication [65]. It is a two-cycle compression test that mimics the first two bites of a food. Then, the results are plotted into graphs to analyze the results, as shown in Fig. 3 [66]. TPA parameters are often related to compression sensory terms, such as firmness and adhesiveness [20]. While this test can be used to build rheology–sensory links, users of TPA should be careful not to violate TPA assumptions (e.g., no sample fracture) and mindfully select parameters for analysis, as not all parameters apply to all samples (e.g., bread is not generally adhesive).

TPA is used for the evaluation of several food products, such as:

1. Baked goods (bread, pound cake, muffins, rolls, etc.)
2. Dairy products (hard and soft cheese).
3. Meat products (hot dogs, hamburgers).
4. Confectionery (jelly beans, candies, chocolate bars, etc.)

During the TPA test, the force and time are recorded, and several key textural parameters are calculated from the resulting force-time curve. These parameters are designed to correlate with specific sensory attributes perceived during mastication [65]:

1. Hardness is the maximum force applied during the first compression, representing the force required to initially deform the food.

2. Cohesiveness describes how well the food holds together during mastication. The product can withstand a second deformation relative to the first compression, indicating the internal bonding strength of the food.
3. Elasticity measures the capacity of the product to spring back after the deformation of the first compression, reflecting its elasticity. It is calculated as the ratio of the height of the food at the start of the second compression to its original height or as the ratio of the corresponding time intervals.
4. Chewiness represents the energy required to chew a solid food until it is ready for swallowing. It is calculated as the product of hardness, cohesiveness, and springiness.
5. Gumminess is the energy required to disintegrate a semisolid food until it is ready to swallow. Analogous to chewiness but applies to semi-solid foods. It is calculated as the product of hardness and cohesiveness.
6. Adhesiveness measures the work required to pull the probe away from the food sample after the first compression, reflecting the stickiness or tackiness of the food's surface.
7. Fracturability is the rate at which an item comes away from the probe (instrumental) or roof of the mouth/teeth during the first compression, indicating its brittleness or crispness.
8. Resilience is the measurement of how a sample recovers from deformation in relation to speed and applied forces and is calculated from the energy recovered during the upstroke of the first compression.

TPA has a wide range of applications in food science [65]:

1. It is extensively used for objectively quantifying the textural properties of a vast array of solid and semi-solid food products, providing a standardized method for texture evaluation.
2. TPA measurements are often correlated with sensory evaluations conducted by trained panels to establish the relationship between instrumental readings and human perception of texture. This allows food scientists to predict how consumers will perceive the texture of a product based on instrumental data.
3. Characterize textural properties of food products in case of production or ingredient changes (e.g., change of flour supplier or formula ratios).
4. TPA serves as a crucial tool for quality control in the food industry, helping manufacturers to characterize textural changes during storage or under different storage conditions.
5. Improve sensory properties of novel or competitive products.

Texture profile analysis is a versatile and widely used empirical technique that offers a multi-parametric assessment of food texture by simulating the process of mastication. While it provides valuable quantitative data on various textural attributes and has been shown to correlate well with sensory perception in many cases, it is important to remember its empirical nature and to carefully consider the test parameters and food system when interpreting the results.

4 Note

Rheometry is still a promising tool for understanding factors that influence food texture. Because the way in which a food deforms, fractures, flows, and breaks down is a key component of textural sensations. By moving beyond simple viscosity measurements, sophisticated methods enable food scientists and product developers to design, optimize, and control the textural attributes of a wide range of food products. Their ability to provide detailed, quantitative information about the mechanical properties of foods at various length scales and under conditions relevant to processing and consumption makes them invaluable tools for food scientists, researchers, and product developers striving to create high-quality, appealing, and innovative food products that meet the ever-evolving demands of consumers.

The field of food rheology continues to evolve, with several emerging trends and future directions promising to further enhance our understanding and application of these techniques. Indeed, there is a growing trend towards combining rheological measurements with other advanced analytical methods, such as microscopy (e.g., confocal laser scanning microscopy) and spectroscopy (e.g., NMR), to gain a more holistic understanding of the relationship between food microstructure and macroscopic rheological properties [48]. This integration allows for the simultaneous investigation of structural and mechanical changes in foods under various conditions. Researchers are continuously working on developing novel rheological techniques and improving existing instrumentation to address specific challenges in food analysis. This includes the development of more sensitive and versatile rheometers, as well as new methodologies for studying complex food systems and processing conditions, such as high-temperature and high-pressure rheology relevant to extrusion [67].

A major area of ongoing research is the development of more accurate and predictive models that link instrumental rheological measurements with the complex sensory experience of eating. This involves not only correlating bulk rheological properties with sensory attributes but also incorporating the dynamics of oral processing through techniques like tribology and mastication studies [68].

Advanced rheological techniques are playing an increasingly important role in the development of innovative food products, such as plant-based meat and dairy alternatives, and foods designed for specific dietary needs (e.g., dysphagia diets) [69]. Rheology helps in understanding the textural requirements for these novel products and in optimizing their formulations and processing to meet consumer expectations.

References

1. Truong VD, Daubert C (2002) Rheological methods for assessment of food freshness and stability. Freshness and Shelf Life of Foods, pp 248–268
2. Khare N, Srivastava T (2021) Rheology's role in food preparation, with a focus on texture: a review. Middle East Res J Engr Technol 1(1): 76–80
3. Morris BA (2017) The science and technology of flexible packaging: multilayer films from resin and process to end use. William Andrew:121–147
4. Mandala I, Apostolidis E (2020) Characterization of nanoencapsulated food ingredients. Nanoencapsul Food Ind 4:529–545
5. Ewoldt RH, Winter P, Maxey J, McKinley GH (2010) Large amplitude oscillatory shear of pseudoplastic and elastoviscoplastic materials. Rheol Acta 49(2):191–212
6. Augusto PED, Rojas ML, Miano AC (2023) Food rheology: a practical guide, 1st edn. CRC Press
7. Ahmed J, Ptaszek P, Basu S (2017) Food rheology: scientific development and importance to food industry. Advances in Food Rheology and Its applications, pp 1–6
8. Abbas KA, Mohammed A, Saleh AM, Ebrahimian M (2010) Suitability of viscosity measurement methods for liquid food variety and applicability in food industry—a review. J Food Agric Environ 8:100–107
9. Wang Y, Ewoldt R (2023) Thixotropy, antithixotropy, and viscoelasticity in hysteresis. J Rheol 67(6):1199–1219
10. Xie F, Halley PJ, Avérous L (2012) Rheology to understand and optimize processibility, structures and properties of starch polymeric materials. Prog Polym Sci 37(4):595–623
11. Mouritsen O, Styrbæk K (2017) Mouthfeel: how texture makes taste. Columbia University Press
12. Tabilo-Munizaga G, Barbosa-Ca'novas GV (2005) Rheology for the food industry. J Food Eng 67(1–2):147–156
13. McClements DJ, Jafari SM (2018) Improving emulsion formation, stability and performance using mixed emulsifiers: a review. Adv Colloid Interf Sci 251:55–79
14. Shamsara O, Muhidinov ZK, Jafari SM, Bobokalonov J, Jonmurodov A, Taghvaei M, Kumpugdee-Vollrath M (2015) Effect of ultrasonication, pH and heating on stability of apricot gum-lactoglobuline two layer nanoemulsions. Int J Biol Macromol 81: 1019–1025
15. McClements DJ (2020) Future foods: is it possible to design a healthier and more sustainable food supply? Nutr Bull 45(3):341–354
16. Grillet AM, Wyatt NB, Gloe LM (2012) Polymer gel rheology and adhesion. Rheology 3: 59–80
17. Ipsen R (2008) Food rheology: a personal view of the past and future. Annu Trans Nord Rheol Soc 16:29–36
18. Athawale GH, Raichurkar SJ, Dagadkhair RA (2022) Pectate as a gelling agent in foods. In: Food processing waste and utilization. CRC Press, pp 201–221
19. Barroso da Silva FL, Carloni P, Cheung D, Cottone G, Donnini S, Foegeding EA et al (2020) Understanding and controlling food protein structure and function in foods: perspectives from experiments and computer simulations. Annu Rev Food Sci Technol 11: 365–387
20. Wang Q, Shi A, Shah F (2019) Rheology instruments for food quality evaluation. Evaluation Technologies for Food Quality, pp 465–490
21. Barbosa-Ca'novas GV, Kokini JL, Ma L, Ibarz A (1996) The rheology of semiliquid foods. Adv Food Nutr Res 39:1–69

22. Kulkarni VS, Shaw C (2016) Essential chemistry for formulators of semisolid and liquid dosages, pp 145–182
23. Yazar G, Duvarci O, Yildirim EM, Kokini J (2019) LAOS (large amplitude oscillatory shear) applications for semisolid foods. Rheology of Semisolid Foods:97–131
24. Samaras G, Bikos D, Vieira J, Hartmann C, Charalambides M, Hardalupas Y, Masen M, Cann P (2020) Measurement of molten chocolate friction under simulated tongue-palate kinematics: effect of cocoa solids content and aeration. Curr Res Food Sci 3:304–313
25. Yang N, Lv R, Jia J, Nishinari K, Fang Y (2017) Application of microrheology in food science. Ann Rev Food Sci Technol 8:493
26. Joyner HS (2018) Explaining food texture through rheology. Curr Opin Food Sci 21:7–14
27. Juszczak L, Witczak T, Ziobro R, Korus J, Cieslik E, Witczak M (2012) Effect of inulin on rheological and thermal properties of gluten-free dough. Carbohydr Polym 90:353–360
28. Sharma M, Kristo E, Corredig M, Duizer L (2017) Effect of hydrocolloid type on texture of pureed carrots: rheological and sensory measures. Food Hydrocoll 63:478–487
29. Paine FA, Paine HY (2012) A handbook of food packaging. Springer Science & Business Media
30. Joyner HS (2019). Rheology of semisolid foods
31. Andrade I, Souki N, Moraes I, Pinho, Samantha (2016) Rheology of emulsion-filled gels applied to the development of food materials. Gels 2:22
32. Simões A, Miranda M, Cardoso C, Vitorino, F. V.A (2020) Rheology by design: a regulatory tutorial for analytical method validation. Pharmaceutics 12(9):820
33. Zheng H (2019) Introduction: measuring rheological properties of foods. Rheology of Semisolid Foods:3–30
34. Tunick MH (2011) Small-strain dynamic rheology of food protein networks. J Agric Food Chem 59(5):1481–1486
35. Egelandsdal B, Fretheim K, Harbitz O (1986) Dynamic rheological measurements on heat-induced myosin gels: an evaluation of the method's suitability for the filamentous gels. J Sci Food Agric 37(9):944–954
36. Kasapis S, Bannikova A (2017) Rheology and food microstructure. Advances in Food Rheology and Its applications:7–46
37. Faber TJ, Jaishankar A, McKinley GH (2017) Describing the firmness, springiness and rubberiness of food gels using fractional calculus. Part II: measurements on semi-hard cheese. Food Hydrocoll 62:325–339
38. Dzadz Ł, Markowski M, Sadowski P, Jakóbczak A, Janulin M (2015) Creep and recovery characteristics of chicken meat frankfurters. J Agric Sci Technol 17:827–835
39. Brito-Oliveira TC, Moraes ICF, Pinho SC, Campanella OH (2022) Modeling creep/recovery behavior of cold-set gels using different approaches. Food Hydrocoll 123:107–183
40. Zhao X, Huebsch N, Mooney D, Suo Z (2010) Stress-relaxation behavior in gels with ionic and covalent crosslinks. J Appl Phys 107:63509
41. Li Z, Yang Z, Zhang Y, Lu T, Zhang X, Qi Y, Wang P, Xu X (2021) Innovative characterization based on stress relaxation and creep to reveal the tenderizing effect of ultrasound on wooden breast. Foods 10(1):195
42. Rukke EO, Abrahamsen RK, Schüller RB (2018) Rheology, texture and meltability of different types of cheese. Ann Trans Nordic Rheol Soc 26:2018
43. Singh H, Rockall A, Martin CR, Chung OK, Lookhart G (2006) The analysis of stress relaxation data of some viscoelastic foods using a texture analyzer. J Texture Stud 37:383–392
44. Gou H, Zhan S, Xu W, Zhang Y, Bao Y (2024) Effects of salt concentration on gel quality and stress relaxation of pig skin gelatin[J]. Meat Res 38(6):19–26
45. Hyun K, Wilhelm M, Klein CO, Cho KS, Nam JG, Ahn KH, Lee SJ, Ewoldt RH, McKinley GH (2011) A review of nonlinear oscillatory shear tests: analysis and application of large amplitude oscillatory shear (LAOS). Prog Polym Sci 36:1697–1753
46. Carmona JA, Ramírez P, Calero N, Muñoz J (2014) Large amplitude oscillatory shear of xanthan gum solutions. Effect of sodium chloride (NaCl) concentration. J Food Eng 126: 165–172
47. Joyner H, Daubert CR (2011) Rheological innovations for characterizing food material properties. Annu Rev Food Sci Technol 2: 153–179
48. Melito H, Daubert C, Foegeding E (2013) Relationships between nonlinear viscoelastic behavior and rheological, sensory and oral processing behavior of commercial cheese. J Texture Stud 44:253–288
49. Różańska S (2017) Extensional rheology in food processing. Advances in Food Rheology and Its applications:125–157

50. Ferguson J, Kembłowski Z (1991) Applied fluid rheology. Elsevier Applied Science, New York, London
51. Evageliou V (2020) Shear and extensional rheology of selected polysaccharides. Int J Food Sci Technol 55(5):1853–1861
52. Chin NL, Campbell GM (2005) Dough aeration and rheology: part 2. Effects of flour type, mixing speed and total work input on aeration and rheology of bread dough. J Sci Food Agric 85(13):2194–2202
53. Dobraszczyk B (2004) The physics of baking: rheological and polymer molecular structure-function relations in breadmaking. J Non-Newtonian Fluid Mech 124:61–69
54. Louw L, Nettesheim F (2024) New insights into stringiness from extensional rheology. Ann Trans Nord Rheol Soc 32:109–117
55. Gamonpilas C, Kongjaroen A, Methacanon P (2023) The importance of shear and extensional rheology and tribology as the design tools for developing food thickeners for dysphagia management. Food Hydrocoll 140: 108603
56. Krstonošić V, Pavlović N, Ćirin D (2023) Principles and applications of interfacial rheology in (pre)formulation development of pharmaceutical preparations. Arhiv za farmaciju 73:337–357
57. Murray B (2002) Interfacial rheology of food emulsifiers and proteins. Curr Opin Colloid Interface Sci 7:426
58. Moschakis T (2013) Microrheology and particle tracking in food gels and emulsions. Curr Opin Colloid Interface Sci 18:311–323
59. Mao Y, Nielsen P, Ali J (2022) Passive and active microrheology for biomedical systems. Front Bioeng Biotechnol 10:916354
60. Xu W, Yu S, Zhong M (2022) A review on food oral tribology. Friction 10:1927–1966
61. Pradal C, Stokes JR (2016) Oral tribology: bridging the gap between physical measurements and sensory experience. Curr Opin Food Sci 9:34–41
62. Prakash S, Tan DDY, Chen J (2013) Applications of tribology in studying food oral processing and texture perception. Food Res Int 54: 1627–1635
63. Fukatsu H, Nohara K, Kotani Y, Tanaka N, Matsuno K, Sakai T (2015) Endoscopic evaluation of food bolus formation and its relationship with the number of chewing cycles. J Oral Rehabil 42:580–587
64. Nazir A, Asghar A, Maan A (2017) Food gels: gelling process and new applications. Woodhead
65. Ganesh K, Ranvare A, Singh A, Suresh C (2024) Texture profile analysis: a comprehensive insight into food texture evaluation
66. Ben Jeddou K, Bouaziz F, Zouari-Ellouzi S, Chaari F, Ellouz-Chaabouni S, Ellouz-Ghorbel R, Nouri-Ellouzi O (2017) Improvement of texture and sensory properties of cakes by addition of potato peel powder with high level of dietary fiber and protein. Food Chem 217:668–677
67. Wittek P, Walther G, Karbstein HP, Emin MA (2021) Comparison of the rheological properties of plant proteins from various sources for extrusion applications. Foods 10(8):1700
68. Shewan HM, Pradal C, Stokes JR (2020) Tribology and its growing use toward the study of food oral processing and sensory perception. J Texture Stud 51(1):7–22
69. Rong L, Shen M, Yu Q, Chen Y, Li J, Xie JH (2023) Rheology, texture and swallowing characteristics of a texture-modified dysphagia food prepared using common supplementary materials. Foods 12:2287

Chapter 17

Lipidomics and Proteomics Approaches in Food Composition Analysis: A Comprehensive Overview

Zimal Ijaz, Zainab Noor, Fakhra Bibi, Shumaila Fayyaz, Mudassar Zafar, Ayesha Baig, Ayesha Siddiqa, and Muhammad Zubair

Abstract

The combination of proteomics and lipidomics has revolutionized food composition analysis in recent years by providing precise instruments for the detection, characterization, and verification of food biomolecules. This chapter offers a comprehensive summary of the most recent analytical developments and methods used to investigate proteins and lipids in diverse food matrices. Proteomics makes it easier to detect allergens, develope functional food, and verify species. Lipidomics makes it possible to profile complex lipid species, contributing to the evaluation of nutritional value, authenticity, and oxidative stability. Various analytical techniques, including chromatography, advanced mass spectrometry, and sample preparation methods, are examined for their roles in proteomics and lipidomics analysis. The chapter further highlights the significance of these methods in food science, such as nutrition, allergen detection, and quality control. The findings demonstrate how well omics-based approaches can identify bioactive peptides, capture processing-induced molecular alterations, and differentiate food origins. The need for research and educational implications, including curriculum integration and funding priorities, are also discussed to advance the area of foodomics. New developments like portable omics systems and AI-driven data interpretation indicate that real-time, precision-based food quality inspection will become widespread in the future. The chapter also emphasizes the regulatory significance of these technologies in ensuring accuracy of labelling, food safety, and traceability, along with the collaborative efforts among nutritionists, bioinformaticians, and food chemists.

Key words Proteomics, Lipidomics, Food composition, Mass spectrometry, Chromatography, Food authentication, Food safety, Omics integration

1 Introduction

1.1 Background

Food composition analysis is the planned process of checking for macronutrients, micronutrients, additives, and possible contaminants in food products. Today, as consumers are more informed, laws are stricter and healthier, and personalized foods are more popular, this expansion is extra important. Getting an in-depth picture of food biomolecules plays a vital role in guaranteeing

Tanmay Sarkar (ed.), *Novel Food Analysis: Techniques and Applications*, Methods and Protocols in Food Science,
https://doi.org/10.1007/978-1-0716-4787-5_17,

food quality and safety as well as in learning how their components affect our health.

1.1.1 Study of Lipidomics

Lipidomics is widely used for lipid composition analysis and detection of the eminence of lipids in food [1]. Lipidomics formulates a system-level analysis of lipids across the board and is employed to identify food adulteration and labeling, along with quantification of discrete lipid molecular types. Numerous analytical techniques based on lipidomics are employed to calculate minute lipid molecules in foods and therefore acquire a complete lipid profile [2].

Lipidomics plays a key role in finding important lipid biomarkers in metabolic parameters. The goal is achieved by comparing shifts in lipid metabolism caused by various conditions. At the end, scientists explore how lipids work. Distinct lipid-metabolizing abilities result from switching out various salts, which can impact the actions of lipolytic enzymes made by microorganisms. So, changes in lipids in meat can change its overall flavor [3].

1.1.2 Study of Proteomics

Exploring proteins from a biological system on a large scale is what experts define as proteomics. The proteome displays the particular state of a biological system at any time and can keep shifting based on various triggers. In addition to studying protein structure and function, proteomics looks at their different types of modifications, the interactions they have, their intracellular locations, and how many there are at any given time. It performs well at exploring how all proteins and their variants work together and change in the body [4]. Proteomics allows us to solve the problems of false positive outcomes that arise in DNA-based and immunoassay techniques. Several techniques to gather proteome information and its link to specific product details have been described in the literature, and these approaches are either targeted or untargeted. At the beginning of biomarker discovery, untargeted proteomic methods are preferred, but validation, implementation and detection of key proteins relies on targeted methods [5, 6].

In the past few years, understanding proteomes has been important in food security to match food quality and safety to health and public interest [1]. UHPLCMS-MS and LESA-MS methods are able to find protein markers. LC–high-resolution mass spectrometry (HRMS) and LC–MS/MS multiple reaction monitoring (MRM) methods are also viewed as complementary for discovering, confirming, and detecting heat-stable peptides from different species in processed meat products [7, 8].

Martakos et al. (2024) presented a new 4D lipidomics method that joined MTBE and methanol microextraction with UPLC-TIMS-QTOF-MS, improving how many of lipid species in pork and chicken meat can be identified. In the course of the analysis, 395 different lipids were discovered in four classes and ten

1.2 Lipidomics and Proteomics Approaches

1.2.1 Lipidomics in Food Composition Analysis

subclasses, with phospholipids making up the most of them. Adding ion mobility separation greatly improved the quality of results for isobaric and isomeric compounds. In particular, the impact was related to finding n-3 fatty acids, suggesting the usefulness of this approach for both nutritional profiling and the development of healthy dietary suggestions [9].

1.2.2 Proteomics in Food Composition Analysis

Luo et al. (2025) designed a new method using 193 nm pulsed UV lasers for photochemical bromination and iodination of peptides. Using UV photo-dissociation as a method gave us comprehensive data with clear patterns, allowing us to recognize many different peptide modifications. High-throughput proteomics applied to food studies highly benefits from this method. Researchers used both DIA and DDA-MS to analyze oxidative and nitrative chemical changes in endothelial cells. The researchers found that nitration was most likely to occur at tyrosine and tryptophan amino acids in proteins. The findings from this study are important for understanding protein oxidation and its effects on antioxidants in food [10]. The studies on fungi discovered that polysaccharides, peptides, and terpenoids have useful properties for both food and pharma sectors, reinforcing the importance of examining fungal biomolecules in making innovative foods. By checking for polyphenols, pesticides, and trace elements in honeydew honeys from Italy, it became possible to label the origin and avoid food fraud. Genomics examine different parts of pancreatic cancer and this approach demonstrates that such detail, common in biomedical research, can be used for foodborne pathogen diagnostics and in precision food safety [11].

1.2.3 Classes of Lipids

Lipidomics looks at groups of lipids, among them fatty acyls, glycerolipids, glycerophospholipids, sphingolipids, sterol lipids, prenol lipids, saccharolipids, and polyketides. Every lipid has its own molecular structure and role in the body, glycerophospholipids help assemble cell membranes, and sphingolipids signal between cells and help with recognition (Fig. 1).

1.3 Importance of Lipidomics and Proteomics in Food Science

Lipidomics and proteomics are now key ways to analyze what foods contain, their safety levels, and their nutritional benefits in food science. Analyzing proteins on a large scale in this field lets us determine important characteristics of food, understand how processing affects these characteristics, and identify factors that influence quality, spoilage, and health.

1.3.1 Role in Nutrition

Lipidomics examines lipids and pathways that use them. It then explains how consumption of omega-3-rich TAGs and PLs keeps the heart healthy. Scientists use LC–MS/MS to compare lipid levels in various meats to discover what affects the meat's nutritional value based on its feeding and age. At the same time, proteomics

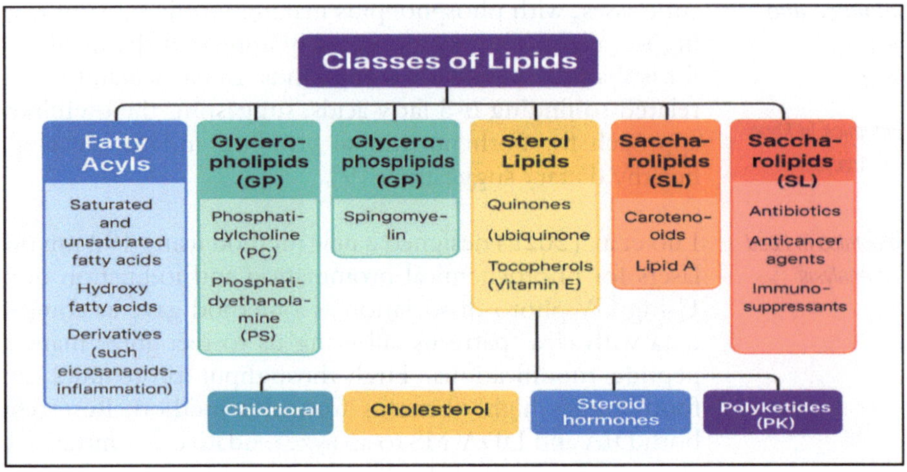

Fig. 1 Hierarchical classification of lipids. It describes the main categories of lipids—fatty acyls, glycerolipids, glycerophospholipids, sphingolipids, sterol lipids, prenol lipids, saccharolipids, and polyketides [12, 13]

investigates proteins and peptides in milk and meat, helping us understand digestion, nutrients, and types of bioactive peptides that play the roles of antioxidants, antihypertensive agents, and have antimicrobial effects, allowing for nutraceutical use [14].

1.3.2 Allergen Detection

Even though lipidomics is not widely used for identifying allergens, it still plays an important role in immune responses by means of lipid signaling. Food processed in ways that promote lipid oxidation may also alter proteins and introduce additional allergens. This way, lipid profiling gives information that can influence protein activity, like changes in allergens would. Proteomics is now a vital approach for finding and counting food allergens. Using the latest mass spectrometry methods, you can spot allergenic proteins in processed foods. Medical studies have found the main allergens in fish, milk, and shellfish. Panchaud and others (2005) stress that this method is vital both for spotting allergens and for keeping track of changes during processing, which is crucial for food safety rules [15].

1.3.3 Food Authentication

The source and type of meat can be determined by using lipidomics, which employs techniques such as LC–MS and GC-FID. Mail profiles present in camel, donkey, and pork meats work as identification and help prevent cheating in meat sales. It also helps by looking for certain peptides in food, for example, in Bolognese sauce and seafood, using methods such as MALDI-TOF MS and tandem mass spectrometry to find out which species are present. It also examines if the milk is real and keeps the product of high quality [14].

1.3.4 Spoilage Detection

Many times, lipid oxidation causes meat to spoil and develop rancid flavors. With lipidomics, it is possible to detect oxidized phospholipids during the freshness assessment of meat. By using UPLC-HRMS and REIMS techniques, researchers located certain oxidized lipid species in irradiated turkey and goat meat, which tried to reduce meat quality. Proteomics can identify the first molecules that indicate meat is beginning to spoil. Its detection of degradation products and microbial protein markers, proteomics can determine the shelf life of products and help put in place solutions to stop spoilage is appreciable.

1.3.5 Relevance for Functional Foods and Nutraceutical Development

Duck and goat meat experiments showed specific lipid subclasses that might be used in the creation of new types of meat products. It is also important that differences in lipids resulting from different diets, such as grass-fed and grain-fed beef, support the production of healthy food for consumers. Functional foods and nutraceuticals work to support health in ways other than regular nutrition, and this field is made possible by proteomics [14]. This chapter aims to present an up-to-date overview of methodologies and technologies in lipidomics and proteomics applied to food analysis. Furthermore, it highlights regulatory aspects, collaborative efforts, educational implications, along with current challenges, and future directions in deploying lipidomic and proteomic tools for comprehensive food characterization.

2 Materials

The materials used in the experimental techniques are selected to optimize the efficiency of both extraction and analytical methods. Solvents with varying polarities are used to facilitate various extraction types [16]. Moreover, to ensure comprehensive lipidomic and proteomic approaches, a wide range of biological matrices is also used. Chromatographic techniques are also used. Conventional and specialized stationary phases allow for the improved resolution of target compounds. Furthermore, highly enzymatic reagents and standards are also used to ensure the precision and reliability in protein analysis. Advanced instrumentation and tools are also used to support the entire analytical sequence [17].

Summary of Materials:

1. *Biological samples* (urine, blood, plasma, serum, saliva, tissue, cerebrospinal fluid, tears, nails).
2. *Solvents* (chloroform, methanol, ethanal, n-hexane, ethyl acetate).
3. *Stationary phases* (alumina, silica gel, silver thiolate).

4. *Standards and reagents* (N-trifluoracetyl-phosphatidylethanol amines, peptides, trypsin enzyme, omega 3-fatty acids).
5. *Analytical tools* (chromatography, spectroscopic techniques).
6. *Storage* equipments.

3 Methodology

3.1 Sample Pretreatment of Lipids

The effective extraction [18] of major and minor lipids (both qualitatively and quantitatively) depends heavily on the use of suitable techniques. The physical state (tissue or fluid), moisture and lipid contents, and the sample's origin (plant or animal) all play a role in choosing the best techniques. Figure 2 shows different methods of lipid extractions. Furthermore, the subsequent needs of the recovered lipids may also serve as the basis for the extraction techniques.

3.1.1 Bligh, Dyer and Folch Method

The effective extraction of major and minor lipids (both qualitatively and quantitatively) depends heavily on the use of suitable techniques. The physical state (tissue or fluid), moisture and lipid contents, and the sample's origin (plant or animal) all play a role in choosing the best techniques. The Bligh, Dyer and Folch that were used to extract lipids from animal tissues are now widely used to extract lipids from a variety of plant and animal materials due to the great efficiency of chloroform and methanol in extracting key lipid classes.

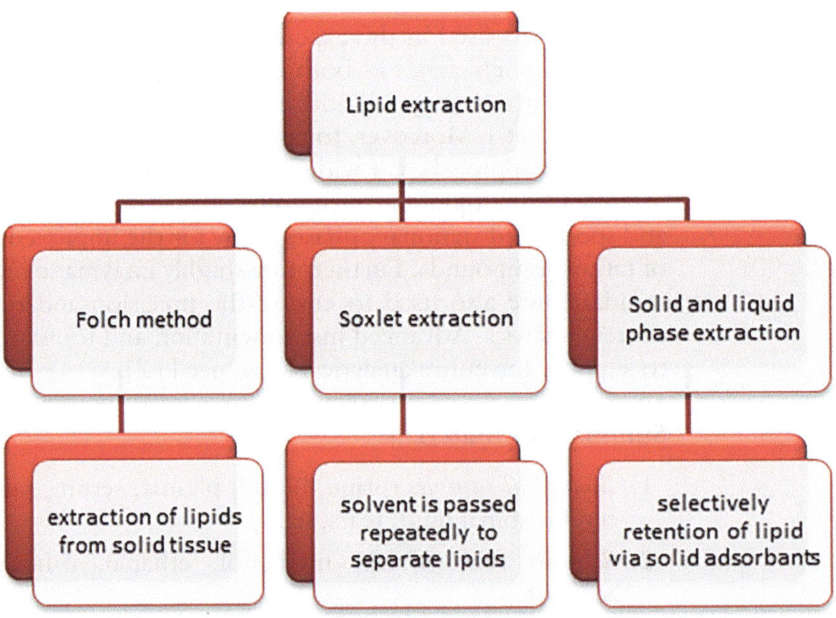

Fig. 2 Methods of lipid extraction

While the Bligh and Dyer method is thought to be beneficial for biological fluids, the Folch method is typically chosen to extract lipids from solid tissue. The multistep nature of the Bligh, Dyer, and Folch Methods restricts their use for screening a large number of samples. By using biomass and dispersing it in a 10 mL solvent system (chloroform/methanol, 2:1, v/v), Axelsson and Gentili created a quicker, one-step method for extracting total lipids from green microalgae. They then added 0.73% NaCl water to create a 2: 1:0.8 system of chloroform/methanol/water (v/v/v/v).

3.1.2 Solid-Phase Extraction (SPE)

In addition to enriching minor lipid classes, SPE is a useful method for the separation and purification of specific lipids. Typically, this is done with small cartridges (columns) filled with ion exchange, normal, or reversed sorbents. Unwanted substances flow through these cartridges, while the desired fractions are selectively held in place by polar (normal phase), hydrophobic (reverse phase), or ionic interactions [18].

3.1.3 Lipid Extraction Utilizing Green Solvents

Lipids from plants, animals, and microorganisms are effectively extracted using the chloroform/methanol-based solvent employed in the conventional Folch, Bligh and Dyer techniques. However, because of health and environmental concerns, researchers have investigated several safer solvents. Supercritical CO_2, plant-derived terpenes (including D-limonene, p-cymene, and α-pinene), and ionic liquids (non-aqueous salt solution) are some of the green solvents that are becoming more popular. Ethanol and ethyl acetate were chosen as possible alternatives to methanol and chloroform for the extraction of lipids from yeast with the aid of the conductor-like screening model for realistic solvatation (COSMO-RS) [16].

3.2 Pretreatment of Proteins

3.2.1 Gel Filtration Chromatography

In size-exclusion-based chromatography, the process is called gel filtration, and molecules are divided into groups according to size, so they do not attach to the media since the molecules are too large to enter the small pores. During the process, smaller molecules tend to go through the column more slowly. Most of the column is open to small molecules because of its many tiny pores. Whereas large-sized proteins are stuck inside the column, small-sized proteins will pass through the holes and leave the column at a reduced speed [4].

3.2.2 Affinity Chromatography

Affinity chromatography relies on a particular interaction between the solid phase and the protein to influence the removal of impurities. The steps involved are identical to those of ion exchange chromatography. It makes it possible to separate proteins according to their unique chemical structures or biological roles. While other proteins flow down the column, proteins with a high affinity for particular chemical groups, such as ligands, will covalently bind to the matrix. The biological interactions between ligands and target proteins include hydrogen bonding, van der Waals forces, and electrostatic or hydrophobic interactions. A solution with a high concentration of the ligand's soluble form will elute the bound proteins from the column [19]. For the proteins to be extracted in an active form, the ligand-target protein molecule binding must be reversible. The linked ligand must maintain its particular binding affinity for the target proteins after the impurities have been removed.

3.3 Analytical Techniques in Lipidomics and Proteomics

3.3.1 Matrix-Assisted Laser Desorption Ionization (MALDI) MS

MALDI, like ESI, is one of the earliest ionization techniques used in lipidomics and proteomics analysis. It is still widely used in MS methods, especially for lipidomics-related disease investigations. The soft ionization method known as Matrix-Assisted Laser Desorption/Ionization (MALDI) embeds analytes in a matrix that absorbs light, such as 2,5-dihydroxybenzoic acid for lipids and α-cyano-4-hydroxycinnamic acid for peptides. Analytes are vaporized and ionized with little fragmentation when pulsed laser irradiation causes matrix ablation. High-throughput analysis is therefore made possible by time-of-flight (TOF) mass analyzers, which distinguish ions based on the mass-to-charge ratio (m/z). It is used to identify lipid types and their structural variations (chain length, saturation), such as phospholipids, sphingomyelins, and fatty acids. Similarly, it also determines the proteins' molecular weights (less than 70 kDa) with an accuracy of $\leq$500 ppm, confirming the integrity of the sequence. When working with a large number of samples to identify new biomarkers, the ability to quickly display the general pattern of changes in both lipidome and peptidome signatures may be of great importance [17]. Given the growing interest in the association between various compounds, the MALDI-MS method's ability to identify many substances simultaneously can be viewed as a benefit. Different MALDI-MS techniques have recently been developed in order to investigate the interactions between various lipids and other substances, such as related proteins. Much more information regarding the relationship has been made available by these simultaneous detection approaches, which could be connected to certain significant biological activity processes [17].

3.3.2 Mass Spectrometry Imaging (MSI)

Despite having a lower sensitivity than other current imaging techniques like fluorescence imaging, MSI is a very popular and effective tool for lipidomics and proteomics studies since it generates images from the individual mass spectra of a biological sample, especially the complex tissues. For the study of numerous biological processes requiring interaction and dynamic geographic distribution, MSI can offer the visualization and distribution data of a single molecule [20].

3.3.3 Chromatography-Based Lipidomics and Proteomics Analysis

Chromatography is a vital method for the thorough examination of complicated samples because of its exceptionally high separation capabilities. Common chromatographic techniques include capillary electrophoresis (CE), gas chromatography (GC), high-performance liquid chromatography (HPLC), thin-layer chromatography (TLC), and supercritical fluid chromatography (SFC). They can meet various lipid and protein analysis requirements. Chromatography techniques can yield a lot of information about the lipidome when hyphenated to MS [21].

3.3.4 Thin Layer Chromatography (TLC)

TLC and its improved form, high-performance thin-layer chromatography (HPTLC), are essential tools in the study of lipidomics because they are quick, easy, and affordable techniques with a respectable resolution of various lipid classes [17]. The resolution of TLC is greatly enhanced by the development of micro-TLC, a multidimensional separation technique, a new planar chromatographic and electro-chromatographic chamber, and a microfluidic paper-based device. This allows for the separation of more than hundreds of spots made up of low-molecular-mass compounds. Alumina, kieselguhr, and silica gel are the most often used stationary phases for lipid separations; silica gel is by far the most dominant phase. Ag-TLC is also a fantastic option for lipidomics analysis. The use of silver-thiolate (AgTCM) as a stationary phase in TLC was initially documented by Dillon et al. in 2012. The AgTCM functions on the same principles as the conventional Ag-TLC, sorting compounds according to their degree of unsaturation, but it has significant advantages in terms of light stability and shelf life. With the help of this AgTCM TLC, a mixture of fatty acid methyl esters (FAMEs) with varying numbers of double bonds and fatty acyl chains can be effectively separated and identified [21].

3.3.5 Spectroscopy

Spectroscopy naturally offers the benefit of visualization analysis, which is crucial for biological process research. It is important to consider the advancement of spectroscopy-based lipidomics in order to accomplish the goal of single-cell imaging detection [21].

3.3.6 Nuclear Magnetic Resonance (NMR)

NMR spectroscopy is not frequently employed in lipidomics, despite being the primary method in metabolomics due to its minimal time consumption and responsiveness to all organic compounds. It can be challenging to distinguish resonances from the different compounds in complex mixtures due to NMR's relatively low sensitivity (especially when compared to MS), high micromolar to millimolar range, and small chemical-shift range, which results in crowded spectra when obtained as a 1D spectrum. Nevertheless, 31P NMR was used to distinguish between several phospholipid groups. Lipids from the liver and plasma of rats with alcoholic liver disease have recently been successfully analyzed by combining 1H and 31P NMR, which can help explain the pathophysiology of ethanol-mediated liver injury. Determining the PL content of cheese and fish was another useful application for the 2D 31P, 1H COSY NMR spectroscopy. Differences in the fatty acid chains affixed to the distinct PLs were shown by variations in the 31P shifts. The detection limit is as high as 4 nmol. NMR in lipidomics analysis may gain new life from 2D NMR [22].

3.3.7 Gel-Free Proteomic Approach

Gel-free methods usually begin by cutting proteins into peptides, followed by separating those peptides according to how hydrophobic they are. After that, the peptides are removed and loaded into a mass spectrometer. After that, all the tandem mass spectra are reviewed using databases, and the original proteins are reconstructed. The method works well when the proteins are part of sequenced organisms. MudPit (multidimensional protein identification technology) allowed researchers to identify proteins from samples that are hard to analyse. One drawback of this method is that it lacks exact number counts, which have been solved by using dilution and isotope labelling. Lacking both measurements and visualization of different proteins and their modifications is an issue with gel-free methods. Cross-species identification is difficult when the sequencing of the genomes is not of high-quality, which is the situation for most plant-based foods. It is done by comparing the peptides of a protein to highly similar proteins found in other species [23]. Some advanced techniques are also mentioned in Fig. 3.

3.4 Lipidomic Analysis

Lipid profiles are extensively affected by various processing methods such as frying, hydrogenation, and freezing [24]. For example, the process of heating causes a reduction in unsaturated phospholipids in chicken meat while an increase in lysophospholipids [25]. Roasting of flaxseed oil significantly modifies the lipid

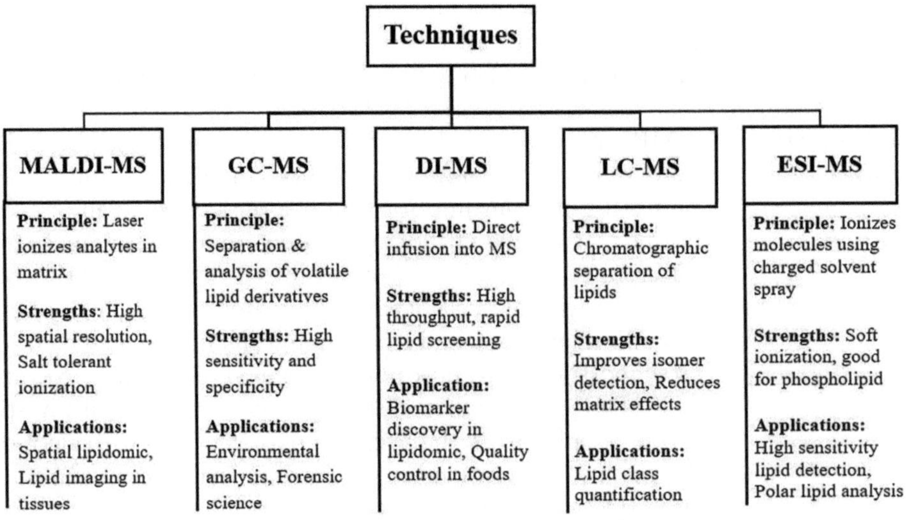

Fig. 3 Techniques for lipidomics and proteomics

Table 1
Key lipid species identified in various food groups

Food category	Subcategory	Key lipid identified	References
Dairy products	Milk	Triglyceride, phospholipid, sphingolipid	[29]
	Cheese	Ceramide, phospholipids, diacylglycerol, triacylglycerol	[30]
Meat and poultry	Beef	Palmitic acid, myristic acid, oleic acid	[31]
	Chicken	Phospholipids, acylglycerols, glycoglycerolipids, sphingolipids	[25]
Fish and seafood	Tuna	Phospholipids, glycerolipids, sphingolipids, fatty acids	[32]
	Mandarin fish	Triglyceride, phosphatidylcholine	[33]
Nuts and oilseeds	Flaxseed oils	Phospholipids, triacylglycerols	[34]
	Olive oil	Triglycerides	[35]
Cereals and grains	Wheat	Monoacylglycerols, wax ester, diacylglycerol, triacylglycerol, o-acylated-ω-hydroxy fatty acids	[36]
	Rice	Free fatty acids, triglycerides	[37]
Fruits and vegetables	Banana	Glycerides, sterols, sphingolipids, glycero-phospholipids	[38]
	Bell pepper	Glycerolipids, glycerophopholipids	[39]

composition. It is characterized by a transient increase in phospholipids and reduction in unsaturated fatty acids and triacylglycerols. The decrease in triacylglycerol levels is connected with the degradation of polyunsaturated fatty acids, which are particularly prone to oxidation [26].

These alterations often result in adverse health outcomes because they affect the nutritional quality of food. Appropriate lipid content is important for food quality and nutritional value. Table 1 highlights the key lipid species identified across a diverse range of commonly consumed food categories. Table 2 shows the lipid contents (%w/w) of some food groups. Changes in lipid composition may cause an increased risk of obesity, chronic disease, and various cardiovascular diseases [24]. Lipidomic profiling has gained importance as a valuable approach to understanding the complex lipid composition of foods. This method provides valuable insights into the functional role of lipids, their metabolic roles, and nutritional value. Lipid species play a critical role in health-related outcomes [27, 28].

3.5 Proteomic Analysis

Proteomic analysis is useful for the identification of allergenic proteins and bioactive peptides. These peptides are useful for human health and are derived from proteins present in foods. Food derived peptides with antioxidant, antihypertensive, immunomodulatory properties have been identified with the help of high-resolution MS techniques [46]. Suitable proteomic techniques are employed on various food matrices to obtain quantitative data. LC–MS/MS

Table 2
Quantitative lipidomics in selected foods

Selected foods	Lipid content (%w/w)	References
Cow milk	3.3–4.7	[40]
Goat milk	4.1–4.5	[40]
Rice	2.94	[41]
Avocado	8.41–57.88	[42]
Chicken nuggets (raw)	32.9	[43]
Chicken nuggets (fried)	38.5	[43]
Egg yolk	33	[44]
Seafood	12.4	[45]

Table 3
Quantitative proteomic analysis in food composition

Food type	Analytical technique	Proteomic markers	Quantitative data	References
Cheese (edam type)	LC-MS/MS	β-Lactoglobulin & α-lactoglobulin,	Protein concentration in β-LG: 0.50–5.31 (%w/w); R2 > 0.9714; CV < 5%; recovery 80–12%	[52]
Seafood (shrimp/crab)	UPLC-MS/MS	Tropomyosin (allergenic protein)	Recovery 77–96%; LOD: 5–5000 microg/L linear; LOQ: 0.1 mg/kg; RSD ≤ 9.97%	[53]
Poultry (chicken)	LC-MS/MS (targeted MRM	Collagen II alpha1 chain peptides	5 markers peptides; detection sensitivity: 96%; detection threshold: 10% MSM (w/w)	[54]
Barley (grain)	LC–MS	Hordeins (storage protein)	3104 proteins identified; variance in protein abundance explained by location: 26.7%, year: 14.3%, genotype: 17.1%; RMSE<2.1%	[55]

technique, used for soybean, identified 550 proteins, including trypsin inhibitors. Comparative profiles studies revealed the degradation and redistribution of bran-specific proteins [47]. Wheat flour analysis through LC–MS/MS identified 564 proteins, and results showed 33% decrease in gluten protein content [48]. MALDI-TOF/TOF mass spectrometry has been effectively used to identify casein and whey as specific protein markers in milk proteomics [49, 50]. Table 3 highlights further proteomic approaches for quantitative food analysis.

Proteomic workflows also characterize bioactive peptides. These peptides are useful for human health and are derived from proteins present in foods. Food-derived peptides with antioxidant, antihypertensive, immunomodulatory properties have been identified with the help of high-resolution MS techniques [46]. Proteomic technologies play significant role in tracing food origin by identifying specific protein markers. These markers are associated with geographic origins and animal breeds. For instance, comparative proteomic analysis of milk samples from cows, goat, and sheep has shown some specific proteins as biomarkers [51].

4 Notes

4.1 Regulatory Aspects

Developments in proteomics and lipidomics are becoming more and more important for food safety, traceability, and food labeling [56]. The U.S. Food and Drug Administration (FDA) and other regulatory agencies are demanding thorough nutritional and compositional statements. To help consumers make better decisions, the FDA's proposed Front-of-Package Nutrition Labeling rule in 2025 calls for packaging to provide clear, consistent nutrient information, including precise quantification of cholesterol, fatty acids, allergenic proteins such as peanuts, lactose, and gluten by using lipidomics and proteomics approaches. Additionally, according to the US Food and Drug Administration, proteomic workflows, in particular targeted LC–MS/MS and high-resolution multiple reaction monitoring (MRM) [57], can even detect the heat-stable peptide markers below 1% (w/w). This allows regulators to verify the presence or absence of particular species in complex matrices like sauces or processed meats. This feature facilitates adherence to the Nutrition Labeling and Education Act's (NLEA) standards for allergy disclosure and "percent juice" or "meat content" statements [58].

Identification of hazards and adherence to evolving regulations depend heavily on lipidomics and proteomics [31]. Lipidomics detects toxic lipid-based oxidation compounds (such as aldehydes and ketones) that are produced during food processing or storage and are associated with inflammation and different types of cancer. By identifying protein-based pollutants like pathogens (such as Salmonella proteins) or allergens (such as casein in dairy) using different proteomics approaches, proteomics makes it possible to comply with food safety laws like the U.S. Food Safety Modernization Act (FSMA) [59]. The advanced lipidomics and proteomics approaches perfectly align with the *Preventive Controls for Human Food* rule of *FSMA,* which made the hazard analysis mandatory for foods such as ready-to-eat salads, which are at high risk of foodborne pathogen. Research based on metabolics has demonstrated that irradiation of food up to 10 kGy can change the levels of glutathione and nucleotide degradation products, although it causes only slightly disruptions in the main lipid types and the pool of free amino acid. Such fingerprinting may be utilized in "process-validation" modules to verify that treatment parameters remain within approved and secure ranges in accordance with FSMA Preventive Controls rules.

Furthermore, modern food regulation relies heavily on traceability in lipidomics and proteomics, which offers irrefutable molecular proof of safety and place of origin. According to the FSMA Food Traceability Final Rule, covered items must have key Data Elements (KDE) records, which include batch IDs and production dates. Species-specific peptide or lipid profiles, which are omics-derived "molecular barcodes" and inherent Critical Tracking Events (CTEs), could link a product's molecular signature to its source, which enables the quick traceback in the context of a contaminated epidemic [60].

4.2 Quality Assurance

Strict QA/QC frameworks are necessary for high-quality lipidomics and proteome analysis to guarantee accurate results for both research and regulatory usage in food composition. Formal SOPs for sample collection, managing, instrument operation, and data processing, extensive staff training, and thorough technique validation are essential components of quality assurance. In targeted workflows, the validation usually adheres to existing regulatory bioanalytical guidelines (such as FDA criteria) in order to establish performance indicators. On the other hand, strong Quality Control (QC) and Quality Assurance (QA) procedures are necessary in untargeted lipidomics and proteomics to guarantee data accuracy, repeatability, and comparability across investigations and labs [61].

In lipidomics and proteomics workflows, strict QA/QC procedures are necessary to guarantee accuracy and reproducibility. Running blanks, pooled quality control (QC) samples, and internal standards with every batch of samples is one of the best practices [62]. Aliquots of every research sample, for instance, are frequently mixed to create pooled QC samples, which are injected at regular intervals (e.g., every ten injections) to track batch effects and instrument drift. It is important to calibrate instruments using recognized reference materials, such as NIST standard mixtures, and to regularly check their performance. The quality of the data is then evaluated using metrics like the coefficient of variation (CV) between QC replicates; peaks with a CV $> 25\%$ are usually examined or eliminated.

Additionally important are proficiency tests and Certified Reference Materials (CRMs) [63]. To measure system performance, NIST has also created proteomic reference mixtures and quality evaluation procedures. Comparability of measurements across various equipment and locations is ensured by the use of such CRMs. The ISO/IEC 17025 guidelines must be followed when evaluating the accuracy, precision, consistency, limit of detection, and durability of targeted proteomics assays during method validation. Even though untargeted workflows are more experimental, they still require document sample preparation standard operating procedures (SOPs) [64], randomize run order, and use reliable data-processing pipelines.

4.3 Collaborative Efforts

The significance of multidisciplinary collaboration is a crucial factor to take into account when using lipidomics and proteomics for food composition analysis. The complexity of food systems and the data-intensive nature of omics technologies have made it impractical for one field to handle every facet of the analytical pipeline. Collaboration among food chemists, bioinformaticians, and nutritionists is necessary to guarantee significant results [65].

Food chemists play a key role in the development and improvement of analytical methods, including mass spectrometry (MS) and sample preparation procedures [66]. Their proficiency guarantees precise extraction of proteins and lipids from a variety of frequently varied dietary matrices [67]. To guarantee consistency, food chemists also validate procedures, especially when examining lipid species that are abundant and scarce in various food kinds. Nevertheless, the analytical outputs produced by these workflows are extensive and multifaceted, demanding specific processing and interpretation skills.

At this point, bioinformaticians are essential. LipidSearch, MZmine, and LIPID MAPS are some of the computational tools and sophisticated statistical techniques they use to organize and analyze high-throughput omics datasets [68]. Furthermore, nutritionists evaluate how lipid and protein profiles translate into nutritional value or illness relevance, contributing to a human-health viewpoint. This involves being aware of how particular food ingredients affect gut flora, lipid metabolism, and long-term health consequences.

The significance of such teams is demonstrated by successful projects. The Dietary Biomarker Consortium (DBDC), for example, brings together nutritionists, metabolomics, epidemiologists, and researchers to use cutting-edge metabolomics to find food intake biomarkers. This is accomplished by utilizing professionals in nutritional biochemistry, bioinformatics, and dietary evaluation. This collaboration serves as an example of how omics analysts and domain specialists work together to develop and evaluate compositional biomarkers [69].

4.4 Educational and Research Implications

The rapid rise of advanced omics technologies in modern food analysis has created an urgent need for experts having specialized training in lipidomics and proteomics. To address this issue, new programs and platforms are specifically created to train food scientists globally. For instance, the Food EDU initiative (by the

Periodic Table of Food) develops fellowships as well as online courses in order to address the important curriculum gaps in food composition and analysis [70]. Similarly, projects funded by the government are integrating workforce development with research. For example, a USDA project in Nebraska is creating a specialized training-and-research hub to increase national capability in food omics [71].

Understanding the data-intensive aspects of modern food research and analysis, incorporating proteomics and lipidomics into mainstream food science training has become extremely important [72]. Both undergraduate and graduate programs should introduce the concept of omics through lab work, case studies, a certified course, and should offer specialized training in advanced omics techniques such as mass spectrometry and research projects based on omics.

On the research side, targeted funding is very important. Public foundations and agencies are increasingly supporting omics research in food science. Projects that connect academia, business, and regulatory bodies should be given priority when it comes to funding to address issues like food authentication, nutritional quality evaluation, and biomarker identification. The benefit of cross-sector partnerships in improving translational research is exemplified by collaborative approaches like as the Clinical Proteomic Tumor Analysis Consortium (CPTAC) [73], which combines proteomics and lipidomics data. The development of affordable technology, such as portable mass spectrometry equipment for field-based food analysis, could also be accelerated by public–private partnerships.

4.5 Challenges and Limitations

Despite being powerful tools, lipidomics and proteomics analysis can be challenging. No single extraction protocol is sufficiently efficient to completely recover all fatty acids and triacylglycerols [74]. Matrix effects can be produced due to complex food matrices, such as proteins, fats, polysaccharides, etc. Ion suppression or enhancement by co-extractives is a main challenge in the analysis of foods in LC–MS [75]. Proteomic analysis may face difficulties due to the occurrence of billions of reactions. In dairy products, these reactions can be triggered during the manufacturing process via heat treatment. It results in chemical changes that cause alterations in protein structures and hinders accurate identifications in proteomic studies. Several proteomic search tools continue to be insufficient for halal food analysis. Such limitations are possibly due to the fact that halal foods typically contain heavily processed meats, which increases the complexity of analysis [76].

Standardization is lacking across platforms, due to which food omics faces considerable challenges. For instance, an analysis of the same dataset using two different widely used software tools showed only 14% agreement in lipid modifications [77]. Obtaining quantitative data accurately can be difficult, since it requires the use of labeled standards for every fatty acid and triglyceride type [78]. In some cases, high-abundance lipids can dominate low-level lipids, making them hard to detect. Overlapped spectral data may be formed due to structural similarities in isomeric and isobaric lipids. Also, the ion source fragmentation may introduce misleading signals, which can distort data interpretation and affect reproducibility. Additionally, poorly ionizable lipid species often go undetected. In this way overall lipidomic profile is affected [79].

The process of exploring food lipidomes can face significant hurdles. Certain specific lipids are highly sensitive, prone to oxidation, and susceptible to light-induced degradation. Thus, there is a need for proper storage and handling conditions, which are not always feasible, compromising lipid stability. Lack of molecular-level studies of some food groups makes lipidomic food analysis. Proteomic strategies may face challenges, notably in the integration of diverse datasets [80]. Integration of multiomics and lipidomics encounters challenges in data interpretation and analysis. To encounter this, advanced computational tools are required [81].

4.6 Future Perspectives

Advances in technology promise the transformation of food omics in the coming years. Artificial intelligence and machine learning are set to transform the landscape of data analysis. [82, 83]. Complex data can be interpreted in an efficient way through artificial intelligence, particularly through machine learning algorithms. With the integration of AI in multiomics, cancer-specific datasets will be analyzed more effectively [84]. Data processing has been advanced by deep learning. It enables accurate predictions and enhances transparency and interpretability. Various intelligent detection systems have been developed for assessing product quality in real time. These technologies play an important role in reducing quality degradation [85]. Smart sensors can also perform real-time data analysis. They enable rapid and accurate monitoring of key parameters. This technology can be used to enhance efficiency in the food industry. Furthermore, real-time quality assurance can be achieved by transitioning from manual processes to artificial intelligence systems for food monitoring. However, this shift demands advanced technologies and skilled expertise [86].

Author Contribution

Conceptualization Muhammad Zubair, Shumaila Fayyaz authored the introduction, highlighting the significance of proteomics and lipidomics. Fakhra Bibi contributed to sample preparation techniques and analytical methods. Zimal Ijaz addressed the materials, quantitative analysis, challenges and future aspects. Zainab Noor highlighted regulatory aspects, quality assurance, collaborative efforts and educational implications. Ayesha Baig and Ayesha Siddiqa reviewed the contents.

References

1. Mao X et al (2023) Overview of omics applications in elucidating the underlying mechanisms of biochemical and biological factors associated with meat safety and nutrition. J Proteome 276:104840
2. Li J et al (2022) LC/MS-based lipidomics to characterize breed-specific and tissue-specific lipid composition of chicken meat and abdominal fat. Lwt 163:113611
3. Guo X et al (2022) UPLC-MS-MS-based lipidomics for the evaluation of changes in lipids during dry-cured mutton ham processing. Food Chem 377:131977
4. Tyers M, Mann M (2003) From genomics to proteomics. Nature 422(6928):193–197
5. Agregán R et al (2023) The potential of proteomics in the study of processed meat products. J Proteome 270:104744
6. Sobsey CA et al (2020) Targeted and untargeted proteomics approaches in biomarker development. Proteomics 20(9):1900029
7. Borràs E, Sabido E (2017) What is targeted proteomics? A concise revision of targeted acquisition and targeted data analysis in mass spectrometry. Proteomics 17(17–18): 1700180
8. Stachniuk A et al (2021) Liquid chromatography–mass spectrometry bottom-up proteomic methods in animal species analysis of processed meat for food authentication and the detection of adulterations. Mass Spectrom Rev 40(1):3–30
9. Martakos IC et al (2024) Food Lipidomics: development and application of a cutting-edge untargeted 4D HRMS workflow for the lipid profiling of food of animal origin. J Food Compos Anal 131:106232
10. Luo P et al (2025) Vacuum ultraviolet laser dissociation and proteomic analysis of halogenated peptides. Se pu= Chin J Chromatogr 43(2):131–138
11. Flores, G.A., et al., Fungal biomolecules for food and pharmaceutical application. eFood, 2025. 6(1): p. e70033
12. Fahy E et al (2011) Lipid classification, structures and tools. Biochim Biophys Acta (BBA) Mol Cell Biol Lipids 1811(11):637–647
13. Shen S et al (2017) A plasma lipidomics strategy reveals perturbed lipid metabolic pathways and potential lipid biomarkers of human colorectal cancer. J Chromatogr B 1068:41–48
14. Sidira M, Smaoui S, Varzakas T (2024) Recent proteomics, metabolomics and lipidomics approaches in meat safety, processing and quality analysis. Appl Sci 14(12):5147
15. Panchaud A, Affolter M, Kussmann M (2012) Mass spectrometry for nutritional peptidomics: how to analyze food bioactives and their health effects. J Proteome 75(12):3546–3559
16. Kim H-Y, Salem N Jr (1990) Separation of lipid classes by solid phase extraction. J Lipid Res 31(12):2285–2289
17. Valdes-Gonzalez T et al (2011) New approach for glyco-and lipidomics–molecular scanning of human brain gangliosides by TLC-blot and MALDI-QIT-TOF MS. J Neurochem 116(5): 678–683
18. Züllig T, Trötzmüller M, Köfeler HC (2020) Lipidomics from sample preparation to data analysis: a primer. Anal Bioanal Chem 412: 2191–2209
19. Cuatrecasas P (1970) Protein purification by affinity chromatography: derivatizations of agarose and polyacrylamide beads. J Biol Chem 245(12):3059–3065
20. Murayama C, Kimura Y, Setou M (2009) Imaging mass spectrometry: principle and application. Biophys Rev 1(3):131
21. Li M, Yang L, Bai Y, Liu H (2013) Analytical methods in lipidomics and their applications. Anal Cham 86:161
22. Serés-Noriega T et al (2023) Nuclear magnetic resonance-based Lipidomics in the assessment of cardiometabolic risk in type 1 diabetes: an exploratory analysis. Diabetes Therapy 14(3): 553–567
23. Abdallah C et al (2012) Gel-based and gel-free quantitative proteomics approaches at a glance. Int J Plant Genomic 2012(1):494572
24. Harlina PW et al (2024) Processing effects on lipid composition in ultra-processed foods: assessing health assumptions and association with blood lipid profiles. Cogent Food & Agric 10(1):2420838
25. Zhang M et al (2023) Lipidomic profile changes of yellow-feathered chicken meat during thermal processing based on UPLC-ESI-MS approach. Food Chem 399:133977
26. Martín-Torres S et al (2023) A comparison of the stability of refined edible vegetable oils under frying conditions: multivariate fingerprinting approach. Foods 12(3):604

27. Lapidot-Cohen T, Rosental L, Brotman Y (2020) Liquid chromatography–mass spectrometry (LC-MS)-based analysis for lipophilic compound profiling in plants. Curr Protocol Plant Biol 5(2):e20109
28. Shi C et al (2023) Development and application of lipidomics for food research. In: Advances in food and nutrition research. Elsevier, pp 1–42
29. Liu Z, Rochfort S (2023) Lipidomics in milk: recent advances and developments. Curr Opin Food Sci 51:101016
30. Tomaiuolo M et al (2023) Untargeted lipidomics and chemometric tools for the characterization and discrimination of irradiated camembert cheese analyzed by UHPLC-Q-Orbitrap-MS. Foods 12(11):2198
31. Zhang S et al (2024) Foodomics as a tool for evaluating food authenticity and safety from field to table: a review. Foods 14(1):15
32. Hu L et al (2022) Lipidomic profiling of Skipjack tuna (Katsuwonus pelamis) by ultrahigh-performance liquid chromatography coupled to high resolution mass spectrometry. Fish Aquat Sci 25(3):140–150
33. Chen J-N et al (2023) Analysis of lipid molecule profiling and conversion pathway in mandarin fish (Siniperca chuatsi) during fermentation via untargeted lipidomics. J Agric Food Chem 71(22):8673–8684
34. Zhang D et al (2021) Lipidomics reveals the changes in lipid profile of flaxseed oil affected by roasting. Food Chem 364:130431
35. Lozano-Castellón J et al (2021) New vacuum cooking techniques with extra-virgin olive oil show a better phytochemical profile than traditional cooking methods: a foodomics study. Food Chem 362:130194
36. Zheng J et al (2021) Lipidomic, transcriptomic, and BSA-660K single nucleotide polymorphisms profiling reveal characteristics of the cuticular wax in wheat. Front Plant Sci 12:794878
37. Zhang D et al (2023) Lipidomics and volatilomics reveal the changes in lipids and their volatile oxidative degradation products of brown rice during accelerated aging. Food Chem 421:136157
38. Su M et al (2024) Lipid dynamics of'Fenjiao'-bananas (Musa ABB Pisang Awak) during post-ripening based on lipidomics analysis. Postharvest Biol Technol 207:112611
39. Sutliff AK et al (2021) Lipidomics-based comparison of molecular compositions of green, yellow, and red bell peppers. Meta 11(4):241
40. Ren W et al (2023) Progress of mass spectrometry-based lipidomics in the dairy field. Foods 12(11):2098
41. Chen G et al (2024) Evaluating rice lipid content, yield, and quality in response to nitrogen application rate and planting density. Front Plant Sci 15:1469264
42. Nasri C et al (2023) Proximate composition, lipid and elemental profiling of eight varieties of avocado (Persea americana). Sci Rep 13(1): 22767
43. Ruiz-Méndez M-V et al (2024) Lipid quality changes in French fries, chicken croquettes, and chicken nuggets fried with high-linoleic and high-oleic sunflower oils in domestic deep fryers. Foods 13(15):2419
44. Abeyrathne EDNS et al (2022) Egg yolk lipids: separation, characterization, and utilization. Food Sci Biotechnol 31(10):1243–1256
45. Tan K et al (2023) Effects of food processing on the lipid nutritional quality of commercially important fish and shellfish. Food Chemistry: X 20:101034
46. Abril AG et al (2022) Proteomics characterization of food-derived bioactive peptides with anti-allergic and anti-inflammatory properties. Nutrients 14(20):4400
47. Molinari MDC et al (2023) Exploring the proteomic profile of soybean bran: unlocking the potential for improving protein quality and quantity. Plants 12(14):2704
48. Nye-Wood MG et al (2021) Proteome analysis and epitope mapping in a commercial reduced-gluten wheat product. Front Nutr 8:705822
49. Shan L et al (2025) Enhancing intact allergen protein detection in dairy products: preventing fragmentation and aggregation with Sinapinic acid salt in MALDI-TOF MS analysis. Food Chem 477:143593
50. Tehrani T, Pont L, Benavente F (2024) Rapid detection and quantification of milk adulteration using MALDI-MS protein profiling and multivariate calibration. J Food Compos Anal 130:106147
51. Martínez A et al (2025) Recent trends in food quality and authentication: the role of omics technologies in dairy and meat production. Int J Mol Sci 26(9):4405
52. von Oesen T et al (2023) Identification of marker peptides for the whey protein quantification in Edam-type cheese. Foods 12(10): 2002
53. Fan S et al (2022) Determination of tropomyosin in shrimp and crab by liquid chromatography–tandem mass spectrometry

based on immunoaffinity purification. Front Nutr 9:848294

54. Wilhelm C, Hofsommer M, Wittke S (2022) Detection of mechanically separated meat from chicken in sausages and cold meat by targeted LC–MS/MS analysis. Food Anal Methods 15(7):1899–1908
55. Ramanan M et al (2024) Barley grain proteome assessment using multi-environment trial data and machine learning. J Agric Food Chem 72(47):26416–26430
56. Bohn K et al (2022) Estimating food ingredient compositions based on mandatory product labeling. J Food Compos Anal 110:104508
57. Wu W-J et al (2022) Qualitative and quantitative analysis of ejiao-related animal gelatins through peptide markers using LC-QTOF-MS/MS and scheduled multiple reaction monitoring (MRM) by LC-QQQ-MS/MS. Molecules 27(14):4643
58. Degnan FH (1997) The food label and the right-to-know. Food Drug Law J 52(1):49–60
59. Barach JT (2017) FSMA and food safety systems: understanding and implementing the rules. John Wiley & Sons
60. Zhang J, Bhatt T (2014) A guidance document on the best practices in food traceability. Compr Rev Food Sci Food Saf 13(5): 1074–1103
61. Beger RD et al (2019) Towards quality assurance and quality control in untargeted metabolomics studies. Metabolomics 15:1–5
62. Broadhurst D et al (2018) Guidelines and considerations for the use of system suitability and quality control samples in mass spectrometry assays applied in untargeted clinical metabolomic studies. Metabolomics 14:1–17
63. Lippa KA et al (2022) Reference materials for MS-based untargeted metabolomics and lipidomics: a review by the metabolomics quality assurance and quality control consortium (mQACC). Metabolomics 18(4):24
64. Bowden JA et al (2018) NIST lipidomics workflow questionnaire: an assessment of community-wide methodologies and perspectives. Metabolomics 14:1–11
65. Sun T et al (2020) Mass spectrometry-based lipidomics in food science and nutritional health: a comprehensive review. Compr Rev Food Sci Food Saf 19(5):2530–2558
66. Gallo M, Ferranti P (2016) The evolution of analytical chemistry methods in foodomics. J Chromatogr A 1428:3–15
67. Tang S, Zhang H, Lee HK (2016) Advances in sample extraction. Anal Chem 88(1):228–249
68. Chelliah R et al (2022) A review on the application of bioinformatics tools in food microbiome studies. Brief Bioinform 23(2):bbac007
69. Chakraborty H et al (2025) The dietary biomarkers development consortium: an initiative for discovery and validation of dietary biomarkers for precision nutrition. Curr Develop Nutr 9(5):107435
70. Association, A.H (2024) Periodic table of food initiative (PTFI). Res Program
71. McCabe-Sellers B et al (2008) Personalizing nutrigenomics research through community based participatory research and omics technologies. OMICS 12(4):263–272
72. Chen H et al (2017) Lipidomics in food science. Curr Opin Food Sci 16:80–87
73. Edwards NJ et al (2015) The CPTAC data portal: a resource for cancer proteomics research. J Proteome Res 14(6):2707–2713
74. Romsdahl TB et al (2022) A lipidomics platform to analyze the fatty acid compositions of non-polar and polar lipid molecular species from plant tissues: examples from developing seeds and seedlings of pennycress (Thlaspi arvense). Front Plant Sci 13:1038161
75. Nasiri A et al (2021) Overview, consequences, and strategies for overcoming matrix effects in LC-MS analysis: a critical review. Analyst 146(20):6049–6063
76. Amir SH et al (2021) Comparative database search engine analysis on massive tandem mass spectra of pork-based food products for halal proteomics. J Proteome 241:104240
77. von Gerichten J et al (2024) Challenges in lipidomics biomarker identification: avoiding the pitfalls and improving reproducibility. Meta 14(8):461
78. Köfeler HC et al (2021) Recommendations for good practice in MS-based lipidomics. J Lipid Res 62:100138
79. Hu C, Duan Q, Han X (2020) Strategies to improve/eliminate the limitations in shotgun lipidomics. Proteomics 20(11):1900070
80. Herráiz-Gil S et al (2023) Multi-omic data integration in food science and analysis. Curr Opin Food Sci 52:101049
81. Abrahams T, Nicholls SJ (2024) Perspectives on the success of plasma lipidomics in cardiovascular drug discovery and future challenges. Expert Opin Drug Discov 19(3):281–290

82. Pu K et al (2022) Integration of non-targeted proteomics mass spectrometry with machine learning for screening cooked beef adulterated samples. J Agric Food Chem 71(4): 2173–2182
83. Sobhan A et al (2025) IoT-enabled biosensors in food packaging: a breakthrough in food safety for monitoring risks in real time. Foods 14(8):1403
84. Biswas N, Chakrabarti S (2020) Artificial intelligence (AI)-based systems biology approaches in multi-omics data analysis of cancer. Front Oncol 10:588221
85. Wang D et al (2024) Intelligent system/equipment for quality deterioration detection of fresh food: recent advances and application. Foods 13(11):1662
86. Kyaw KS et al (2024) Toward in-process technology-aided automation for enhanced microbial food safety and quality assurance in milk and beverages processing. Crit Rev Food Sci Nutr 64(6):1715–1735

Chapter 18

Artificial Intelligence and Machine Learning in Food Analysis

Homa Behmadi, Elham Azarpazhooh, Hosahalli S. Ramaswamy, and Khaled Ahmadauli

Abstract

Artificial Intelligence (AI) and machine learning (ML) are updating food analysis, improving quality control, safety assurance, and authenticity verification. AI/ML replaces traditional, labor-intensive methods with data-driven automation, enabling fast, non-destructive, high-throughput food analysis. For example, convolutional neural networks (CNNs) with hyperspectral imaging achieve over 95% accuracy in detecting defects in fruits such as apples, while support vector machines (SVMs) with Raman spectroscopy identify adulterants in honey with 98% precision, meeting regulatory standards. Advanced preprocessing, such as spectral normalization and dimensionality reduction, improves models, and supervised learning, like random forests, for tasks such as predicting meat spoilage with an error margin of less than one day. Unsupervised techniques, including autoencoder-based anomaly detection (a type of neural network that learns compressed representations of data to identify outliers), further empower laboratories to uncover unknown contaminants or spoilage markers without relying on labeled datasets. Beyond technical innovations, AI/ML covers important industry needs: Real-time pathogen detection in meat production reduces contamination risk, automated sorting systems reduce packaging labor costs, and metabolomics traces high-value product origins (e.g., olive oil) to combat fraud. Challenges such as data heterogeneity, computational costs, and ethical concerns around algorithmic bias remain hurdles, but standardization (e.g., spectral library harmonization, ISO 17025 workflows) enables scalable, reproducible solutions. Developing trends, including federated learning for secure multi-lab collaboration and edge AI on IoT devices, enable decentralized, real-time field analysis. Explainable AI tools like LIME (Local Interpretable Model-agnostic Explanations) and SHAP (SHapley Additive exPlanations Support Vector Machine) are associate with FDA and EFSA guidelines by clarifying model decisions, promoting trust among regulators and stakeholders. In conclusion, while AI/ML offers significant opportunities for food systems, challenges like data fragmentation, regulatory gaps, and adoption barriers must be resolved to bridge the gap between technological potential and real impact.

Key words Artificial Intelligence, Edge Computing, Explainable AI, Food Authenticity, Food Safety, Machine Learning, Predictive Modeling

Tanmay Sarkar (ed.), *Novel Food Analysis: Techniques and Applications*, Methods and Protocols in Food Science,
https://doi.org/10.1007/978-1-0716-4787-5_18,

1 Introduction

The integration of artificial intelligence (AI)-systems that mimic human cognitive processes and Machine Learning (ML)-algorithms that learn patterns from data-is revolutionizing food analysis by addressing long-standing challenges in safety, quality control, and authenticity verification. Traditional methods such as manual inspection and laboratory assays are time-consuming, subjective, and often too slow to meet the demands of modern, global food supply chains [1]. For instance, manual fruit grading in packing facilities often suffers from inconsistency, with human inspectors missing minor defects due to fatigue or variability in judgment. Similarly, conventional chromatographic techniques used to detect pesticide residues are destructive and can require several hours, delaying necessary safety decisions [2].

By contrast, AI and ML offer fast, objective, and non-destructive alternatives that can process vast datasets derived from spectroscopic, imaging, and sensor technologies. Convolutional neural networks (CNNs)—a type of deep learning model optimized for image analysis- paired with hyperspectral imaging, have demonstrated a training accuracy of 98% and a testing accuracy of 85–90% in detecting internal blueberry defects, which are often invisible to the human eye [3].

AI-enhanced electronic nose (e-nose) and tongue (e-tongue) systems enable rapid, non-destructive food quality analysis. E-noses detect volatile organic compounds (VOCs) to monitor freshness, spoilage, and adulteration, while e-tongues evaluate liquid taste profiles for authenticity and contamination. These systems, integrated with Internet of Things (IoT) devices and supported by nanotechnology, offer scalable, real-time solutions that reduce reliance on traditional, time-intensive testing procedures, thereby improving regulatory compliance and operational efficiency [4].

AI/ML also excels at managing complex, high-dimensional datasets. Traditional chemometric tools such as Principal Component Analysis (PCA) often struggle with non-linear relationships common in food matrices [5]. In contrast, models such as Support Vector Regression (SVR) and LASSO (See Table 1) have achieved high performance in detecting honey adulteration, with R^2 values exceeding 0.99 when trained on product-specific datasets. These tailored models, especially when paired with ultra-fast gas chromatography (UF-GC), offer highly accurate and rapid food authentication [6]. Deep learning autoencoders further expand analytical capabilities by enabling unsupervised anomaly detection and identification of unknown contaminants, for proactive food safety management.

To improve readability and accessibility for interdisciplinary audiences, Table 1 defines the most important abbreviations,

Table 1
Essential AI and machine learning terms in food analysis

Acronym	Full term	Brief explanation	Example application in food analysis	Reference
AI	Artificial Intelligence	Systems that mimic human intelligence to perform tasks (e.g., pattern recognition).	Automated grading of fruits using computer vision.	[7]
ANN	Artificial Neural Network	Computational model inspired by biological neural networks for pattern recognition.	Predicting meat spoilage from spectral data.	[8]
CNN	Convolutional Neural Network	Deep learning model optimized for image and spatial data analysis.	Detecting bruises in apples via hyperspectral imaging.	[9]
DL	Deep Learning	Subset of ML using multi-layered neural networks to model complex data patterns.	Anomaly detection in untargeted metabolomics datasets.	[10]
FTIR	Fourier-Transform Infrared Spectroscopy	Analytical technique identifying chemical bonds via infrared absorption spectra.	Quantifying lipid oxidation in edible oils.	[11]
GC-MS	Gas Chromatography-Mass Spectrometry	Hybrid method for separating and identifying volatile compounds.	Detecting pesticide residues in fruits.	[12]
HACCP	Hazard Analysis Critical Control Point	Systematic preventive approach to food safety.	AI-driven monitoring of CCPs in meat processing.	[13]
HPLC	High-Performance Liquid Chromatography	Technique for separating, identifying, and quantifying compounds in liquid mixtures.	Authenticating olive oil using phenolic compound profiles.	[14]
IoT	Internet of Things	Network of interconnected devices for real-time data exchange.	IoT-enabled e-noses for spoilage monitoring in seafood.	[15]
LIME	Local Interpretable Model-agnostic Explanations	Explains single predictions by fitting interpretable models (e.g., linear regression) to localized subsets of complex model outputs.	Interpreting a deep learning model's classification of tomato ripeness	[16]
LOD	Limit of Detection	Minimum concentration of an analyte detectable by an analytical method.	Validating AI models for trace contaminant detection.	[17, 18]

(continued)

Table 1
(continued)

Acronym	Full term	Brief explanation	Example application in food analysis	Reference
ML	Machine Learning	Algorithms that learn patterns from data to make predictions or decisions.	Classifying honey authenticity using Raman spectra.	[19]
NIR	Near-Infrared Spectroscopy	Non-destructive technique analyzing molecular overtone and combination vibrations.	Predicting grain moisture content in real time.	[20]
PCA	Principal Component Analysis	Dimensionality reduction method for identifying key variables in datasets.	Simplifying hyperspectral imaging data for defect detection.	[21]
PLS-R	Partial Least Squares Regression	Chemometric method modeling relationships between predictors and responses.	Correlating FTIR spectra with olive oil quality parameters.	[22]
RMSE	Root Mean Square Error	Statistical measure of prediction accuracy (lower values indicate better performance).	Validating shelf-life prediction models for dairy products.	[23]
ROI	Region of Interest	Selected subset of data (e.g., image area) for focused analysis.	Extracting bruise regions from apple hyperspectral images.	[24]
SHAP	SHapley Additive exPlanations	Framework for interpreting ML model outputs by assigning feature importance scores.	Identifying key spectral bands in honey adulteration models.	[25]
SVM	Support Vector Machine	Supervised learning algorithm for classification/regression using hyperplanes.	Detecting melamine adulteration in dairy products.	[26]
VOC	Volatile Organic Compound	Organic chemicals emitted as gases, often indicators of food quality or spoilage.	Monitoring seafood freshness via e-nose sensor arrays.	[27]
XAI	Explainable Artificial Intelligence	Techniques to make AI/ML model decisions transparent and interpretable.	Justifying olive oil authenticity predictions for regulators.	[28]
ISO/IEC 17025	International Standard for Testing/ Calibration Labs	Quality management system ensuring technical competence and reliable results.	Validating AI models for regulatory compliance.	[29]
GDPR	General Data Protection Regulation	EU regulation governing data privacy and security.	Anonymizing geospatial data in AI-driven supply chains.	[30]

machine learning (ML) techniques, and analytical methods referenced throughout the chapter. For expanded definitions, see the cited references.

Recent AI/ML advances also improve Hazard Analysis Critical Control Point (HACCP) systems for animal-source foods (ASFs). ML enhances predictive analytics via advanced feature extraction, while technologies like hyperspectral imaging and smartphone-integrated sensors provide scalable, real-time data acquisition. Advancing tools such as hybrid models and AI-driven computer vision are being applied for automated quality control. However, persistent challenges include algorithmic bias, data heterogeneity, and the interpretability of deep learning models, which hinder widespread regulatory acceptance. Explainable AI (XAI) tools, such as SHAP, are increasingly used to clarify model decisions-such as identifying specific wavelengths that indicate adulterants-thereby improving regulatory trust and transparency [17, 25].

Ethical considerations, particularly related to algorithmic bias, also demand attention. Models trained on region-specific datasets may misclassify products from underrepresented areas, potentially leading to false claims of fraud. This issue can arise from biased training data, inadequate algorithm design, or a lack of diversity in sample representation. To ensure fairness, it is essential to diversify training datasets, implement fairness-aware algorithms, and conduct routine bias audits throughout the AI development lifecycle [31, 32].

Despite these challenges, AI and ML technologies are reshaping the food sector into a more transparent, efficient, and resilient system. Ethical principles that prioritize transparency, accountability, and inclusivity are key to ensuring that technological advancements benefit all participants [33, 34].

This introduction outlines the high potential of AI and ML in food analysis, emphasizing improvements in precision, efficiency, and decision-making for researchers, industry professionals, and policymakers. It also highlights the evolution from basic chemometric tools to modern deep learning models, and introduces the foundational protocols for successful AI/ML implementation-including data collection, preprocessing, model optimization, deployment, and interpretation. The section concludes by covering ongoing challenges such as data heterogeneity and ethical risk, and by introducing technologies like federated learning, edge computing, and quantum ML [35–37], which promise to reshape food analysis in the years ahead. For clarity and ease of reference, the introduction section is organized into the following sections:

1. Role of AI/ML in Modern Food Analysis; reviews the historical role and recent evolution of AI/ML in food analysis.
2. Evolution of AI/ML in Food Science; summarizes the key applications, from pathogen detection to shelf-life prediction.

3. Key Applications of AI/ML in Food Analysis; outlines current materials including data sources, hardware, software, and reference standards which are needed for AI/ML workflows.
4. Challenges in Implementing AI/ML; examines major challenges (technical, ethical, regulatory) and proposed mitigation strategies.
5. Chapter Objectives; presents step-by-step methods for integrating AI/ML protocols into food analysis pipelines.

1.1 Role of AI/ML in Modern Food Analysis

The traditional analytical toolbox in food testing-chromatography, microbial culturing, and sensory panels has long struggled to keep pace with the demands of today's global supply chains. Manual grading in packing facilities is slow, subjective, and can miss subtle defects such as internal bruising or early-stage rot. Likewise, microbial plating for pathogen detection may take 48–72 hours, delaying critical safety interventions. These limitations highlight an urgent need for automated, high-throughput methods that deliver real-time, objective insights.

AI and ML directly address these bottlenecks by automating feature extraction and decision-making on large, complex datasets [1]. Supervised learning algorithms-such as support vector machines (SVMs) and random forests-excel at classification tasks essential to food safety and authenticity. For example, Raman spectroscopy data classified by an SVM model achieved 98% accuracy in detecting honey adulteration, outperforming conventional chemometric approaches [38]. Similarly, random forests trained on multispectral images have predicted meat spoilage within a one-day margin of error [39].

Beyond classification, unsupervised learning techniques uncover hidden patterns in unlabeled data. Autoencoders can detect unknown contaminants by learning the normal "fingerprint" of a food matrix and flagging deviations, while clustering methods applied to metabolomic profiles reveal natural versus fraudulent geographical signatures in olive oil [40].

Predictive analytics powered by regression models-such as LSTM networks-further transform supply-chain management. By integrating time-series sensor data (e.g., temperature, humidity) with Fourier-transform infrared (FTIR) spectra, these models forecast shelf life and dynamically adjust expiration dates, reducing waste by up to 30% in pilot trials [41].

Finally, edge AI and IoT integration are decentralizing analysis, enabling real-time testing outside centralized labs. Portable NIR spectrometers running lightweight ML models on devices like the NVIDIA Jetson Nano can quantify grain moisture on-site in minutes rather than days [42]. This shift not only accelerates decision-making but also extends advanced testing capabilities to low-resource and field settings.

1.2 Evolution of AI/ML in Food Science

The application of AI and ML methods in food science has progressed through distinct phases, each driven by technological advances and analytical needs.

1.2.1 1990s: Chemometrics and Dimensionality Reduction

Early work focused on linear multivariate methods to extract meaningful information from complex spectral and chromatographic datasets. Principal component analysis (PCA) and partial least squares regression (PLS-R) became staples for interpreting near-infrared (NIR) and Fourier-transform infrared (FTIR) spectra. For example, PCA reduces thousands of spectral wavelengths to a few principal components, facilitating moisture and fat predictions in grains and edible oils [5].

1.2.2 2000s: Kernel Methods and Ensemble Learning

The introduction of kernel-based algorithms and ensemble models addressed the limitations of linear approaches. Support Vector Machines (SVMs) mapped data into high-dimensional feature spaces, improving the classification of adulterated versus authentic samples. Random forests, with their aggregations of decision trees, handled noisy, high-dimensional inputs such as electronic-nose sensor arrays. SVM models have been effectively used to distinguish extra-virgin from adulterated olive oil, achieving high accuracy through methods like Fourier Transform Infrared (FTIR) spectroscopy [43, 44]. For meat spoilage detection, random forests have been applied using techniques such as FTIR spectroscopy, enabling rapid identification of microbial spoilage [45, 46].

1.2.3 2010s: Deep Learning Revolution

Deep neural networks transformed image- and signal-based analyses. CNNs automatically learned hierarchical features from hyperspectral cubes or RGB images, vastly outperforming manual feature engineering. ResNet and ResNeXt convolutional neural networks achieved 88.4% and 87.8% accuracy (93.3% and 89.4% recall, respectively) in detecting internal damage in blueberries using hyperspectral transmittance data, outperforming traditional machine learning models such as SMO (80.8% accuracy) and RF (73.1% accuracy). With preprocessing and rapid classification times (5.2 ms for ResNet, 6.5 ms for ResNeXt per sample), the approach shows promise for real-time agricultural sorting, emphasizing high recall to minimize costly misclassification of damaged fruit [3]. Autoencoders and recurrent architectures have further enabled unsupervised anomaly detection in metabolomic and time-series datasets, opening new avenues for fraud and contamination screening.

1.2.4 2020s and Beyond: Transformers, Self-Supervision, and Explainability

Recent advances employ self-supervised pretraining to reduce reliance on labeled data, particularly in machine learning applications [47, 48]. Vision transformers (ViTs) are being adapted for sequential spectral inputs, capturing long-range dependencies critical for shelf-life modeling. Federated learning frameworks allow secure, collaborative model training across laboratories without raw data sharing, addressing privacy and heterogeneity concerns [49, 50]. Meanwhile, Explainable AI (XAI) tools such as SHAP and LIME are incorporated by design to demystify model decisions—essential for regulatory compliance [51, 52]. Near-infrared hyperspectral imaging (NIR-HSI) combined with machine learning (e.g., PLS-DA, ANN, SVM, RF) and spectral preprocessing (Savitzky-Golay, SNV/MSC) effectively detects adulteration in extra-virgin olive oil, achieving up to 100% accuracy in identifying adulterants like hazelnut, refined olive, and pomace oils at concentrations as low as 1% (m/m). PLS-DA and ANN models outperformed other algorithms, with key variable selection showing no significant impact on performance. This demonstrates a robust, non-destructive approach for olive oil authentication and fraud prevention [53]. Table 2 is a compact table that maps each era to its hallmark techniques, impact, and a key example.

Figure 1 illustrates the chronological progression of key machine learning methods applied in food analysis, highlighting their foundational techniques.

1.3 Key Applications of AI/ML in Food Analysis

AI and ML have found transformative uses across the food value chain. They can be highlighted in four core domains (Fig. 2):

1.3.1 Food Safety: Pathogen Detection and Spoilage Monitoring

AI-enhanced hyperspectral imaging and e-nose systems deliver rapid, non-destructive detection of contaminants. For example, CNNs trained on hyperspectral data from chicken fillets achieved 97.62% sensitivity for Salmonella detection, significantly reducing outbreak risk [54]. E-noses coupled with SVM classifiers monitor VOCs in seafood, attaining F1-scores above 0.93 for real-time freshness assessment [55].

Table 2
Evolutionary milestones

Era	Key techniques	Impact on food analysis	Example study	Reference
1990s	PCA, PLS-R	Dimensionality reduction for spectral data	Moisture prediction in grains	[5]
2000s	SVM, random forests	Non-linear classification, ensemble robustness	Olive oil adulteration detection	[43, 44]
2010s	CNNs, autoencoders, RNNs	Automated feature learning; unsupervised anomaly detection	Blueberry bruise detection	[3]
2020s +	Transformers, self-supervised learning, federated learning, XAI	Reduced labeling needs; privacy-preserving collaboration; explain ability	ViTs for spectral time series	[47, 49, 50]

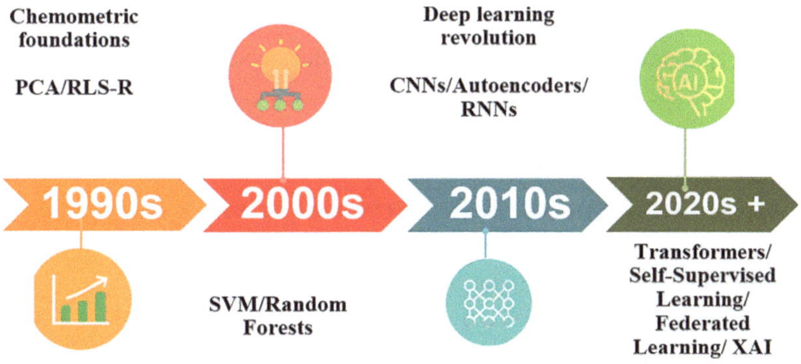

Fig. 1 Evolution of machine learning techniques in food analysis

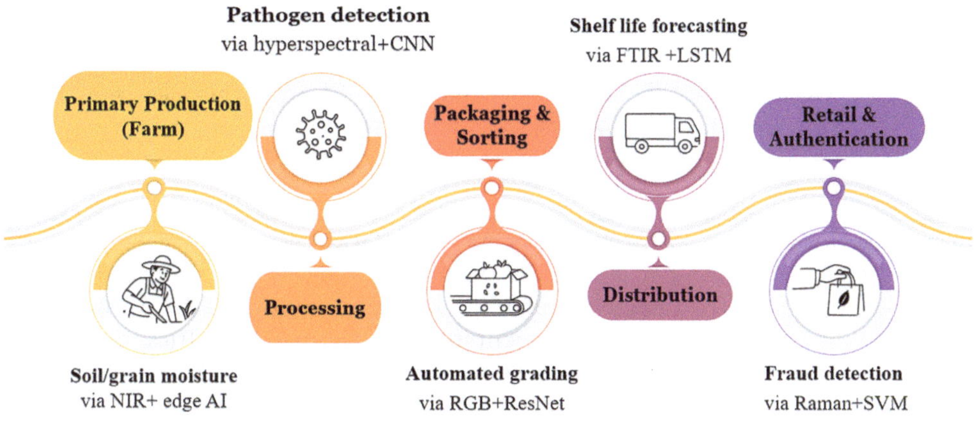

Fig. 2 AI/ML applications across the food supply chain

1.3.2 Quality Control: Automated Grading & Defect Detection

Deep-learning models replace subjective manual inspections. A ResNet-50 architecture fine-tuned on RGB images of tomatoes reached 96.35% accuracy in ripeness classification [56]. Similar systems are applied to detect apple bruises and potato blemishes, employing data augmentation techniques to improve robustness under variable lighting conditions.

1.3.3 Authenticity Verification: Metabolomic Fingerprinting

Random forests and SVMs parse high-resolution chromatographic or spectroscopic profiles to expose fraud. In olive oil, ML models analyzing fatty acid ratios via GC–MS achieved 100% accuracy in distinguishing extra-virgin from adulterated blends [57]. Raman-SVM pipelines detect honey adulteration (e.g., corn syrup) with 98% precision [58].

1.3.4 Shelf-Life Prediction: Dynamic Expiration Management

Time-series models-such as LSTM networks-combine sensor data (temperature, humidity) with FTIR spectra to forecast spoilage. LSTMs applied to dairy FTIR data reduced prediction error to under 5%, enabling real-time adjustment of expiration dates and cutting waste by up to 30% [59].

Table 3
AI/ML applications in food analysis: techniques, performance, and impact

Domain	Data modality	AI/ML Technique	Key performance	Real-world benefit	Reference
Food Safety	Hyperspectral Imaging	CNN	97.62% sensitivity	Early Salmonella detection in poultry	[60]
Spoilage Monitoring	E-nose VOC Profiles	SVM	F1-score > 0.93	Real-time seafood freshness alerts	[61]
Quality Control	RGB/ Multispectral Imaging	ResNet-50 (CNN)	96.35% accuracy	Automated tomato ripeness sorting	[56]
Authenticity	Raman Spectroscopy	SVM	98% precision	Honey adulteration detection	[58]
Authenticity	GC-MS	Random Forest	100% accuracy	Olive oil fraud detection	[57]
Shelf-Life Prediction	FTIR + Time Series	LSTM	<5% error	30% dairy waste reduction	[62]

CNN convolutional neural network, *SVM* support vector machine, *LSTM* long short-term memory

Table 3 summarizes key applications of AI/ML in food analysis, highlighting techniques, performance metrics, real-world benefits, and references for each domain, from pathogen detection to waste reduction.

1.4 Challenges in Implementing AI/ML

Despite their transformative potential, AI and ML systems in food analysis face several critical hurdles, which can be organized into three main categories and outline strategies to address each (Table 4).

1.4.1 Data Heterogeneity

Challenge

Spectral, imaging, and sensor data often vary across instruments, laboratories, and environmental conditions. For example, near-infrared (NIR) spectra from different devices can differ by 15–20% due to calibration drift or light-source variability [5].

Impact

Models trained on one dataset may perform poorly when applied to data from other sources, undermining reliability.

Strategies

Standardized preprocessing: Apply baseline correction, Standard Normal Variate (SNV) normalization, and Savitzky–Golay smoothing to harmonize spectra.

Data augmentation: Simulate instrument and environmental variability during training.

Table 4
Key challenges and mitigation strategies for implementing AI/ML in food analysis

Challenge	Description	Mitigation strategy
Data heterogeneity	Instrument- and lab-to-lab variability in spectral/imaging/ sensor data	Standardized preprocessing; data augmentation; federated learning
Model interpretability	"Black-box" nature of deep learning models	Explainable AI (SHAP, LIME); surrogate models; model cards
Ethical and regulatory	Algorithmic bias; compliance with lab and data standards	Diverse data gathering; bias audits; compliance with ISO/IEC 17025 and FDA AI/ML standards

SHAP Shapley Additive exPlanations; *LIME* Local Interpretable Model-agnostic Explanations

Federated learning: Train shared models across multiple laboratories without exchanging raw data, preserving privacy and improving generalization [63].

1.4.2 Model Interpretability

Challenge

Deep learning models (e.g., CNNs, transformers) are often "black boxes," making it difficult for regulators and end-users to understand why a prediction was made.

Impact

Opacity hinders regulatory approval and erodes user trust, especially for safety-critical decisions.

Strategies

Explainable AI (XAI) Integrate tools such as SHAP and LIME during development to generate feature-importance explanations [25]. Opacity in deep learning models remains a major barrier to regulatory approval and user adoption, particularly in high-stakes contexts like food safety. To address this, explainable AI (XAI) tools such as SHAP and LIME are increasingly integrated during model development. SHAP provides global and local interpretability by attributing feature importance based on game-theoretic principles, while LIME builds simplified surrogate models around individual predictions. These tools help uncover why a model classified a food sample as adulterated or unsafe, offering transparency essential for compliance with FDA and ISO/IEC 17025 standards.

Simplified Surrogate Models Use decision-tree approximations of complex models for high-level auditing.

Model Cards Publish concise documentation of model inputs, training data, performance metrics, and known limitations.

1.4.3 Ethical and Regulatory Concerns

Challenge

Algorithmic bias can arise from unbalanced training datasets, leading to unfair outcomes-e.g., misclassifying products from under-represented regions as adulterated.

Impact

Bias undermines equity, damages producer livelihoods, and may trigger legal or reputational risks.

Strategies

Diverse data collection Deliberately include samples from different geographies, seasons, and production methods.

Bias audits Regularly evaluate performance across subgroups using confusion matrices and fairness metrics.

Meet regulatory requirements Follow standards such as ISO/-IEC 17025 for laboratory competence and the FDA's AI/ML Action Plan for ongoing model monitoring.

To ensure fairness and reliability in AI systems, Fig. 3 outlines essential mitigation strategies such as conducting regular bias audits, diversifying data collection, and adhering to ethical standards for interpretability.

1.5 Chapter Objectives

This chapter is designed to equip researchers, industry professionals, and policymakers with the knowledge and protocols necessary to deploy AI/ML effectively in food analysis. Specifically, it aims to:

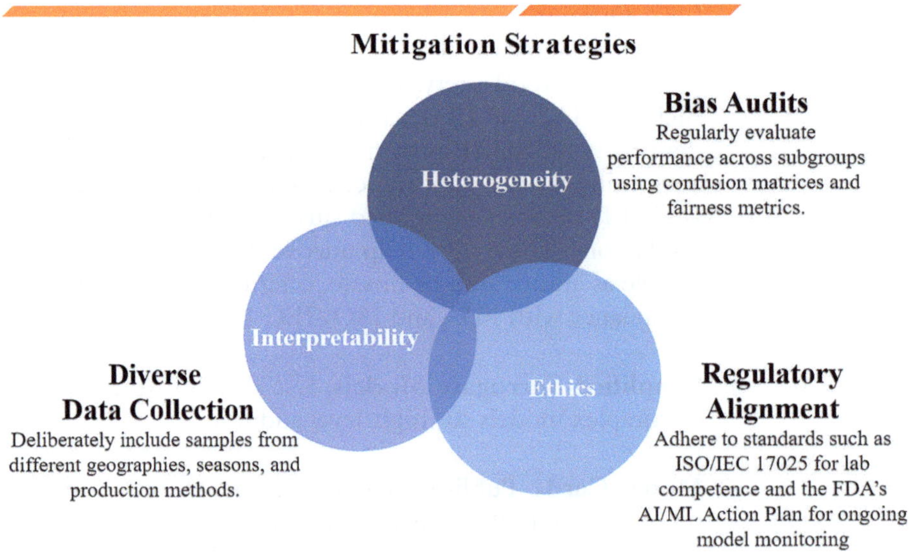

Fig. 3 Key strategies for mitigating bias in AI models

1.5.1 Outline Best Practices for Data Acquisition and Preprocessing

Demonstrate standardized protocols for collecting and cleaning spectral, chromatographic, and imaging data to maximize model robustness and reproducibility.

1.5.2 Present Step-by-Step Modeling Workflows

Provide clear, hands-on guidance for training supervised (e.g., SVM, CNN) and unsupervised (e.g., autoencoder) models, including hyperparameter tuning and cross-validation strategies.

1.5.3 Detail Regulatory and Ethical Considerations

Explain compliance requirements (e.g., ISO/IEC 17025, FDA AI/ML Action Plan) and bias-mitigation techniques to ensure transparent, fair, and legally sound AI deployments.

1.5.4 Showcase Deployment and Validation Strategies

Describe methods for model optimization (e.g., quantization, ONNX conversion), edge-device integration, and continuous performance monitoring to maintain accuracy over time.

1.5.5 Highlight Trends and Future Directions

Advancing trends including self-supervised pretraining, federated learning, quantum ML, and secure multi-party collaboration are expected to significantly expand access to advanced analytical capabilities. Quantum ML, though still in its early stages, holds potential for accelerating pattern recognition and dimensionality reduction in high-dimensional spectral datasets, offering promising future applications in food analysis.

2 Materials

This section details the essential components-data sources, hardware, software tools, and reference standards-needed to develop robust, reproducible AI/ML workflows in food analysis.

2.1 Data Sources

High-quality, well-annotated datasets are the foundation of any AI/ML model (Table 5). Key modalities include:

Table 5
Key data sources for AI/ML in food analysis

Data type	Examples / repositories	Typical applications	Reference
Spectral	NIR, FTIR, Raman; USDA FoodData	Moisture, fat, adulterant quantification	[5, 64, 65]
Chromatographic	GC–MS, LC–MS; MassBank, GNPS	Pesticide screening, metabolomics	[66, 67]
Imaging	Hyperspectral (400–1000 nm); RGB	Bruise, mold, ripeness detection	[68, 69]
Omics	LC-MS/MS; MetaboLights	Origin tracing, biomarker discovery	[70, 71]

2.1.1 Spectral Data

Near-infrared (NIR), Fourier-transform infrared (FTIR), and Raman spectra capture molecular vibrations for non-destructive compositional analysis. Public repositories such as USDA Food Data Central and in-house libraries support tasks like moisture prediction in grains and lipid-oxidation monitoring in oils.

2.1.2 Chromatographic Data

Gas chromatography–mass spectrometry (GC-MS) and liquid chromatography–mass spectrometry (LC-MS) can generate detailed profiles of volatile and non-volatile compounds. Databases like Mass Bank and GNPS facilitate pesticide-residue screening and metabolomic fingerprinting.

2.1.3 Imaging Data

Hyperspectral cubes (400–1000 nm) and RGB image sets (e.g., Fruit Net, UNIMIB2016) enable defect detection and grading via computer vision pipelines.

2.1.4 Omics Data

Metabolomic and proteomic profiles (e.g., MetaboLights) underpin geographic traceability and fraud detection in high-value products like olive oil and honey.

Best practice: Prior to modeling apply baseline correction, Standard Normal Variate (SNV) normalization, and Savitzky–Golay smoothing to reduce instrument and environmental variability.

2.2 Hardware

The combined use of specialized sensors and optimized compute platforms creates a closed-loop system for perishables monitoring: sensors capture high-fidelity physicochemical data, while compute resources transform it into predictive insights. This hardware stack enables end-to-end traceability, from field measurements to cloud-based analytics, while edge deployments reduce latency for time-sensitive applications like spoilage alerts.

Advancing hardware trends-such as miniaturized multispectral sensors, neuromorphic chips for low-power inference, and quantum-accelerated hyperspectral processing-promise further gains in speed, accuracy, and energy efficiency. Scalability remains critical, with modular architectures allowing easy upgrades as sensing and computational demands change.

Effective AI/ML workflows require both advanced sensors for data capture and compute platforms for model development and deployment (Table 6).

2.2.1 Sensors

Modern AI/ML-driven quality control relies on advanced sensor technologies to capture precise, high-dimensional data from perishable goods. These sensors provide the foundational inputs for defect detection, freshness assessment, and compositional analysis, bridging the gap between raw agricultural produce and actionable insights.

Table 6
Hardware components for AI/ML in food analysis

Category	Example device	Function	Key metric
Hyperspectral	Specim IQ	Image & spectral capture	400–1000 nm; 5 nm resolution
Electronic nose	MOX-array e-nose	VOC profiling	F1-score > 0.9
Portable NIR	Tellspec AgriFood scanner	Moisture/protein analysis	Minutes per sample
GPU cluster	NVIDIA A100	Deep-learning training	624 TFLOPS; 40 GB memory
Edge device	NVIDIA Jetson Nano	On-device inference	128-core GPU; 4 GB RAM

1. Hyperspectral Cameras (400–1000 nm, ~5 nm resolution) capture spectral and spatial information, enabling pixel-level detection of defects such as bruises or mold.
2. Electronic Noses (12-sensor VOC arrays) profile volatile organic compounds for spoilage monitoring, achieving F1-scores >0.9 in seafood freshness assays.
3. Portable NIR Spectrometers (e.g., Tellspec AgriFood Scanner) perform on-site moisture and protein analysis, reducing lab turnaround from days to minutes.

2.2.2 Compute Platforms

The computational backbone of AI/ML systems ranges from high-performance clusters for model training to compact edge devices for decentralized inference. These platforms ensure rapid processing of sensor-derived data, enabling both large-scale analytics and real-time decision-making in supply chain environments.

1. GPU Clusters (e.g., NVIDIA A100 with 40 GB memory) accelerate training of deep networks on large hyperspectral and chromatographic datasets.
2. Edge Devices (e.g., NVIDIA Jetson Nano) enable real-time inference in field settings, powering portable spectrometers and e-nose deployments.

2.3 Software and Tools

The interoperability of these tools—such as exporting PyTorch models to ONNX for edge deployment or combining OpenCV with scikit-learn for hybrid image/spectral workflows—enables tailored solutions for perishables analytics. Future toolkits must prioritize lightweight containerization (e.g., Docker for sensor-fused models) and federated learning support to address data privacy constraints in multi-stakeholder supply chains.

2.3.1 Preprocessing and Modeling

1. Python: scikit-learn for traditional ML (SVM, random forest); TensorFlow & PyTorch for deep learning architectures.
2. R: pls and caret for chemometric techniques (PCA, PLS-R).

3. OpenCV: ROI extraction and augmentation in image workflows.
4. MATLAB: Spectral smoothing and baseline correction routines.

2.3.2 Deployment and Integration

1. TensorFlow Lite & ONNX: Model compression and cross-platform export for edge deployment.
2. Flask APIs are used to develop cloud-based inference services and support automated reporting pipelines.

Table 7 summarizes key software tools and their specialized roles across the AI/ML pipeline, from spectral preprocessing to cloud deployment. The frameworks were selected based on benchmarking studies and industry adoption rates. Open-source dominance reflects the field's emphasis on reproducibility and modular integration. (Table 7).

2.4 Reference Standards

Follow certified materials and protocols ensures both analytical accuracy and regulatory compliance [72, 73] (Table 8).

2.4.1 Certified Reference Materials (CRMs)

1. NIST SRM 3257 (olive oil) for fatty-acid calibration under ISO/IEC 17025 guidelines [73].
2. ERM-BD600 (pesticide residues in wheat flour) for validating LC-MS workflows [74].

2.4.2 Adulterant Libraries

1. Raman spectra of high-fructose corn syrup and beet sugar (AOAC 994.12) for honey authentication [72].
2. GC-MS profiles of palm and sunflower oils for olive-oil fraud detection [74].

Table 7
Software frameworks and their roles

Stage	Tools / libraries	Main use case
Spectral preprocess	MATLAB signal processing toolbox	Savitzky–Golay smoothing, SNV
Chemometrics	R (pls, caret)	PCA, PLS-R modeling
Machine learning	Python (scikit-learn)	SVM, random forest
Deep learning	TensorFlow, PyTorch	CNN, RNN, autoencoder architectures
Image processing	OpenCV	ROI extraction, augmentation
Edge deployment	TensorFlow lite, ONNX	Quantized model inferencing
Cloud integration	Flask	REST APIs for remote inference

Table 8
Reference standards and guidelines

Standard type	Example	Application
Calibration CRM	NIST SRM 3257 (olive oil)	Fatty-acid quantification
Validation CRM	ERM-BD600 (wheat flour)	Pesticide-residue testing
Adulterant Library	AOAC 994.12 (HFCS, beet sugar spectra)	Honey authenticity
Lab Accreditation	ISO/IEC 17025	Testing & calibration quality management
Data Governance	FDA AI/ML Action Plan, GDPR	Model transparency & data privacy

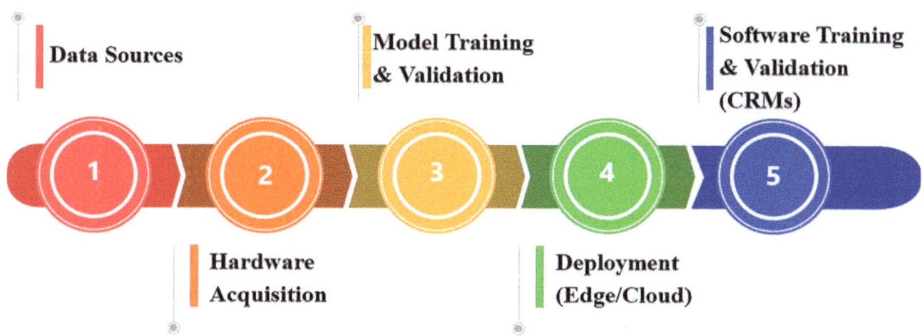

Fig. 4 End-to-end workflow from data acquisition to actionable insights

2.4.3 Regulatory Frameworks

1. ISO/IEC 17025: Quality-management standard for testing laboratories [75].
2. FDA AI/ML Action Plan & GDPR: Guidelines for data governance, model transparency, and ongoing monitoring [76].

Figure 4. presents a schematic illustrating the flow from data sources to actionable insights.

3 Methods

Analytical methods are vital in food analysis, particularly through the application of AI and ML. Techniques such as spectroscopic analysis (NIR, FTIR, Raman), chromatographic analysis (GC-MS, LC-MS), and imaging (hyperspectral, RGB) applies AI/ML algorithms like SVM, random forests, and CNNs to improve accuracy and efficiency [77, 78]. These methods enable precise identification, quantification, and authentication of food products, ensuring safety and quality [79]. The following sections provide an overview of these analytical methods, highlighting their key algorithms, performance metrics, and applications in the food industry. Table 9 is a comprehensive overview of analytical methods.

Table 9
Overview of analytical methods

Method category	Data type	Key algorithms	Typical performance	Section
Spectroscopic analysis	NIR, FTIR, Raman spectra	SVM, Random Forest	$R^2 > 0.95$; RMSE <2%	3.1
Chromatographic analysis	GC–MS, LC–MS	Random Forest, ANOVA/VIP	Accuracy > 95%	3.2
Imaging and computer vision	Hyperspectral, RGB images	CNN (ResNet-50), Grad-CAM	Accuracy > 98%	3.3
Sensor-based VOC profiling	E-nose arrays	SVM, k-NN	F1-score > 0.9	3.4
Omics-driven authentication	LC–MS/MS metabolomics	LASSO, k-means clustering	Silhouette >0.8; $R^2 > 0.9$	3.5
Validation and deployment	All	Cross-validation; TensorFlow lite	AUC > 0.97; inference <200 ms	3.6

3.1 Spectroscopic Analysis

Spectroscopic analysis is a powerful technique used to identify and quantify chemical compounds based on their interaction with electromagnetic radiation. This section outlines the various spectroscopic methods, including Near-Infrared (NIR), Fourier Transform Infrared (FTIR), and Raman spectroscopy. Each method has unique advantages in terms of sensitivity and specificity, making them suitable for diverse applications in fields such as food quality assessment, pharmaceuticals, and environmental monitoring. The subsequent subsections detail the processes involved in data acquisition, preprocessing, feature extraction, and model training, which are critical for achieving accurate and reliable results.

3.1.1 Data Acquisition

NIR, FTIR, and Raman spectroscopy are established techniques for non-destructive analysis in various fields, including food safety and medical diagnostics. They provide reliable spectral data for quality control and analysis [80, 81]

1. Instrument: NIR (400–2500 nm), FTIR (4000–400 cm^{-1}), Raman (400–3500 cm^{-1}).
2. Sample Prep: Grind to uniform particle size; maintain constant path length.

3.1.2 Preprocessing

Baseline correction using methods like Asymmetric Least Squares (ALS) is effective for removing scatter and improving data quality [82, 83]. Normalization techniques such as Standard Normal Variate (SNV) are widely used to adjust for variations in path length

[84]. The Savitzky–Golay filter is a recognized method for smoothing spectral data [85].

1. Baseline Correction: Asymmetric least squares (ALS) remove scatter [86].
2. Normalization: Standard Normal Variate (SNV) adjusts for path-length variations [87].
3. Smoothing: Savitzky–Golay filter (2nd order, window 11 points) reduces noise [88].

3.1.3 Feature Extraction and Dimensionality Reduction

PCA is a robust method for dimensionality reduction and is commonly used in spectroscopy to capture significant variance in data [89, 90].

1. PCA to capture ≥95% variance [91].
2. Wavelet Transforms (Daubechies-4) for localized feature isolation.

3.1.4 Model Training

SVM and Random Forest algorithms are effective for classification tasks in spectroscopy, with studies showing their high accuracy in various applications [92, 93]

1. Algorithms: SVM with RBF kernel; Random Forest (500 trees).
2. Hyperparameter Tuning: Grid search on C and γ (SVM), tree depth (RF).
3. Validation: 10-fold cross-validation; report RMSE and R^2 [94]

3.2 Chromatographic Analysis

Chromatographic analysis is essential for separating and analyzing complex mixtures of compounds. Techniques such as gas chromatography–mass spectrometry (GC–MS) and liquid chromatography–mass spectrometry (LC–MS) are widely used in analytical chemistry for their high resolution and sensitivity. This section provides an overview of the methodologies employed in chromatographic analysis, including data acquisition, preprocessing, feature selection, and model development. Understanding these processes is necessary for optimizing the analysis of samples in various applications, from clinical diagnostics to environmental testing.

3.2.1 Data Acquisition

1. Techniques: GC–MS (electron impact ionization); LC–MS/MS (electrospray).
2. Settings: Column temp gradient 50–300 °C (GC), mobile phase A/B gradient for LC.

3.2.2 Data Preprocessing

1. Peak Detection & Alignment: XCMS for retention-time alignment [95]
2. Annotation: CAMERA for adduct/isotope removal.

3.2.3 Feature Selection

ANOVA ($p < 0.01$) or VIP scores from PLS-DA for high-importance peaks [96]

3.2.4 Model Development

1. Random Forest classifiers on peak intensities; optimize tree number via out-of-bag error.
2. Validation: External test set; accuracy and confusion matrix.

3.3 Imaging and Computer Vision

Imaging and computer vision techniques have revolutionized the way we analyze visual data, enabling the extraction of meaningful information from images. This section focuses on the use of hyperspectral and RGB imaging for applications such as agricultural monitoring and quality control. The methodologies discussed include image acquisition, preprocessing, model training, and validation, which are vital for developing robust models that can accurately classify and interpret visual data. The integration of advanced algorithms, such as CNNs, increases the capability to analyze complex image datasets effectively.

3.3.1 Image Acquisition

1. Hyperspectral: Line-scan cameras, 5 nm spectral resolution; integrate over 224 × 224 px ROIs [97].
2. RGB: Standardized lighting; capture multi-angle views.

3.3.2 Preprocessing

1. ROI Extraction: OpenCV to segment fruits/vegetables.
2. Augmentation: Rotation, flipping, brightness jitter to improve robustness.

3.3.3 Model Training and Interpretation

1. Architecture: Fine-tune ResNet-50 pretrained on ImageNet [98]
2. Transfer Learning: Replace the final fully connected layer with n-class softmax.
3. Explainability: Grad-CAM to visualize salient regions.

3.3.4 Validation

1. Metrics: Accuracy, Precision, Recall, F1.
2. Cross-Validation: Stratified 5-fold on balanced datasets.

3.4 Sensor-Based VOC Profiling

Sensor-based profiling of volatile organic compounds (VOCs) is an innovative approach to monitoring air quality and detecting specific chemical signatures. This section describes the use of electronic nose (E-nose) technology and various sensor types, including metal-oxide and quartz crystal microbalance sensors. The methodologies for data collection, drift correction, and classification are outlined, emphasizing the importance of accurate VOC profiling in applications such as food safety, environmental monitoring, and industrial processes. The integration of ML algorithms further improves the classification accuracy of VOCs.

3.4.1 Data Collection

1. Devices: Metal-oxide (MOX) and quartz crystal microbalance (QCM) sensors.
2. Sampling: Headspace VOC capture at 40 °C for 60 s.

3.4.2 Drift Correction

1. Algorithms: Component Correction (CC) to stabilize baseline over time [99].

3.4.3 Classification

1. Models: SVM (RBF) and k-NN ($k = 5$).
2. Validation: Leave-one-out CV against GC-MS ground truth; report F1-score [100].

3.5 Omics-Driven Authentication

Omics-driven authentication employs advanced analytical techniques to authenticate products based on their molecular profiles. This section discusses the application of metabolomics, particularly using LC–MS/MS, to identify biomarkers that can distinguish between different origins or qualities of products. The processes of data processing, biomarker selection, clustering, and classification are detailed, highlighting the significance of these methods in ensuring product integrity and safety in industries such as food and pharmaceuticals. The use of robust statistical techniques ensures the reliability of the authentication process.

3.5.1 Data Processing

1. Feature Extraction: XCMS deconvolution; TIC and PQN normalization [101].

3.5.2 Biomarker Selection

1. LASSO regression to penalize non-informative ions, selecting top 20 features [102].

3.5.3 Clustering & Classification

1. k-means (k chosen by silhouette score > 0.8) for provenance grouping.
2. Validation: Bootstrapped cluster stability and external dataset testing.

3.6 Model Validation and Deployment

Model validation and deployment are critical steps in ensuring that analytical methods are not only accurate but also practical for real-world applications. This section outlines the performance metrics and validation strategies necessary for assessing model efficacy. It emphasizes the importance of nested cross-validation and external validation on independent datasets to provide unbiased performance estimates. Additionally, the deployment strategies, including edge optimization and model monitoring, are discussed to ensure that models remain effective over time, adapting to changes in data and maintaining high performance in operational settings.

3.6.1 Performance Metrics

1. Classification: ROC-AUC > 0.97, F1 > 0.9.
2. Regression: RMSE <2%, $R^2 > 0.9$.

3.6.2 Validation Strategies

1. Nested Cross-Validation for hyperparameter tuning and unbiased performance estimates [94].
2. External Validation on independent datasets and CRMs (NIST SRMs).

3.6.3 Deployment

1. Edge Optimization: Quantization via TensorFlow Lite; inference time < 200 ms [36].
2. Model Monitoring: Data-drift detection using Kolmogorov–Smirnov tests; periodic retraining.

4 Notes

Schedule Periodic Retraining

Ensuring high quality data starts with accurate instrument calibration, such as regular recalibration of spectrometers, chromatographs and imaging devices with certified reference materials (e.g. NIST SRM 3257) to reduce data drift. Consistency in sample preparation, such as standardization of grinding, separation or dilution protocols is equally important to minimize variability in spectral, chromatographic or imaging results. Each pre-processing step, from normalization to noise removal, should be carefully documented and version controlled (e.g. via Git or DVC) to ensure reproducibility and transparency. When developing models, start with simpler approaches such as PLS-R or SVM to establish a baseline before exploring complex architectures, as this will clarify whether additional complexity really improves performance. Nested cross-validation is essential to prevent overfitting by separating hyperparameter tuning from the final performance evaluation. Also, routinely check models for bias by evaluating the performance of subgroups (e.g., by geographic origin or variety) using confusion matrices and fairness metrics to ensure equitable results. For deployment, optimize models for target hardware through quantization and pruning, especially for edge devices, and validate latency for real-world inference (ideally below 200 ms). Implement robust data drift detection mechanisms, such as Kolmogorov-Smirnov tests or control charts, to detect deviations in input distributions. Schedule regular retraining, quarterly or after major process changes (e.g. new instruments or seasonal shifts) to maintain model accuracy. Ethical and regulatory compliance requires maintaining detailed model maps that summarize training data, use cases, performance metrics and constraints for

Table 10
Practical notes at a glance

Category	Key note
Data quality	Regular calibration; consistent sample prep; version-controlled preprocessing
Model development	Benchmark simple models; nested CV; bias audits
Deployment and maintenance	Optimize for hardware; data-drift alerts; periodic retraining
Ethics and compliance	Model cards; follows ISO/IEC 17025 & FDA guidelines; data privacy safeguards

each AI/ML system. associate with practices with standards such as ISO/IEC 17025 for laboratory proficiency and the FDA's AI/ML Transparency Action Plan. When handling sensitive supply chain or personal data, adhere to GDPR and local privacy regulations; federated learning can help avoid raw data sharing while preserving privacy. Table 10 summarizes the most important practical notes at a glance.

Conclusion

This chapter has reviewed the transformative impact of AI and ML on food analysis, spanning foundational chemometric techniques to advanced methodologies such as deep learning, federated learning, and edge-AI deployments. The necessary materials, methods, and practical considerations for developing robust and reproducible workflows have been outlined, along with key challenges related to data heterogeneity, interpretability, and ethical implications. By following standardized protocols, using explainable AI tools, and meeting regulatory requirements (like ISO/IEC 17025 and the FDA's AI/ML Action Plan), we can speed up innovation while maintaining trust and compliance. Advancing trends-including self-supervised pretraining, quantum ML, and secure multi-party collaboration are anticipated to further democratize advanced analytical capabilities, enabling real-time, decentralized testing even in resource-limited settings. Ultimately, the integration of technical rigor with ethical and regulatory vigilance will be critical to fully realizing the potential of AI/ML for safer, more transparent, and sustainable food systems.

References

1. Lin Y, Ma J, Wang Q, Sun D-W (2023) Applications of machine learning techniques for enhancing nondestructive food quality and safety detection. Crit Rev Food Sci Nutr 63: 1649–1669
2. Pawar TJ, Ramos-Cruz SX, Bonilla-Landa I, Muteeb G, Delgado-Alvarado E, Patil SV et al (2025) Quantitative NMR for detection of spinosad residues in agricultural soils. RSC Adv 15:5547–5557

3. Wang Z, Hu M, Zhai G (2018) Application of deep learning architectures for accurate and rapid detection of internal mechanical damage of blueberry using hyperspectral transmittance data. Sensors 18:1126
4. Vanaraj R, I.P B, Mayakrishnan G, Kim IS, Kim S-C. (2025) A systematic review of the applications of electronic nose and electronic tongue in food quality assessment and safety. Chem 13:161
5. Ballabio D, Consonni V (2013) Classification tools in chemistry. Part 1: linear models. PLS-DA. Anal Methods 5:3790
6. Punta-Sánchez I, Dymerski T, Calle JLP, Ruiz-Rodríguez A, Ferreiro-González M, Palma M (2024) Detecting honey adulteration: advanced approach using UF-GC coupled with machine learning. Sensors 24:7481
7. Sharma S, Gahlawat VK, Rahul K, Mor RS, Malik M (2021) Sustainable innovations in the food industry through artificial intelligence and big data analytics. Logistics 5:66
8. Bhagya Raj GVS, Dash KK (2022) Comprehensive study on applications of artificial neural network in food process modeling. Crit Rev Food Sci Nutr 62:2756–2783
9. Kagaya H, Aizawa K, Ogawa M (2014) Food detection and recognition using convolutional neural network. In: Proceedings of the 22nd ACM International Conference on Multimedia [Internet]. ACM, Orlando [cited 2025 May 17], pp 1085–1088. Available from: https://dl.acm.org/doi/10.1145/2647868.2654970
10. Zhou L, Zhang C, Liu F, Qiu Z, He Y (2019) Application of deep learning in food: a review. Comp Rev Food Sci Food Safe. 18:1793–1811
11. Rodriguez-Saona LE, Allendorf ME (2011) Use of FTIR for rapid authentication and detection of adulteration of food. Annu Rev Food Sci Technol 2:467–483
12. Putri SP, Ikram MMM, Sato A, Dahlan HA, Rahmawati D, Ohto Y et al (2022) Application of gas chromatography-mass spectrometry-based metabolomics in food science and technology. J Biosci Bioeng 133:425–435
13. Savage RA (1995) Hazard analysis critical control point: a review. Food Rev Intl 11: 575–595
14. Chen X, Wang L (2021) Analysis of the application of high performance liquid chromatography. IOP Conf Ser: Earth Environ Sci 769: 032020
15. Zhang L, Dabipi IK, Brown WL (2018) Internet of things applications for agriculture. In: Hassan Q (ed) Internet of things A to Z [Internet]. 1st ed. Wiley [cited 2025 May 17], pp 507–528. Available from: https://onlinelibrary.wiley.com/doi/10.1002/9781119456735.ch18
16. Ribeiro MT, Singh S, Guestrin C (2016) "Why should I trust you?": Explaining the predictions of any classifier. In: Proceedings of the 22nd ACM SIGKDD International Conference on Knowledge Discovery and Data Mining [Internet]. ACM, San Francisco [cited 2025 May 17], pp 1135–1144. Available from: https://dl.acm.org/doi/10.1145/2939672.2939778
17. ISO/IEC (2017) IEC 17025: 2017; general requirements for the competence of testing and calibration laboratories. I. ISO, Geneva
18. FAO (2023) Guidelines on measurement uncertainty [Internet]. FAO. Available from: http://www.fao.org/documents/card/en/c/cc6556en
19. Wang X, Bouzembrak Y, Lansink AO, Van Der Fels-Klerx HJ (2022) Application of machine learning to the monitoring and prediction of food safety: a review. Comp Rev Food Sci Food Safe 21:416–434
20. Xu Y, Kong T, Ma Y, Zhao Y, Chu L, Zheng M (2025) Near-infrared spectroscopy: application in ensuring food quality and safety. Anal Methods 17:3381–3406
21. Cozzolino D, Power A, Chapman J (2019) Interpreting and reporting principal component analysis in food science analysis and beyond. Food Anal Methods 12:2469–2473
22. Helland I (2005) Partial least squares regression. In: Kotz S, Read CB, Balakrishnan N, Vidakovic B (eds) Encyclopedia of statistical sciences [Internet], 2nd edn. Wiley [cited 2025 May 17]. Available from: https://onlinelibrary.wiley.com/doi/10.1002/0471667196.ess6004.pub2
23. Hodson TO (2022) Root-mean-square error (RMSE) or mean absolute error (MAE): when to use them or not. Geosci Model Dev 15:5481–5487
24. Poldrack RA (2007) Region of interest analysis for fMRI. Soc Cogn Affect Neurosci 2:67–70
25. Lundberg SM, Lee S-I (2017) A unified approach to interpreting model predictions. In: Advances in neural information processing systems, p 30. NIPS
26. Liang Y, Xu Q-S, Li H-D, Cao D-S (2016) Support vector machines and their application in chemistry and biotechnology [Internet], 0th edn. CRC Press [cited 2025 May 17]. Available from: https://www.taylorfrancis.com/books/9781439821282

27. Yang H, Wu M, Shen X, Lai Y, Wang D, Ma C et al (2025) A comprehensive review of VOCs as a key indicator in food authentication. eFood 6:e70057
28. Adak A, Pradhan B, Shukla N (2022) Sentiment analysis of customer reviews of food delivery services using deep learning and Explainable artificial intelligence: systematic review. Foods 11:1500
29. Vlachos NA, Michail C, Sotiropoulou D (2002) Is ISO/IEC 17025 accreditation a benefit or hindrance to testing laboratories? The Greek experience. J Food Compos Anal 15:749–757
30. Voigt P, Von Dem Bussche A (2024) The EU general data protection regulation (GDPR): a practical guide [Internet]. Springer Nature Switzerland, Cham [cited 2025 May 17]. Available from: https://link.springer.com/10.1007/978-3-031-62328-8
31. Katoch N (2025) Addressing bias in AI: ethical concerns, challenges, and mitigation strategies. IJSREM 09:1–9
32. Jain LR, Menon V (2023) AI algorithmic bias: understanding its causes, ethical and social implications. In: 2023 IEEE 35th International Conference on Tools with Artificial Intelligence (ICTAI) [Internet]. IEEE, Atlanta [cited 2025 May 15], pp 460–467. Available from: https://ieeexplore.ieee.org/document/10356540/
33. Sreerama J, Krishnamoorthy G (2022) Ethical considerations in AI addressing bias and fairness in machine learning models. JKLST 1: 130–138
34. Bahangulu JK, Owusu-Berko L (2025) Algorithmic bias, data ethics, and governance: ensuring fairness, transparency and compliance in AI-powered business analytics applications. World J Adv Res Rev 25:1746–1763
35. Thirupathi L, Vasundara B, Rajan P, Arella T (2025) Optimization tools and techniques for AI and ML: Exploring future directions and emerging trends. In: Rai HM, Razaque A (eds) Advances in systems analysis, software engineering, and high performance computing [Internet]. IGI Global [cited 2025 May 16], pp 125–150. Available from: https://services.igi-global.com/resolvedoi/resolve.aspx?doi=10.4018/979-8-3693-6885-5.ch005
36. Liakos K, Busato P, Moshou D, Pearson S, Bochtis D (2018) Machine learning in agriculture: a review. Sensors 18:2674
37. Javeed D, Saeed MS, Ahmad I, Adil M, Kumar P, Islam AKMN (2024) Quantum-empowered federated learning and 6G wireless networks for IoT security: concept, challenges and future directions. Futur Gener Comput Syst 160:577–597
38. Ahmed E (2024) Detection of honey adulteration using machine learning. Barage S, editor. PLOS Digit Health 3:e0000536.
39. Khaled AY, Parrish CA, Adedeji A (2021) Emerging nondestructive approaches for meat quality and safety evaluation—a review. Comp Rev Food Sci Food Safe 20:3438–3463
40. González-Pereira A, Otero P, Fraga-Corral M, Garcia-Oliveira P, Carpena M, Prieto MA et al (2021) State-of-the-art of analytical techniques to determine food fraud in olive oils. Foods 10:484
41. Masengu R, Muchenje C, Ruzive B (2025) Leveraging predictive analytics to enhance food safety risk management in supply chains: a conceptual framework. J Infras Policy Dev 9: 10114
42. Dakhia Z, Russo M, Merenda M (2025) AI-enabled IoT for food computing: challenges, opportunities, and future directions. Sensors 25:2147
43. Gurdeniz G, Ozen B (2009) Detection of adulteration of extra-virgin olive oil by chemometric analysis of mid-infrared spectral data. Food Chem 116:519–525
44. Downey G, McIntyre P, Davies AN (2002) Detecting and quantifying sunflower oil adulteration in extra virgin olive oils from the eastern Mediterranean by visible and near-infrared spectroscopy. J Agric Food Chem 50:5520–5525
45. Ellis DI, Broadhurst D, Kell DB, Rowland JJ, Goodacre R (2002) Rapid and quantitative detection of the microbial spoilage of meat by Fourier transform infrared spectroscopy and machine learning. Appl Environ Microbiol 68:2822–2828
46. Nychas G-JE, Skandamis PN, Tassou CC, Koutsoumanis KP (2008) Meat spoilage during distribution. Meat Sci 78:77–89
47. Ahmed SAA, Awais M, Kittler J. SiT: self-supervised vision transformer. ArXiv, abs/210403602 [Internet]. Available from: https://www.semanticscholar.org/paper/SiT%3A-Self-supervised-vIsion-Transformer-Ahmed-Awais/166e98317ed9c4687e71bef55a6800431e00b8fa
48. Yang X, He X, Liang Y, Yang Y, Zhang S, Xie P (2020) Transfer learning or self-supervised learning? A tale of two pretraining paradigms [Internet] [cited 2025 May 18]. Available from: https://www.techrxiv.org/doi/full/10.36227/techrxiv.12502298.v1

49. Liu L, Zhang J, Song SH, Letaief KB (2020) Client-edge-cloud hierarchical federated learning. In: ICC 2020–2020 IEEE International Conference on Communications (ICC) [Internet]. IEEE, Dublin [cited 2025 May 18]. pp 1–6. Available from: https://ieeexplore.ieee.org/document/9148862/
50. Kairouz P, McMahan HB, Avent B, Bellet A, Bennis M, Nitin Bhagoji A et al (2021) Advances and open problems in federated learning. FNT Mach Learn 14:1–210
51. Gupta N (2025) Explainable AI for regulatory compliance in financial and healthcare sectors: a comprehensive review. IJAEM 7:489–494
52. Duggal B, Explainable AI (2025) For fraud detection in financial transactions. IJSREM 09:1–9
53. Malavi D, Raes K, Van Haute S (2024) Integrating near-infrared hyperspectral imaging with machine learning and feature selection: detecting adulteration of extra-virgin olive oil with lower-grade olive oils and hazelnut oil. Curr Res Food Sci 9:100913
54. Bai Z, Du D, Zhu R, Xing F, Yang C, Yan J et al (2024) Establishment and comparison of in situ detection models for foodborne pathogen contamination on mutton based on SWIR-HSI. Front Nutr 11:1325934
55. Wang T, Zhang H, Wu Y, Chen X, Chen X, Zeng M et al (2022) Classification and concentration prediction of VOCs with high accuracy based on an electronic nose using an ELM-ELM integrated algorithm. IEEE Sensors J 22:14458–14469
56. Kumar S, Pal S, Singh VP, Jaiswal P (2023) Performance evaluation of ResNet model for classification of tomato plant disease. Epidemiol Methods 12:20210044
57. Malavi D, Nikkhah A, Raes K, Van Haute S (2023) Hyperspectral imaging and Chemometrics for authentication of extra virgin olive oil: a comparative approach with FTIR, UV-VIS, Raman, and GC-MS. Foods 12:429
58. Oroian M, Ropciuc S, Paduret S (2018) Honey adulteration detection using Raman spectroscopy. Food Anal Methods 11:959–968
59. Oliver Muncharaz J (2020) Comparing classic time series models and the LSTM recurrent neural network: an application to S&P 500 stocks. fmv 6:137–148
60. Nikzadfar M, Rashvand M, Zhang H, Shenfield A, Genovese F, Altieri G et al (2024) Hyperspectral imaging aiding artificial intelligence: a reliable approach for food qualification and safety. Appl Sci 14:9821
61. Sonwani E, Bansal U, Alroobaea R, Baqasah AM, Hedabou M (2022) An artificial intelligence approach toward food spoilage detection and analysis. Front Public Health 9:816226
62. Shehzad K, Predictive AI (2025) Models for food spoilage and shelf-life estimation. Glob Trends Food Sci Technol 1:75–94
63. Revelou P-K, Tsakali E, Batrinou A, Strati IF (2025) Applications of machine learning in food safety and HACCP monitoring of animal-source foods. Foods 14:922
64. Guo M, Wang K, Lin H, Wang L, Cao L, Sui J (2024) Spectral data fusion in nondestructive detection of food products: strategies, recent applications, and future perspectives. Comp Rev Food Sci Food Safe 23:e13301
65. Kharbach M, Alaoui Mansouri M, Taabouz M, Yu H (2023) Current application of advancing spectroscopy techniques in food analysis: data handling with chemometric approaches. Foods 12:2753
66. Gilbert-López B, Valdés A, Acunha T, García-Cañas V, Simó C, Cifuentes A (2017) Foodomics: LC and LC-MS-based omics strategies in food science and nutrition. Liquid Chromatography [Internet] Elsevier [cited 2025 May 18], pp 267–299. Available from: https://linkinghub.elsevier.com/retrieve/pii/B9780128053928000104
67. Zhu M (2024) Application of gas chromatography in food safety testing. fq [Internet] [cited 2025 May 18];2. Available from: https://ojs.scineer-pub.com/index.php/FSQ/article/view/3871
68. Huang L, Huang R (2024) Study of progress on application of hyperspectral imaging combined with deep learning approaches in detecting foods content. In: Zhou P, editor. Advanced Fiber Laser Conference (AFL2023) [IInternet]. SPIE, Shenzhen [cited 2025 May 18]. p 319. Available from: https://www.spiedigitallibrary.org/conference-proceedings-of-spie/13104/3023637/Study-of-progress-on-application-of-hyperspectral-imaging-combined-with/10.1117/12.3023637.full
69. Dai Q, Sun D, Xiong Z, Cheng J, Zeng X (2014) Recent advances in data mining techniques and their applications in hyperspectral image processing for the food industry. Comp Rev Food Sci Food Safe 13:891–905
70. Ahmed S, De La Parra J, Elouafi I, German B, Jarvis A, Lal V et al (2022) Foodomics: a data-driven approach to revolutionize nutrition and sustainable diets. Front Nutr 9:874312

71. Walsh AM, Crispie F, Claesson MJ, Cotter PD (2017) Translating omics to food microbiology. Annu Rev Food Sci Technol 8:113–134
72. Ahn S-H, Kim BJ, Hwang Y-J (2011) Stability monitoring of pesticide residues in a Chinese cabbage certified reference material. Bull Korean Chem Soc 32:1365–1367
73. Sander L, Schantz M, Sharpless K, Wise S (2009) Standard reference materials to support measurement of fatty acids. Lipid Technol 21:7–12
74. Wiest LA, Hepner JR, Fisher JE, Risha KM, Lidgett JH, Ballarotto VN et al (2023) Stability study and handling recommendations for multiresidue pesticide mixes under diverse storage conditions for LC–MS/MS and GC–MS/MS. J AOAC Int 106:1550–1563
75. De Bièvre P (2006) On (certified) reference materials. Accred Qual Assur 10:459–460
76. Bates J, Bahadoor A, Cui Y, Meija J, Windust A, Melanson JE (2019) Certification of ochratoxin a reference materials: calibration solutions OTAN-1 and OTAL-1 and a mycotoxin-contaminated Rye flour MYCO-1. J AOAC Int 102:1756–1766
77. Workman J, Mark H (2023) Artificial intelligence in analytical spectroscopy, Part II: Examples in spectroscopy. Spectroscopy, pp 10–15
78. Rodriguez-Lazaro D, Cook N, Bertheau Y, Bosch A, Godefroy SB, Magnuson B et al (2008) Editorial: "Food analytical methods", an integrated approach for food analysis. Food Anal Methods 1:1–1
79. Delahaut P, Marega R (2022) Novel analytical methods in food analysis. Foods 11:1512
80. Naumann D (1998) Infrared and NIR Raman spectroscopy in medical microbiology. In: Mantsch HH, Jackson M (eds). San Jose [cited 2025 May 19]. pp 245–257. Available from: http://proceedings.spiedigitallibrary.org/proceeding.aspx?articleid=933885
81. Fakayode SO, Baker GA, Bwambok DK, Bhawawet N, Elzey B, Siraj N et al (2020) Molecular (Raman, NIR, and FTIR) spectroscopy and multivariate analysis in consumable products analysis[1]. Appl Spectrosc Rev 55: 647–723
82. Chen T, Son Y, Dong C, Baek S-J (2025) Baseline correction of Raman spectral data using triangular deep convolutional networks. Analyst:10.1039.D5AN00253B
83. Devos O, Mouton N, Sliwa M, Ruckebusch C (2011) Baseline correction methods to deal with artifacts in femtosecond transient absorption spectroscopy. Anal Chim Acta 705:64–71
84. Kumar N, Panchariya PC, Patel SS, Kiranmayee AH, Ranjan R (2018) Application of various pre-processing techniques on infrared (IR) spectroscopy data for classification of different ghee samples. In: 2018 Fourth International Conference on Computing Communication Control and Automation (ICCUBEA) [IInternet]. IEEE, Pune [cited 2025 May 19]. pp 1–6. Available from: https://ieeexplore.ieee.org/document/8697787/
85. Eilers PHC (2003) A perfect smoother. Anal Chem 75:3631–3636
86. Rinnan Å, Berg FVD, Engelsen SB (2009) Review of the most common pre-processing techniques for near-infrared spectra. TrAC Trends Anal Chem 28:1201–1222
87. Nicolaï R (2007) Contacts des langues et contact dans la langue : heterogeneite, construction de l'homogene et emergence du "linguistique". J Lang Contact 1:198–221
88. dell'Isola F, Steigmann DJ (eds) (2020) Discrete and continuum models for complex metamaterials [Internet], 1st edn. Cambridge University Press [cited 2025 May 19]. Available from: https://www.cambridge.org/core/product/identifier/9781316104262/type/book
89. Zou Y, Sun P, Liu W (2015) Classification of materials using terahertz spectroscopy with principal components analysis. In: 2015 40th International Conference on Infrared, Millimeter, and Terahertz waves (IRMMW-THz) [Internet]. Hong Kong, China: IEEE [cited 2025 May 19]. pp 1–2. Available from: http://ieeexplore.ieee.org/document/7327794/
90. Cao LJ, Chua KS, Chong WK, Lee HP, Gu QM (2003) A comparison of PCA, KPCA and ICA for dimensionality reduction in support vector machine. Neurocomputing 55:321–336
91. Berglund A, Kettaneh N, Uppgård L, Wold S, Bendwell N, Cameron DR (2001) The GIFI approach to non-linear PLS modeling. J Chemom 15:321–336
92. Ahmed Khan T, Sadiq R, Shahid Z, Alam MM, Mohd Su'ud M (2024) Sentiment analysis using support vector machine and random forest. JIWE 3:67–75
93. Azhari M, Situmorang Z, Rosnelly R (2021) Perbandingan Akurasi, Recall, dan Presisi Klasifikasi pada Algoritma C4.5, Random Forest, SVM dan Naive Bayes mib 5:640

94. Kohavi R (1995) A study of cross-validation and bootstrap for accuracy estimation and model selection. Available from: https://www.semanticscholar.org/paper/8c70a0a39a686bf80b76cb1b77f9eef156f6432d
95. Smith CA, Want EJ, O'Maille G, Abagyan R, Siuzdak G (2006) XCMS: processing mass spectrometry data for metabolite profiling using nonlinear peak alignment, matching, and identification. Anal Chem 78:779–787
96. Worley B, Halouska S, Powers R (2013) Utilities for quantifying separation in PCA/PLS-DA scores plots. Anal Biochem 433:102–104
97. Sun C, Feng L, Zhang Z, Ma Y, Crosby T, Naber M et al (2020) Prediction of end-of-season tuber yield and tuber set in potatoes using in-season UAV-based hyperspectral imagery and machine learning. Sensors 20:5293
98. He K, Zhang X, Ren S, Sun J (2016) Identity mappings in deep residual networks. In: Leibe B, Matas J, Sebe N, Welling M (eds) Computer vision—ECCV 2016 [Internet]. Springer International Publishing, Cham [cited 2025 May 19]. pp 630–645. Available from: http://link.springer.com/10.1007/978-3-319-46493-0_38
99. Sanaeifar A, Mohtasebi SS, Ghasemi-Varnamkhasti M, Siadat M (2014) Application of an electronic nose system coupled with artificial neural network for classification of banana samples during shelf-life process. In: 2014 International Conference on Control, Decision and Information Technologies (CoDIT) [Internet]. IEEE, Metz [cited 2025 May 19]. pp 753–757. Available from: http://ieeexplore.ieee.org/document/6996991/
100. Ghasemi-Varnamkhasti M, Tohidi M, Mishra P, Izadi Z (2018) Temperature modulation of electronic nose combined with multi-class support vector machine classification for identifying export caraway cultivars. Postharvest Biol Technol 138:134–139
101. Katajamaa M, Orešič M (2007) Data processing for mass spectrometry-based metabolomics. J Chromatogr A 1158:318–328
102. Tibshirani R (1996) Regression shrinkage and selection via the Lasso. J R Statis Soc Ser B 58:267–288

Chapter 19

Tunable Resistive Pulse Sensing for Nanoparticle-Based Food Analysis

Emmanuel Ibukun Aduloju, Emmanuel Ayorinde Ajiboye, and Bosede Alice Omachi

Abstract

Nanotechnology is a quickly growing field in the modern innovative world with significant applications in the modern food industry, enhancing many aspects of food production, safety, and quality. Tunable Resistive Pulse Sensing (TRPS) has emerged as a powerful analytical technique for the high-resolution characterization of nanoparticles in food matrices, offering single-particle resolution for size, concentration, and real-time surface charge measurements. This technique operates by detecting changes in electrical resistance as nanoparticles traverse a tunable nanopore ranging from ~50 nm to cellular scales, generating unique resistive pulses corresponding to each particle. Unlike conventional methods such as dynamic light scattering or electron microscopy, TRPS enables label-free, real-time analysis of polydisperse samples with unparalleled accuracy, enhanced sensitivity, and specificity. This makes it ideal for detecting and quantifying engineered nanoparticles, additives, nutrient delivery systems such as liposomes, protein carriers, color enhancers, protein aggregates, or lipid vesicles in complex food systems. It is also useful for assessing naturally occurring nanoparticles and nanoplastics in food products and packaging, while providing simultaneous zeta potential analysis to assess particle stability and interactions. This chapter explores the principles, advantages, and applications of TRPS in Nanoparticle-Based Food Analysis, emphasizing its ability to monitor nanoparticle behavior under physiologically relevant conditions (e.g., pH, ionic strength). It also highlights the advancements of TRPS in food nanotechnology, ensuring the safety, authenticity, and efficacy of nanoparticle applications in modern food systems.

Key words Tunable Resistive Pulse Sensing (TRPS), Nanoparticle characterization, Single-particle analysis, Size distribution analysis, Zeta potential measurement, Nanopore sensing, Real-time monitoring

1 Introduction

Nanoparticlesplay an increasingly critical role in food science, from enhancing nutrient delivery and food safety to enabling innovative packaging solutions. The increasing application of nanotechnology in the food industry has raised significant interest in the characterization and monitoring of nanoparticles within food matrices. The

Tanmay Sarkar (ed.), *Novel Food Analysis: Techniques and Applications*, Methods and Protocols in Food Science,
https://doi.org/10.1007/978-1-0716-4787-5_19,

most paramount nanomaterials used in food industries are usually organic and inorganic particles, including metals and metal salts such as copper, gold, zinc, silicon dioxide, and titanium dioxide. While organic nanoparticles are proteins (casein micelle and zein nanoparticle), phospholipids, lipids (such as liposomes), carbohydrates, including nanocellulose nanogels and nanochitin.

Synthetic nanoparticles and extracellular vesicles (EVs) [1, 2], have been widely employed in therapeutic, food production, packaging, preservation, and diagnostic applications, including as biomarkers, drug carriers, and tools for disease detection [3, 4], due to their unique physicochemical properties. The key attributes of these nanomaterials, such as their surface characteristics, shape, and size, play a critical role in determining their biodistribution, interactions, and pharmacokinetics within biological systems, whether at the cellular or organ level [5]. However, their potential toxicity, unintended ingestion raise concerns about food safety. To ensure their safe and effective use in food applications, it is crucial to accurately characterize their properties, such as individual particle size (particle-by-particle), concentration, and surface charge. Reliable methods for detecting and measuring nanoparticles are important not only for maintaining food quality and safety but also for gaining insight into their behavior and possible interactions within biological systems. Traditional techniques such as dynamic light scattering (DLS) and nanoparticle tracking analysis (NTA) provide valuable insights but often lack the resolution and specificity required for complex food matrices.

Tunable resistive pulse sensing (TRPS), also known as a modified form of the Coulter counter [6, 7], is an advanced analytical technique, a promising solution used to detect and quantify transient changes in ionic current caused by particles translocating through a nanopore. This method enables precise determination of *real-time, high-resolution analysis* of nanoparticle size, charge (zeta potential) [8], concentration, surface charge, and overall nanoparticle characterization [9], even in complex food matrices. This technique enables the simultaneous counting and sizing of individual nanoparticles (typically between 50 nm and 10 μm [10] by monitoring fluctuations in electrical current (zeta potential) as each particle passes through a calibrated nanopore while suspended in an electrolyte solution. The size and concentration of biological colloidal particles are determined using thermoplastic polyurethane (TPU) nanopores as the sensing element [11]. In biosensing applications, these elastic, size-adjustable pores are a relatively recent innovation and are typically created by mechanically puncturing a stretchable polyurethane membrane with a micro-scale tungsten needle [12]. TRPS technology operates by applying a transmembrane DC voltage, which drives electrophoretic transport of particles [13] (Fig. 1). This voltage establishes a stable ionic baseline current through the nanopore, and any disturbance to this current

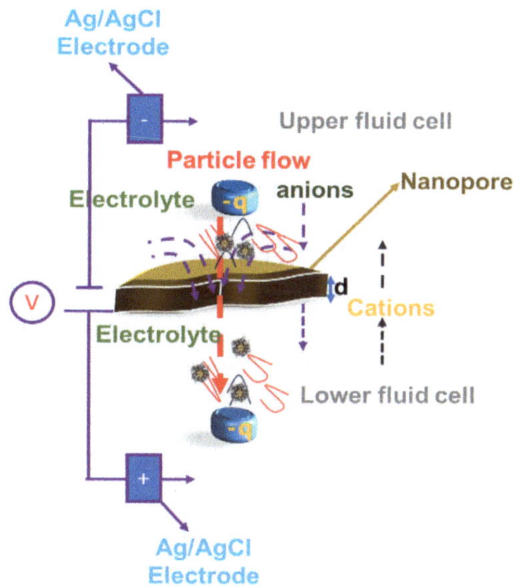

Fig. 1 Schematic illustration of TRPS, showing the arrangement of the nanopore, electrodes, and fluid chambers, along with the flow direction of ions and particles

indicates particle translocation. TRPS is particularly advantageous for analyzing heterogeneous samples in complex media due to its high sensitivity, single-particle resolution, and wide dynamic range. TRPS presents several notable advantages, including minimal sample volume requirements, customizable pore sizes, and high-resolution, particle-by-particle analysis. This versatile technique has been widely applied across various fields for the characterization of biomarkers [14], DNA, bacteria, and viruses [15]. It has proven especially valuable in the analysis of extracellular vesicles [16], as well as in the study of synthetic and biological colloidal particles. TRPS has also been instrumental in drug delivery research [17], the investigation of nanomaterials [12], and the detection of metal ions and nanorod aggregation induced by specific analytes [18]. The resistive pulse sensing (RPS) approach continues to evolve, as demonstrated in recent works by [19, 20]. With the ability to tune the nanopore diameter and control experimental conditions, TRPS provides a versatile platform for real-time, label-free particle analysis across a wide range of applications, including biomedicine, environmental science, and food technology.

1.1 Challenges in Detecting and Characterizing Nanoparticle-Based FoodAnalysis

Nanoparticles in food analysis pose significant challenges related to safety, detection, and regulatory gaps. Key concerns include nanoparticle migration from packaging into food, unknown interactions with food components, and potential toxicity risks. The variability in nanoparticle size and concentration, combined with complex food matrices (proteins, lipids), complicates accurate detection.

Current techniques, such as *dynamic light scattering (DLS)*, lack resolution for polydisperse samples, *electron microscopy* is slow and non-scalable, and *uv-vis spectroscopy* fails to distinguish nanoparticle populations effectively.

Toxicity, bioaccumulation, and safety remain poorly understood due to insufficient data on nanoparticle behavior in food and their long-term health effects. Migration from packaging raises exposure risks, yet standardized toxicity assessment methods are lacking. *Regulatory frameworks* are underdeveloped, leading to inconsistent monitoring and limited transparency in nanoparticle usage. *Economic and technological barriers* further hinder progress, with high costs for advanced characterization tools and a pressing need for innovation in nanoparticle synthesis and safety testing. Unlike conventional methods, TRPS enables *single-particle detection*, overcoming polydispersity challenges while supporting regulatory compliance and risk assessment.

Addressing these challenges requires *standardized detection protocols*, presenting comprehensive exposure data, *stricter regulations, and investment in advanced technologies* such as TRPS to ensure food safety and foster consumer trust in nanoparticle applications.

1.2 The Mechanistic and Key Parameters of TRPS

In 1996, Kasianowicz and colleagues reignited interest in Coulter-based resistive pulse sensing by demonstrating the detection of single-stranded DNA using a biological α-hemolysin nanopore [21]. The capacity of the method to distinguish subtle variations in the sizes of the four DNA bases [22] has since sparked significant interest in nanopore-based DNA sequencing technologies [23]. The straightforward design, high sensitivity, and ability to analyze individual particles make resistive pulse sensors highly appealing for precise characterization applications.

1.3 Principles of TRPS

TRPS is based on the Coulter principle, where individual particles are detected as they pass through a nanopore, generating resistive pulses (Fig. 1). A steady ionic current is maintained between two electrodes placed in separate reservoirs filled with electrolyte and divided by a nanopore (Fig. 1). When nanoparticles from one reservoir move through the nanopore, they momentarily block the flow of ions, causing a detectable change in the current [24]. This phenomenon is known as blockade occurs when a particle traverses the nanopore, temporarily reducing the magnitude of the baseline ionic current and generating a brief resistive pulse. These blockade (lasting milliseconds) enable single-particle detection and counting. The pulse magnitude correlates with particle size, while the blockade frequency reflects sample (nanoparticle) concentration under consistent experimental conditions, translocation frequency is directly proportional to analyte concentration [25]. Additionally, the characteristics magnitude of the current pulses and each analyte's translocation duration are influenced by particle size and surface charge [26]. The system's detection limit hinges on the smallest particle that produces a discernible current change above background noise [27].

The core mechanism of TRPS involves:

1.3.1 Nanopore System

A tunable elastomeric membrane with a single, size-adjustable pore separates two electrolyte-filled chambers. The nanopore membrane is mounted on a mechanically adjustable stage, allowing in situ stretching or relaxation of the material over millimeter-scale displacements [24]. These macroscopic adjustments enable fine-tuning of pore diameter at the submicrometer level size. Each pore features a conical geometry, narrowing from a broad entrance to a tight constriction. This design ensures maximum sensitivity for resistive pulse measurements at the narrowest point, where changes in ionic current are most pronounced during particle translocation.

1.3.2 Electrodes and Voltage

An electric field is applied across the pore using electrodes in each chamber, creating a current flow through the pore. In RPS experiments, the nanopore and its surrounding area were filled with an electrolyte solution (~40 μL in the upper fluid cell and ~100 μL in the lower fluid cell). A voltage was applied across the membrane using Ag/AgCl electrodes placed in both compartments [28, 29]. The resulting ionic current flowing through the nanopore was recorded and analyzed with the instrument's proprietary electronics and software. This analysis provided measurements of the baseline current (I_0) observed in the absence of particle translocation, the pulse magnitude (ΔI) representing the change in current caused by particle translocation, and the full width at half maximum (FWHM), which denotes the duration of each resistive pulse. Using coulter principle, the pore Size & Calibration can be calculated as shown in Eq. (1),

$$d_p = \frac{d_b . \Delta I}{I_0} \tag{1}$$

where, d_p is the effective pore diameter, d_b is the known bead diameter, ΔI is the blockage current, and I_0 is the baseline current. Usually, a computer-integrated software such as Izon control suite can automate this calculation.

1.3.3 Particle Detection

As a particle moves through the pore (driven by electrophoresis and/or pressure), it temporarily displaces electrolyte, increasing electrical resistance. This causes a measurable resistive pulse (drop in current). This electrical resistance is calculated for all pores, including cylindrical pores, as shown in Eq. (2)–(4)

Resistance is denoted by R, and is given by

$$\mathcal{R} = \rho \int_L^0 \frac{dz}{A\,(z)} \tag{2}$$

The effective resistive pore length in a cylindrical pore is given as

$$L_C = L + 0.8D \tag{3}$$

where L is the pore lenght and D is the pore diameter, z is the direction along pore length axis and *d is the* particle diameter. Also, resistance in a pore is given as in Eq. (4) when there is no blockage.

$$\mathcal{R} = \frac{4L_c\rho}{\pi D^2} \tag{4}$$

However, the total resistance in a conical pore is given in Eq. (5)

$$\mathcal{R} = \frac{4\ L_C\rho}{\pi D_S D_L} \tag{5}$$

where D_L and L is replaced with D_z and z. Considering the for instance, a cylindrical pore containing a spherical particle, Eqs. (2) and (6) are combined together as shown in Eq. (7).

1.3.4 Pulse Characteristics

The *amplitude* of the pulse correlates with the *particle volume (size)*. The pulse magnitude, denoted by $(\frac{\Delta\mathcal{R}}{\mathcal{R}})$ is given in Eq. (6)

$$\frac{\Delta\mathcal{R}}{\mathcal{R}} = f\frac{v}{V}\,S\left(\frac{d}{D}\right) \tag{6}$$

where, the particle shape factor f, for the spherical particle is $3/2$ and $d \ll D(S = 1)$. *This is the bases the Maxwell's theory, as given in* Eq. (7).

$$\frac{\Delta R}{R} = \frac{d^3}{D^2 L_c}\,S\left(\frac{d}{D}\right) \tag{7}$$

The ΔR is given in Eq. (8), where ρ is the resistivity of the electrolyte used in filling the pore, and $A(z)$ is the cross-sectional area perpendicular to the pore's axis in z (same as in Eq. (4)), and R is the pore resistance.

$$\Delta R = \rho \int_0^L \frac{dz}{A(z)} - R \tag{8}$$

The particle factor for a spherical particles is given in Eq. (9)

$$f = f_\perp + (f_\sqcap - f_\perp)\cos^2 \propto \tag{9}$$

where, $f_\sqcap$ *and* $f_\perp$ are the shape factors for spheroid with the longest dimension parallel and perpendicular to the pore wall, respectively. Both for an irregularly shaped flattened and elongated spheroids as they tumble through a pore. Where the correction factor is given in Eq. (10) based on $(\frac{d}{D})$

$$S\left(\frac{d}{D}\right) = \left[1 - 0.8\left(\frac{d}{D}\right)^3\right]^{-1} \quad (10)$$

The frequency of pulses relates to particle concentration. Pulse frequency is given in Eq. (11)

$$J = \frac{\pi v_s D^2}{4} C \quad (11)$$

where, $_s$ is the Object velocity through pore, (meanwhile the velocity contributions are from external pressure, diffusion, and electro-osmotic and electrophoretic flow), C is the Particle concentration.

The duration and shape of the pulse can provide information on the particle's shape and surface charge. Pulse duration is denoted as Δt and is given by Eq. (12)

$$\frac{1}{\Delta t} \propto f(\nu_\mu) + f(\nu_\rho) + f(\nu_\epsilon) \quad (12)$$

where ν_ρ and ν_ϵ are electrosmostic velocity ($m^2/V{\cdot}s$) and fluid velocity, if these are negligible then the particle charge μ can be calculated. The particle velocity through a cylindrical pore combines both fluid flow and electrokinetic forces. The combined velocities are inversely proportional to the pulse duration, while the diffusion velocity is negligible. Electrophoretic mobility is given in Eq. (13) as,

$$\mu = \frac{L_c^2}{E_M \Delta t} \quad (13)$$

where E_M is electric field strength (V/m), and L_c is the effective length or the distance the particle travelled.

1.4 Comparison with Other Particle Analysis Methods for Nanoparticle-Based Food Analysis

There are various single particle analysis techniques applied for the characterization of nanoparticle-based food analysis including resistive pulse sensing (RPS) [30]), nanoflow cytometry, dynamic light scattering (DLS), and nanoparticle tracking analysis (NTA) [31].

1.4.1 Nanoparticle Tracking Analysis (NTA)

Nanoparticle tracking analysis (NTA) is one of the prominent technologies used for high-throughput analysis of individual EVs. Nanoparticle tracking analysis (NTA), which was first commercialized in 2006, is an innovative system for sizing particles from about 30 to 1000 nm, with the lower detection limit being dependent on the refractive index of the nanoparticles (Table 1). This technique combines laser light scattering microscopy with a charge-coupled device (CCD) camera (Fig. 2), which enables the visualization and recording of nanoparticles in solution.

Table 1
Comparative summary of NTA, DLS, and TEM for nanoparticle characterization

Aspect	NTA (nanoparticle tracking analysis)	DLS (dynamic light scattering)	(TRPS) tunable resistive pulse sensing	TEM (transmission electron microscopy)	References
Size resolution	High resolution for unimodal/bimodal samples. Size range is ~1 nm—6 μm	Lower resolution; averages over population Size range is ~50 nm	High resolution; capable of distinguishing small differences in nanoparticle size individually	Ultra-high resolution at nanometer scale Size range is ~10 nm–2 μm	[64–66]
Multimodal capability	Limited in resolving 3+ populations	It can suggest multiple populations but lacks detail	Moderate; can analyze polydisperse samples but may have challenges in resolving closely overlapping populations	Its excellent for visualizing distinct population	[67]
Measurement principle	It can track Brownian motion and light scattering	It measures intensity fluctuation of scattered light	Coulter principle: detection of resistive pulses as individual particles pass through a tunable nanopore	Direct imaging using electron beam	[31]
Concentration measurement	It can estimate particle concentration (with variability)	N/A (Yes (directly))	Accurate quantification of particle concentration on a particle-by-particle basis	N/A	[68]
Sample preparation	It requires dilute / minimal suspensions	It requires dilution; sensitive to impurities	Minimal; samples need to be in liquid suspension and filtered to avoid clogging	Extensive; samples must be dry, thin, and conductive	[69]
Visualization	It tracks particles optically, no direct image	No direct visualization	No direct imaging; data is obtained as electrical signals (resistive pulses)	Provides detailed morphological images	[46]
Detection sensitivity	It is influenced by size, refractive index, camera settings (CL, DT)	It is size-intensity bias; less sensitive to polydispersity	High sensitivity to particle size, charge, and concentration	Excellent for small particles <10 nm	[70]
Cost and accessibility	Moderate instrument cost, real-time data	Widely available and affordable	Moderate to high; specialized equipment and trained personnel required	Expensive and requires skilled operation	[71]

Standardization efforts	SOPs improve reproducibility and regulatory use	Standard protocols exist	Protocols vary by manufacturer, but efforts are increasing toward harmonization	Standardization based on imaging protocols and contrast agents	[72]
Sample type	Aqueous suspensions	Aqueous suspensions	Liquid-based nanoparticle suspensions; suitable for synthetic, biological, and food matrix samples	Aqueous suspensions	[71]
Quantification accuracy	Moderate (biased by larger particles))	High (single-particle detection	High; counts and sizes individual particles, reducing reliance on indirect estimations	High	[71]
Advantages	Fast Simple to use Suitable for monodisperse samples	High sensitivity Single-particle readout Tunable size range	Single-particle analysis Real-time size and zeta potential Label-free No assumption of shape/optics	Tracks individual particles Provides size distribution and concentration	[73]
Limitations	Not ideal for polydisperse or heterogeneous samples.	Requires pore tuning. Sensitive to clogging.	Limited throughput Potential for pore clogging, requires calibration and Skilled handling.	Lower resolution for very small particles.	[74]

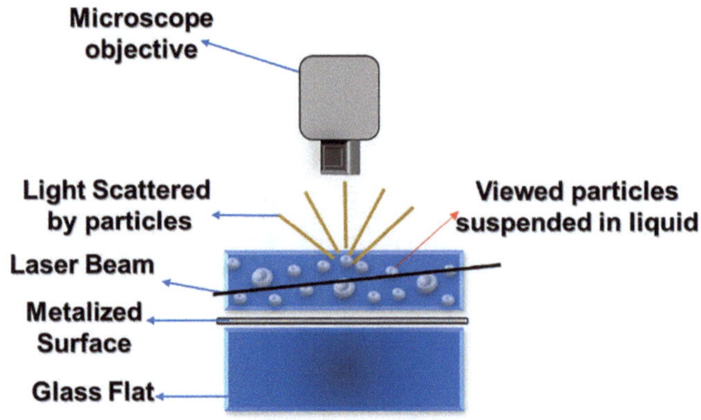

Fig. 2 Schematized diagram for nanoparticle tracking analysis (NTA) techniques

Principles and Mode of Operation of NTA

The principle of NTA is based on the characteristic movement of (nano)particles in solution according to the Brownian motion, relates the movement to a particle size according to the following formula derived from the Stokes-Einstein. The trajectory of the particles in a defined volume is documented by a camera that captures the scattered light upon illumination of the particles with a laser. Applying the Stokes–Einstein equation results in size determination for each tracked particle. In addition, single particle count in relation to the cell volume allows determining particle concentration of samples (Table 1).

Briefly, during NTA measurement, EVs are visualized by the light scattered when they are irradiated with a laser beam. The scattered light is focused by a microscope onto a camera, which records the particle movement. The NTA software tracks the random thermal motion of each particle, known as Brownian motion, to determine the diffusion coefficient, which is used to calculate the size of each particle using the Stokes–Einstein equation, as shown in Eq. (14).

$$(x, y)^2 = \frac{2k_B T}{3r_h \pi \eta} \tag{14}$$

where, r_h is the hydrodynamic radius, T is the temperature, $(x, y)^2$ is the mean-squared speed of a (nano)particle and k_B is Boltzmann constant.

NTA was first applied to the measurement of EVs in a biological sample in 2011. NTA provides a higher resolution for multimodal samples. NTA offers a unique ability to directly visualize and size nanoparticles in a liquid suspension, allowing for the observation of aggregation behavior through visual data. In the NTA analysis, a liquid sample is introduced into a chamber and exposed to laser illumination. The movement of individual particles is monitored by capturing the scattered light using a camera, which operates either in scientific complementary metal–oxide–semiconductor (sCMOS) mode or Electron Multiplying charge-coupled device (emCCD) mode.

The Brownian motion of each individual particle is monitored in real-time through the camera, allowing each particle to be individually visualized and tracked using specialized image analysis. The collected data is then processed and interpreted using dedicated software. NTA reports mean size (average particle size), as well as the modal size (frequent particle size).

The visualization of the particles allows each particle to be simultaneously and individually sized, overcoming inherent problems associated with techniques such as PCS, which is also referred to as dynamic light scattering (DLS). PCS observes the time-dependent fluctuations in scattering intensity caused by constructive and destructive interference resulting from the relative Brownian movements of the particles within a sample. Other techniques used in place of NTA include optical flow cytometry [32], rheology [33], high-speed photography [34], optical microscopy [35], atomic force microscope [36], and dynamic light scattering (DLS) [37, 38]. These methods can be categorized as dynamic approaches and static techniques based on their operating characteristics.

Applications and Uses of NTA

1. Used for determining size and concentration of particles in various matrices [63].
2. Applied in therapeutic protein formulations [39], gold nanoparticles [40], extracellular vesicles (EVs) [1, 2], liposomes, viruses [41], and drug delivery systems.
3. To support nanotoxicology, protein aggregation studies, biomedical research, and vaccine development [42].
4. To measure zeta potential, fluorescence, and relative refractive index (rRI).
5. To estimate refractive index using relative light scattering intensity (rLSI) and Mie scattering theory.
6. It differentiates particles of similar size but different compositions (e.g., metallic vs. lipid-based).

1.4.2 Dynamic Light Scattering (DLS)

Dynamic light scattering (DLS) measures the correlation of the time-dependent fluctuations in the light scattered by a suspension of nanoparticles (autocorrelation function), which is determined by their Brownian motion. The nanoparticle diffusion coefficient(s) is then calculated by fitting the autocorrelation function and used to determine the hydrodynamic diameter (or radius) of the particles via the Stokes-Einstein equation. The analysis assumes that the object measured is of spherical shape and requires that the viscosity of the solution is known. The intensity of light scattered from a nanoparticle as a function of particle radius follows a power law and increases with the sixth power for a Rayleigh particle. Hence, the average particle size produced by PCS (which measures total light scattered from an ensemble of particles) is heavily weighted towards small numbers of large, perhaps contaminant, particles. it a major drawback is that dust particles or small amounts of large aggregates can impede the accurate size determination if the main component exhibits a distinctly smaller size.

Principles and Mode of Operation of DLS

DLS is a very fast and inexpensive analytical technique that does not require highly specialized personnel to operate. DLS measurements can be conducted using either a dynamic or static compact goniometer, such as DLS/SLS-5000. For example, a helium–neon (He–Ne) laser with a power output of 22 mW emitting a polarized light at a wavelength (λ) of 6328 Å, was used as the incident beam. DLS measurements were typically carried out at 25 °C at various scattering angles, including 90°. In this analysis, the time-dependent correlation of scattered light intensity can be calculated based on the function, $g^{(2)(}\tau)$, as given in Eq. (15), which is linked to the scattered field correlation function, $g^{(1)(}\tau)$, as expressed in Eq. (16).

$$g^{(2)}\ (\tau) = 1 + \beta \left| g^{(1)}(\tau)^2 \right| \tag{15}$$

$$g^{(1)}\ (\tau) = \int_0^\infty G\ (\Gamma) \exp(-\Gamma_\tau)\ d\Gamma \tag{16}$$

where β is the instrumental coherence factor close to unity, τ is the decay time, while $g^{(1)}\ (\tau)$ is the decay rate distribution function. Γ is the relaxation rate, which can be calculated by Laplace inversion of $g^{(2)(}\tau)$ on the basis of equations both Eqs. (15 and 16). $G(\Gamma)$ can be obtained from $g^{(2)}\ (\tau)$ using CONTIN (constrained regularization method) [43].

DLS offers several advantages such as minimal sample preparation, short measurement times. However, its disadvantages include the inability to determine the distribution details and non-existence or poor concentration information due to the ill-posed nature of converting data to distribution.

In summary, resistive pulse sensing (RPS) is a single-particle analytical method capable of measuring the particle size distribution (PSD) and concentration of synthetic and polymer-based nanoparticles. While RPS has been effectively utilized to classify various blood cell types, its sensitivity has traditionally been limited to particles in the micrometer range.

1.4.3 Flow Cytometry (FCM)

As an alternative, flow cytometry (FCM) has been employed to detect extracellular vesicles, analyzing eukaryotic cells and sub-micron nanoparticle populations. However, traditional systems often lack the sensitivity needed to detect nanoscale biological particles and organelles. Recent developments in high-sensitivity flow cytometry have enabled accurate, sensitive, and targeted analysis of these smaller entities, offering significant advantages for studying bacteria, mitochondria, viruses, and extracellular vesicles. A key benefit of flow cytometry is its ability to determine the particle concentration directly, as the volume of the sample being analyzed is known, eliminating the need for bead-based calibration.

Principles and Mode of Operation of FCM

Flow cytometry operates by directing particles through a hydrodynamically focused fluid stream, where each particle is individually illuminated by a laser beam [44] (Fig. 3). This interaction causes particles to scatter light and emit fluorescence. Forward-scattered light (FSC) and side-scattered light (SSC), along with fluorescence, are detected by separate sensors to determine particle size, complexity, and fluorescent labeling (Fig. 3). Fluorescence-activated cell sorting (FACS) can also isolate specific subpopulations (Table 1).

A key challenge in using conventional flow cytometry for nanoscale biological particle and organelle analysis is distinguishing particle signals from background noise (Fig. 4). Light scattering, which occurs due to differences in refractive index between particles and their surrounding medium, is vital for analyzing size and structure. However, the technique has a detection limit of 200–500 nm for high-refractive-index polystyrene beads. Since biological particles like bacteria (1.38–1.45) [45], extracellular vesicles (EVs)(1.32–1.42) [46], four-modal silica nanosphere

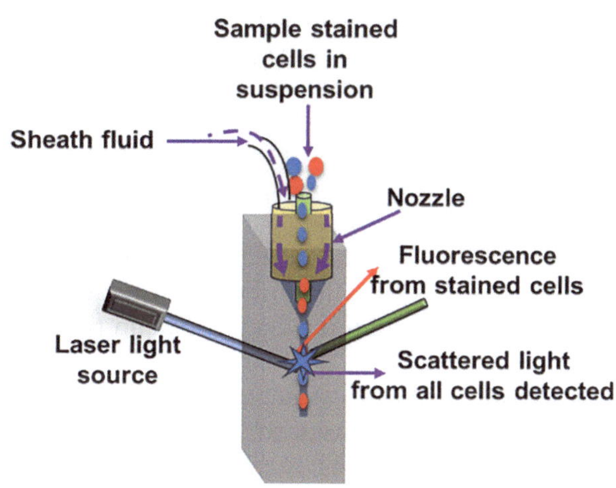

Fig. 3 Schematic illustration of flow cytometry

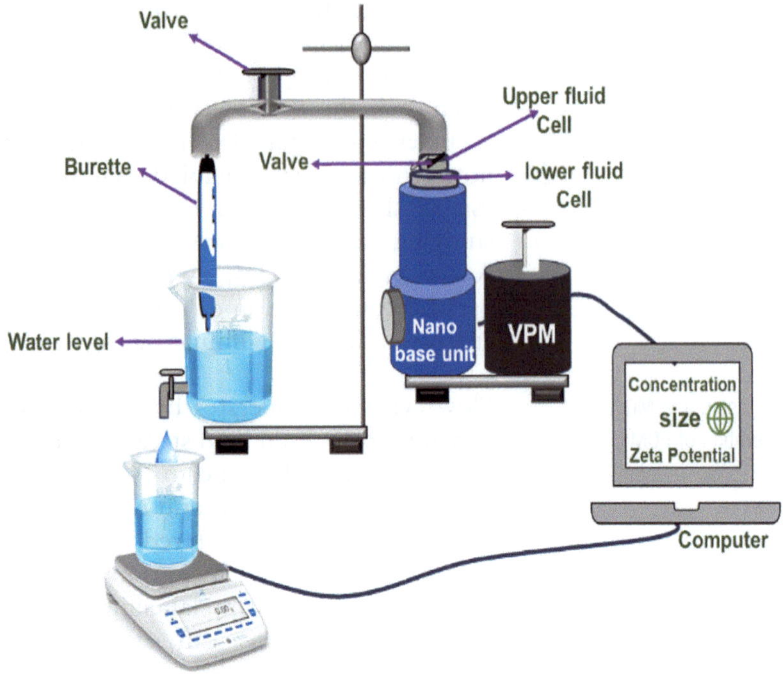

Fig. 4 Schematic illustration of components of tunable resistive pulse sensing TRPS instrumentation

mixture (1.45) and viruses (1.45–1.46) [47] have lower refractive indices, their detection threshold is higher, making small particle analysis more difficult. High-resolution side scatter detection enables label-free identification of all particles, including non-EV materials. To distinguish specific subpopulations, additional fluorescent staining techniques are employed [48]. Sample concentration is determined by calibrating the flow rate using 250 nm silica nanoparticles with a known particle concentration.

Flow cytometry demonstrates the highest accuracy when sizing standardized reference beads; however, its precision tends to decrease when analyzing extracellular vesicles due to their variable refractive indices and heterogeneous composition. Despite this limitation, specialized flow cytometry remains a rapid and efficient technique, particularly valued in clinical research for its capability to perform multiplex fluorescence detection. This enables simultaneous identification of multiple biomarkers, making it especially useful for diagnostic and therapeutic monitoring applications.

Applications and Uses of FCM

Flow cytometry has been successfully employed to assess bacterial responses to antimicrobial treatments by measuring parameters such as membrane potential, membrane permeability, esterase activity, respiratory function, and glucose uptake [49–52]. However, conventional flow cytometers still face significant limitations in detecting small bacteria or low-abundance biochemical components (Table 1). Recent advancements in mitochondrial research have highlighted their functional and structural diversity, including differences in metabolic activity, reactive oxygen species (ROS) generation, membrane potential, lipid levels, protein expression, mtDNA copy number, and calcium uptake [53–56]. These variations are linked to critical biological, physiological, and pathological processes like cancer progression, cell differentiation, drug resistance, and aging, making single-organelle flow cytometric analysis of isolated mitochondria highly valuable [57].

Flow cytometry's ability to perform multiparametric analysis has enabled the exploration of complex interrelationships within mitochondria. Additionally, flow virometry, also known as flow cytometric analysis of viruses or virus-like particles, has gained momentum [58, 59]. It reveals virus heterogeneity in terms of aggregation [60], size [61], protein conformation [62], and composition [61] integrity, and receptor interactions, offering insights essential for virus sorting [63], assessing infectivity, and supporting vaccine development. Fluorescence-based triggering, using nucleic acid stains, antibodies, or membrane dyes, is commonly applied due to the small size of viruses.

For obtaining both size data and detailed morphological information, electron microscopy techniques such as scanning electron microscopy (SEM) and transmission electron microscopy (TEM) are commonly used. However, these techniques are not well-suited for high-throughput analysis due to their low automation and time-intensive nature.

Table 1 summarizes the comparison between the various particle analysis methods, as well as their advantages and disadvantages

2 Materials

2.1 Fundamental Materials and Components Needed for TRPS Analysis of Food Matrices

TRPS System (e.g., qNano or qViro-X by Izon Science), Control Software (e.g., Izon Control Suite), Non-conductive elastic membrane (NP100, NP200, NP400), Calibration particles (carboxylated polystyrene (CO-psty) beads), Filtration (0.45 μm syringe filter or peristaltic pump), centrifugation devices or ultrasonication device, Suspension Buffer (SEB—Standard Electrolyte Buffer) such as PBS, Tris-HCl, *0.1 M KCl* (potassium chloride), *15 mM Tris* (2-amino-2-hydroxymethyl-propane-1,3-diol), *0.01% v/v Triton X-100* (nonionic surfactant), *3 mM EDTA* (ethylenediaminetetraacetic acid), *pH adjustment:* HCl (hydrochloric acid) to pH 8 and Surfactants (such as Tween 20, Triton X-100).

2.2 TRPS Instrumentation, Set-Up, and Workflow

The TRPS is a simple procedure that can be easily set up. The system consists of the following essential components, such as:

analyzed by multi-point pressure method, the accuracy of measurement would be enhanced and electrokinetic effects corrected.

To ensure optimal nanoparticle dispersion and prevent aggregation, samples should be gently vortexed and subjected to brief sonication (1–2 minutes). For real-world sample matrices (such as biological fluids, food, or environmental samples), pre-filtration is recommended due to the presence of proteins, undissolved solids, and fats, using either a 0.22 μm or 0.45 μm syringe filter to remove large debris or aggregates that could lead to clogging of the pores. If the nanoparticles are aggregated and persists, short sonication with 1–2 minutes may be applied to disperse the particles, although prolonged sonication must be avoided to prevent structural degradation.

For samples in non-conductive or incompatible media, such as milk or organic solvents, the samples can undergo centrifugation followed by resuspension in a suitable TRPS buffer. Subsequently, the sample is gently introduced into the upper fluid cell with a pipette, ensuring that critical parameters such as pressure and voltage are continuously monitored throughout the process. Depending on the pressure control technique applied, pressures typically range between 0.3 and 2.0 kPa, [25]. Pressure control must be optimized for consistent particle translocations without clogging or excessive noise. Sample volumes of approximately 35–40 μL are commonly used for each measurement. To ensure reproducibility, it is also advisable to perform measurements in triplicate.

3.2 Calibration and Standardization Procedures in Tunable Resistive Pulse Sensing

TRPS relies on rigorous calibration and standardization protocols to ensure high-resolution, accurate measurements of particle size, concentration, and zeta potential.

3.2.1 Calibration with Known Standards

Firstly, concentrations are verified using calibration beads (polystyrene or silica beads with predefined size and concentration) with the application of large multiple pressures $(P_1 \gg P_2)$ (0.3–2.0 kPa) to ensure consistency across varying conditions. The beads are usually analyzed before and after sample measurements to validate nanopore performance. Once the concentration has been measured, a sample of the concentration bead before and after the modification of the same is prepared to the same concentration. Subsequently, both samples are run at a range of voltages with no additional pressure to produce a plot of pulse frequency against voltage. The resistance change (ΔR) caused by a calibration particle passing through the nanopore is proportional to its volume (V_c). This relationship is used to calculate sample particle volume (V_s) in shown in Eq. (17),

$$\frac{\Delta R_s}{\Delta R_c} = \frac{V_s}{V_c} \tag{17}$$

Particle concentration is determined by comparing the translocation rate of sample particles to calibration particles of known concentration. Additional sample preparation steps, such as pre-filtration, brief sonication, and buffer compatibility assessments, are crucial and should be performed in accordance with ICH guidelines and ISO 13319-2:2023 standards, as previously outlined.

Following this, appropriate nanopore selection is carried out, often involving the use of individual or combined nanopores to ensure accurate size detection. Throughout the experimental process, measures are implemented to minimize errors, including precautions to prevent pore blockage, air bubble formation, and excessive sonication. Adherence to ISO 13319-2:2023 and robust calibration protocols ensures TRPS delivers reliable, standardized data for applications in nanomedicine, environmental monitoring, and biotechnology. Its single-particle approach and multi-pressure calibration make it indispensable for high-stakes characterization where precision is paramount.

4 Tunable Resistive Pulse Sensing (TRPS) for Characterizing Food Matrices

4.1 Tunability and Sensitivity

Tunable Resistive Pulse Sensing (TRPS), formerly known as Scanning Ion Occlusion Spectroscopy, is an advanced technique for analyzing nanoparticles in complex food matrices. It provides high sensitivity, precision, and exceptional size resolution, enabling the measurement of zeta potential, particle size, and concentration [76]. TRPS's main strength lies in its tunability by adjusting nanopore size and membrane stretch, it can accurately detect a wide range of particle sizes [77], making it ideal for diverse, polydisperse food systems such as emulsions and protein aggregates [42]. It has been successfully used to analyze viruses, bacteria [77], exosomes, dextran particles [65], liposomes, emulsions [78], lyophilisomes [79], magnetic beads [80], and other nanoscale materials relevant to food and biomedical applications.

Researchers such as Vogel et al. [81] have employed TRPS technique combined with custom built manometer, to measure diverse range of carboxylate polystyrene nanoparticles. Other researchers, including Somerville et al. [78], Kozak et al. [82], and Weatherall et al. [83] used the same method for the detection of ζ-potential in water-in-oil emulsion, in tunable pores to determine the shape of individual pulses, and in calculating ζ-potential in place of geometric parameters, respectively.

Generally, the ζ-potential of a particle is given as shown in Eq. (18), when the particle size is greater than the Debye length and the ζparticle is measured based on electrophoretic mobility using the Smoluchowski approximation. Using the Nernst–Planck equation, the distinct contributions of the particle flux through the pore are summed up as given in Eq. (18) [77].

$$\frac{J}{C} = \left(\frac{\varepsilon\left(\zeta_{\text{Particle}} - \zeta_{\text{pore}}\right)}{\eta}\right) E + \frac{Q_p}{A} \tag{18}$$

where ζpore is the ζ-potential of the pore wall, E is the applied electrical field ε is the permittivity of the solution and η is the kinematic viscosity, respectively. A is the cross-sectional area of the pore, Q_p is the volumetric rate of pressure driven through the pore. In any given experiment where an external pressure (P applied) is applied across the fluid cell, the evaluation of Po at which the transport mechanism is balanced, so that there is no net motion of the particles through the pore ($J = 0$). To calculate ζ_{Particle} in a semi-analytical model is given in Eq. (19)

$$\zeta_{\text{Particle}} = -\frac{\eta Q_p}{EA} - \zeta_{\text{pore}} \tag{19}$$

As shown in Eq. (20), Smoluchowski approach define zeta potential as

$$\zeta = \frac{\eta \mu}{\varepsilon} \tag{20}$$

The Smoluchowski model suggests that a particle's zeta potential can be calculated based on its velocity, considering the effects of convective flow, electro-osmotic forces, and the particle's electrophoretic mobility. The ζ can be validated using Henry's equation as given in Eq. (21)

$$\zeta = \frac{3\eta \mathcal{U}_e}{2 \in_T \in_0 f\ (\kappa a)} \tag{21}$$

where $f\ (\kappa a)$ is the Henry's function, which is approximately equal to 1.5 for Smoluchowski approximation.

J can be related to the concentration of the analyte (Cs) and velocity of the transversing particles (ν_ρ), as shown in Eq. (22).

$$J = C_s \times \nu_\rho \tag{22}$$

ν_ρ *is the* sum of the electroosmotic, fluidic, and electrophoretic velocities ($\nu_\rho = \nu_O + \nu_F + \nu_E$). However, due to the magnitude of the other forces, these contributory effects from diffusion can be ignored, and the end effects not considered. Therefore, ν_ρ is given in Eq. (23) as,

$$\nu_\rho = \frac{Q}{\pi\left(\frac{D_s}{2}\right)} + \frac{\varepsilon \zeta_{\text{Particle}}}{\eta} E - \frac{\varepsilon \zeta_{\text{pore}}}{\eta} E \tag{23}$$

The pulse frequency, J, where, ΔP is the pressure across the zeta potential of the channel surface (ζ_{pore}) and particle (ζ_{Particle}). Q is the volumetric rate of pressure driven through the pore, and is given in Eq. (24),

$$Q = \frac{3\pi D_s^{3} \Delta P}{128\eta\left(\frac{L}{D_L - D_s}\right)} \tag{24}$$

5 Notes

Tunable resistive pulse sensing (TRPS) is a precise analytical method used to characterize nanoparticles within complex food systems. It operates by employing a tunable nanopore membrane to measure key particle properties such as size, concentration, and surface charge (zeta potential) based on brief disruptions in ionic current as particles pass through the nanopore. This technique is especially useful in ensuring food safety and quality, where accurate detection and analysis of engineered nanoparticles (such as nutrient carriers, antimicrobial agents) or unintended particles (including contaminants) are essential. This system includes the qNano System by IZON Science, a commercial Tunable Resistive Pulse Sensing (TRPS) platform equipped with adjustable nanopores ranging from 50 to 2000 nm, allowing it to analyze a wide variety of nanoparticles between 10 and 2000 nm in size. It features precise pressure control using a syringe pump and a manometer, maintaining fluid flow with an accuracy of ±5 Pa, which ensures consistent particle translocation behavior. Additionally, the system offers customizable membranes, where the nanopore size and surface properties can be modified to selectively detect specific particle types, such as lipid nanoparticles in nutraceutical applications or metal oxide particles in environmental contaminant analysis.

This system effectively measures nanoparticle size distribution, concentration, and zeta potential, making it ideal for analyzing polydisperse and low-abundance samples in complex food matrices such as dairy and juices. It offers clear advantages over dynamic light scattering and electron microscopy by minimizing sample preparation and interference, while providing real-time results. In food science, it supports safety monitoring by detecting pathogens, toxins, and microplastics, ensures quality assurance of nanoencapsulated nutrients and additives, and aids regulatory compliance by identifying nanoparticle migration from packaging into food. These features make it a valuable tool for advancing food safety, quality, and regulatory standards. TRPS faces challenges like matrix interference in complex foods, reliance on standard calibration beads, and high costs of equipment and consumables. AI-enhanced TRPS enables automated nanoparticle profiling and on-site food testing, bridging lab research and industry with precise, regulatory-friendly tools for food safety and innovation.

5.1 Application of TRPS in Food Nanotechnology

Recent progress in synthetic biology and nanofabrication has led to the development of a wide variety of resistive pulse sensors, featuring diverse materials, structures and configurations. These sensors are generally classified into three categories: biological, synthetic, and microfluidic-based systems.

1. *Characterization of Food-Grade Nanostructures*
   (a) TRPS is used to analyze nanoemulsions [83], liposomes, nanogels [7], exosomes in human and bovine milk, lacto-globulin aggregates [78], and air microbubbles.
   (b) It is essential for evaluating particle size, concentration, and zeta potential.
2. *Encapsulation and Controlled Release.*
   (a) It helps in designing nanocarriers for encapsulation of nutrients and bioactives.
   (b) It enables assessment of controlled release profiles for improved nutrient delivery.
3. *Enhanced Bioavailability*
   (a) It supports the development of systems that increase bio-availability of poorly soluble compounds.
   (b) It ensures consistent dosage and absorption in functional foods.
4. *Food Safety and Quality Control*
   (a) It is used to detect micro- and nanoscale contaminants, such as microplastics or nanoparticles from packaging.
   (b) It assists in verifying formulation stability during storage and processing.
5. *Real-Time Analysis and Process Monitoring*
   (a) It enables on-site and rapid measurement of particle dynamics during food production.
   (b) It also helps to optimize manufacturing conditions for nanostructured food products.
6. *Zeta Potential Measurements*
   (a) It is used to evaluates colloidal stability to predict aggre-gation behavior.
   (b) It is crucial for shelf-life prediction and consumer safety.
7. *High-Throughput Screening.*
   (a) Recent TRPS platforms allow rapid classification of food particles at high flow rates (e.g., 1 mL/min) [84].
   (b) Facilitates screening of diverse particle types including algae-based additives.

8. *Material Property Analysis.*
   (a) It helps to differentiates particles based on porosity, shape, and conductivity.
   (b) It used in quality assurance and material verification for food-grade nanosystems.

5.2 Detection Limitations in Resistive Pulse Sensing (RPS)

5.2.1 Detection Speed and Size Range

The sensitivity of resistive pulse sensing (RPS) depends on instrument bandwidth and particle translocation speed. Faster translocation or lower bandwidth can reduce accuracy, leading to underestimated particle sizes. Experimental conditions like voltage, salt concentration, viscosity, and pore modifications, can adjust translocation speed. While fixed pores limit detectable size ranges, tunable pores expand versatility.

Signal-to-noise ratio (SNR) also impacts detection sensitivity. Low-frequency noise (e.g., from nanobubbles) and high-frequency noise (e.g., Johnson noise) hinder performance. Innovations like MOSFET sensors and differential amplification have enhanced SNR, allowing detection of ultra-small particles.

Off-axial translocation: when particles pass through a pore off-center causes a sizing error up to 15%. Hydrodynamic focusing can minimize this by guiding particles centrally through the pore.

Electrolyte conductivity is vital; low-conductivity media limit RPS effectiveness. However, microfluidic capacitance sensors enable detection in such conditions by measuring AC capacitance instead of DC resistance.

Pore design innovations, including tunable elastic pores and lithography-based systems, now support real-time pore resizing and parallel measurements. These advances improve sensitivity, expand size range, and increase throughput, enabling detection from nanoparticles to whole cells.

5.3 Future Directions of TRPS and Its Integration with Other Analytical Tools

Tunable Resistive Pulse Sensing (TRPS) has shown remarkable versatility in nanoparticle characterization, particularly for measuring size, concentration, and zeta potential of particles ranging from a few nanometers to several micrometers. As the field advances, future research is expected to focus on enhancing the sensitivity, throughput, and multiplexing capabilities of TRPS. One promising direction lies in miniaturizing and automating TRPS systems to allow for high-throughput analysis suitable for clinical and industrial settings. Additionally, improving the robustness of nanopore materials and the reproducibility of measurements across different pore types will be essential for standardization.

Moreover, the integration of TRPS with other analytical platforms will significantly broaden its application scope. For example, coupling TRPS with mass spectrometry (MS) could enable both physical and compositional profiling of single particles, such as extracellular vesicles or protein aggregates. Further demonstrating

nanodroplets and nanoparticles. Curr Opin Colloid Interface Sci 53:101427
33. Yu K, Zhang H, Biggs S, Xu Z, Cayre OJ, Harbottle D (2018) The rheology of polyvinylpyrrolidone-coated silica nanoparticles positioned at an air-aqueous interface. J Colloid Interface Sci 527:346–355
34. Tanimura Y, Yoshida K, Watanabe Y (2010) A study on cleaning ability of oscillating bubbles driven by low-frequency ultrasound. Jpn J Appl Phys 49:07HE20
35. Yu K, Zhang H, Hodges C, Biggs S, Xu Z, Cayre OJ, Harbottle D (2017) Foaming behavior of polymer-coated colloids: the need for thick liquid films. Langmuir 33:6528–6539. https://doi.org/10.1021/ACS.LANGMUIR.7B00723
36. Phan KK, Truong T, Wang Y, Bhandari B (2020) Nanobubbles: Fundamental characteristics and applications in food processing. https://www.sciencedirect.com/science/article/pii/S0924224419301414
37. Qiu J, Zou Z, Wang S, Wang X, Wang L, Dong Y, Zhao H, Zhang L, Hu J (2017) Formation and stability of bulk nanobubbles generated by ethanol–water exchange. ChemPhysChem 18:1345–1350. https://doi.org/10.1002/CPHC.201700010
38. Ulatowski K, Sobieszuk P, Mróz A, Ciach T (2019) Stability of nanobubbles generated in water using porous membrane system. Chem Eng Process 136:62–71. https://doi.org/10.1016/j.cep.2018.12.010
39. Bai K, Barnett GV, Kar SR, Das TK (2017) Interference from proteins and surfactants on particle size distributions measured by nanoparticle tracking analysis (NTA). Pharma Res 34(4):800–808. https://doi.org/10.1007/s11095-017-2109-3
40. Zhang L, Mazouzi Y, Salmain M, Liedberg B, Boujday S (2020) Antibody-gold nanoparticle bioconjugates for biosensors: synthesis, characterization and selected applications, Biosens Bioelectron 165:112370. https://www.sciencedirect.com/science/article/pii/S095656632030364X
41. Kramberger P, Ciringer M, Štrancar A, Peterka M (2012) Evaluation of nanoparticle tracking analysis for total virus particle determination. Virol J 9:265. https://doi.org/10.1186/1743-422X-9-265
42. Guerrini G, Magrì D, Gioria S, Medaglini D, Calzolai L (2022) Characterization of nanoparticles-based vaccines for COVID-19. Nat Nanotechnol 17:570–576. https://doi.org/10.1038/s41565-022-01129-w
43. Liu J, Yan T, Li Y, Ren H, Wang Q, Guan F, Jiao Q (2020) Dual-mode response behavior of a graphene oxide implanted energetic system under different thermal stimuli. RSC Adv 10:10789–10798. https://doi.org/10.1039/D0RA00857E
44. Shapiro HM (2005) Practical flow cytometry. https://books.google.com/books?hl=en&lr=&id=JhSyimPKuJwC&oi=fnd&pg=PR7&ots=OTCYCWSJSB&sig=EXRrdL4KdFjck93n_IOGYS0G7n8
45. Foladori P, Quaranta A, Ziglio G (2008) Use of silica microspheres having refractive index similar to bacteria for conversion of flow cytometric forward light scatter into biovolume. Water Res 42(14):3757–3766. https://www.sciencedirect.com/science/article/pii/S0043135408002741
46. van der Pol E, Coumans FAW, Grootemaat AE, Gardiner C, Sargent IL, Harrison P, Sturk A, van Leeuwen TG, Nieuwland R (2014) Particle size distribution of exosomes and microvesicles determined by transmission electron microscopy, flow cytometry, nanoparticle tracking analysis, and resistive pulse sensing. J Thromb Haemost 12:1182–1192. https://doi.org/10.1111/jth.12602
47. Ma L, Zhu S, Tian Y, Zhang W, Wang S, Chen C, Wu L, Yan X (2016) Label-free analysis of single viruses with a resolution comparable to that of electron microscopy and the throughput of flow cytometry. Angew Chem Int Ed 55:10239–10243. https://doi.org/10.1002/ANIE.201603007
48. Zhu S, Ma L, Wang S, Chen C, Zhang W, Yang L, Hang W, Nolan JP, Wu L, Yan X (2014) Light-scattering detection below the level of single fluorescent molecules for high-resolution characterization of functional nanoparticles. ACS Nano 8:10998–11006. https://doi.org/10.1021/NN505162U
49. Kramer B, Thielmann J (2016) Monitoring the live to dead transition of bacteria during thermal stress by a multi-method approach. J Microbiol Methods 123:24–30. https://www.sciencedirect.com/science/article/pii/S0167701216300264
50. Lee H, Hwang JS, Lee J, Kim JI, Lee DG (2015) Scolopendin 2, a cationic antimicrobial peptide from centipede, and its membrane-active mechanism. Biochim Biophys Acta 1848:634–642. https://www.sciencedirect.com/science/article/pii/S0005273614004143
51. Morishige Y, Fujimori K, Amano F (2015) Use of flow cytometry for quantitative analysis of metabolism of viable but non-culturable (VBNC) Salmonella. Biol Pharma Bull 38(9):